Lecture Notes in Physics

The Editorial Policy for Proceedings

The series Lecture Notes in Physics reports new developments in physical research and teaching – quickly, informally, and at a high level. The proceedings to be considered for publication in this series should be limited to only a few areas of research, and these should be closely related to each other. The contributions should be of a high standard and should avoid lengthy redraftings of papers already published or about to be published elsewhere. As a whole, the proceedings should aim for a balanced presentation of the theme of the conference including a description of the techniques used and enough motivation for a broad readership. It should not be assumed that the published proceedings must reflect the conference in its entirety. (A listing or abstracts of papers presented at the meeting but not included in the proceedings could be added as an appendix.)

When applying for publication in the series Lecture Notes in Physics the volume's editor(s) should submit sufficient material to enable the series editors and their referees to make a fairly accurate evaluation (e.g. a complete list of speakers and titles of papers to be presented and abstracts). If, based on this information, the proceedings are (tentatively) accepted, the volume's editor(s), whose name(s) will appear on the title pages, should select the papers suitable for publication and have them refereed (as for a journal) when appropriate. As a rule discussions will not be accepted. The series editors and Springer-Verlag will normally not interfere with the detailed editing except in fairly obvious cases or on technical matters.

Final acceptance is expressed by the series editor in charge, in consultation with Springer-Verlag only after receiving the complete manuscript. It might help to send a copy of the authors' manuscripts in advance to the editor in charge to discuss possible revisions with him. As a general rule, the series editor will confirm his tentative acceptance if the final manuscript corresponds to the original concept discussed, if the quality of the contribution meets the requirements of the series, and if the final size of the manuscript does not greatly exceed the number of pages originally agreed upon. The manuscript should be forwarded to Springer-Verlag shortly after the meeting. In cases of extreme delay (more than six months after the conference) the series editors will check once more the timeliness of the papers. Therefore, the volume's editor(s) should establish strict deadlines, or collect the articles during the conference and have them revised on the spot. If a delay is unavoidable, one should encourage the authors to update their contributions if appropriate. The editors of proceedings are strongly advised to inform contributors about these points at an early stage.

The final manuscript should contain a table of contents and an informative introduction accessible also to readers not particularly familiar with the topic of the conference. The contributions should be in English. The volume's editor(s) should check the contributions for the correct use of language. At Springer-Verlag only the prefaces will be checked by a copy-editor for language and style. Grave linguistic or technical shortcomings may lead to the rejection of contributions by the series editors. A conference report should not exceed a total of 500 pages. Keeping the size within this bound should be achieved by a stricter selection of articles and not by imposing an upper limit to the length of the individual papers. Editors receive jointly 30 complimentary copies of their book. They are entitled to purchase further copies of their book at a reduced rate. As a rule no reprints of individual contributions can be supplied. No royalty is paid on Lecture Notes in Physics volumes. Commitment to publish is made by letter of interest rather than by signing a formal contract. Springer-Verlag secures the copyright for each volume.

The Production Process

The books are hardbound, and the publisher will select quality paper appropriate to the needs of the author(s). Publication time is about ten weeks. More than twenty years of experience guarantee authors the best possible service. To reach the goal of rapid publication at a low price the technique of photographic reproduction from a camera-ready manuscript was chosen. This process shifts the main responsibility for the technical quality considerably from the publisher to the authors. We therefore urge all authors and editors of proceedings to observe very carefully the essentials for the preparation of camera-ready manuscripts, which we will supply on request. This applies especially to the quality of figures and halftones submitted for publication. In addition, it might be useful to look at some of the volumes already published. As a special service, we offer free of charge LaTeX and TeX macro packages to format the text according to Springer-Verlag's quality requirements. We strongly recommend that you make use of this offer, since the result will be a book of considerably improved technical quality. To avoid mistakes and time-consuming correspondence during the production period the conference editors should request special instructions from the publisher well before the beginning of the conference. Manuscripts not meeting the technical standard of the series will have to be returned for improvement.

For further information please contact Springer-Verlag, Physics Editorial Department II, Tiergartenstrasse 17, D-69121 Heidelberg, Germany

Gülen Aktaş Cihan Saçlıoğlu
Meral Serdaroğlu (Eds.)

Strings and Symmetries

Proceedings of the Gürsey Memorial Conference I
Held at Istanbul, Turkey, 6-10 June 1994

Springer

Editors

Gülen Aktaş
Cihan Saçlıoğlu
Meral Serdaroğlu
Physics Department, Bogaziçi University
TR-80815 Bebek, Turkey

ISBN 978-3-662-14032-1 ISBN 978-3-540-49204-7 (eBook)
DOI 10.1007/978-3-540-49204-7

CIP data applied for

Originally published by Springer-Verlag Berlin Heidelberg New York in 1995
Softcover reprint of the hardcover 1st edition 1995

Typesetting: Camera-ready by the editors
SPIN: 10481096 55/3142-543210 - Printed on acid-free paper

Foreword

Feza Gürsey was one of the most respected members of the physics community and his untimely death on April 13, 1992 was a great loss to theoretical physics. He will always be remembered for his many seminal and deep contributions to theoretical physics as well as for his kindness, civility and scholarship. For those of us who knew him he epitomized a style of physics and an epoch in the history of physics.

Feza's scientific work is marked with remarkable originality and elegance as well as intellectual courage. He never hesitated to pick problems that were not fashionable. He worked at them in depth, planting seeds that in some cases developed into whole branches of our discipline. Outstanding examples would include his conception of the pion in terms of spontaneously broken chiral symmetry, and his contributions to the introduction of exceptional gauge groups for grand unification. To the end of his life he was tackling the most difficult problems, planting new seeds in unknown soil.

In the early part of his career, Gürsey studied the conformal group and conformally invariant quantum field theories, concepts whose role in physics are now central. This developed into his long and multifaceted interest in the unitary representations of non-compact groups and their applications to space-time. In the late fifties he did his work on Pauli-Gürsey transformations and later introduced the non-linear chiral Lagrangian, one of his most seminal contributions to theoretical physics. Chiral symmetry and non-linear realizations of symmetry groups have since become an integral part of theoretical physics. In the 1960 s, Feza became well known for his work on the $SU(6)$ symmetry that combines the unitary spin $SU(3)$ of the eight-fold way with non-relativistic spin degrees of freedom of quarks. Subsequent attempts to understand the origin of $SU(6)$ symmetry led to the introduction of the color degrees of freedom of quarks. Feza's introduction in the mid-1970s of the grand unified theory based on the exceptional group E_6 -which has continued to fascinate theoretical physicists ever since- was one facet of his long interest in the possible role of quaternions and octonions in physics.This interest also led to Feza's work on quaternion analyticity, which continued practically to the end of his life.

Feza was an exceptionally inspiring teacher. He trained many Ph.D. students who now hold academic positions in numerous countries of the world. Throughout his life he retained a youthful spirit and was always enthusiastic about learning new things. He had a special rapport with the young people and enjoyed their company.

Reminiscing only about Feza Gürsey the physicist would not do full justice to him. He was a very cultured man who distilled the essential and sublime elements of Western and Turkish cultures and synthesized them into a singularly unique whole in hispersonality and wisdom. One could have deep and penetrating discussions with him on the music of Franz Schubert and Dede Efendi, on the poetry of Yunus Emre and Goethe, on the novels of Thomas Mann and Marcel Proust, on the paintings of Van Gogh and Giotto, in short, on essentially any subject of depth and beauty.

The Gürsey Memorial Conferences that are to be held biannually are hopefully a fitting way to pay tribute to his memory. We hope that these will be conferences that Gürsey himself would have enjoyed! His memory will always be with those of us who were his friends and colleagues.

February 1995

Murat Günaydın
Edward Witten

Feza Gürsey
7 April 1921-13 April 1992

Preface

This volume is intended to be the first in a biennial conference series endowed by TÜBİTAK, the Turkish Scientific and Technical Research Council, and dedicated to Feza Gürsey's memory. Feza (Feza Bey to his Turkish compatriots) had just retired from Yale University (but by no means from Physics) where he had occupied the J. Willard Gibbs chair since 1977, in order to take up a position at Boğaziçi University; thus it is both sad and fitting that the first conference was held at Boğaziçi.

In addition to the outstanding scientific talks, the conference also afforded an opportunity for Feza's friends and colleagues to reminisce about him at length. In what follows, we will liberally quote or paraphrase some of these contributions in order to convey aspects of Feza's life, work and personality to younger physicists who did not have the good fortune of knowing him personally. After all, as Alan Chodos put it, Feza himself remained a young man in preserving his idealism and openness to new ideas and thus so inspired his merely chronologically younger students that not a single one of them later opted out of physics. We feel an effort should be made to keep his memory alive so that he may continue to serve as a gentleman-physicist role model in spite of the tendency, observed by Antony Zee, of the breed to become extinct.

Charlie Sommerfield aptly noted the cosmopolitan influences in Feza's childhood and schooling. This was in considerable part due to the remarkable careers of his parents: the father, a late Ottoman army doctor with a scientific *Wanderlust,* took his young wife and the toddler Feza to Paris, ostensibly for the purpose of studying the use of X-rays in medicine; but once there, he promptly took off in search of instruction in quantum mechanics from Schrödinger himself. The mother, left to her own devices, managed to put Feza through elementary school as well as earning a Doctorate in chemistry from the Sorbonne.

The French connection was reasserted in Feza's secondary education at the venerable Galatasaray Lycée, the first school set up by the Ottoman government as a window on the West. Afterwards, following an undergraduate education in physics and mathematics at Istanbul University, where Maxwell's equations were considered mere speculation, Einstein a charlatan and quantum mechanics an unspeakable heresy, Feza left for graduate study in England in 1945. Because of his already well-developed interest in "bohemian intellectual life" as he put it, he chose Imperial College, London, rather than Edinburgh, where Max Born reigned. He later came to regard this as one of the missed opportunities in his life, compounded by his failure to appreciate Born's advice in 1945 to learn about Lie groups and their representations with a view to applying them in physics. As we all know, the "mistake" was amply made up for later.

It was apparently in 1946 that Feza and Freeman Dyson met in England. In a beautiful essay, "Unfashionable Pursuits", delivered on Feza's 60th birthday, Dyson writes: "When I met Feza in London long, long ago, I think he was probably the first physicist I ever met. I had the idea all physicists were like that. That had a lot to do with my becoming a physicist myself; of course I found out afterwards I was under something of a misapprehension."

Feza partook of the lively intellectual atmosphere of London. He also turned out a number of papers on relativity and statistical mechanics and obtained his Ph.D. in 1950. The 1950 paper "Classical Statistical Mechanics of a Rectilinear Assembly" was re-examined at length much later in Münster's two-volume treatise on statistical mechanics. After spending 1950-51 as a postdoctoral fellow at Cambridge, where the seeds of this lifelong interest in general relativity and the conformal group were apparently sown, he returned to Istanbul University. The years 1951-57 were spent in Istanbul University and in part in the Army, due to the compulsory military service requirement. His university colleagues describe how Feza organized a quantum field theory seminar, where they worked their way through the papers of Schwinger and Feynman. It was during this period that he met and married his beloved Suha, who was also a junior faculty member at the Physics Department.

In spite of the cold shoulder treatment he received from most of his other colleagues and the administration for disturbing the peace in a provincial physics department, the publication list reveals that these were not unproductive years: papers on the classical spinning electron, the conformal group and the use of quaternions in relativity give hints of what were going to become recurring themes in his research. John Wheeler once wrote, "As a man who was in many ways self-educated, Gürsey has a most valuable independence of point of view and originality". Perhaps the Istanbul period was an important part of this process of self-education.

The academic year 1957-58, which he spent at Brookhaven on an Atomic Energy Commission grant, turned out to be the bifurcation point between two entirely different world lines. Had he stayed in Istanbul, he almost certainly would have continued in the same semi-amateurish vein, given the weakness of stimulating contacts with the mainstream. At Brookhaven, on the other hand, he found himself hurled into the middle of it at the age of 36. He used to recount an encounter with Dyson in which he was bluntly told the Conformal Group was for crackpots and that if he wished to re-establish his credentials he had better turn out a respectable calculation in QED. After heeding the advice and finishing the paper in half the time Dyson expected, Feza's subsequent paper and his resulting brief collaboration with Pauli caused a major upheaval: as C. Sommerfield put it, "first raising and then demolishing Pauli's hopes about the Heisenberg non-linear spinor theory". This led to an estrangement between Pauli and Heisenberg; the episode is obliquely referred to in the book "Physics and Beyond" by Heisenberg.

The next important contribution from the period 1958-1961, spent partly at IAS, Princeton, and Columbia, and partly in Istanbul, is the nonlinear sigma model with chiral symmetry. This was perhaps the first model in particle physics which featured spontaneous symmetry breaking and the attendant massless scalar fields. Versions of the model surfaced later in electroweak theory, string theory or in supersymmetric contexts.

It is remarkable that with all the attractive international career opportunities that now opened up in front of him, Feza chose to go back to Turkey to continue with his efforts to create a school of theoretical physics. Having become convinced it was hopeless to fight the old guard at Istanbul University, he took part in the enterprise of setting up the new Middle East Technical University in the middle of the Central Anatolian steppe just outside Ankara. With similarly idealistic colleagues and energetic administrators such as Erdal Inönü and Kemal Kurdaş the new University flourished quickly. For us students it was not an uncommon event in those days to hear a lecture or a seminar by a physicist of the stature of Coleman, Chew, Dashen, Gell-Mann or Wightman. During the same period, quite unbeknownst to us, on the floor above the Theoretical Physics department, Robert Langlands, who happened to be visiting METU for a year, was effecting a major breakthrough in number theory through his correspondence with H. Jacquet. We are very grateful that he also kindly contributed to this volume.

Feza realized that the newly founded Department of Theoretical Physics would turn into a stagnant backwater if connections with the outside academic world were not kept wide open. He thus arranged extended visits abroad for graduate students and young Ph.D. students. He knew also that unless he himself revisited leading physics centers, what he could impart to his younger colleagues would quickly become outdated. This meant the Gürsey family had to uproot themselves to spend 1963-64 at the Institute for Advanced Study, Princeton followed by Brookhaven, and 1965-67 at Yale. It was in the summer of 1964 that the SU(6) paper with L. Radicati was written. This clinched Feza's scientific reputation as well as the reality of quarks. The tension between the phenomenological success of SU(6) versus the wrong statistics of quark wave functions on the one hand and no-go theorems forbidding the mixing of spin and internal symmetry on the other was only satisfactorily resolved with the advent of a gauged color degree of freedom and its consequence, asymptotic freedom.

The same years also saw a number of papers on general relativity and group theory. It is worth remembering that unlike nowadays, in the sixties, general relativity was considered a distant and esoteric subject by most high energy physicists.

In 1967, Gregory Breit was to retire from the Yale Physics Department. As Vernon Hughes explained, Breit himself, as well as glowing reference letters from Dyson, Chew, Wick, Goldberger, Serber, Pais, T.D. Lee, Radicati, Wigner Wheeler, supported the idea of Feza being his successor.

Feza accepted the appointment with the understanding that he could keep his ties to METU through roughly biennial trans-Atlantic commutes. It is ironic that it was the METU administration in power in 1974 that went out of its way to put an end to Feza's efforts. Feza was told by the Vice President that he should make a choice between up-to-date, high caliber research and "research in line with Turkey's realities". The former would mean staying abroad and the latter in Turkey, both on a permanent basis. When he refused to resign, he was finally fired in 1974. Not surprisingly, the administrator in question was promoted to positions of higher academic responsibility in subsequent years.

Feza's output from his Yale years (1969-1992) shows amazing variety: the papers range from particle physics phenomenology to supersymmetry and superstrings, use of group theory in nuclear physics, a very early work on conformal invariance in 2-dimensional field theories, use of exceptional Lie algebras as GUT gauge groups, investigation of Kerr-Schild geometries, Skyrmions, and proposals for possible uses for quaternions, octonions and Jordan algebras in fundamental physics. While some of these such as conformal field theory and exceptional GUT groups were eventually adopted by mainstream researchers, many of the others are in the class of the unfashionable mathematical physics that Dyson characterizes as being likely to be very important in the long run. Given Feza's earlier record, this is probably not at all an unrealistic expectation.

It took little over a year from Feza's retirement from Yale and the diagnosis to his death from cancer. Even the illness did not slow down Feza, the irrepressible optimist: undergoing treatment and knowing he had only about an even chance for survival, Feza moved his belongings to Istanbul, attended all the talks at a conference in Edirne, Turkey, and kept on doing research at an accelerated pace, learning and applying new approaches such as A. Connes noncommutative geometry.

The talks at the conference also dwelt upon Feza's broad range of interests in history, literature and the arts. It is perhaps not terribly rare to find persons of immense and wide erudition in academic circles; what distinguished Feza was his original and penetrating insights and exquisite taste in these fields in addition to his learning. He was also unusual in discerning and appreciating the best in both Western and Turkish culture in an unselfconscious way. The dichotomy and the often resulting identity crisis that afflicts many individuals exposed to both influences were transcended in a seamless synthesis in his case. Thus he almost regarded himself as a modern-day version of the Turkish sufi dervishes and was fond of quoting the medieval dervish-poet Yunus Emre in his Turkish public speeches. While the sufis preached ascetic self-discipline and renouncement of worldly ambition as the path to mystical revelation, Feza allowed for a mild Epicurean enjoyment of worldly pleasures and substituted the rational search for scientific truth and beauty for the ultimate aim.

Finally, it was not possible to have known Feza and Suha Gürsey and not be the beneficiary of some very exceptional generosity and kindness. Their house at 1066 Whitney Avenue was a home away from home for generations of Turkish students and young scholars from diverse fields and nations. As Vernon Hughes put it, if Feza had any weakness, it was perhaps over generosity. Suha definitely shared this weakness.

Feza received many honors for his work. Among them one might mention the J.R. Oppenheimer Medal (shared with S. Glashow) in 1977, the Einstein Medal in 1979, the Medal of the College de France and the A. Cressy Morrison Prize (with R. Griffiths) in 1981, the Wigner Medal for the applications of Group theory to Physics in 1986 and the TÜBITAK Science Award in 1969.

Feza's lifelong dream was to bring Turkish physics into the international mainstream. He made considerable personal sacrifices in order to turn this into reality and eloquently defended the indispensability of fundamental science for the material and cultural development of a country. Now that he is gone with his hopes only partially realized, we are profoundly grateful to TÜBITAK for providing the main support for this conference, which will contribute significantly to keeping his dream alive if it indeed becomes a permanent institution. It is a pleasure to acknowledge additional support from The Boğaziçi University Foundation, Burçelik A.Ş. and The Turkish Atomic Energy Authority. We are indebted to the distinguished members of the Advisory Committee who either contributed to the conference by invited talks or by suggesting other outstanding speakers; our special thanks go to Murat Günaydın who did both. We thank Üstün Ergüder, the Rector of Boğaziçi University, who graciously put the campus facilities at the disposal of the conference participants, as well as our colleagues in the Local Organization Committee and the Physics Departments of Boğaziçi University and Eastern Mediterranian University. Aydın Akkaya deserves to be singled out for his tireless involvement in every aspect of the organization.

It is a matter of profound sadness to us that Prof. Asım Barut, who contributed his article to the Proceedings and also presided over the after-dinner talks at the Conference Banquet died after a brief illness in January, 1995. The Turkish Physics Community is very unfortunate in having lost two leaders of international stature in the past two years.

January 1995
Istanbul

Gülen Aktaş
Cihan Saçlıoğlu
Meral Serdaroğlu

Contents

Superstrings: The View from Below

P.Ramond

Institute for Fundamental Theory, University of Florida, Gainesville, FL 32611

Abstract. We review the Standard Model in a form conducive to formulating its possible short distance extensions. This depends on the value of the Higgs mass, the only unknown parameter of the model. We suggest methods to reproduce many of the small numbers in the model in terms of scale ratios, applying see-saw like ideas to the breaking of chiral symetries. We then investigate how the N=1 Standard Model extrapolated to or near the Planck scale can fit superstring models, emphasizing the use of some non-renormalizable operators generic to superstrings.

I. Introduction

The Standard Model is a remarkably compact description of all fundamental matter, described in terms of only three gauge groups and nineteen parameters. Yet, it hardly looks like a fundamental theory; it probably is the low energy manifestation of a more integrated, more satisfying, and less broken-up theory. If it is indeed to be viewed as an effective low energy theory, there must exist an ultraviolet cut-off. The *raison d'être* and the value of this cut-off are the central question of fundamental theory. Since there is no experimental indication of its existence, it must be at least of the order of hundreds of GeVs. At the higher end, Nature provides us with its own cut-off, the Planck scale, the largest cut-off we can presently imagine to the standard model. However, it is far removed from present experimental scales.

Local field theories of gravity certainly break down at the Planck scale, and the only known possible cure is to formulate theories which deviate from space-time locality, superstring theories. Of these, the heterotic string seems to contain all the necessary ingredients needed to reproduce the low energy world. Unfortunately, there is no detailed matching between the standard model and string theory. Yet if our world has its origin in a superstring theory, there ought to be satisfactory matching of the two at some scale below the Planck scale. The knowledge of the standard model alone, may not be sufficient to identify this match. Still, effective theories derived from superstrings typically reproduce not only renormalizable interactions, such as those found in the standard model, but also a pattern of non-renormalizable interactions, some of which may provide low energy signatures for superstring models. In addition, these theories contain not only the observed chiral families, but also a number of vector-like particles, some with electroweak quantum numbers, but with hitherto undetermined $\Delta I_W=0$ masses. Supersymmetry at low energy provides another hint, although superstring theories do not yet predict its breaking scale.

We start by reviewing the standard model, and present arguments for low energy supersymmetry. The N = 1 standard model is perturbative until the Planck scale, where, we can hope to match it with superstring models. We present a mode of attack which, although based mostly on our incomplete knowledge of the Yukawa sector, hints at the presence of certain types of non-renormalizable terms, some of which are generic to superstrings.

II. The N = 0 Standard Model

Extension of the standard model predicts new phenomena at shorter distances, although none so far have distinguished themselves either by reproducing the *values* of the parameters, or even their multiplicity. Thus it is time to review the types of extensions which might generically explain the observed patterns.

The N = 0 standard model is described by three Yang-Mills groups, each with its own dimensionless gauge coupling, α_1 for the hypercharge U(1), α_2 for the weak isospin SU(2), and α_3 for QCD. QCD itself predicts strong CP violation, with strength proportional to a fourth dimensionless parameter $\bar{\theta}$.

The electroweak symmetry breaking Higgs sector contains two unknowns, a dimensionless Higgs self-coupling, and the Higgs mass. The "measured" value of the Fermi coupling accounts for one parameter, and the other is the value of the Higgs mass, the only parameter of the model yet to be determined from experiment. The Yukawa interactions between the fermions and the Higgs yields the nine masses of the elementary fermions. This sector also contains three mixing angles which monitor interfamily decays, and one phase which describes CP violation.

It also contains two dimensionful parameters, the Higgs mass, and the QCD confinement scale, obtained by dimensional transmutation. The QCD scale is a tiny number in Planck units $\Lambda_{QCD} \sim 10^{-20}\ M_{P\ell}$. This small number has a natural explanation due to the logarithmic variation of the QCD coupling with scale.

The Higgs mass is unknown, but the electroweak order parameter is determined by the Fermi constant. In terms of the Planck mass it is also very small $G_F^{-1/2} \sim 10^{-17}\ M_{P\ell}$. The origin of this small number is a matter of much speculation. In perturbation theory the Higgs mass is of the same order of magnitude as the electroweak order parameter. The most natural idea is to relate this number to dimensional transmutation associated with new strong forces just beyond electroweak scales, called technicolor. It yields a satisfying natural explanation of this value, but these models fail to reproduce the values of the fermion masses.

Another class of theories postulates supersymmetry [1] at TeV scales. There, the electroweak order parameter is related to that of supersymmetry breaking. While not at first sight very economical, the breaking of supersymmetry automatically generates electroweak breaking [2] in a wide class of theories. The beautiful ideas of technicolor can then be applied to supersymmetry breaking, without encountering the problem of fermion masses of technicolor applied to electroweak breaking.

There are many other numbers to explain, notably in the Yukawa sector of the theory. Quark and charged lepton masses break weak isospin by half a unit, along $\Delta I_W = 1/2$, and hypercharge by one unit, the same quantum numbers as the electroweak order parameter, which also gives the W-boson its mass. In this sense charged fermion masses should be of the same order as the W mass. This happens only for the top quark. The others are unnaturally small

$$\frac{m_{u,d}}{M_W} \sim O(10^{-4})\,; \quad \frac{m_s}{M_W} \sim O(10^{-3})\,; \quad \frac{m_c}{M_W} \sim O(10^{-2})\,; \quad \frac{m_b}{M_W} \sim .05.$$

Similarly for the charged leptons,

$$\frac{m_e}{M_W} \sim O(10^{-5})\,; \quad \frac{m_\mu}{M_W} \sim O(10^{-3})\,; \quad \frac{m_\tau}{M_W} \sim .02,$$

which range from the tiny to the small. Neutrino masses are predicted to be exactly zero in the standard model only because of the global chiral lepton number symmetries. However there is mounting experimental evidence that neutrinos have masses. In the absence of new degrees of freedom they are of the Majorana kind, and break weak isospin by one unit, as $\Delta I_W = 1$. Direct experimental limits on neutrino masses indicate that they are at most extremely small: $m_{\nu_e} < 10^{-17} M_W$.

The values of the three gauge parameters are known to great accuracy from measurements at low energy, although because of endemic problems associated with strong QCD, the color coupling is the least well known. Given these parameters, we can extrapolate the standard model to shorter distances, using the renormalization group perturbatively. The most interesting effect occurs in the extrapolation of the three gauge couplings. These hypercharge and weak isospin couplings meet at a scale of 10^{13} GeV, with a value $\alpha^{-1} \approx 43$, but at that scale, the QCD coupling is much larger, $\alpha_3^{-1} \approx 38$. Thus, although the quantum numbers indicate possible unification into a larger non-Abelian group, the gauge coupling do not follow suit in this naive extrapolation. Historically, before the couplings were measured to such accuracy, it was believed that all three did indeed unify in the ultraviolet. In the ultraviolet, the values of these couplings is less disparate than at experimental scales. Similarly, nothing spectacular occurs to the Yukawa couplings. For instance, the bottom quark and τ lepton Yukawa couplings meet around 10^9 GeV, and part in the deeper ultraviolet .

The situation is potentially more extreme in the Higgs sector because of the renormalization group behavior of the Higgs self coupling [3]. We can consider two cases, depending on the value of the Higgs mass. If it is below 150 GeV, the self-coupling turns negative somewhere below Planck scale. This results in a loss of perturbation theory, with a potential unbounded from below. Using the recently announced value of the top quark mass, a Higgs mass of 120 GeV means that "instability" sets in at 1 TeV, indicating some new physics at that scale. When operative, this bound provides a low (with respect to Planck mass) energy cut-off for the standard model.

If the Higgs mass is above 200 GeV, its self-coupling rises dramatically towards its Landau pole at a relatively low energy scale. It means loss of perturbative control of the theory, and sets an upper bound on the Higgs mass since there is no evidence of any strong electroweak coupling at experimental scales. Strong coupling must happen, meaning that the Higgs is a composite. An example of this view is the technicolor scenario where the Higgs is a condensate of techniquarks.

Within a tiny range of intermediate values for the Higgs mass, the instability and triviality bounds are pushed to scales beyond the Planck length, and there is no standard model prediction of new physics; the cut-off is indeed the Planck scale.

The dependence of the various standard model parameters on the cut-off is very different. Quantum fluctuations *additively* renormalize the Higgs mass with a term linearly proportional to the cut-off. Thus even if the Higgs mass is in a region that does not *technically* require new physics below Planck mass, its value is unnaturally small, if Planck mass is the cut-off. On the other hand, the cut-off dependence of any chiral fermion mass is only logarithmic. The reason is chiral symmetry, which is recovered by setting the fermion mass to zero. It affords a protection mechanism which results in this weak cut-off dependence.

III The N = 1 Standard Model

Supersymmetry avoids the *technical* naturalness problem by linking any fermion to a boson of the same mass. With exact supersymmetry, the boson mass is then protected by the chiral symmetry of the fermion. As long as supersymmetry is broken at energies in the range of TeV, this is enough protection to produce a low Higgs mass. This might seem to be small progress, since a new symmetry has been introduced to relax the strong cut-off dependence, a symmetry which has to be broken itself at a small scale, $V_{SUSY} \sim 10^{-15} M_{P\ell}$

In the N = 1 standard model, there are only gauge and Yukawa coupling constants. None of these couplings blow up below Planck mass. In particular, the perky Higgs self-coupling is replaced by the square of gauge and Yukawa couplings, which allows for the perturbative extrapolation all the way to Planck scale, opening the way for comparison with fundamental theory!

There are tantalizing hints of simplicity in the extrapolation of the couplings. Firstly the gauge couplings seem to be much closer to unification, and at a scale large enough not to be invalidated by proton decay bounds. The hypercharge and weak isospin couplings meet at a scale of the order of 10^{16} GeV, with a value $\alpha^{-1} \approx 25$, and the QCD coupling is much closer to, if not right on the same value [4]. It may still be a shade higher than the others, with $(\alpha^{-1} - \alpha_3^{-1}) \leq 1.5$

The second remarkable thing is that with simple boundary conditions at or near Planck mass, inspired by a simple picture of supersymmetry breaking, the renormalization group drives one of the Higgs masses to imaginary values in the infrared. This in turns triggers electroweak breaking [2], made possible by the large top quark mass!

The Higgs mass is not arbitrarily high in the minimal extension. At tree-level, it is predicted to be below the Z-mass, but it suffers large radiative corrections due to the top Yukawa coupling, raising it above the Z, but not by an arbitrarily large amount [5].

This general scheme allows us to study the pattern of fermion masses at these shorter distances; there are more regularities with supersymmetry. For instance, the bottom quark and t masses seem to unify at or around 10^{16-17} GeV [6], the same scale where the gauge couplings converge.

The most striking aspect of the fermion masses is that only one chiral family has large masses, leading us to consider theories where the tree-level Yukawa matrices are simply of the form.

$$\mathbf{Y}_{u,d,e} = y_{t,b,\tau} \begin{pmatrix} 0 & 0 & 0 \\ 0 & 0 & 0 \\ 0 & 0 & 1 \end{pmatrix}$$

The matrices imply a global chiral symmetry, $U(2)_L \times U(2)_R$, in each charged sector. The hierarchy between the bottom and top quark masses requires explanation. In the N = 1 model, it is linked to another parameter which comes from the Higgs sector, the ratio of the *vev* of the two Higgs. Hence it may not pertain to properties of the Yukawa matrices. Why are the other two families so light? Starting from the rank two Yukawa matrices, we must find a scheme by which the zeros get filled, presumably in higher orders of perturbation theory. In order to see how this might come about, let us examine one well-known case in which small numbers are naturally generated, the see-saw mechanism [7].

In the standard model, the neutrino Majorana mass matrix is zero at tree-level. A detailed examination shows that these zeros are protected from quantum corrections by conservation of chiral global lepton number for each species. In the see-saw mechanism, the usual neutrinos are mixed with new electroweak singlet fields (neutral leptons), by $\Delta I_W = 1/2$ terms, of electroweak breaking strength, to give them the same lepton numbers. These new particles can acquire $\Delta I_W = 0$ Majorana masses, M, of any magnitude, in particular well above the electroweak scale. Upon diagonalization, this generates a mass for the familiar neutrinos, depressed from typical electroweak values by the ratio of scale m/M, where m is the electroweak order parameter. A scale ratio between electroweak and chiral lepton number breaking is used to generate a small number.

A similar analysis may apply to the charged Yukawa matrices, where the zeros are also protected by chiral symmetries. We first couple the massless fermions of the first two families to new fermions with similar quantum numbers, thereby extending the chiral symmetries to them. Unlike the neutral case, these new fermions have electroweak charges, and cannot have Majorana masses. Breaking of chiral symmetry is done by their Dirac masses, which requires the presence of vector-like partners (this differs from the neutral sector), along the $\Delta I_W = 0$ direction at a new undetermined scale M.

One may also take the point of view that these non-renormalizable operators come from physics beyond the Planck scale, in which case, the question is relegated to one of classifying the possible non-renormalizable operators.

Now that low energy supersymmetry allows its perturbative extrapolation deep in the ultraviolet, we may ask how it can be made to match with more fundamental theories, one type being superstrings.

IV. Matching to Superstrings

Superstring theories are not understood in detail, but some of their generic features are evident. We are interested in the effective theory they generate at or near Planck scale. An estimate of string effects indicates that this scale is related to the gauge coupling through the formula [8]

$$M_U \approx 2.5 \sqrt{\alpha_U} \times 10^{18}\,\mathrm{GeV}.$$

With $M_X = 10^{16}$ GeV, and $\alpha_U^{-1} < \alpha_X^{-1} \approx 25$, this implies that contact with the superstring can be made provided that $M_U / M_X > 50$: there is a discrepancy in the matching of scales.

The second feature of superstring theories is to produce at lower energies remnants of **27** and $\overline{\mathbf{27}}$ representations of E_6. In the effective low energy theory, these yield the three chiral families, together with many vector-like particles, capable of sharing quantum numbers with the chiral particles. These particles may be used in see-saw like mechanisms to generate small numbers in the Yukawa matrices. It also means that there may be many intermediate thresholds between the supersymmetry and unification scales.

A third feature of superstring theories is that the gauged group at the string scale is larger than the standard model group. This implies the existence of more gauge bosons at intermediate scales and many vector-like particles with electroweak quantum numbers. The values of their masses to be determined by the flat directions in the superpotential, and the discrete symmetries of the particular model.

A fourth generic feature is the existence of a local U(1) symmetry, with anomaly cancelled through the Green-Schwarz mechanism. This symmetry is however broken close to the Planck scale. Does any trace of this symmetry appear in the extrapolated low energy standard model? Ibàñez [9] has argued that this symmetry can be used to fix the Weinberg angle in superstring theories. Following Ibàñez and Ross [10], we argue [11] that this Abelian symmetry sets the dimensions of the Froggart and Nielsen [12] Yukawa operators. Are any of these features present in the extrapolated low energy theory?

Consider first the unification of the gauge couplings. It is predicated on two assumptions: that the weak hypercharge coupling is normalized to its unification into a higher rank Lie group, such as SU(5), SO(10) or E_6 [13], and on the absence of intermediate thresholds with matter carrying strong or electroweak quantum numbers between 1 TeV and 10^{16} GeV. The gauge couplings may not exactly unify at M_X, and we may want to alter this simple picture by requiring at least one intermediate threshold between the SUSY scale and the illusory unification scale at M_X to obtain unification at the string scale M_U [14].

At one-loop, the couplings $\alpha_i^{-1}(t)$ for the three gauge groups, (i = 1,2,3 for $U(1)_Y$, $SU(2)_L$, and $SU(3)^c$, respectively) run with scale according to

$$\alpha_i^{-1}(t) = \alpha_i^{-1}(t_X) + \frac{b_i}{2\pi}(t - t_X),$$

where

$$t = \ln(\mu / \mu_0), \qquad t_X = \ln(M_X / \mu_0),$$

and μ_0 is an arbitrary reference energy. For the three families and two Higgs doublets of the minimal supersymmetric standard model, we have

$$b_1 = -\frac{33}{5} \; ; \qquad b_2 = -1; \, b_3 = 3.$$

Since the low energies value of α_1 and α_2 are known with the greatest accuracy, we use their trajectories to define t_X as the scale at which they meet:

$$\alpha_X^{-1} \equiv \alpha^{-1}(t_X) = \alpha_2^{-1}(t_X).$$

The extrapolated data show that $\alpha_X^{-1} \approx 25$, with $M_X \approx 10^{16}$ GeV. We do not assume precisely the same value for $\alpha_3(t_X)$ at that scale; rather we set

$$\alpha_X^{-1} = \alpha_3^{-1}(t_X) + \Delta.$$

Present uncertainties on the QCD coupling suggest that

$$|\Delta| \leq 1.5.$$

Suppose there is an intermediate threshold above supersymmetry at

$$t_I = \ln(M_I / \mu_0); \quad t_I < t_X,$$

caused by new vector-like particles with electroweak singlet masses at M_I. Their effect is to alter the b_i coefficients:

$$b_i \to b_i - \delta_i, \; i = 1,2,3.$$

By requiring unification at M_U, we find the constraints

$$\frac{r}{14} = \frac{t_U - t_X}{t_U - t_I}$$

and

$$\frac{q}{4} = \frac{t_U - t_X - \pi\Delta/2}{t_U - t_I} ,$$

written in terms of

$$q \equiv \delta_3 - \delta_2 \quad \text{and} \quad \frac{2}{5} r = \delta_2 - \delta_1 .$$

For vector-like matter generated from superstrings, q and r are integers. The value of the gauge coupling at unification is now

$$\alpha_U^{-1} = \alpha_X^{-1} - \frac{1}{2\pi} \left[\delta_2(t_U - t_I) + t_U - t_X\right].$$

These equations have solutions for non-exotic matter. For instance when $\Delta = 0.82$ with $r = 5$, $q = 1$, we get

$$M_U = 7.5 \times 10^{17}\ \text{GeV}; \qquad M_I = 4.4 \times 10^{12}\ \text{GeV}; \qquad \alpha_U^{-1} = 11.$$

However most solutions do not allow large M_X/M_I.

In realistic superstring models, the assumption of one intermediate scale is probably not justified. For several intermediate thresholds, by applying these equations repeatedly, we obtain similar equations, with q and r replaced by average quantities which are no longer integers. Take for instance the interesting example of the 3-family Calabi-Yau superstring model of ref. [15]. After flux breaking, the surviving gauge group is

$$SU(3)_L \times SU(3)^c \times SU(3)_R .$$

There are at least two *a priori* distinct intermediate scale order parameter associated with each reduction in rank. Many chiral superfields survive flux breaking: 9 leptons, 6 mirror leptons, 7 quarks, 4 mirror quarks, 7 antiquarks, 4 mirror antiquarks. With all these particles concentrated at one mass scale, there is no solution, but this is hardly realistic. It is convenient to define the effective intermediate scale as the average intermediate scale weighted by δ_2, i.e.,

$$t_I \equiv \frac{\sum_{a=1}^N t_{Ia}\, \delta_{2a}}{\sum_{a=1}^N \delta_{2a}} = \frac{1}{29} \sum_{a=1}^N t_{Ia}\, \delta_{2a} .$$

Taking 5 x 10^{17} GeV as a minimum for M_U, we find the high vaule.

$$M_{\bar{I}} > 3 \times 10^{15} \text{ GeV}$$

Gauge coupling unification can be attained in this example in a calculably perturbative way, but it requires that many of the vector-like particle which survive flux breaking be very close to the string scale, and that electroweak-doublet vectorlike particles be heavier on average than the strongly-interacting electroweak singlet vectorlike particles.

It might seem rather surprising that in the MSSM the gauge couplings should appear to be nicely headed for unification at M_X, only to be redirected to a new meeting place at M_U, but the apparent perverseness of this situation allows us put some non-trivial constaints on the scenario.

Let us now turn to the last topic, the possibility of an Abelian gauge symmetry, with anomaly cancelled by the Green-Schwarz mechanism. Such can be recognized if it plays a role in determining the dimensions of the entires of the Yukawa matrices [10,11]. The most general Abelian charge that can be assigned to the particles of the Minimal Supersymemtric Standard Model, with μ term, can be written as

$$X = X_0 + X_3 + \sqrt{3}\, X_8 \ ,$$

where X_0 is the family independent part, X_3 is along λ_3, and X_8 is along λ_8. We set

$$X_{3,8} = (a_{3,8}, b_{3,8}, c_{3,8}, d_{3,8}, e_{3,8}),$$

where the entries correspond to the components in the family space of the fields **Q, $\bar{\mathbf{u}}$, $\bar{\mathbf{d}}$**, L, and $\bar{e}$, respectively. Both Higgs doublets have the same zero X-charge, without loss of generality, since an imbalance can be created by mixing in the hypercharge Y.

With the tree-level Yukawa coupling *only* to the third family, we obtain the constraints.

$$\frac{m+n}{3} = 2(a_8 + c_8), \quad \frac{m+p}{3} = 2(a_8 + c_8), \quad \frac{q+r}{3} = 2(d_8 + e_8).$$

The excess X-charge at each of their entries is made up by powers of an electroweak singlet field, resulting in operators of higher dimensions. A typical term would be of the form

$$Q_i \bar{u}_j H_u \left(\frac{\theta}{M}\right)^{n_{ij}}$$

where θ is some field with charge x, M is some large scale, and n_{ij} is the excess X–charge listed above for the Yukawa matrices. The exponents are determined by X–symmetry. In order to produce a small coefficient, the *i*th and *j*th fermions need to go through a number of intermediate steps to interact. The larger the number steps, the larger n_{ij}, and the smaller the entry in the effective Yukawa matrix. This approach was advocated long ago by Froggart and Nielsen [12]. This yields approximate zeros in the matrices, creating textures [16]. For example, in the charge 2/3 sector,

$$n_{12}x = 3(a_8 + b_8) + a_3 - b_3.$$

Since θ may have a large expectation value, it is likely accompanied by its vector-like partner $\bar{\theta}$, with opposite charge, showing that the exponents n_{ij} need not be positive, but if all the n_{ij} are positive, several interesting phenomenological consequences follow [11]. First the n_{ij} exponents are not all independent, resulting in order of magnitude estimates among the Yukawa matrix elements

$$(Y)_{11} \sim \frac{(Y)_{13}(Y)_{31}}{(Y)_{33}},$$

$$(Y)_{22} \sim \frac{(Y)_{23}(Y)_{32}}{(Y)_{33}},$$

valid for each of the three charge sectors. These relations are consistent with many of the allowed textures. Another important consequence is that the X-charge of the determinant in each charge sector is *independent* of the texture coefficients that distinguish between the two lightest families

$$\text{charge } \frac{2}{3}: 6(a_8 + b_8) \equiv U, \text{ charge } -\frac{1}{3}: 6(a_8 + c_8) \equiv D, \text{ charge } -1: 6(d_8 + e_8) \equiv L.$$

Let the value of $\frac{\theta}{M}$ be a small parameter λ. In the simplest case, this parameter would be the same for all three charge sectors. Then we have

$$\frac{m_d m_s d_b}{m_e m_\mu m_\tau} \sim O(\lambda^{(D-L)/x})$$

It is more difficult to compare the up and down sectors in this way since we do not know the value of tan β, which sets the normalization between the two sectors

$$\frac{m_u m_c d_t}{m_d m_s m_b} \sim \tan^3 \beta \text{ x } O\,(\lambda^{(U-D)/x}).$$

Since this ratio is much larger than one, it means either that tan β is itself large, with U close to D, or that tan β is not large, but D > U.

In general, the X symmetry is anomalous. The three chiral families contribute to the mixed gauge anomalies as follows

$$C_3 = 2m + n + p,$$

$$C_2 = 3m + q + 2s,$$

$$C_1 = \frac{1}{3}\,m + \frac{8}{3}\,n + \frac{2}{3}\,p + q + 2r + 2s.$$

The subcript denotes the gauge group of the Standard Model, i.e. 1 ~ U(1), 2 ~ SU(2), and 3 ~ SU(3). The X-charge also has a mixed gravitational anomaly, which is simply the trace of the X-charge,

$$C_g = (6m + 3n + 3p + 2q + r + 4s) + C'_g,$$

where C'_g is the contribution from the particles that do not appear in the minimal N = 1 model. The last anomaly coefficient is that of the X-charge itself, C_X, which is the sum of the cubes of the X-charge.

It was suggested by Ibàñez [9], that an anomalous U(1) symmetry, with its anomalies cancelled through the Green-Schwarz mechanism, is capable of relating the ratio of gauge couplings to the ratios of anomaly coefficients

$$\frac{C_i}{k_i} = \frac{C_X}{k_X} = \frac{C_g}{k_g},$$

which relates the Weinberg angle to the anomaly coefficients, without the use of Grand Unification. The k_i are the Kac-Moody levels; are integers for the non-Abelian factors only. The mixed Y X X anomaly, however, must vanish by itself.

We demand that the non-Abelian gauge groups have the same Kac-Moody levels, which means that

$$C_2 = C_3 \quad \text{or} \quad q = n + p - m - 2s.$$

Secondly we require that at or near the unification or string scale, the Weinberg angle have the value

$$\sin^2\theta_W = \frac{3}{8} ,$$

which translates into the further constraint

$$5C_2 = 3C_1 \quad \text{or} \quad r = 2m - n.$$

These equations are sufficient to infer that L = D, which implies, remarkably enough, that the products of the charged lepton masses is of the same order of magnitude as that of the down-type quarks [11]. It is satisfying to note that extrapolation of the masses to short distances indicates that these two products are in fact roughly equal in the deep ultroviolet.

This formalism has been used [10] to generate symmetric textures, of the kind found to be allowed by experiment [16]. Work is in progress to determine how these equations constrain possible textures. One result is that it appears to be difficult to generate acceptable constraints, without invoking Green–Schwarz cancellation. In that case, this particular way of generating textures would require the type of mechanism that is generic to superstrings!

The following examples have shown how several problems with the standard model might require a superstring explanation. While it is clearly too soon to claim to have made the connection, it is a fruitful path to take, as we must try to match the apparent unruliness and ugliness of the world we observe to the more beautiful and satisfying contructs of our imagination. Feza would have liked that.

I wish to thank Professor Serdaroğlu for her kind hospitality during the conference. This work was supported in part by the United States Department of Energy under contract No. DEFG05-86-ER-40272.

References

[1] For reviews, see H.P. Nilles, Phys. Rep. **110** (1984) 1 and H.E.Haber and G.L.Kane, Phys. Rep. **117** (1985) 75.

[2] L.E.Ibáñez and G.G. Ross, Phys. Lett. **110B** (1982) 215; K.Inoue, A.Kakuto, H. Komatsu, and S.Takeshita, Prog.Theor. Phys. **68** (1982) 927; L. Alvarez-Gaumé, M. Claudson, and M. Wise, Nucl. Phys. **B207** (1982) 16; J.Ellis, J.S. Hagelin, D.V.Nanopoulos, and K.Tamvakis, Phys. Lett. **125B** (1983) 275.

[3] M.Sher, Phys. Rep. **179** (1989), 273.

[4] U.Amaldi, W. de Boer, and H.Furstenau, Phys. Lett **B260** (1991) 447; J.Ellis, S. Kelley and D.Nanopoulos, Phys. Lett. **260B** (1991) 131; P.Langacker and M.Luo, Phys. Rev. **D44** (1991) 817.

[5] G.L.Kane, C.Kolda, and J.D.Wells, Phys. Rev. Lett. **70** (1993) 2686.

[6] H.Arason, D.J. Castano, B.Keszthelyi, S. Mikaelian, E.J.Piard, P.Ramond, and B.D.Wright, Phys. Rev. Lett. **67** (1991) 2933; A.Givenon, L.J.Hall, and U.Sarid, Phys. Lett. **271B** (1991) 138.

[7] M.Gell-Mann, P.Ramond, and R.Slansky in Sanibel Talk, CALT-68-709, Feb 1979, and in *Supergravity* (North Holland, Amsterdam 1979). T.Yanagida, in *Proceedings of the Workshop on Unified Theory and Baryon Number of the Universey* KEK, Japan, 1979.

[8] J.P. Derendinger, S.Ferrara, C.Kounnas, and F.Zwirner, Nucl. Phys. **B372** (1992) 145.

[9] L.Ibáñez, Phys. Lett. **B303** (1993) 55.

[10] L.Ibáñez and G.G.Ross, Phys. Lett. **B332** (1994) 100.

[11] P.Binétruy and P.Ramond, in preparation.

[12] C.Froggart and H.B. Nielsen Nucl. Phys. **B147** (1979) 277.

[13] J.C.Pati and A.Salam, Phys. Rev. **D10** (1974) 275; H.Georgi and S.Glashow, Phys. Rev. Lett. **32** (1974) 438; H.Georgi, in *Particles and Fields* 1974, edited by C.E.Carlson, AIP Conference Proceedings No.23 (American Institute of Physics, New York, 1975) p.575; H.Fritzsch and P.Minkowski, Ann. Phys. NY **93** (1975) 193; F.Gürsey, P.Ramond, and P.Sikivie, Phys. Lett. **60B** (1975) 177.

[14] S.Martin and P.Ramond, in preparation.

[15] B.R.Greene, K.H.Kirklin, P.J.Miron, and G.G. Ross, Phys. Lett, **B180** (1986) 69; Nucl. Phys. **B278** (1986) 667; Nucl. Phys. B292 (1987) 606.

[16] P.Ramond, R.G. Roberts and G.G. Ross, Nucl. Phys. **B406** (1993) 19.

Note on Holomorphic Anomalies in Topological Field Theories

Hirosi Ooguri

Research Institute for Mathematical Sciences, Kyoto University
Kyoto 606-01, Japan
and
Department of Physics, University of California at Berkeley
Berkeley, CA 94702, U.S.A.

1. Introduction

A model in quantum field theory usually has several parameters, and it is important to understand how various quantities one can compute in such a model depend on the parameters. In the case of the non-linear-sigma-model, the moduli of the target space can be regarded as such parameters. Thus we are interested in how amplitudes in the non-linear sigma-model behave as we change the geometry of the target space.

In the case of the N=2 supersymmetric sigma-model in two dimensions, and when the target space is a Ricci-flat Kähler manifold (Calabi-Yau manifold) M, the model becomes ultra-violet finite and the N=2 superconformal algebra is realized as symmetry of the model. One can make *topological string theory* from such a model. Recently we have derived a set of differential equations for vacuum loop amplitudes of the topological string theory when M is complex three-dimensional [1] [2].

Before describing the equations, I should explain some (local) properties of the moduli space of M. The moduli space locally splits into two subspaces; one parametrizes the complex structure of M and the other the Kähler class. Since the two subspace have similar structure (due to the mirror symmetry [3]), here I concentrate on the moduli space of the complex structure of M and call it $\mathcal{M}$. When dim M=3, it is known that [4] [5]:

(1) $\mathcal{M}$ is complex manifold with dimensions given by dim $\mathcal{M}$ =dim $H^{2,1}(M)$

(2) $\mathcal{M}$ is a Kähler manifold, i.e. its Weil-Peterson metric $G_{i\bar{j}}(i, \bar{j}=1,\ldots, dim\, H^{2,1}(M))$ can be locally written as $G_{i\bar{j}}=\partial_i\bar{\partial}_{\bar{j}}K$ for some function K which is defined in a neighbourhood of each point on $\mathcal{M}$. This (locally defined) function K is called a Kähler potential.

(3) $\mathcal{M}$ is a restricted Kähler manifold, i.e. there is a holomorphic line bundle $\mathcal{L}$ on $\mathcal{M}$ whose metric on the fiber is given by e^K.

(4) $\mathcal{M}$ is a special Kähler manifold, i.e. there is a global holomorphic section C_{ijk} of $\mathcal{L}^2 \mathrm{x}\, Sym^{\otimes 3} T^{*(1,0)}\mathcal{M}$ and the Riemann curvature computed from the metric $G_{i\bar{i}}$ is expressed as

$$R_{i\bar{j}k\bar{l}} = G_{i\bar{j}}G_{k\bar{l}} + G_{i\bar{l}}G_{k\bar{j}} - e^{2K}G^{m\bar{n}}C_{ikm}\bar{C}_{\bar{j}\bar{k}\bar{l}}. \tag{1.1}$$

The holomorphic section C_{ijk} is called *Yukawa coupling*. Once we know C_{ijk}, we can use this equation to compute the metric $G_{i\bar{j}}$. For the complex moduli space of the Calabi-Yau 3-fold M, C_{ijk} is given by

$$C_{ijk} = \int_M \Omega\partial_i\partial_j\partial_k\Omega,$$

where Ω is the unique holomorphic 3-form on M.

We can now write down the equations for the g-loop vacuum amplitude F_g of the topological string theory on M. The one-loop amplitude F_1 is a scalar function on $\mathcal{M}$ and it is shown in [1] that it satisfies

$$\partial_i\bar{\partial}_{\bar{j}}F_1 = \frac{1}{2}e^{2K}G^{k\bar{k}}G^{l\bar{l}}C_{ikl}\bar{C}_{\bar{j}\bar{k}\bar{l}} - \left(\frac{\chi}{24} - 1\right)G_{i\bar{j}} \tag{1.2}$$

where χ is the Euler characteristic of M. For $g \geq 1$, Fg is a section of $\mathcal{L}^{2-2g}$ rather than a function, and it is shown in [2] that it solves

$$\bar{\partial}_{\bar{i}}F_g = \frac{1}{2}\bar{C}_{\bar{i}\bar{j}\bar{k}}e^{2K}G^{j\bar{j}}G^{k\bar{k}}\left(D_jD_kF_{g-1} + \sum_{r=1}^{g-1}D_jF_rD_kF_{g-r}\right) \tag{1.3}$$

where D_i is a covariant derivative. These equations are called the holomorphic anomaly equations since they are related to a subtle breakdown of the N=2 supersymmetry in the topological string.

The one-loop equation can be easily integrated. By using the relation, we can rewrite (1.2) as

$$\partial_i\bar{\partial}_{\bar{j}}F_1 = -\frac{1}{2}R_{i\bar{j}} + \frac{1}{2}\left(3 + \dim\mathcal{M} - \frac{\chi}{12}\right)G_{i\bar{j}}.$$

Since $G_{i\bar{j}}$ is Kähler, its Ricci curvature is given by $R_{i\bar{j}} = \partial_i\bar{\partial}_{\bar{j}}\,log\,det\,G$. Therefore we can integrate this equation to obtain

$$F_1 = \frac{1}{2}\log\left[\exp\left[(3 + \dim\mathcal{M} - \frac{\chi}{12})\right][\det G]^{-1}|f|^2\right],$$

where f is some holomorphic object on $\mathcal{M}$. In ref. [1] we were able to determine f for some class of Calabi-Yau manifolds, and thus obtained explicit expressions for F_1 in these cases.

In this note, I would like to explain how we can integrate the equation (1.3) for $g \geq 1$ We shall see that the solution for the equations can be constructed using a set of rules similar to the Feynman rule.

2. Feyman rules at g = 2,3

As a warm-up, let us start with the genus-2 case.

$$\bar{\partial}_{\bar{i}}F_2 = \frac{1}{2}\bar{C}_{\bar{i}\bar{j}\bar{k}}e^{2K}G^{j\bar{j}}G^{k\bar{k}}\left(D_j\partial_kF_1 + \partial_jF_1\partial_kF_1\right). \tag{2.1}$$

Interestingly enough, a key to solving this equation lies in a genus-0 object. Because the Yukawa-coupling $\overline{C}_{\bar{i}\bar{j}\bar{k}}$ is totally symmetric in its indices and satisfies

$$D_{\bar{i}}\overline{C}_{\bar{j}\bar{k}\bar{l}} = D_{\bar{j}}\overline{C}_{\bar{i}\bar{k}\bar{l}},$$

we can always *integrate* the Yukawa coupling locally as

$$\overline{C}_{\bar{i}\bar{j}\bar{k}} = e^{-2K} D_{\bar{i}} D_{\bar{j}} \bar{\partial}_{\bar{k}} S. \tag{2.2}$$

where S is a local section of $\mathcal{L}^{-2}$. In fact, in all the examples we will discuss later, it is possible to construct S globally on the moduli space of the topological theories. We will present such constructions later in this note. To simplify the expressions below, we use the following notation.

$$\begin{aligned} S_{\bar{i}} &\equiv \bar{\partial}_{\bar{i}} S \\ S_{\bar{i}}^{j} &\equiv \bar{\partial}_{\bar{i}} S^{j}, \quad \text{where } S^{j} \equiv G^{j\bar{j}} S_{\bar{j}} \end{aligned} \tag{2.3}$$

In this notation,

$$\overline{C}_{\bar{i}}^{jk} = \bar{\partial}_{\bar{i}} S^{jk} \tag{2.4}$$

where

$$\overline{C}_{\bar{i}}^{jk} \equiv \overline{C}_{\bar{i}\bar{j}\bar{k}} e^{2K} G^{j\bar{j}} G^{k\bar{k}}, \quad S^{jk} \equiv G^{j\bar{j}} S_{\bar{j}}^{k}.$$

We now solve the genus-2 equation (2.1) by "integration-by-parts". We first rewrite (2.1) using (2.4) as

$$\bar{\partial}_{\bar{i}} \left[F_2 - \frac{1}{2} S^{jk} (D_j \partial_k F_1 + \partial_j F_1 \partial_k F_1) \right] = -\frac{1}{2} S^{jk} \bar{\partial}_{\bar{i}} (D_j \partial_k F_1 + \partial_j F_1 \partial_k F_1)$$

The r.h.s. can be evaluated using the holomorphic anomaly of F_1 and the special geometry relation for $[\,\bar{\partial}_{\bar{i}}, D_j\,]$ as

$$\begin{aligned} &-\frac{1}{2} S^{jk} \bar{\partial}_{\bar{i}} (D_j \partial_k F_1 + \partial_j F_1 \partial_k F_1) = \\ &= -\frac{1}{2} \overline{C}_{\bar{i}}^{mn} S^{jk} \left(\frac{1}{2} C_{nmjk} + C_{mnj} \partial_k F_1 + C_{jkm} \partial_n F_1 \right) + \frac{\chi}{24} S_{\bar{i}}^{j} \partial_j F_1 \end{aligned}$$

Now we repeat the integration-by-parts.

$$\begin{aligned} \bar{\partial}_{\bar{i}} \Bigg[F_2 &- \frac{1}{2} S^{jk} (D_j \partial_k F_1 + \partial_j F_1 \partial_k F_1) + \\ &+ \frac{1}{4} S^{mn} S^{jk} \left(\frac{1}{2} C_{nmjk} + 2 C_{mnj} \partial_k F_1 \right) - \frac{\chi}{24} S^{j} \partial_j F_1 \Bigg] = \\ &= \frac{1}{4} S^{mn} S^{jk} \bar{\partial}_{\bar{i}} \left(\frac{1}{2} C_{nmjk} + 2 C_{mnj} \partial_k F_1 \right) - \frac{\chi}{24} S^{j} \bar{\partial}_{\bar{i}} \partial_j F_1. \end{aligned}$$

It turns out that r.h.s. of this equation can also be written as a total derivative with respect to $\bar{t}^{\bar{i}}$. By using the genus-1 anomaly (1.2) and the special geometry relation (1.1), we find

$$\frac{1}{4}S^{mn}S^{jk}\overline{\partial}_{\bar{i}}\left(\frac{1}{2}C_{nmjk}+2C_{mnj}\partial_k F_1\right)-\frac{\chi}{24}S^j\overline{\partial}_{\bar{i}}\partial_j F_1=$$

$$=\overline{\partial}_{\bar{i}}\Bigg[S^{jk}S^{pq}S^{mn}(\frac{1}{8}C_{jkp}C_{mnq}+\frac{1}{12}C_{jpm}C_{kqn})-$$

$$-\frac{\chi}{48}S^jC_{jkl}S^{kl}+\frac{\chi}{24}(\frac{\chi}{24}-1)S\Bigg].$$

Thus the iteration stops here. We have converted the genus-2 anomaly equation (2.1) into the following form.

$$\overline{\partial}_{\bar{i}}F_2=$$

$$=\overline{\partial}_{\bar{i}}\Bigg[\frac{1}{2}S^{jk}D_j\partial_k F_1+\frac{1}{2}S^{jk}\partial_j F_1\partial_k F_1-\frac{1}{8}S^{jk}S^{mn}C_{jkmn}-$$

$$-\frac{1}{2}S^{jk}C_{jkm}S^{mn}\partial_n F_1+\frac{\chi}{24}S^j\partial_j F_1+$$

$$+\frac{1}{8}S^{jk}C_{jkp}S^{pq}C_{qmn}S^{mn}+\frac{1}{12}S^{jk}S^{pq}S^{mn}C_{jpm}C_{kqn}-$$

$$-\frac{\chi}{48}S^jC_{jkl}S^{kl}+\frac{\chi}{24}(\frac{\chi}{24}-1)S\Bigg]. \tag{2.5}$$

Now one can easily integrate this equation as

$$F_2=\frac{1}{2}S^{ij}C^{(1)}_{ij}+\frac{1}{2}C^{(1)}_iS^{ij}C^{(1)}_j-\frac{1}{8}S^{jk}S^{mn}C_{jkmn}-$$

$$-\frac{1}{2}S^{ij}C_{ijm}S^{mn}C^{(1)}_n+\frac{\chi}{24}S^iC^{(1)}_i+$$

$$+\frac{1}{8}S^{ij}C_{ijp}S^{pq}C_{qmn}S^{mn}+\frac{1}{12}S^{ij}S^{pq}S^{mn}C_{ipm}C_{jqn}-$$

$$-\frac{\chi}{48}S^iC_{ijk}S^{jk}+\frac{\chi}{24}(\frac{\chi}{24}-1)S+f_2(t) \tag{2.6}$$

where we used the notation $C^{(g)}_{i_1 \ldots i_n}=D_{i_1}\ldots D_{i_n}F^{(g)}$.

Here $f_2(t)$ is some meromorphic object which is not fixed at this stage. Since both F_2 and S are section of $\mathcal{L}^{-2}$ and C_{ijk} is a section of $\mathcal{L}^2\, x Sym^{\otimes 3}T^{(1,0)}\ \mathcal{M}$ on the moduli space, f_2 must be a meromorphic section of $\mathcal{L}^{-2}$. Although we cannot determine f_2 from the holomorphic anomaly alone, the holomorphicity gives rather stringent constraints on f_2 and, in many cases, almost uniquely determines it. In the case of the sigma-model, we can exploit the geometric meaning of F_2 studied to fix f_2.

This method also works in the case of $g=3$. After six iterations of integration-by-parts, we obtain

$$
\begin{aligned}
F_3 =& \frac{1}{2}S^{ij}C_{ij}^{(2)} + C_i^{(1)}S^{ij}C_j^{(2)} + (\frac{\chi}{24}+2)S^iC_i^{(2)} + \\
&+ 2F_2S^iC_i^{(1)} - \frac{1}{2}S^{ij}C_{ijk}S^{kl}C_l^{(2)} - \frac{1}{4}S^{ij}S^{kl}C_{ijkl}^{(1)} - \\
&- \frac{1}{2}S^{ij}C_{ijk}^{(1)}S^{kl}C_l^{(1)} - \frac{1}{4}S^{ij}S^{kl}C_{ik}^{(1)}C_{jl}^{(1)} + \\
&+ \cdots \text{(it would take five more pages to write them all)} \cdots + f_3(t)
\end{aligned}
\tag{2.7}
$$

Here $f_3(t)$ is a meromorphic section of $\mathcal{L}^{-4}$.

One may observe that the equations (2.6) and (2.7) have a strong resemblance to the Feynman rule. Consider a finite dimensional quantum system with $(-S^{ij})$ as a propagator connecting the indices i and j, C_{ijk}, C_{ijkl} ...as classical vertices, $C_i^{(1)}$, $C_{ij}^{(1)}$,... as one-loop corrected vertices etc, and compute two- and three-loop partition functions according to the Feynman rule. If we multiply an overall factor of (–1) after the computation, we reproduce all the terms in (2.6) and (2.7) including all the symmetry factors, except for those containing S^i, S and the holomorphic sections f_2 and f_3.

The terms involving S^i and S can also be recovered if we introduce one more degree of freedom φ and extend the Feynman rule as follows. The propagators are given as

$$
K^{ij} = -S^{ij}, \quad K^{i\varphi} = -S^i, \quad K^{\varphi,\varphi} = -2S
\tag{2.8}
$$

and the vertices are given by

$$
\begin{aligned}
&\tilde{C}^{(g)}_{i_1\cdots i_n,\varphi^{m+1}} = (2g-2+n+m)\tilde{C}^{(g)}_{i_1\cdots i_n,\varphi^m} \\
&\tilde{C}^{(g)}_{i_1\cdots i_n} = C^{(g)}_{i_1\cdots i_n}, \qquad \tilde{C}^{(1)}_{\varphi} = \frac{\chi}{24} - 1. \\
&\tilde{C}^{(0)}_{\varphi^m} = 0, \quad \tilde{C}^{(0)}_{i,\varphi^m} = 0, \quad \tilde{C}^{(0)}_{ij,\varphi^m} = 0, \quad \tilde{C}^{(1)} = 0.
\end{aligned}
\tag{2.9}
$$

Compute two- and three-loop partition functions using this Feynman rule and multiply the overall factor of (–1) after the computation. By adding the meromorphic sections f_2 and f_3, we recover the expressions (2.6) and (2.7). The definition (2.9) of the vertices reminds us of the puncture equation in the topological gravity. In fact, we will now identify the variable φ with the dilaton which is the first topological descendant of the puncture operator [6] $\sigma_1(P)$. All the other topological descendants decouple from the correlation functions simply by the $U(1)$ charge conservation and thus *the only non-vanishing correlation functions involve those of marginal fields and the dilaton field.* So far we have only discussed the marginal fields. To properly discuss the dilaton field coupling we need to enlarge the field space from that of pure topological theory. However luckily the correlation for the dilaton field can quite generally be eliminated from correlation functions by the recursion relations. In fact the first equation in (2.9) is precisely the general recursion relation of [6] and so φ is indeed the dilaton field.

3. Feynman rules for arbitrary g

The emergence of the Feynman rule is rather mysterious from the way we discovered it at g = 2 and 3. It would be extremely difficult to prove the Feynman rule for g ≥ 4 by using the method in the above since the number of iterations would grow exponentially. Thus we will develop another technique which enables us to prove the Feynman rule directly for all g. We will do so by reducing the Feynman rule for F_g to the Schwinger-Dyson equation of the finite dimensional system. Let us consider a generating function $W(\lambda, x, \varphi, t, \bar{t})$ for the vertices $\tilde{C}^{(g)}_{i_1\ldots i_n,\varphi^m}$ of the Feynman rule as

$$W(\lambda,x,\varphi;t,\bar{t})=\sum_{g=0}^{\infty}\sum_{n,m=0}^{\infty}\frac{1}{n!m!}\lambda^{g-1}\tilde{C}^{(g)}_{i_1\cdots i_n;\varphi^m}x^{i_1}\cdots x^{i_n}\varphi^m$$

By using this generating function, the holomorphic anomaly equations, (1.2) and (1.3) are expressed in the following compact form.

$$\frac{\partial}{\partial\bar{t}^i}\exp(W)=\left[\frac{\lambda^2}{2}\bar{C}_{\bar{i}}^{jk}\frac{\partial^2}{\partial x^j\partial x^k}-G_{\bar{i}j}x^j\frac{\partial}{\partial\varphi}\right]\exp(W). \tag{3.1}$$

It turns out that there is another function of x^i and φ which satisfies almost the same equations as (3.1). It is given as follows

$$Y(\lambda,x,\varphi;t,\bar{t})=-\frac{1}{2\lambda^2}\left(\Delta_{ij}x^ix^j+2\Delta_{i\varphi}x^i\varphi+\Delta_{\varphi\varphi}\varphi^2\right)+\frac{1}{2}\log\left(\frac{\det\Delta}{\lambda^2}\right) \tag{3.2}$$

Here Δ is an inverse of the propagator K defined by (2.8), i.e.

$$\begin{aligned}
&S^{ij}\Delta_{jk}+S^i\Delta_{k\varphi}=-\delta^i_k\\
&S^{ij}\Delta_{j\varphi}+S^i\Delta_{\varphi\varphi}=0\\
&S^i\Delta_{ij}+2S\Delta_{j\varphi}=0\\
&S^i\Delta_{i\varphi}+2S\Delta_{\varphi\varphi}=-1.
\end{aligned} \tag{3.3}$$

Thus Y may be regarded as a kinetic term for the finite dimensional system of x^i and φ. The most important properties of these inverse propagators are

$$\begin{aligned}
&\bar{\partial}_{\bar{i}}\Delta_{jk}=\bar{C}_{\bar{i}}^{mn}\Delta_{mj}\Delta_{nk}+G_{\bar{i}j}\Delta_{k\varphi}+G_{\bar{i}j}\Delta_{j\varphi}\\
&\bar{\partial}_{\bar{i}}\Delta_{j\varphi}=\bar{C}_{\bar{i}}^{mn}\Delta_{mj}\Delta_{n}+G_{\bar{i}j}\Delta_{\varphi\varphi}\\
&\bar{\partial}_{\bar{i}}\Delta_{\varphi\varphi}=\bar{C}_{\bar{i}}^{mn}\Delta_{m\varphi}\Delta_{n\varphi}
\end{aligned}$$

which we can derive from (2.3) and (2.4). Just as the anomaly equations for $C^{(g)}_{i_1\ldots i_n}$ are encoded in (3.1), the above equations for Δ's can be written as a differential equation for Y.

$$\frac{\partial}{\partial\bar{t}^i}\exp(Y)=\left[-\frac{\lambda^2}{2}\bar{C}_{\bar{i}}^{jk}\frac{\partial^2}{\partial x^j\partial x^k}-G_{\bar{i}j}x^j\frac{\partial}{\partial\varphi}\right]\exp(Y) \tag{3.4}$$

Now we consider the following integral.

$$Z = \int dx d\varphi \exp(Y + W) \tag{3.5}$$

Although this integral itself may be divergent, we can compute its perturbative expansion with respect to λ. The integral Z may be regarded as a partition function of a finite dimensional quantum system with dynamical variables x^i and φ, and the perturbative expansion of Z can be evaluated using the standard technique of the Feynman rule as

$$\begin{aligned} \log Z = \lambda^2 &\left[F_2 - \frac{1}{2}S^{ij}C_{ij}^{(1)} - \frac{1}{2}C_i^{(1)}S^{ij}C_j^{(1)} + \cdots\right] + \\ + \lambda^4 &\left[F_3 - \frac{1}{2}S^{ij}C_{ij}^{(2)} - C_i^{(1)}S^{ij}C_j^{(2)} + \cdots\right] + \\ + \lambda^6 &\left[F_4 - \frac{1}{2}S^{ij}C_{ij}^{(3)} - C_i^{(1)}S^{ij}C_j^{(3)} - \frac{1}{2}C_i^{(2)}S^{ij}C_j^{(2)} + \cdots\right] + \\ + \cdots + \lambda^{2g-2} &\left[F_g - \frac{1}{2}S^{ij}C_{ij}^{(g-1)} - \frac{1}{2}\sum_{r=1}^{g-1} C_i^{(r)}S^{ij}C_j^{(g-r)} + \cdots\right] + \cdots \end{aligned} \tag{3.6}$$

where (...) in the coefficient of λ^2 represents the terms in the r.h.s. of (2.6), (...) in the coefficient of λ^4 represents those in (2.7), and so on.

Previously we found, by the iterative method, that the coefficients of λ^2 and λ^4 in the perturbative expansion of Z are holomorphic in t. We can now prove the holomorphicity of Z to all order in the perturbation as the Schwinger-Dyson equation of the finite dimensional system. By using (3.1) and (3.4), we obtain

$$\begin{aligned} \frac{\partial}{\partial \bar{t}^i} Z = &\int dx d\varphi \, e^Y \left[\frac{\lambda^2}{2}\bar{C}_{\bar{i}}^{jk} \frac{\partial^2}{\partial x^j \partial x^k} - G_{\bar{i}j} x^j \frac{\partial}{\partial \varphi}\right] e^W + \\ + &\int dx d\varphi \, e^W \left[-\frac{\lambda^2}{2}\bar{C}_{\bar{i}}^{jk} \frac{\partial^2}{\partial x^j \partial x^k} - G_{\bar{i}j} x^j \frac{\partial}{\partial \varphi}\right] e^Y \\ = &\frac{\lambda^2}{2}\bar{C}_{\bar{i}}^{jk} \int dx d\varphi \left[\frac{\partial}{\partial x^j}\left(e^Y \frac{\partial}{\partial x^k}(e^W)\right) - \frac{\partial}{\partial x^j}\left(e^W \frac{\partial}{\partial x^k}(e^Y)\right)\right] - \\ &- G_{\bar{i}j} \int dx d\varphi \, \frac{\partial}{\partial \varphi}\left[x^j e^{Y+W}\right]. \end{aligned}$$

The point is that the integrand in the r.h.s. of this equation is total derivative with respect to x^i and φ. In the perturbative expansion, we are free to perform the integration-by-part and drop boundary terms since integrals involved in the perturbation are all Gaussian. Thus we have derived

$$\frac{\partial}{\partial \bar{t}^i} Z = 0$$

As is evident from the expansion (3.6), the holomorphicity of Z means that we can express F_g as a meromorphic section f_g of $\mathcal{L}^{2-2g}$ minus a sum over the Feynman graphs constructed from the propagators (2.8) and the vertices (2.9).

4. Construction of propagators

So far we have assumed that there is a global section S of $\mathcal{L}^{-2}$ which satisfies (2.2). Now we are going to construct such an object. The important ingredient is again the special geometry relation

$$R_{i\bar{j}l}{}^{k} = -\bar{\partial}_{\bar{j}}\Gamma^{k}_{il} = G_{i\bar{j}}\delta^{k}_{l} + G_{k\bar{j}}\delta^{k}_{i} - C_{ilm}\overline{C}^{km}_{\bar{j}}. \tag{4.1}$$

Since $G_{i\bar{j}} = \partial_i\bar{\partial}_{\bar{j}}K$, this can be rewritten as

$$\bar{\partial}_{\bar{i}}[S^{jk}C_{klm}] = \bar{\partial}_{\bar{i}}\left[\partial_l K\delta^{j}_{m} + \partial_m K\delta^{j}_{l} + \Gamma^{j}_{lm}\right].$$

This can be easily integrated as

$$S^{ij}C_{jkl} = \delta^{i}_{l}\partial_k K + \delta^{i}_{k}\partial_l K + \Gamma^{i}_{kl} + f^{i}_{kl} \tag{4.2}$$

where f^{i}_{kl} is some meromorphic object which should compensate for the non-covariance of $\partial_k K$ and Γ^{i}_{kl} in the r.h.s. side. We can express f^{i}_{kl} as

$$f^{i}_{kl} = \delta^{i}_{l}\partial_k \log f + \delta^{i}_{k}\partial_l \log f - \sum_{a=1}^{n} v_{l,a}\partial_k v^{i,a} + \tilde{f}^{i}_{kl},$$

where f is a meromorphic section of $\mathcal{L}$, $\{v^{i,a}\}_{a=1,\dots,n}$ (n is the dimensions of the moduli space) are meromorphic tangent vectors which are linearly independent almost everywhere on the moduli space, $v_{i,a}$ are inverse of $v^{i,a}$ ($\sum_a v_{i,a}v^{j,a} = \delta^{j}_{i}$) and $\tilde{f}^{i}_{kl}$ is a meromorphic section of $T \times Sym^{\otimes 2}T^{(1,0)*}\mathcal{M}$. In general, (4.2) has $\frac{1}{2}n^2(n+1)$ equations for $\frac{1}{2}n(n+1)$ variables S^{ij} and it is over-determined when $n > 1$. Thus we should make an appropriate choice of $\tilde{f}^{i}_{kl}$ to ensure that (4.2) is solvable with respect to S^{ij}.

The situation is much simpler in the one-modulus case since there is only one equation in (4.2) and there is no constraint on $\tilde{f}^{1}_{11}$. In order to construct F_g by using the Feynman rule, (2.6) and (2.7) for example, we do not need the most general solution to $\bar{\partial}S^{11} = \overline{C}^{11}_{\bar{1}}$ since any holomorphic ambiguity in S^{11} is absorbed into the holomorphic section f_g which we add to F_g at the end of the computation. Thus we can, for example, choose $\tilde{f}^{1}_{11} = 0$. With this choice, S^{11} becomes

$$S^{11} = \frac{1}{C_{111}}\left[2\partial\log(e^{K}|f|^2) - (G_{1\bar{1}}v)^{-1}\partial(vG_{1\bar{1}})\right] \tag{4.3}$$

To find S^i, we need to integrate

$$\bar{\partial}_{\bar{i}}S^{j} = G_{i\bar{k}}S^{jk}.$$

Substituting (4.3) into this, we obtain

$$\begin{aligned}\bar\partial S^1 =&\frac{1}{C_{111}}\left[2\partial\log(e^K|f|^2)G_{1\bar 1} - v^{-1}\partial(vG_{1\bar 1})\right]\\ =&\frac{1}{C_{111}}\bar\partial\left[(\partial\log(e^K|f|^2))^2 - v^{-1}\partial(v\partial K)\right].\end{aligned}$$

A special solution to this equation can be easily found as

$$S^1 = \frac{1}{C_{111}}\left[\left(\partial\log(e^K|f|^2)\right)^2 - v^{-1}\partial\left(v\partial\log(e^K|f|^2)\right)\right] \tag{4.4}$$

Finally we need to find S which satisfies

$$\bar\partial S = G_{1\bar 1}S^{\bar 1}. \tag{4.5}$$

A special solution to this equation is given by

$$\begin{aligned}S =& \left[S^1 - \frac{1}{2}D_1S^{11} - \frac{1}{2}(S^{11})^2C_{111}\right]\partial\log(e^K|f|^2)+\\ &+\frac{1}{2}D_1S^1 + \frac{1}{2}S^{11}S^1C_{111}.\end{aligned} \tag{4.6}$$

Let us check that this indeed satisfies (4.5). We first note that the following special combination of S^1 and S^{11} is holomorphic

$$\begin{aligned}&\bar\partial\left[S^1 - \frac{1}{2}D_1S^{11} - \frac{1}{2}(S^{11})^2C_{111}\right] =\\ &= G_{1\bar 1}S^{11} - \frac{1}{2}[\bar\partial, D_1]S^{11} - \frac{1}{2}(G^{1\bar 1})^2\partial\bar C_{\bar 1\bar 1\bar 1} - \bar C_{\bar 1}^{11}S^{11}C_{111}\\ &= 0,\end{aligned}$$

where we used the special geometry relation[1] (4.1), the definitions of S^1 and S^{11} and $\partial\bar C_{\bar 1\bar 1\bar 1} = 0$. Now it is straightforward to check the equation (4.5) as

$$\begin{aligned}\bar\partial S =&G_{1\bar 1}S^1 - \frac{1}{2}G_{1\bar 1}D_1S^{11} - \frac{1}{2}G_{1\bar 1}(S^{11})^2C_{111}+\\ &+\frac{1}{2}[\bar\partial, D_1]S^1 + \frac{1}{2}D_1S^1_{\bar 1} + \frac{1}{2}\bar C_{\bar 1}^{11}S^1C_{111} + \frac{1}{2}S^{11}S^1_{\bar 1}C_{111}\\ =&G_{1\bar 1}S^1 + \frac{1}{2}[\bar\partial, D_1]S^1 + \frac{1}{2}\bar C_{\bar 1}^{11}S^1C_{111}\\ =&G_{1\bar 1}S^1.\end{aligned}$$

Here we once again used the special geometry relation[2].

To summarize, in the one-modulus case, the propagators S^{11}, S^1, and S are given as

$$\begin{aligned}S^{11} =& \frac{1}{C_{111}}\partial\log\left[2\partial\log(e^K|f|^2) - (G_{1\bar 1}v)^{-1}\partial(vG_{1\bar 1})\right]\\ S^1 =& \frac{1}{C_{111}}\left[\left(\partial\log(e^K|f|^2)\right)^2 - v^{-1}\partial\left(v\partial\log(e^K|f|^2)\right)\right]\\ S =& \left[S^1 - \frac{1}{2}D_1S^{11} - \frac{1}{2}(S^{11})^2C_{111}\right]\partial\log(e^K|f|^2)+\\ &+\frac{1}{2}D_1S^1 + \frac{1}{2}S^{11}S^1C_{111}.\end{aligned}$$

---

1 $D_1S^{11} = (\partial - 2\Gamma_1^{11} - 2\partial K)S^{11}$. Therefore $[\bar\partial, D_1]S^{11} = 2G_{1\bar 1}S^{11} - 2C_{111}\bar C_{\bar 1}^{11}S^{11}$.

2 $D_1S^1 = (\partial - \Gamma_1^{11} - 2\partial K)S^1$. Therefore $[\bar\partial, D_1]S^1 = -C_{111}\bar C_{\bar 1}^{11}S^1$.

In the multi-moduli case, (4.2) gives

In the multi-moduli case, (4.2) gives

$$S^{ij}C_{jkl} = (\delta^i_l\partial_k + \delta^i_k\partial_l)\log(e^K|f|^2) - \sum_{a=1}^{n} v_{l,a}G^{i\bar{i}}\partial_k(v^{m,a}G_{m\bar{i}}) + \tilde{f}^i_{kl}. \tag{4.7}$$

In order to obtain an expression for S^{ij} from this equation, we need to "invert" the Yukawa coupling. Although we do not know if it is possible to do so in general, it is certainly possible for the A-model discussed in section 4. In this model, each chiral field corresponds to a Kähler form in the target space and, in the large volume limit, the Yukawa coupling C_{ijk} is given as an intersection of the three Kähler forms. There is a distinguished Kähler modulus t^1 in this model corresponding to an overall scaling of the target space metric. In the large volume limit $t^1 \to \infty$, the Yukawa coupling C_{ij1} then gives the inner product of the two Kähler forms k_i and k_j, and it is non-degenerate as an n x n matrix, $\det(C_{ij1})_{i,j=1,\ldots,n} \neq 0$. Since $\det(C_{ij1})$ is holomorphic in t, this means that $\det(C_{ij1})$ should be non-zero almost everywhere on the moduli space. Therefore we can invert C_{ij1} in (4.7) to find an expression for S^{ij}, provided we made an appropriate choice of $\tilde{f}^i_{kl}$.

As in the one-modulus case, we substitute (4.7) into $\bar{\partial}_{\bar{i}}S^j = G_{\bar{i}i}S^{ij}$ to obtain

$$\begin{aligned}\bar{\partial}_{\bar{i}}[S^j]C_{jkl} =& G_{\bar{i}l}\partial_k\log(e^K|f|^2) + G_{\bar{i}k}\partial_l\log(e^K|f|^2) + \\ & -\sum_{a=1}^{n} v_{l,a}\partial_k(v^{m,a}G_{\bar{i}m}) + \tilde{f}^i_{kl}G_{\bar{i}i}.\end{aligned}$$

This can be easily integrated as

$$\begin{aligned}S^iC_{ijk} =& \partial_j\log(e^K|f|^2)\partial_k\log(e^K|f|^2) - \sum_{a=1}^{n} v_{k,a}\partial_j\left[v^{l,a}\partial_l\log(e^K|f|^2)\right] + \\ & + \tilde{f}^l_{jk}\partial_l\log(e^K|f|^2) + \tilde{f}_{jk}.\end{aligned} \tag{4.8}$$

Here $\tilde{f}_{jk}$ is a meromorphic section of $Sym^{\otimes 2}T^{(1,0)*}\mathcal{M}$.. As in the case of S^{ij} in (4.7), with an appropriate choice of $\tilde{f}_{jk}$, we can invert the Yukawa coupling in the above and obtain an exression for S^i.

To complete the Feynman rule, we need S which satisfies

$$\bar{\partial}_{\bar{i}}S = G_{\bar{i}i}S^i. \tag{4.9}$$

A special solution to this equation is given by

$$\begin{aligned}S =& \frac{1}{2n}\left[(n+1)S^i - D_jS^{ij} - S^{ij}S^{kl}C_{jkl}\right]\partial_i\log(e^K|f|^2) + \\ & + \frac{1}{2n}\left(D_iS^i + S^iS^{jk}C_{ijk}\right).\end{aligned}$$

Let us check that this satisfies (4.9). As in the case of one-modulus, the following combination of S^i and S^{ij} is holomorphic due to the special geometry relation and $\bar{\partial}_{\bar{i}}C_{jkl} = 0$

$$
\begin{aligned}
&\overline{\partial}_{\bar{i}}\left[(n+1)S^{j}-D_{k}S^{jk}-S^{jk}S^{mn}C_{jmn}\right]=\\
&=(n+1)G_{\bar{i}k}S^{jk}-[\overline{\partial}_{\bar{i}},D_{k}]S^{jk}-G^{j\bar{j}}G^{k\bar{k}}\partial_{k}\overline{C}_{\bar{i}\bar{j}\bar{k}}-\\
&\qquad-\overline{C}_{\bar{i}}^{jk}S^{mn}C_{kmn}-\overline{C}_{\bar{i}}^{mn}S^{jk}C_{kmn}\\
&=0.
\end{aligned}
$$

We can then compute $\overline{\partial}_{\bar{i}}S$ as

$$
\begin{aligned}
\overline{\partial}_{\bar{i}}S=&\frac{n+1}{2n}S_{\bar{i}}+\\
&+\frac{1}{2n}\left([\overline{\partial}_{\bar{i}},D_{j}]S^{j}+D_{j}S_{\bar{i}}^{j}+S_{\bar{i}}^{j}S^{kl}C_{jkl}+S^{j}\overline{C}_{\bar{i}}^{kl}C_{jkl}-S_{\bar{i}}^{j}S^{kl}C_{jkl}\right)\\
=&G_{\bar{i}j}S^{j}.
\end{aligned}
$$

Here again, we used the special geometry relation for $[\overline{\partial}_{\bar{i}}, D_j]$.

We have found that the topological string amplitudes are computable using the set of Feynman rules. This suggests that there is some finite dimensional system behind the topological string theory. It would be extremely interesting to understand an origin of such a system.

References

[1] M. Bershadsky, S. Cecotti, H. Ooguri and C. Vafa, Nucl. Phys. B (1993) 1.

[2] M. Bershadsky, S. Cecotti, H. Ooguri and C. Vafa, *Kodaira-Spencer Theory of Gravity and Exact Results for Quantum String Amplitudes*, preprint hep-th/9309140, to be published in Commun. Math. Phys.

[3] *Essays on Mirror Manifold*, ed. S.-T. Yau, International Press (1992).

[4] R. Bryant and P. Griffiths, in *Arithmetic and Geometry*, papers dedicated to I.R. Shafarevitch, eds. M. Artin and J. Tate, (Boston, Birkhäuser, 1983), vol.2, p.77.

[5] B. de Wit and A. van Proeyen, Nucl. Phys. B245 (1984) 89;
B. de Wit, P.G. Lauwers and A. van Proyen, Nucl. Phys. B255 (1985) 569;
E. Cremmer, C. Kounnas, A. van Proeyen, J.-P. Derendinger, S.Ferrara, B. de Wit and L. Girardello, Nucl. Phys. B250 (1985) 385;
S. Cecotti, Comm. Math. Phys. 131 (1990) 517.

[6] E. Witten, Comm. Math. Phys. 117 (1988) 353;
E. Witten, Comm. Math. Phys. 118 (1988) 411;
T. Eguchi and S.-K. Yang, Mod. Phys.Lett. A5 (1990) 1693.

Folded Strings

Itzhak Bars

Department of Physics and Astronomy
University of Southern California
Los Angeles, CA 90089-0484

ABSTRACT Recent progress on the complete set of solutions of two dimensional classical string theory in any curved spacetime is reviewed. When the curvature is smooth, the string solutions are deformed folded string solutions as compared to flat spacetime folded strings that were known for 19 years. However, surprizing new stringy behavior becomes evident at singularities such as black holes. The global properties of these solutions require that the "bare singularity region" of the black hole be included along with the usual black hole spacetime. The mathematical structure needed to describe the solutions include a recursion relation that is analogous to the transfer matrix of lattice theories. This encodes lattice properties on the worldsheet on the one hand and the geometry of spacetime on the other hand. A case is made for the presence of folded strings in the quantum theory of non-critical strings for $d \geq 2$.

1.1 Introduction

Feza Gürsey was a great master and an artist in finding connections between Mathematics and Physics. In this conference we have the pleasure to hear from many of his friends that have admired his leadership in several areas of Physics. I am very appreciative for having been given the opportunity to express my gratitude to Feza for the inspiration he has provided to me as my teacher as well as my colleague and friend.

Among his first discoveries was the *non-linear sigma models*, which he applied to pion physics[1]. Nowdays, sigma models are at the basis of string theory in the form of conformal field theories. In recent years, through the use of gauged Wess-Zumino-Witten models based on non-compact groups, it has been possible to construct *exact conformal field theories* that describe (super)strings propagating

*Research partially supported by DOE grant No. DE-FG03-84ER40168.

[1]Feza was the first to introduce the idea of the sigma models. Many people think of the paper by Gell-Mann and Lévy in connection with sigma models, but it is important to recall that Gell-Mann and Lévy refer to Feza's paper.

in curved space-time in 2D to 4D. These models combine several fields that deeply interested Feza: non-linear sigma models, conformal invariance, classical and quantum gravity, non-compact groups, unification of forces, string theory. In honoring Feza today, I would like to highlight recent progress made in this field.

1.2 Motivation

The original physical motivations for studying string theory were: (1) understanding unification of forces including quantum gravity, and (2) understanding the Standard Model. In recent years it has become more and more evident that these goals should be examined in the presence of curved 4D space-time string backgrounds. The construction of 4D curved space-time string theories that correspond to exact conformal theories have provided models in which various questions can be investigated [1][2].

The usual scenario of flat 4D plus extra curved dimensions may not be the right approach for making predictions about the Standard Model. The gauge symmetries and spectrum of quark + lepton families, which are the main ingredients of the Standard Model, were probably fixed during the early times in the evolution of the Universe. At such times 4D space-time was curved. Since curvature contributes to the central charge and other topological aspects of String Theory, it is likely that the predictions of String Theory under such conditions may be quite different than the flat 4D approach. Therefore, String Theory in curved space-time must be better understood before attempting to make connections to low energy physics. One should consider all kinds of curved backgrounds, not only the traditional cosmological backgrounds, since the passage from curved space-time to flat space-time may involve various phase transitions, including inflation of a small region of the original curved universe to today's universe that is essentially homogeneous and flat. The gauge bosons, and chiral families of quarks and leptons in a small region of the early curved universe would become the ones observed in today's inflated flat universe. The possibility of such a scenario suggests that curved space-time string theory deserves intensive study. In addition, the issues surrounding gravitational singularities should be answered in the context of curved space-time string theory, as it is the only known theory of quantum gravity.

1.3 Some Results in 1-time G/H

With these questions in mind, we have been pursuing a program of building and analyzing exactly solvable models of string theory in curved spacetime based on conformal field theory. The main tool is the G/H gauged WZW model based on non-compact groups, such that the coset contains a single time coordinate. A lot of progress was made on the construction of exact conformal field theory models for bosonic, supersymmetric and heterotic strings in curved spacetime, and some exact results were derived. These include :

- Classification of G/H models with 1-time + $(d-1)$-space coordinates [1]. G is non-compact and H can be non-compact or compact. After identifying the simple cases for G/H that yield a single time coordinate, the classification is easily extended to semi-simple, with Abelian factors, and their contractions to solvable groups [3][4]. There may also exist other exact conformal models which may not be G/H models.

- The non-linear sigma model geometries for these models have been derived [6][7][1][2], giving the metric $G_{\mu\nu}(x)$, the anti-symmetric tensor $B_{\mu\nu}(x)$, and the dilaton $\Phi(x)$. These automatically solve Einstein's equations for dilaton gravity. The global spaces for these manifolds have been constructed, and rich duality symmetries have been identified. Furthermore, the exact point particle geodesics in the global space have been obtained through group theoretical methods [6][7][1].

- Quantum corrections to these geometries have been computed to all orders in the sigma model interactions, thanks to the group theoretical construction, mainly by using algebraic methods[7]. This led to an exact quantum effective action [8] with the quantum corrected metric, antisymmetric tensor and dilaton. The results show that for type-II superstrings, thanks to the supersymmetry, there are no corrections to the classical expressions. In bosonic or heterotic cases the corrections show that certain singularities of the metric get shielded by quantum corrections in parts of spacetime.

- In these models the exact spectrum of the Laplacian can be obtained through unitary representation theory of non-compact groups. This provides the method for extracting the spectrum of quarks and leptons, but more work is needed along these lines.

1.4 Classical Solutions

More recently, it became apparent that a physical interpretation of the models as well as further progress will be accomplished through the study of the classical equations of such models. Therefore, we have turned to the classical theory. This is relevant to fundamental questions of singularities in gravitational physics, as well as stringy questions about the early universe and its influence on the low energy spectrum of quarks and leptons. The classical string solution for any gauged WZW model was obtained in general terms in [9], and its specialization to particle solutions was given explicitly in [6].

A more detailed exploration of the general 2D classical string theory in any curved spacetime (i.e. not only WZW models) was done in [5]. In 2D the only non-trivial stringy solutions turn out to be necessarily folded strings, and therefore they are the only path toward analyzing stringy questions in a toy model. In addition to the interest in singular gravitational behavior (such as black holes) there has also been a long-standing interest in exploring consistent generalizations of non-critical strings with the hope that they may be relevant for some branch of physics. Folded strings fall into this category, especially in the area of string-QCD relations. Therefore two aspects of string theory were investigated: (i) strings in curved space-time and (ii) folded strings.

In papers [5][10][11] the complete set of solutions of two dimensional classical string theory were constructed for *any 2D curved spacetime*. The classical action is $\int d^2\sigma\, G_{\mu\nu}(x)\partial_+ x^\mu \partial_- x^\nu$. In 2D $B_{\mu\nu}(x)$ can be eliminated since it produces a total derivative in the action, and in the classical theory the dilaton is absent. The most general metric can always be transformed into the conformal form $G_{\mu\nu} = \eta_{\mu\nu}G(x)$. Then the most general 2D *classical* string equations of motion and conformal (Virasoro) constraints take the form

$$\begin{array}{c} \partial_+(G\,\partial_- u) + \partial_-(G\,\partial_+ u) = \frac{\partial G}{\partial v}(\partial_+ u\partial_- v + \partial_+ v\partial_- u) \\ \partial_+(G\,\partial_- v) + \partial_-(G\,\partial_+ v) = \frac{\partial G}{\partial u}(\partial_+ u\partial_- v + \partial_+ v\partial_- u) \\ \partial_+ u\partial_+ v = 0 = \partial_- u\partial_- v\ , \end{array} \tag{1.1}$$

where we have used the target space lightcone coordinates $u(\sigma^+,\sigma^-) = \frac{1}{\sqrt{2}}(x^0 + x^1)$, $v(\sigma^+,\sigma^-) = \frac{1}{\sqrt{2}}(x^0 - x^1)$, and the world sheet lightcone coordinates $\sigma^\pm = (\tau \pm \sigma)/\sqrt{2}$, $\partial_\pm = (\partial_\tau \pm \partial_\sigma)/\sqrt{2}$.

In flat space-time the solutions are given in terms of arbitrary left-moving and right-moving functions $x_L^\mu(\sigma^+), x_R^\mu(\sigma^-)$

$$x^\mu(\tau,\sigma) = x_L^\mu(\sigma^+) + x_R^\mu(\sigma^-).$$

As shown by BBHP [12][13], the constraints are also satisfied provided[2]

$$\begin{array}{c} u(\sigma^+,\sigma^-) = u_0 + \frac{p^+}{2}\left[(\sigma^+ + f(\sigma^+)) + (\sigma^- - g(\sigma^-))\right] \\ v(\sigma^+,\sigma^-) = v_0 + \frac{p^-}{2}\left[(\sigma^+ - f(\sigma^+)) + (\sigma^- + g(\sigma^-))\right] \end{array} \tag{1.2}$$

where $f(\sigma^+)$ and $g(\sigma^-)$ are any two *periodic functions,* $f(\sigma^+) = f(\sigma^+ + \sqrt{2})$, $g(\sigma^-) = g(\sigma^- + \sqrt{2})$, with slopes $f'(\sigma^+) = \pm 1$ and $g'(\sigma^-) = \pm 1$. The slope can change discontinuously any number of times at arbitrary locations σ_i^+, σ_j^- within the basic intervals $-1/\sqrt{2} \le \sigma^\pm \le 1/\sqrt{2}$ (and then repeated periodically), but the functions f, g are continuous at these points. The discontinuities in the slopes are allowed since the equations of motion are first order in either ∂_+ or ∂_-. The number of times the slope changes in the basic interval corresponds to the number of folds for left movers and right movers respectively. The simplest BBHP solution is the so called yo-yo solution given by $f = |\sigma^+|_{per}$ and $g = |\sigma^-|_{per}$ which are the periodically repeated absolute value. These solutions describe folded strings, with the folds oscillating against each other, and moving

[2] Although the original BBHP solutions were for open strings, the same solutions also apply to closed strings by simply taking independent functions f, g for left movers and right movers.

at the speed of light. Examples are plotted in Figures 1,2. In Fig.1 one sees the yo-yo solution with equal periods for $|\sigma^+|_{per}$, and $|\sigma^-|_{per}$. Fig. 2 is generated by taking the period of $|\sigma^-|_{per}$ to be half of that of $|\sigma^+|_{per}$.

As discovered in [5][10], the complete set of classical solutions in curved spacetime are classified by their behavior in the asymptotically flat region of spacetime $G(u,v) \to 1$, where they tend to the folded string solutions of BBHP given in (1.2) as boundary conditions. The curved space solutions are given in the form of a map from the world sheet to target spacetime, where (as a mathematical convenience) the world sheet is divided into lattice-like patches corresponding to different maps. The world-sheet lattice structure is determined by the sign patterns of $(f', g') = (\pm, \pm)$ inherent in the BBHP solutions, thus the lattice is dictated by the boundary conditions in the asymptotically flat region of spacetime $G(u,v) \to 1$. The lattice is on the world-sheet, not in curved spacetime, it is only a mathematical tool to keep track of patches, and the world sheet is not at all discretized. In each patch of the lattice one set of signs holds, hence there are 4 types of patches called A, B, C, D. For each such patch there is a solution of the equations of motion that is valid within the patch. The forms of the solutions in patches labelled by an integer k are (see eq.(1.4) for an example of a pattern of patches)

$$\begin{array}{ll} A: u = U_k(\sigma^+) & v = V_k(\sigma^-) \\ B: u = U_k(\sigma^-) & v = V_k(\sigma^+) \\ C: u = u_k & v = W[\alpha_k(\sigma^+) + \beta_k(\sigma^-),\, u_k] \\ D: u = \bar{W}[\alpha_k(\sigma^-) + \beta_k(\sigma^+),\, v_k] & v = v_k\ , \end{array} \tag{1.3}$$

where the constants u_k, v_k and the functions $U_k(\sigma^\pm), V_k(\sigma^\pm),\ \alpha_k(\sigma^\pm),\ \beta_k(\sigma^\pm)$ are given by a recursion relation whose form depends on the metric G. It is easy to verify that, independently of the recursion relation, the forms listed in (1.3) solve the differential equations for any $U_k(\sigma^\pm), V_k(\sigma^\pm),\ \alpha_k(\sigma^\pm),\ \beta_k(\sigma^\pm)$ provided the functions $W, \bar{W}$ are defined by inverting the following functions

$$\int^W dv' G(u_k, v') = \alpha + \bar{\beta}, \qquad \int^{\bar{W}} du' G(u', v_k) = \bar{\alpha} + \beta.$$

By construction, in flat spacetime the recursion reproduces the BBHP solutions given above.

The recursion relation, which is analogous to a "transfer matrix", connects the maps in different patches into a single continuous map. It is derived by demanding continuity accross the boundaries of each patch (see below for an example). Thus, the functions in the various patches get related to each other. This "transfer matrix" encodes the properties of the world sheet lattice on the one hand and the geometry of spacetime on the other hand. Thus, lattices on the world-sheet plus geometry in space-time lead to "transfer matrices". Recall that the lattice is dictated by the nature of the solution (1.2) in the asymptotically flat region of target spacetime. This seems to be a rich area of mathematical physics to explore in more detail in the future.

As an example we consider the simplest yo-yo solution as a boundary condition. This defines the sign patterns according to the slopes of the periodic functions $|\sigma^+|_{per}$ and $|\sigma^-|_{per}$, and the following lattice emerges from the periodic behavior of these functions . The world sheet is labelled by σ horizontally and by τ vertically. Periodicity in σ is imposed, hence the world sheet is a cylinder. It is sliced by equally spaced 45° lines that form a light-cone lattice in $\sigma^\pm$. The crosses in the diagram represent the corners of the cells on the world sheet.

$\sigma = 0$	$\sigma = 1$	$\sigma = 2$	$\sigma = 3$	$\sigma = 4 \equiv 0$
$\vdots$	$\vdots$	$\vdots$	$\vdots$	$\vdots$
$\times$	$U_{k+2}(\sigma^+)$ $V_{k+2}(\sigma^-)$	$\times$	$U_{k+2}(\sigma^-)$ $V_{k+2}(\sigma^+)$	$\times$
$\cdots$ u_{k+1} $W_{k+1}(\sigma^+,\sigma^-)$	$\times$	$\bar{W}_{k+1}(\sigma^+,\sigma^-)$ v_{k+1}	$\times$	u_{k+1} $W_{k+1}(\sigma^+,\sigma^-)$ $\cdots$
$\times$	$U_{k+1}(\sigma^-)$ $V_{k+1}(\sigma^+)$	$\times$	$U_{k+1}(\sigma^+)$ $V_{k+1}(\sigma^-)$	$\times$
$\cdots$ $\bar{W}_k(\sigma^+,\sigma^-)$ v_k	$\times$	u_k $W_k(\sigma^+,\sigma^-)$	$\times$	$\bar{W}_k(\sigma^+,\sigma^-)$ v_k $\cdots$
$\times$	$U_k(\sigma^+)$ $V_k(\sigma^-)$	$\times$	$U_k(\sigma^-)$ $V_k(\sigma^+)$	$\times$
$\cdots$ u_{k-1} $W_{k-1}(\sigma^+,\sigma^-)$	$\times$	$\bar{W}_{k-1}(\sigma^+,\sigma^-)$ v_{k-1}	$\times$	u_{k-1} $W_{k-1}(\sigma^+,\sigma^-)$ $\cdots$
$\vdots$	$\vdots$	$\vdots$	$\vdots$	$\vdots$

(1.4)

The transfer matrix for this "yo-yo lattice" was derived in [5][11][10] for any metric G. Here we give only the results for the $SL(2,R)/R$ black hole space-time $ds^2 = du\,dv(1-uv)^{-1}$ and for the cosmological deSitter space-time $ds^2 = dt^2 - R^2(t)\frac{dr^2}{1-kr^2} = \frac{4}{H^2}(u+v)^{-2}\,du\,dv$, for $|R(t)| = e^{Ht}$, $k=0$.

For the black hole metric the "transfer matrix" is

$$\begin{aligned} \bar{W}_k &= \frac{1}{v_k}\left[1 - \frac{\left(1-U_k(\sigma^+)v_k\right)\left(1-U_k(\sigma^-)v_k\right)}{1-u_{k-1}v_k}\right] \\ U_{k+1}(z) &= \frac{1-u_k v_k}{1-u_{k-1}v_k}\left[U_k(z) + \frac{u_k - u_{k-1}}{1-u_k v_k}\right] \\ u_{k+1} &= \frac{2u_k - u_{k-1} - u_k^2 v_k}{1-u_{k-1}v_k}, \end{aligned} \tag{1.5}$$

and similarly W_k, V_k, v_k are obtained from the above by interchanging $U \leftrightarrow V$ and $u \leftrightarrow v$. The constants u_k, v_k are the values of the functions $U_k(z), V_k(z)$ at the boundaries of the cell labelled by k :

$$u_{k-1} = U_k(-1/\sqrt{2}), \quad u_k = U_k(1/\sqrt{2}), \quad etc.$$

These constants describe the motion of folds that move at the speed of light. Note that for $u, v \to 0$ or ∞ the metric approaches the flat metric.

Remarkably, there is an invariant of this "transfer matrix" that corresponds to the "lattice area" swept by the string [for comparison, recall the form of the action density whose meaning is area $dA = (1-uv)^{-1}(\partial_+ u \partial_- v + \partial_+ v \partial_- u)$]

$$dA_k = \frac{(u_k - u_{k-1})(v_k - v_{k-1})}{1 - \frac{1}{4}(u_k + u_{k-1})(v_k + v_{k-1})}. \tag{1.6}$$

It can be easily verified that $dA_{k+1} = dA_k$, implying that this quantity remains a constant even in the vicinity of singularities $uv \approx 1$. This observation leads to new interesting phenomena as discussed below.

The "transfer matrix" for the deSitter spacetime is

$$\begin{aligned} \bar{W}_k(\sigma^+,\sigma^-) &= \left[\frac{1}{U_k(\sigma^+)+v_k} + \frac{1}{U_k(\sigma^-)+v_k} - \frac{1}{u_{k-1}+v_k}\right]^{-1} - v_k \\ U_{k+1}(z) &= \left[\frac{1}{U_k(z)+v_k} + \frac{1}{u_k+v_k} - \frac{1}{u_{k-1}+v_k}\right]^{-1} - v_k \\ u_{k+1} &= \left[\frac{2}{u_k+v_k} - \frac{1}{u_k+v_{k-1}}\right]^{-1} - v_k \end{aligned} \tag{1.7}$$

Similar formulas hold for W_k, V_k, v_k respectively. In this case too there is an invariant area

$$dA_k = \frac{4}{H^2}\frac{(u_k - u_{k-1})(v_k - v_{k-1})}{(u_k + v_{k-1})(u_{k-1} + v_k)}$$

The 2D deSitter space can be embedded in 3D as the surface of a hyperboloid described by

$$\begin{gathered} x_0^2 - x_1^2 - x_2^2 = -H^{-2} \\ x_0 = \frac{uv - H^{-2}}{u+v}, \quad x_1 = \frac{uv + H^{-2}}{u+v}, \quad x_2 = \frac{1}{H}\frac{u-v}{u+v} \end{gathered} \tag{1.8}$$

Then the deSitter metric takes the flat form

$$ds^2 = dx_0^2 - dx_1^2 - dx_2^2. \tag{1.9}$$

The motion is more easily visualized in this parametrization.

The recursion relations are solved in terms of two functions $U_0(z), V_0(z)$ that are associated with the initial cell $k = 0$. The remaining conformal invariance may be used to fix the form of the functions $U_0(z)$, $V_0(z)$ in the initial cell (although this is not necessary). For the yo-yo solution the initial functions $U_0(z)$, $V_0(z)$ need not contain more than 4 constants that are related to the initial positions and velocities of the two folds. However, there is a physical requirement: the time coordinate $x^0(\tau,\sigma)$ constructed from $U_0(\sigma^{\pm})$, $V_0(\sigma^{\pm})$ must be an increasing function of the proper time τ for any value of σ, so that physically every point on the string moves forward in time (no bits of anti-strings). Therefore, the simplest physical gauge fixed form is

$$\begin{aligned} U_0(z) &= \tfrac{1}{2}(u_0 + u_{-1}) + \tfrac{1}{\sqrt{2}}(u_0 - u_{-1})\, z_{per} \\ V_0(z) &= \tfrac{1}{2}(v_0 + v_{-1}) + \tfrac{1}{\sqrt{2}}(v_0 - v_{-1})\, z_{per}, \end{aligned} \tag{1.10}$$

where z_{per} is the linear function $z_{per} = z$ in the interval $-1/\sqrt{2} \leq z \leq 1/\sqrt{2}$, and then repeated periodically. This form reproduces the BBHP yo-yo solution in flat spacetime from the recursion relations, provided one uses $G(u,v) = 1$. However, any other function with the same 4 boundary constants and general increasing character will produce the same, gauge independent, physical motion for the folds in flat or curved spacetime, since their motion is given by gauge independent equations involving only the gauge independent constants u_k, v_k.

Evidently, the motion of the intermediate points of the string is gauge dependent, as expected.

The constants (u_k, v_k) are sufficient to describe the physical motion of the folds (or end points), as well as the whole string. The trajectories of the folds are plotted in Figs.4,5,6 by feeding the recursion relations to a computer. As in Fig.3, the string performs oscillations that are similar to those of flat spacetime around a center of mass that follows on the average the geodesic of a massive point particle. This is expected intuitively. A detailed discussion of the black hole case was given in [5]. The main surprize is the tunelling of the string into the forbidden region in Fig.5 (the bare singularity region of the black hole), where particles cannot go. This behavior cannot be avoided since it follows from minimal area *conservation laws* that were given above [5][10][11]. It could be compared to diffraction, or light illuminating the wall around a corner, that can happen with waves, but not with particles. In addition, the *massive* point particle geodesic (as well as the string geodesic) does not stop at the black hole, rather it reaches the black hole in a finite amount of proper time, and then it continues into a second sheet of spacetime that is glued to the first sheet at the black hole singularity. The observers on the second sheet see it as if the massive particle or the string is coming out of a white hole. The motion may continue from white hole to black hole singularities, each time moving into a new sheet, interpreted as a new world, like in the Reissner-Nordstrom spacetime. For more details see [5][10] [11].

1.5 Quantum Folded String

Given the fact that the string in 2D is quite non-trivial classically, we expect that there is a consistent quantization procedure that includes the non-trivial folded states. Therefore we should try to make a case for folded strings in the quantum theory.

As pointed out many times in our past work, folded 2D-string states are present in the $d = 2$ and $c \leq 25$ sector of the quantum theory in flat as well as curved spacetime. In simple string models, when it has been possible to compute the spectrum, their norm is positive and is proportional to $(c-26)$. Only if $d = 2$ and $c = 26$ simultaneously (e.g. $d = 2$ flat space-time with linear dilaton such that $c = 26$) the folded string states become zero norm states and then the special discrete momentum states survive as the only stringy states. A simple model in which these properties may be easily seen is the *covariant* quantization of the 2D string theory, in which the physical states are identified as the subset that satisfies the Virasoro constraints, i.e. $L_0 - \frac{d-2}{24} = L_{n\geq 1} = 0$ applied on states. For example, it has been known for a long time that the $d \leq 25$ sector of the flat string theory has non-trivial positive norm states (including for $d = 2$) that satisfy the Virasoro constraints and that there are no ghosts [14]. A similar covariant quantization can be carried out for the 2D black hole string by using the Kac-Moody current algebra formulation, and relaxing the $c = 26$ condition (i.e. $k < 9/4$) to include the folded strings.

Why $c = 26$? There are several approaches to the quantization of strings that converge on the requirement of $c = 26$. These include the light-cone gauge, the Polyakov path integral and the BRST quantization. However, they each involve certain steps that seem to inadverdently exclude the $c < 26$ string. We can point out that

(i) The usual light-cone approach throws away the folded states from the beginning by assuming a uniform momentum density $P^+(\tau,\sigma) = p^+$, a statement that is not true for the BBHP solutions even in flat spacetime.

(ii) The Polyakov approach assumes a certain measure for the path integral, thus locking into a *definition* of a quantum theory. A different measure that takes into account folds can be considered as in [15] mentioned below.

(iii) the BRST approach requires $Q^2_{BRST} = 0$ as an operator. This is a stronger requirement than the Virasoro constraints satisfied only in the physical subspace $< phys|L_n - \alpha_0\delta_{n0}|phys >= 0$. An analogous statement would be $< phys|Q_{BRST}|phys >= 0$, which does not lead to $c = 26$. Actually, the fact that there exists a consistent covariant quantization of the flat free string in $d < 26$ is already proof that the $Q^2_{BRST} = 0$ approach is too strong.

Therefore, it appears that a more general quantization of string theory for $c < 26$, that would permit folded string states, seems possible. What would also be interesting is to find the correct formulation for interacting folded strings. The path integral approach discussed in [15] seems to be promising, and it may be possible to make faster progress by reformulating it in the conformal gauge and relating it to our classical solutions Note that the definition of fold in ref.[15] does not take into account that the map from the world sheet to spacetime may be many to one (i.e. a region maped to a segment, as is the case for our solutions). This feature may be important in the formulation of folds and their interactions in the path integral approach. In particular, the description of folds in the conformal gauge, as in our papers, may eventually prove to be a more convenient mathematical formulation than the one used in [15].

1.6 Higher Dimensions

Folded strings exist in higher dimensions as well. One can display the general solution in flat space-time in the temporal gauge

$$\begin{gathered} x^0 = p^0\tau,\ \mathbf{x}(\tau,\sigma) = \mathbf{x}_L(\sigma^+) + \mathbf{x}_R(\sigma^-),\ (\partial_+\mathbf{x}_L)^2 = p_0^2 = (\partial_-\mathbf{x}_R)^2 \\ \partial_+\mathbf{x}_L = p^0\left(\frac{2\mathbf{f}}{1+\mathbf{f}^2}, \frac{1-\mathbf{f}^2}{1+\mathbf{f}^2}\varepsilon_L\right),\quad \partial_-\mathbf{x}_R = p^0\left(\frac{2\mathbf{g}}{1+\mathbf{g}^2}, \frac{1-\mathbf{g}^2}{1+\mathbf{g}^2}\varepsilon_R\right) \end{gathered} \tag{1.11}$$

where $\mathbf{f}(\sigma^+)$, $\mathbf{g}(\sigma^-)$ are arbitrary periodic vectors in $d-2$ dimensions, *which could be discontinuous*, and $\varepsilon_L(\sigma^+)$, $\varepsilon_R(\sigma^-)$ take the values ± 1 in patches of the corresponding variables such that the sign patterns repeat periodically (as in the 2D string). When $\mathbf{f}, \mathbf{g}$ are both zero the solution reduces to the 2 dimensional BBHP case. In general, the presence of discontinuous $\varepsilon_L, \varepsilon_R$, and the discontinuities in $\mathbf{f}(\sigma^+)$, $\mathbf{g}(\sigma^-)$ gives a larger set of solutions, which include strings that are partially or fully folded. Discontinuities are allowed since the differential equations are first order in the derivatives ∂_+ and ∂_-. Such solutions are usually missed in the lightcone gauge even in the flat classical theory (therefore, the lightcone "gauge" is not really a gauge).

The curved space-time analogs of such solutions in higher dimensions are presently under investigation.

1.7 Comments and Conclusions

We have solved generally the classical 2D string theory in any curved space-time. All stringy solutions correspond to folded strings. All solutions tend to the BBHP solutions (as boundary conditions) in the asymptotically flat region of the curved space-time. Therefore, the BBHP solutions of eq.(1.2) serve to classify all the solutions for any curved space-time. In fact, the sign patterns of the BBHP solutions provide the method for dividing the world-sheet into patches, thus defining the lattices associated with the A, B, C, D solutions. The matching of boundaries for these functions gives the general solution in curved space-time in the form of a "transfer matrix". Thus, lattices on the world-sheet plus geometry in space-time lead to transfer matrices. This seems to be a rich area to explore in more detail.

The general physical motion of the string is: oscillations around a center of masss that follows on the average a geodesic of a massive particle, consistent with intuition. The oscillations are deformed by curvature as compared to the BBHP solutions in flat spacetime, but they maintain the same general character as long as the curvature is smooth. However, new stringy behavior becomes evident in the vicinity of singularities where new phenomena, such as tunneling (similar to diffraction), take place. There is also the continuation of the motion into new worlds, in a finite amount of proper time, that the string as well as the massive particle geodesics do (but not the massless particle! - see [10][11]). Because of the tunelling and the new worlds, the global space of the $SL(2,R)/R$ black hole is not just the usual black hole space, $uv < 1$. Rather, it must include also the $uv > 1$ "bare singularity" region even for the classical description of strings (actually this region is not really singular, as argued in [10][11]). We conjecture that the inclusion of the bare singularity region is a more general requirement than the $SL(2,R)/R$ case for the correct description of string motion. Of course, by duality, the quantum theory must include all the regions.

Folded string are also of interest in a string-QCD relation. Gluons are expected to behave just like the folds, since only at the location of a gluon the color flux tube can fold. Some recent discussion if this point can be found in [16] and [10][11].

We suspect that the inclusion of the quantum states corresponding to folded strings may lead to a consistent quantum theory in less than 26 dimensions. As already emphasized earlier in the paper, the free string is perfectly consistent as a quantum theory for $c < 26$, including the folded states. The interacting quantum string with folds remains as an open possibility.

1.8 References

[1] For a recent review, see: I. Bars, "Curved Space-time Geometry for Strings ...", hep-th 930942, in *Perspectives in Mathematical physics" Vol.III, Eds.* R. Penner and S.T.Yau, International Press (1994), page 51.

[2] A.A. Tseytlin, reviews, hep-th/9408040, 9410008. G.T.Horowitz and A.A. Tseytlin, hep-th/9409021.

[3] C. Nappi and E. Witten, hep-th/9310112.

[4] K. Sfetsos, hep-th/9311093, K. Sfetsos and A.A. Tseytlin hep-th/9404063.

[5] I. Bars and J. Schulze, "Folded Strings Falling into a Black Hole", hep-th/9405156 or USC-94/HEP-B1, Phys. Rev. D., to appear

[6] I. Bars and K. Sfetsos, Phys. Rev. **D46** (1992) 4495.

[7] I. Bars and K. Sfetsos, Phys. Rev. **D46** (1992) 4510, and other articles refered in [1][2].

[8] I. Bars and K. Sfetsos, Phys. Rev. **D48** (1993) 844. A.A. Tseytlin, Nucl. Phys. **B399** (1993) 601. A.A. Tseytlin and K. Sfetsos, Phys. Rev. **D49** (1994) 2933.

[9] I. Bars and K. Sfetsos, Mod. Phys. Lett **A7** (1992) 1091.

[10] I. Bars, "Folded Strings in Curved Spacetime", USC-94/HEP-B2 or hep-th/9411078.

[11] I. Bars, "Classical Solutions of String Theory in any 2D Curved Spacetime", USC-94/HEP-B3 or hep-th/9411nnn.

[12] W.A.Bardeen, I.Bars, A.Hanson and R.Peccei, Phys.Rev.**D13** (1976) 2364.

[13] W.A.Bardeen, I.Bars, A.Hanson and R.Peccei, Phys.Rev.**D14** (1976) 2193.

[14] C. Thorn, Nucl. Phys. **B248** (1974) 551, and lecture in *Unified String Theory*, Eds. D. Gross and M. Green, page 5.

[15] O. Ganor, J. Sonnenschein and S. Yankielowicz, "Folds in 2D String Theories", TAUP-2152-94.

[16] I. Bars, "Strings and QCD in 2D", hep-th 9312018, in Proc. Strings 93, World Scientific (1994).

Figures

Fig.1.-- Minimal surface of flat string with 2 critical points that move at 45 degrees. The paths of different points along the string are marked with different symbols.

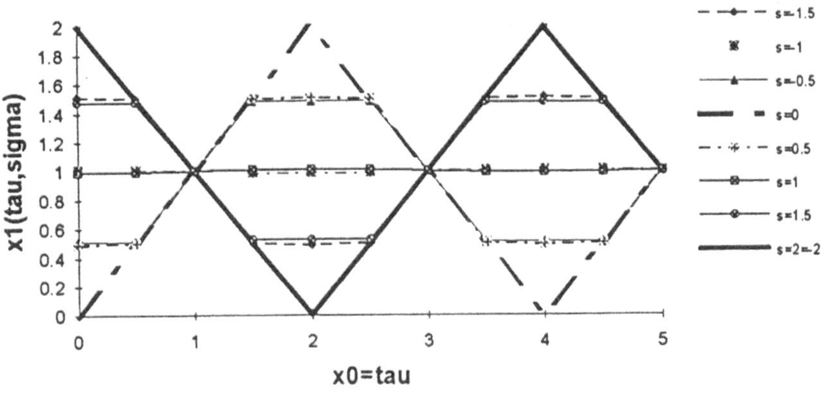

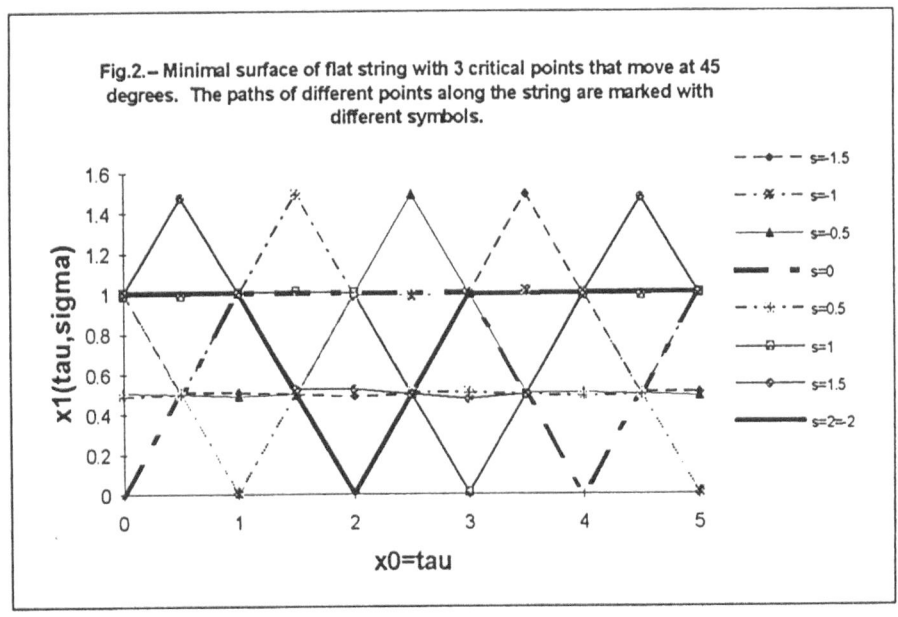

Fig.2.– Minimal surface of flat string with 3 critical points that move at 45 degrees. The paths of different points along the string are marked with different symbols.

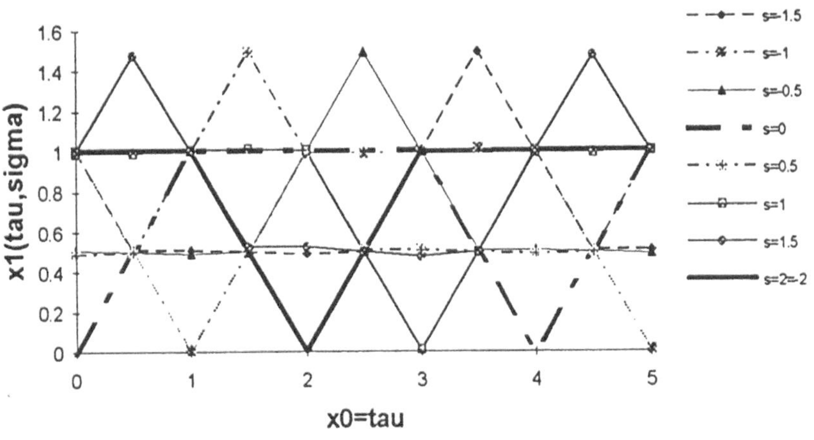

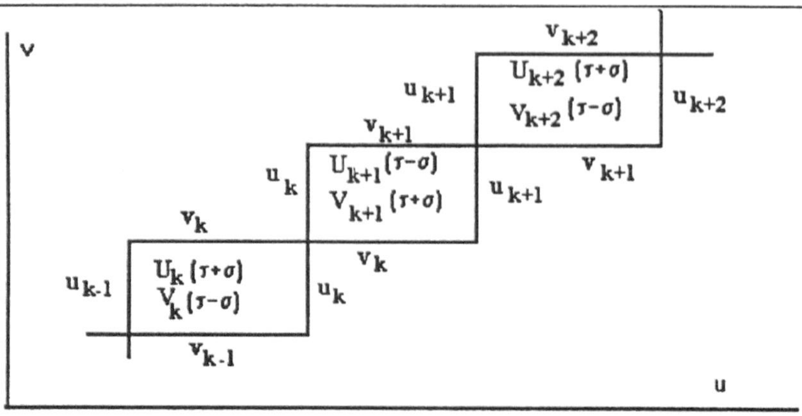

Fig.3 – Minimal area in curved spacetime. The sizes of the rectagles change depending on the curvature.

Fig.4. Ingoing string on 1st sheet meets black hole, moves out to 2nd sheet.

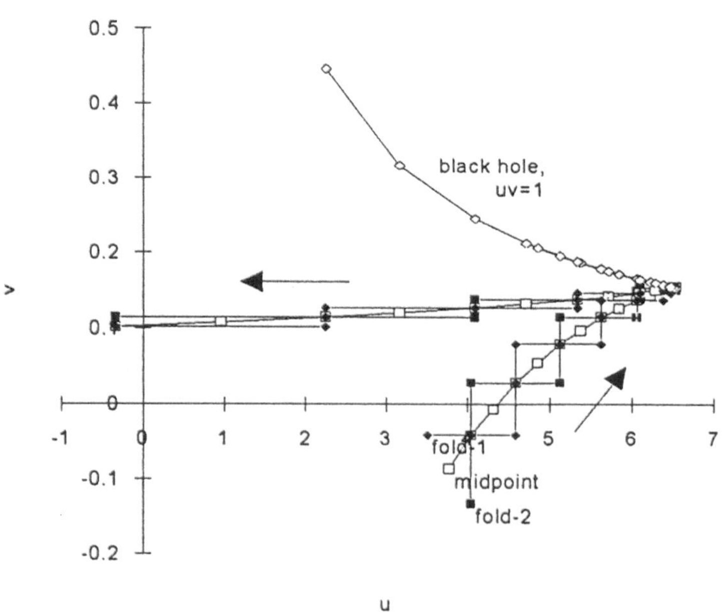

Fig.5. String minimal area tunnels to forbidden region beyond black hole. Arrows along trajectories of midpoint.

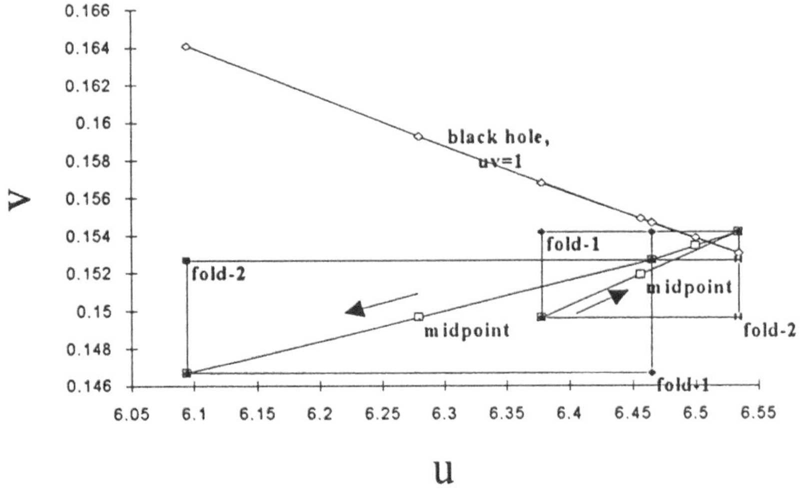

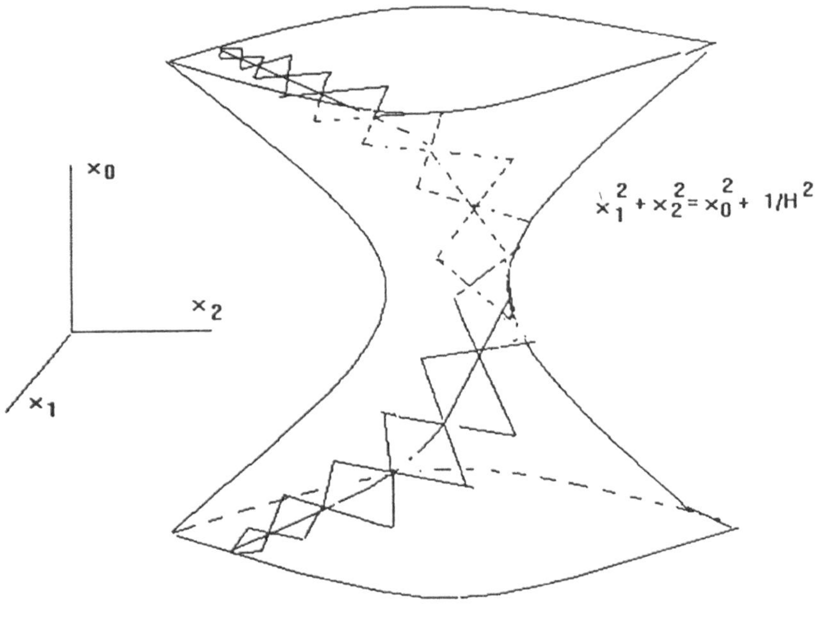

Fig.6 – String moving in deSitter space

Algebro-geometric Aspects of the Bethe Equations

Dedicated to the memory of Feza Gürsey

Robert P. Langlands[1] and Yvan Saint-Aubin[2]

1 School of Mathematics, Institute for Advanced Study, Princeton, N.J., 08540

2 Centre de recherches mathématiques, Université de Montréal, C.P. 6128A, succ. centre-ville, Montréal, Qc, Canada H3C 3J7

Although the Ansatz introduced by Bethe in 1931 ([B]) has been exploited repeatedly by physicists, who have adapted it successfully to a variety of problems, it has never been given a careful mathematical treatment. As a result there is often a disquieting imprecision in its formulation that discourages a resolute pursuit of its analytical consequences; moreover, and more to the point here, its algebraic charm has been little appreciated. Two years ago, the present authors undertook a study of the equations with standard techniques from algebraic geometry. The enterprise, rewarding as it has been, has taken more time and energy than expected. Complete proofs, even adequate understanding, have cost a great deal of effort and patience, and there are still gaps, but the project is nearing completion, and in this paper we describe, albeit in a somewhat provisional form, the principal features of the treatment. Details will appear in [BL].

There is no need here to recall the physical origins of the eigenvalue problem treated by Bethe. The mathematical problem is that of finding the eigenvalues and eigenvectors of an operator on a space of dimension 2^N. This space is

$$\mathfrak{X} = \otimes_1^N \mathbb{C}^2.$$

We take as basis of $\mathbb{C}^2$ two vectors u^+ and u^- and as basis of $\mathfrak{X}$ the vectors

$$u_{m_1,\ldots,m_r} = \otimes u_i, \qquad \{m_1,\ldots,m_r\} \subset \{1,\ldots,N\}$$

where $u_i = u^+$ if $i \in \{m_1,\ldots,m_r\}$ and $u_i = u^-$ otherwise. Thus the index attached to an element of the basis is a sequence of N signs. It is to be thought of as a cyclic sequence, so that each sign has two neighbors, those of the sign at position 1 being those at positions 2 and N. A typical vector will be written

$$\sum_{r=0}^{N} \sum a_{m_1,\ldots,m_r} u_{m_1,\ldots,m_r} = \sum x_r,$$

the inner sum running over all sequences with r positive signs. There is a corresponding decomposition $\mathfrak{X} = \oplus_{r=0}^N \mathfrak{X}_r$.

The operator H whose eigenvalues are to be calculated leaves each of the spaces $\mathfrak{X}_r$ invariant and on $\mathfrak{X}_r$ is given by

$$H_r: \quad x_r \to x_r'$$

with

$$a'_{m_1,\ldots,m_r} = \sum(a_{m_1,\ldots,m_r} - a_{m'_1,\ldots,m'_r}).$$

The sum runs over all sequences $\{m'_1,\ldots,m'_r\}$ that can be obtained from the sequence $\{m_1,\ldots,m_r\}$ by interchanging two adjacent and opposite signs. For example, $- - - - - + + - -$ allows just two possibilities: $- - - - + - + - -$ and $- - - - - + - + -$. Recall that adjacency is to be interpreted in the cyclic sense. For H_r there are $\binom{N}{r}$ eigenvalues and eigenvectors to be found.

If $z = (z_1,\ldots,z_r)$ with z_i complex and $m = (m_1,\ldots,m_r)$ we set $z^m = \prod_k z_k^{m_k}$. If P is a permutation of the set $\{1,\ldots,r\}$ we set $Pm = m'$, $m'_k = m_{P^{-1}k}$. The Bethe Ansatz is to search for eigenvectors of the form

$$a_{m_1,\ldots,m_r} = \sum_P z^{Pm} w_P \tag{1}$$

The sum runs over all permutations of $\{1,\ldots,r\}$. The complex number w_P is obtained from a collection of complex numbers $w_{k,l}$, $k \neq l$, with $w_{k,l} = w_{l,k}^{-1}$:

$$w_P = \prod_{\substack{k>l \\ P^{-1}k<P^{-1}l}} w_{k,l}.$$

Bethe, less preoccupied with the algebro-geometric aspects of the equations, chose as variables f_k and $\varphi_{k,l}$ with $z_k = \exp if_k$, $w_{k,l} = \exp i\varphi_{k,l}$. The formula (1) for the eigenvectors is moreover not quite that of Bethe; we have multiplied his eigenvectors by appropriate constants to obtain more symmetric formulas.

The vector (1) will be an eigenvector (or zero) if the points z and $w = (w_{k,l})$ satisfy the equations

$$\begin{aligned} w_{k,l} &= -\frac{z_k z_l - 2z_k + 1}{z_k z_l - 2z_l + 1}, \qquad k \neq l, \\ z_k^N &= \prod_{l\neq k} w_{k,l}. \end{aligned} \tag{2}$$

If it is not zero, the associated eigenvalue is

$$2\epsilon = \sum_1^r 1 - \cos f_k = \sum_1^r 1 - \tfrac{z_k+z_k^{-1}}{2}.$$

It turns out that these equations evince in algebro-geometrical respects an inconvenient degeneracy. Fortunately this degeneracy is absent for a more general eigenvalue problem, a simple variant of that associated in [TΦ] to the six-vertex model. The equations (2) are replaced by

$$(3)\qquad \begin{aligned} w_{k,l} &= -\frac{z_k z_l - 2\Delta z_k + 1}{z_k z_l - 2\Delta z_l + 1}, \qquad k \neq l, \\ R(z_k) &= \prod_{l\neq k} w_{k,l}. \end{aligned}$$

Δ is a parameter whose value will at first be chosen to be generic. The function R is a rational function of degree N with zeros $\alpha_1, \ldots, \alpha_N$ and poles $\beta_1, \ldots, \beta_N$. The α_i are arbitrary but

$$(4)\qquad \beta_i = 2\Delta - 1/\alpha_i = A(\alpha_i).$$

This equation defines the fractional linear transformation A. The equations (3) can be studied for generic values of Δ and generic (with respect to the constraints imposed) R.

We recall that the term generic simply means that the pertinent parameters (at present Δ and the $N+1$ parameters needed to specify R, for example $\alpha_1, \ldots, \alpha_N$ and $R(\infty)$, thus $N+2$ in all) are required to lie outside a countable collection of algebraic subvarieties of dimension less than $N+2$. If we can calculate eigenvectors and eigenvalues for generic values of the parameters then, taking limits, we can calculate them for any values. Generic values of the parameters are thus values that do not satisfy any of some countable collection of non-trivial equations, the equations themselves to be determined (explicitly or implicitly) in the course of analyzing the problem. For example, in the present problem the zeros and poles of R are to be distinct and the matrix of A

$$(5)\qquad \begin{pmatrix} 2\Delta & -1 \\ 1 & 0 \end{pmatrix}$$

may not have eigenvalues that are roots of unity. Thus Δ may not take any of the values $\cos(a\pi/b)$, a/b rational, although these are far from the sole constraints.

There are several trivial, yet basic, observations to be made about the equations (3). First of all, if Q is a permutation and

$$Q(z)_k = z_{Q^{-1}k}, \qquad Q(w)_{k,l} = w_{Q^{-1}k, Q^{-1}l}$$

then $(Q(z), Q(w))$ is a solution whenever (z, w) is. Since

$$\sum Q(z)^{Pm} Q(w)_P = w_Q^{-1} \sum z^{Pm} w_P,$$

the associated eigenvalue is not changed and the associated eigenvector not changed in any essential way. Moreover if Q is simply an interchange of two integers k and l then $w_Q = w_{k,l}$. Thus whenever two coordinates z_k and z_l are equal $w_Q = -1$; the vector defined by (1) is zero; and the solution of (3) is not admissible because it leads to nothing. Finally, there are solutions of (3) for which $z = \alpha_k$ or $z = \beta_k$. For these solutions, some $w_{k,l}$ is zero or infinity and these solutions are also not admissible because (1) is then not well defined.

A solution will therefore be called *admissible* if:

(1) all the coefficients z_k are different;

(2) no z_k is equal to zero (or infinity) or to a zero or pole of the function R and no $w_{k,l}$ is zero or infinite.

The admissible solutions come in sets with $r!$ elements, any two elements in these subsets differing by a permutation. Since we need $\binom{N}{r}$ vectors we need at least $N(N-1)\dots(N-r+1)$ solutions. The principal theorem of [BL] is easily stated.

Theorem. *For generic Δ and R there are exactly $N(N-1)\dots(N-r+1)$ admissible solutions (z,w) of the equations (3). The vector (1) attached to such a solution is not zero and the collection of vectors obtained in this way generate the space $\mathfrak{X}_r$.*

It will become clear that it is easy to find generically at least $N(N-1)\dots(N-r+1)$ admissible solutions whose associated eigenvectors generate the full space $\mathfrak{X}_r$. The difficult assertion is that there are *exactly* this many solutions, and we concentrate on it.

To count the number of admissible solutions we have to count the number of all solutions and then subtract the number of inadmissible solutions. The equations (3) can be regarded as defining the fixed points of an at first imprecisely defined *algebraic correspondence*. Let

$$X = \prod_{k=1}^{r} \mathbb{P}^1 \times \prod_{1 \le k < l \le r} \mathbb{P}^1 = Z \times W.$$

The coordinates are z_k and $w_{k,l}$ with redundant coordinates $w_{l,k} = w_{k,l}^{-1}$. For the moment take $C = X$ and consider the two mappings φ and ψ of C into X defined (inadequately) by

$$\varphi(z,w) = (z',w'), \quad z'_k = R(z_k), \quad w'_{k,l} = w_{k,l},$$

$$\psi(z,w) = (z'',w''), \; z''_k = \prod_{l \ne k} w_{k,l}, \; w''_{k,l} = Q(z_k,z_l) = -F(z_k,z_l)/F(z_l,z_k), \tag{6}$$

with $F(z,z') = F_\Delta(z,z') = zz' - 2\Delta z + 1$. The equations (3) define the points p for which $\varphi(p) = \psi(p)$.

Before discussing the failings of the definition of the correspondence (φ,ψ) we observe that the inadmissible solutions can be defined as the fixed points of similar correspondences. Suppose that $\{A_1,\dots,A_s,B'_1,\dots,B'_t,B''_1,\dots,B''_t\}$ (denoted more compactly (A,B)) is a disjoint decomposition of $\{1,\dots,r\}$ into non-empty subsets and that to each l, $1 \le l \le t$, there is associated a zero $\alpha(l)$ of R. Set $i \equiv j$ if i and j belong to the same A_l or if $i \in B'_l$, $j \in B'_m$, and

$\alpha(l) = \alpha(m)$, or finally if $i \in B_l''$, $j \in B_m''$, and $\alpha(l) = \alpha(m)$. Define a sub-variety $X^{A,B}$ of X by the conditions

(1) $z_i = z_j$ and $w_{i,m} = w_{j,m}$ if $i \equiv j$;
(2) if $i \equiv j$ then $w_{i,j} = -1$;
(3) if $i \in B_l'$ then $z_i = 0$;
(4) if $i \in B_l''$ then $z_i = \infty$;
(5) $i \in B_{l'}'$ and $j \in B_{l''}''$ with $\alpha(l') = \alpha(l'')$ then $w_{i,j} = 0$ and $w_{j,i} = \infty$;
(6) if $\alpha' = \alpha(l') \neq \alpha(l'') = \alpha''$ and $\beta' = A(\alpha')$, $\beta'' = A(\alpha'')$ then
 (a) $w_{i,j} = Q(\alpha', \alpha'')$ if $i \in B_{l'}'$ and $j \in B_{l''}'$,
 (b) $w_{i,j} = Q(\alpha', \beta'')$ if $i \in B_{l'}'$ and $j \in B_{l''}''$,
 (c) $w_{i,j} = Q(\beta', \alpha'')$ if $i \in B_{l'}''$ and $j \in B_{l''}'$,
 (d) $w_{i,j} = Q(\beta', \beta'')$ if $i \in B_{l'}''$ and $j \in B_{l''}''$.

If $t = 0$ and $s = r$ then $X^{A,B}$ is X itself. It is convenient to set

$$\mathbb{B}_\alpha' = \cup_{\{l|\alpha=\alpha(l)\}} B_l', \qquad \mathbb{B}_\alpha'' = \cup_{\{l|\alpha=\alpha(l)\}} B_l'',$$

and $\mathbb{B}_\alpha = \mathbb{B}_\alpha' \cup \mathbb{B}_\alpha''$. It is not excluded that one or the other of $\mathbb{B}_\alpha'$ and $\mathbb{B}_\alpha''$ is empty but then both are.

The variety C is replaced, again for a provisional definition of the correspondence, by a sub-variety $C^{A,B}$ defined by the conditions (1) (2) (5) and (6) together with the following modifications of (3) and (4):

(3) if $i \in B_l'$ and $\alpha = \alpha(l)$ then $z_i = \alpha$;
(4) if $i \in B_l''$ and $\alpha = \alpha(l)$ then $z_i = \beta = A(\alpha)$.

It is clear how to restrict (φ, ψ) to $C^{A,B}$ to obtain $(\varphi^{A,B}, \psi^{A,B})$.

There is an obvious order on the decompositions $\{A, B\}$. The decomposition $\{\tilde{A}, \tilde{B}\}$ is deeper than $\{A, B\}$ if it imposes more conditions. We write $\{\tilde{A}, \tilde{B}\} \prec \{A, B\}$. We attach to a fixed point $p = (z, w)$ the decomposition $\{A(p), B(p)\}$ in which each B_l is of the form $B_l' = \{i | z_i = \alpha\}$, $B_l'' = \{i | z_i = \beta = A(\alpha)\}$, $\alpha(l) = \alpha$, and for which i and j are in a common A_l if and only if $z_i = z_j$ and z_i is neither a zero nor a pole of R. It is pretty clear, apart from the unresolved ambiguities in the correspondences, that a fixed point in X lies in $X^{A,B}$ if and only if $\{A(p), B(p)\}$ is deeper than $\{A, B\}$.

To each set A_l or $B_l = B_l' \cup B_l''$ of a decomposition we attach the weight $(-1)^{n-1}(n-1)!$, n being the number of integers in the set. To the decomposition itself we attach the weight $\omega(A, B)$ equal to the product of the weights of all terms in the decomposition. Thus the weight of the decomposition $\{1,2\}$ into $\{\{1\},\{2\}\}$ is 1 and of that into $\{\{1,2\}\}$ is -1. That of $\{1,2,3,4,5,6,7\}$ into $\{\{1,2,3\},\{4,5,6,7\}\}$ is -12. The following lemma is easily verified.

Lemma. *For each fixed point p in X the sum*

$$\sum_{(A(p),B(p)) \prec (A,B)} \omega(A, B)$$

is 0 unless p is admissible, but then it is 1.

Thus if $n(A,B)$ is the number of fixed points on $X^{A,B}$ it suffices to verify that

$$N(N-1)\dots(N-r+1)=\sum_{(A,B)}\omega(A,B)n(A,B). \tag{7}$$

Long ago, in the twenties, Lefschetz ([L]) introduced a topological formula not for the number of fixed points of a correspondence but for the number of fixed points counted with multiplicities. If all multiplicities are one, or even if a given fixed point p has the same multiplicity for all correspondences $(\varphi^{A,B},\psi^{A,B})$ of which it is a fixed point and this multiplicity is one if the point is admissible, then it is legitimate to substitute for $n(A,B)$ in (7) the number $\lambda(A,B)$ of fixed points counted with multiplicity. Since part of our strategy is to show that all these multiplicities are one for generic values of the parameters, we shall use this new form of (7).

There is a disagreeable complication. The Lefschetz formula expresses the number of fixed points in terms of topological data associated to the correspondence. These data are relatively easy to determine, at least in a combinatorial form, from the inadequately defined correspondences with which we have dealt so far because the underlying topological spaces are products of the Riemann sphere with itself, so that their homology has a simple structure. The true correspondences are obtained by blow-ups that are yet to be described. One of our principles has been to make the count, in a manner that cannot be completely correct, with the ill-defined correspondences, leaving the complete justification, which should pose no problems, until the more serious algebraic difficulties are out of the way.

The homology of Z is spanned by products of an arbitrary number of the factors $\mathbb{P}^1$ appearing in Z. An element of this basis is therefore given by a subset of $\{1,\dots,r\}$. An element of the analogous basis of the homology of W is given by a collection of unordered pairs $\{k,l\}$, $k\neq l$. A basis for the homology of X is obtained by taking all products of a basis element for Z with one for W. In particular, the homology groups of X are non-zero only in even dimensions. There are maps $\varphi_i : H_i(C)\to H_i(X)$, and by Poincaré duality an associated map $\varphi^i : H_i(X)\to H_i(C)$, as well as a map $\psi_i : H_i(C)\to H_i(X)$. The composition $\psi_i\varphi^i$ takes $H_i(X)$ to itself and the Lefschetz formula is

$$\lambda(X)=\sum_{i=0}^{\dim X} Tr\,(\psi_{2i}\varphi^{2i}). \tag{8}$$

There is of course a similar formula for $\lambda(X^{A,B})=\lambda(A,B)$.

The map φ is so simple that it is easy to determine φ_{2i}. It multiplies a basis element whose Z component is the product of s factors $\mathbb{P}^1$ by N^s. From this it is easy to determine φ^{2i}. The map ψ on the other hand interchanges the two factors Z and W. According to the formulas (6), $w_{k,l}$ is for fixed z_l a fractional linear function of z_k and for fixed z_k a fractional linear function of z_l. As a consequence the class in $H_2(Z)$ associated to the index k is mapped to the sum of the classes in $H_2(W)$ associated to those indices $\{l,m\}$ for which $m = k$ or $l = k$. In the same way z_k is a linear function of $w_{k,l}$ for each l. As only the classes of the product $X = Z \times W$ that appear in the representation of their own image contribute to the trace, we obtain a combinatorial expression for this trace. It is the sum over a collection of oriented graphs G on the set $\{1,\ldots,r\}$ of $N^{\pi(G)}$, $\pi(G)$ being the number of connected components of G.

The graphs are subject to two constraints:

(1) they are trees;

(2) there is at most one bond issuing from each point.

The meaning of these conditions is easily explained. Suppose S is a subset of $R = \{1,\ldots,r\}$ and T a subset of unordered pairs $\{k,l\}$. The basis element

$$\eta = \prod_{k\in S} \mathbb{P}^1 \times \prod_{\{l,m\}\in T} \mathbb{P}^1$$

will contribute either 0 or 1 to the trace. To each basis element contributing 1 we associate a graph G. For such a basis element the cardinality of S must be that of T. Moreover it must be possible to assign to each $\{l,m\} \in T$ a k in $\{l,m\}$ and in S so that as z_k moves so does $w_{l,m}$. Thus $k \in \{l,m\}$ and $\{l,m\}$ is the bond issuing from k. Since a given z_k can not be responsible for the movement of two factors in W, we attach different vertices to different bonds, and the equality $|S| = |T|$ assures us that there is exactly one bond issuing from a vertex in S. The vertices of $\{1,\ldots,r\}$ not in S are the final points of G and some may be isolated.

Suppose there were a cycle in G with bonds $\{k_1,k_2\},\ldots,\{k_p,k_1\}$. Then the number of remaining bonds is equal to the number of remaining vertices in S and each of the remaining bonds is responsible for the movement of one of the remaining vertices. Thus the effective movement in $(z_{k_1},\ldots,z_{k_p})$ is achieved by the bonds in the cycle. However if $P = \{k_1,\ldots,k_p\}$ and Q is its complement in R

$$\prod_{k\in P} z_k'' = \prod_{k\in P}\prod_{l\in Q} w_{k,l}. \tag{9}$$

Thus the product is independent of the variables $w_{k_i,k_{i+1}}$ and the image of the class attached to η is zero. Equation (9), valid for any set P, is basic and will be used again, but without comment.

The Lefschetz numbers attached to $(\varphi^{A,B},\psi^{A,B})$ are calculated in a similar way with a similar result. The vertices of the graphs are now $A_1,\ldots,A_s$ and $B_1,\ldots,B_t$. Each vertex carries a weight, the cardinality of the respective set; so does each graph, the product over the directed bonds of the weights of the vertices at which the bond ends. Moreover no bond may issue from a vertex B_i. Thus B_i is *forbidden* as the *source* of a bond.

Although it is possible to carry out a good part, but not all, of the combinatorial discussion without them, it turns out that these numbers defined combinatorially can be expressed by simple algebraic formulas that, together with their proofs, are due to Fan Chung. It is sufficient to give a formula for the number of connected graphs of the described type. The number of forbidden vertices is then at most one. Suppose there are n vertices with weights $\omega_1, \ldots, \omega_n$.

Lemma(F. Chung).

(1) *If there are no forbidden vertices the number of connected graphs counted with weight is* $(\omega_1 + \cdots + \omega_n)^{n-1}$.

(2) *If there is one forbidden vertex of weight* ω *the number of connected graphs counted with weight is* $\omega(\omega_1 + \cdots + \omega_n)^{n-2}$.

This lemma simplifies, for example, the proof of formula (7). The remaining obstacle is the proof that the multiplicites of the fixed point are generically one, and it is surprisingly difficult to overcome. The bulk of our effort has had to be devoted to this.

A closer examination of the description of the effect of ψ, or more generally $\psi^{A,B}$, on cycles reveals the inadequacy of the formulas (6). It is possible that $F(z_k, z_l)$ and $F(z_l, z_k)$ vanish simultaneously so that $w''_{k,l}$ is not well defined, not even as infinity, or that $w_{k,l} = 0$ while $w_{k,m} = \infty$ so that z''_k is not well defined. The second possibility has to be dealt with, but its consequences are largely technical, while an adequate understanding of the consequences of the first is central to the argument.

It is necessary to blow up C by introducing new coordinates $(u_{k,l}, u_{l,k})$ and $\lambda_{k,l}$. The first of these are projective coordinates, so that only their ratio matters, and to fix $\lambda_{k,l}$ it can be supposed that one or the other is 1. They are subject to one basic relation and to all relations that can be deduced from it: $u_{k,l}F(z_l, z_k) = u_{l,k}F(z_k, z_l)$. Of course these new coordinates are redundant except where $F(z_l, z_k) = F(z_k, z_l) = 0$. Additional blow-ups are necessary at $\Delta = 0$ when one of the z_k is a zero or a pole of the function $R(z)$. The blow-ups are carried out in a systematic way, so that the result is a family of correspondences parametrized by Δ with generically smooth C_Δ, or more generally $C_\Delta^{A,B}$.

The fiber is not smooth at $\Delta = 0$. We proceed, however, by first analyzing the fixed points of the correspondence at this fiber, and then studying its deformations. The first point, less evident than we at first thought, is that in a neigborhood of $\Delta = 0$ and thus generically the fixed points of the modified, well-defined correspondences $(\varphi_\Delta^{A,B}, \psi_\Delta^{A,B})$ are isolated, although perhaps with multiplicity greater than one. This is not true at $\Delta = 0$ where the variety of fixed points has a large number of components of positive dimension.

The next point is to show that appropriate fixed points at $\Delta = 0$ deform to give at least $\lambda(A, B)$ distinct fixed points in its neighborhood. It then follows immediately from the Lefschetz formula for $(\varphi_\Delta^{A,B}, \psi_\Delta^{A,B})$ that these are all the fixed points and that they have multiplicity one. The constraints imposed by (A, B) do not affect the form of the equations. They need only be taken into account when counting the number of solutions.

At $\Delta = 0$

$$F(z_k, z_l) = F(z_l, z_k) = z_k z_l + 1$$

so that unless $z_k z_l = -1$ at a fixed point the value of $w_{k,l}$ there is -1. As a consequence at a fixed point p the collection $\{1, \ldots, r\}$ breaks up naturally into subsets $E_1, F_1, \ldots, E_m, F_m$, two indices k and l lying in the same subset if and only if $z_k = z_l$ and lying in *opposed* subsets E_i and F_i if and only if $z_k z_l = -1$. (For a generic choice of the function R the values of the z_k at a fixed point are finite and not zero, and $z_k^2 \neq -1$.) Moreover when k and l do not lie in opposed subsets then $w_{k,l} = -1$.

Thus the equations decouple and as a first approximation it is convenient to deform into a neighborhood of $\Delta = 0$ as though $w_{k,l}$ continued to be -1 when k and l are not opposed.

Let γ_i and δ_i be the values of the coordinates in E_i and F_i, of which one can be empty but not both. There are two types to distinguish: either $\{\gamma_i, \delta_i\}$ is not a pair $\{\alpha, \beta\}$ for some zero and matching pole of R or it is. In the second case neither set is empty.

The first case is, however, the easiest to treat. If one of the two sets is empty then for all k in the other

$$R(z_k) = (-1)^{r-1}.$$

For generic R this equation has N distinct solutions (of which no two are opposed) and they deform uniquely to a whole neighborhood of $\Delta = 0$. Choosing distinct $z_1, \ldots, z_r$ among these solutions we obtain $N(N-1)\ldots(N-r+1)$ admissible fixed points. (In each pair (E_i, F_i) of the associated configuration one set is then empty and the other contains a single element.) Thus it is to be expected that all other fixed points will deform to inadmissible fixed points. The existence of these solutions means that the existence generically of admissible solutions sufficient to generate all eigenvectors is rather simple. The significant mathematical assertion is that there is no redundancy. It is unlikely that this is without consequences for the set of eigenvalues, but we have not had an opportunity to pursue the matter very far.

Continuing to suppose that γ_i is neither a zero nor a pole of R we suppose that neither E_i nor F_i is empty. Let their cardinalities be μ and ν. The number $\rho = R(z)R(-1/z)$ is independent of z at $\Delta = 0$ and may be supposed generic. Multiplying the equations $R(z_k) = \prod_l w_{k,l}$, $k \in E$ and $k \in F$, together we find that

$$(-1)^{\mu(r-\nu)} R(\gamma_i)^{\mu} \rho^{\nu} = (-1)^{\nu(r-\mu)} R(\gamma_i)^{\nu}. \tag{10}$$

Since ρ is generic we conclude that $\mu \neq \nu$ and that there are $|\mu - \nu| N$ possibilities for γ_i.

At $\Delta = 0$ and for $k \in E_i$, $l \in F_i$ it is best not to use the projective coordinates $(u_{k,l}, u_{l,k})$ but new coordinates $(t_{k,l}, v_{k,l})$ such that

$$\frac{u_{k,l}}{u_{l,k}} = \frac{v_{k,l} - 2a_{k,l}t_{k,l}}{v_{k,l} - 2t_{k,l}}, \qquad a_{k,l} = z_k/z_l.$$

When $z_k = \gamma_i$ and $z_l = \beta_i = -1/\gamma_i$ then $a_{k,l} = a_i = -\gamma_i^2$, and for generic parameters this is never 1 at a fixed point. Disregarding the possibility that there may be solutions with some $t_{k,l} = 0$, we take all $t_{k,l}$, $k \in E_i$, $l \in F_i$ equal to 1. Then the $v_{k,l}$ are not independent but may be written as $v_{k,l} = r_k + s_l$. There is an indeterminacy in these new parameters; all the r_k may be replaced by $r_k + t$ and the s_l by $s_l - t$.

Upon decoupling, the equations that refer to the coordinates in E_i and F_i are at $\Delta = 0$

$$
\begin{aligned}
c_i &= \prod_{l \in F_i} \frac{r_k + s_l - 2a_i}{r_k + s_l - 2}, \qquad & d_i &= \prod_{k \in E_i} \frac{r_k + s_l - 2a_i}{r_k + s_l - 2}, \\
c_i &= (-1)^{r-\mu} R(\gamma_i), & d_i &= (-1)^{r-\nu} R^{-1}(\delta_i).
\end{aligned} \tag{11}
$$

The values of c_i and d_i in (11) are generic, and in contrast to the Bethe equations themselves it is possible to show directly that each solution of these equations is of multiplicity one, an assertion that remains meaningful even when the equations are recoupled in a neighborhood of $\Delta = 0$.

The solutions of (10) are counted differently for the different correspondences $(\varphi^{A,B}, \psi^{A,B})$ because there are different constraints imposed. For given (A, B) the set E_i and the set F_i must each be the union of the A_l it contains, for at present γ_i is neither an α_k nor a β_k. If E_i contains e elements of weights $\mu_1, \dots, \mu_e$ and F_i contains f elements of weights $\nu_1, \dots, \nu_f$ then $\mu = \mu(E_i) = \sum \mu_i$, $\nu = \mu(F_i) = \sum \nu_j$ and the number of solutions of (11), counted by determining the degrees of the maps involved – not forgetting that (11) implies (10) – is $|\mu - \nu| \mu^{f-1} \nu^{e-1} N$.

Before commenting on the difficult case that $\{\gamma_i, \delta_i\}$ is a pair $\{\alpha, \beta\}$, we examine the nature of the combinatorial problem that remains. The Lefschetz formula yields the number of fixed points with multiplicities as a polynomial in N. The coefficient of a given power N^s is the number of graphs of a certain type counted with weights but with s connected components. This number is thus the sum over all decompositions of $\{1, \dots, r\}$ into s subsets of the product from 1 to s of the number, counted with weights, of *connected* graphs on the elements in the corresponding subset. This structure is shared by the other number to which it is to be compared. The set $\{1, \dots, r\}$ is decomposed into the sets $D_i = E_i \cup F_i$ and the equations, to a first approximation that then has to be improved, decoupled. For each element of the decomposition there are equations to be solved, and the total number of solutions is obtained by summing over all partitions of the product of the number of solutions attached to the elements of the partition.

For each collection of vertices D and all possible decompositions $D = E \cup F$ (the elements defining the decomposition of $\{1, \dots, r\}$ are now themselves decomposed) we have to count the solutions of each type observing that if D contains forbidden vertices only solutions of the second type are possible. In all cases, it is essential that the result is a combinatorial factor, independent of N, times N. This is so for solutions of the first type. In all there are

$$
N \sum_{E \cup F = D} |\mu(E) - \nu(F)| \mu(E)^{f-1} \mu(F)^{e-1} \tag{12}
$$

where e and f are, as before, the numbers of elements in E and F. The sum runs over disjoint partitions of D without regard to order. It is not difficult to show that

$$\sum_{E\cup F=D}(\mu(E)+\nu(F))\mu(E)^{f-1}\mu(F)^{e-1}=\mu(D)^{d-1},$$
(13)
$$d=|D|,\quad \mu(D)=\sum_{A_l\in D}\omega_l$$

so that, if D contains no forbidden vertices, there are N times

(14)
$$2\sum_{E\cup F=D}\min\{\mu(E),\nu(F)\}\mu(E)^{f-1}\mu(F)^{e-1}$$

deformations of solutions of the second type to be found. If D contains forbidden vertices there are only solutions of the second type and the pertinent analogue of (14) is

(15)
$$\omega\mu(D)^{d-2}=\sum\{(\mu(E)+\omega')^f(\mu(F)+\omega'')^e-\mu(E)^f\mu(F)^e\}$$

if $\omega=\omega'+\omega''$ is the weight of the forbidden vertex of D and the sum is over decompositions of the set $\{A_1,\dots,A_s\}$ of the set of remaining vertices into two subsets E and F. Recall that the weight $\omega=|B|$ of a forbidden vertex is the sum of $\omega'=|B'|$ and $\omega''=|B''|$.

For pairs (E_i,F_i) of the second type the source of the factor N is clear for there are N possibilities for α. The combinatorics are, however, less transparent because a crude count of possibilities does not distinguish between two disjoint pairs (E_i,F_i) and (E_j,F_j) associated to the same α_k and the pair $(E_i\cup E_j,F_i\cup F_j)$. Although we have yet to deal with this problem in general, it cannot be major. A second difficulty that we also expect to overcome is more serious. The decoupled equations of the first type define at $\Delta=0$ a variety of positive dimension. Moreover, although only a finite number of points on this variety admit a deformation to a neighborhood of $\Delta=0$, there may be several such deformations, so that we have to count not the number of such points but the number of curves (parametrized by Δ) passing through them.

We do not attempt to describe what we know at present, for our knowledge is incomplete. Yet a few words to indicate the source of the difficulties are appropriate. At $\Delta=0$ and when γ_i is a zero α of R and $\delta_i=A(\alpha)$ it is necessary, in order to define the correspondences completely, to blow C up further by the introduction of projective coordinates (p_k,q_k), $k\in E_i\cup F_i$, such that

$$z_k-\alpha=\xi_kp_k,\quad \Delta=\xi_kq_k,\quad k\in E_i,$$
$$\tilde z_k-\tilde\beta=\xi_kp_k,\quad \Delta=\xi_kq_k,\quad k\in F_i.$$

The coordinates ξ_k have to be introduced at the same time as the supplementary projective coordinates, and (for convenience) in equations containing $\tilde z_k=1/z_k$ rather than z_k.

In a form that should be taken as a first symbolic approximation the equations for the variables attached to $E_i \cup F_i$ are

$$
(16) \qquad
\begin{aligned}
A\Delta p_k &= (-1)^{r-\mu} \prod_{l \in F_i} \frac{p_k + p_l}{p_k + p_l - 2b}, \\
B\Delta p_l &= (-1)^{r-\nu} \prod_{k \in E_i} \frac{p_k + p_l}{p_k + p_l - 2b}.
\end{aligned}
$$

The number $b = 1 + \alpha^2$ is not 0. Moreover A is the derivative of $R(z)$ at α and $B = -A/\rho$. As a result of the presence of Δ on the left side these equations can be solved at $\Delta = 0$ simply be demanding that for each k in E_i or F_i there be an l in the other set such that $p_k + p_l = 0$. To find the solutions that deform and their deformations one looks not for numerical solutions in p_m but for solutions that are Puiseux expansions in Δ, not all constant.

The strategy is to find as many solutions as demanded by the combinatorics and the Lefschetz formula, and therefore to deduce indirectly that we have all. We have no idea whether this might be done directly; it certainly is not promising. We also observe that it is by no means obvious that the deformations of solutions inadmissible at $\Delta = 0$ should remain inadmissible. Yet this is so – or rather our analysis confirms it so far! There are now two combinatorial elements that arise in the algebra and that must be taken into account in the combinatorial comparison with (14) and (15). Not only must the number of solutions of various equations be found, but also, as the first step, the possible leading exponents of the Puiseux expansions calculated. Even for very low values of e and f the analysis is far from transparent.

The ultimate purpose of the Bethe Ansatz is useful insight into the behavior of the eigenvalues of the operator H. To this end Bethe introduced the notion of *Wellenkomplex* for which the customary terminology is now the less formal *string*. At $\Delta = 1$, the value of the parameter that Bethe treated, this notion has limitations, so that clarifications are necessary, but near $\Delta = \infty$ there is a precise, general definition. Since admissible solutions remain admissible as the parameters are deformed, the strings can be transported to 1, but their relation with those of Bethe will not be examined here.

For large $|\Delta|$, set $\lambda_\pm = 1 \pm \sqrt{1 - 1/\Delta^2}$. The matrix

$$V = \begin{pmatrix} \kappa\lambda_+ & \lambda_- \\ \kappa/\Delta & 1/\Delta \end{pmatrix}$$

diagonalizes the matrix (5),

$$V^{-1}AV = \Delta \begin{pmatrix} \lambda_+ & 0 \\ 0 & \lambda_- \end{pmatrix}.$$

As a consequence the Bethe equations take a simpler form when the unknowns z_i are replaced by $u_i = V^{-1}(z_i)$. Define κ so that $\kappa^2 = \lambda_-/\lambda_+$ and so that κ behaves like $1/2\Delta$ when $\Delta \to \infty$.

Adapting the ideas of [B] to our purposes we fix a decomposition of $\{1,\ldots,r\}$ into ordered subsets $S_1 = \{u_{1,1},\ldots,u_{1,L_1}\}, S_2, \ldots$ of lengths $L_1, L_2, \ldots,$ and look for solutions that are Laurent series in κ,

$$u_{l,i} = \kappa^{L_l+1-2i} v_{l,i} + \ldots, 1 \le i \le L_l.$$

The zeros α_i have to be so chosen that $\alpha_i = \epsilon_i \kappa$, where ϵ_i is itself a power series in κ with generic initial term, and

$$R(z) = c \prod \frac{z - \alpha_i}{z/\beta_i - 1},$$

where c is also a power series in κ with generic initial term. Once again, if we find enough solutions of this type generically admissible in a neighborhood of $\Delta = \infty$ then we have them all. It turns out that the correct number of solutions is obtained by taking all possible decompositions, and by choosing the leading coefficients $v_{l,i} = v_l$ to be independent of i, to be different if the associated strings have the same length ($v_l \neq v_{l'}$ if $L_l = L_{l'}$), and to satisfy the system of equations

$$d_l v_l^N = \prod_{k \neq l} \left(\frac{v_l}{v_k}\right)^{X(L_l, L_k)} \tag{17}$$

with

$$X(l,m) = \begin{cases} 2l - 1, & \text{if } l = m \\ 2\min(l,m), & \text{if } l \neq m. \end{cases}$$

The number d_l is determined by the leading coefficients ϵ_i^0 and c_0 of the series for ϵ_i and c,

$$d_l = c_0^{L_l} (-1)^{N + L_l(r - L_l)} \prod_{i=1}^{N} (\epsilon_i^0)^{L_l - 1}.$$

That there are enough solutions of (17) is a simple combinatorial problem; the implicit function theorem can then be used to establish that the desired solutions in series exist. We observe that the notion of strings is only useful when $2r \le N$ and that there is a duality in the equations that permits the replacement of r by $N - r$, adding that our argument does not yet deal with the case $2r = N$.

We have explicit and complete sets of admissible solutions near 0 and near ∞ that can be analytically continued along paths, either closed, starting near 0 or ∞ and ending where they began, or passing from 0 to ∞. It may well be of considerable mathematical interest to examine the effect on the solutions and the partitions into strings of different choices of path for there is very likely ramification at some non-generic points. We have, however, only begun to experiment.

References

[B] H. Bethe, *Zur Theorie der Metalle*, Zeit. f. Physik, **71**, 205-226 (1931)

[BL] R. P. Langlands and Y. Saint-Aubin, *Les sommes de Bethe et la formule de Lefschetz*, in preparation.

[L] S. Lefschetz, *Intersections and transformations of complexes and manifolds*, Trans. A. M. S., **28**, 1-49 (1926)

[ТФ] Л. А. Тахтаджян и Л. Д. Фаддеев, Квантовой метод обратной задачи у XYZ модель Гейзенберга, Успехи Мат. Наух, **34**, вып. 5, 13-63 (1979)

Extended Superconformal Symmetry , Freudenthal Triple Systems and Gauged WZW Models

Murat Günaydin

Physics Department, 104 Davey Lab.
Penn State University ,University Park, PA 16802

Abstract: We review the construction of extended ($N = 2$ and $N = 4$) superconformal algebras over triple systems and the gauged WZW models invariant under them. The $N = 2$ superconformal algebras (SCA) realized over Freudenthal triple systems (FTS) admit extension to "maximal" $N = 4$ SCA's with $SU(2) \times SU(2) \times U(1)$ symmetry. A detailed study of the construction and classification of $N = 2$ and $N = 4$ SCA's over Freudenthal triple systems is given. We conclude with a study and classification of gauged WZW models with $N = 4$ superconformal symmetry.

1 Introduction

It is a singular honor for me to give a talk in the First of the Gürsey Memorial Conferences which are to be held biannually. I regard Feza Gürsey as my mentor and a role model I try to emulate. He has had a great influence on my style of physics.

In this talk I will discuss extended superconformal algebras and their connection with triple systems, in particular the Freudenthal Triple Systems. I will also discuss a class of Lagrangian field theories, namely the gauged WZW models, that are invariant under extended superconformal groups. I learned about Freudenthal triple systems more than twenty years ago when I was working with Feza on the questions of physical implications of extending the underlying number system of quantum mechanics from complex numbers to octonions and what possible role exceptional groups may play in physics.

Infinite conformal algebra in two dimensions and its supersymmetric extensions have been studied extensively in recent years. They underlie string and superstring theories as their local gauge symmetries[1]. The classical vacua of string theories are described by conformal field theories. For example, the heterotic string vacua with $N = 1$ space-time supersymmetry in four dimensions are described by "internal" $N = 2$ superconformal field theories with central charge $c = 9$ [2, 3, 4, 5, 6, 7]. The $N = 2$ space-time supersymmetric vacua of the heterotic string are described by an internal superconformal field theory with four supersymmetries [8]. The extended superconformal algebras have important applications to integrable systems [9] and to topological field theories in two dimensions as well [10]. The conformal and superconformal algebras have been studied in detail via the coset space method of Goddard, Kent and Olive (GKO) [11] which is a generalization of Sugawara-Sommerfield construction of the Virasoro algebra in terms of bilinears of the generators of a current algebra [12]. In the first part of my talk I will review a novel ternary algebraic approach to the construction and study of extended superconformal algebras. The construction of N=2 SCA's over Jordan Triple Systems (JTS) was given in [13]. This construction was generalized to the realization of N=2 SCA's over more general Kantor triple systems (KTS) in [14, 15]. The coset spaces associated with the KTS's are, in general, not symmetric spaces. For a particular subclass of Kantor triple systems, namely the Freudenthal triple systems (FTS) the N=2 SCA's admit extensions to "maximal" $N = 4$ SCA's with the gauge group $SU(2) \times SU(2) \times U(1)$ [16]. The construction of $N = 2$ and $N = 4$ superconformal algebras over FTS's and their classification will be discussed in detail following [14, 15, 17, 18]. The last part of my talk is devoted to the study of gauged WZW models that are invariant under $N = 4$ superconformal symmetry [19].

2 Construction of Extended Superconformal Algebras over Triple Systems

The realization of $N = 2$ superconformal algebras over hermitian Jordan triple systems given in [13] is equivalent to their realization over hermitian symmetric spaces of Lie groups a la Kazama and Suzuki [5] which can be compact or non-compact . The connection between hermitian symmetric spaces and hermitian Jordan triple systems arises as follows. If G/H is a hermitian symmetric space then the Lie algebra g of G can be given a 3-graded decomposition with respect to the Lie algebra g^0 of H:

$$g = g^{-1} \oplus g^0 \oplus g^{+1} \tag{2.1}$$

where $\oplus$ denotes vector space direct sum and g^0 is a subalgebra of maximal rank. We have the formal commutation relations of the elements of various grade subspaces

$$[g^m, g^n] \subseteq g^{m+n} \; ; m, n = -1, 0, 1 \tag{2.2}$$

where $g^{m+n} = 0$ if $|m + n| > 1$. The Tits-Kantor-Koecher (TKK) construction [20] of the Lie algebra g establishes a mapping between the grade +1 subspace of g and the underlying Hermitian JTS V.

The exceptional Lie algebras G_2, F_4 and E_8 do not admit a TKK type construction. A generalization of the TKK construction to more general triple systems was given by Kantor [21]. All finite dimensional simple Lie algebras admit a construction over these generalized triple systems. The Kantor's construction of Lie algebras was generalized to a unified construction of Lie algebras and Lie superalgebras in [22] which we shall review briefly restricting our discussion to Lie algebras only.

This more general construction starts from the fact that every simple Lie algebra g with the exception of $SU(2)$ admits a 5-grading (Kantor structure) with respect to some subalgebra g^0 of maximal rank [21, 22]:

$$g = g^{-2} \oplus g^{-1} \oplus g^0 \oplus g^{+1} \oplus g^{+2} \tag{2.3}$$

One labels the elements of g^{+1} subspace with the elements of a vector space V [21, 22]:

$$U_a \in g^{+1} \Longleftrightarrow a \in V \tag{2.4}$$

g admits a conjugation under which the grade $+m$ subspace gets mapped into the grade $-m$ subspace, which allows one to label the elements of the grade -1 subspace by the elements of V as well :

$$U^a \equiv U_a^\dagger \in g^{-1} \Longleftrightarrow U_a \in g^{+1} \tag{2.5}$$

One defines the commutators of U_a and U^b as

$$\begin{aligned} &[U_a, U^b] = S_a^b \in g^0 \\ &[U_a, U_b] = K_{ab} \in g^{+2} \\ &[U^a, U^b] = K^{ab} \in g^{-2} \\ &[S_a^b, U_c] = U_{(abc)} \in g^{+1} \end{aligned} \tag{2.6}$$

where (abc) is the triple product under which the elements of V close. The remaining non-vanishing commutators of g can all be expressed in terms of the triple product (abc):

$$\begin{aligned} &[S_a^b, U^c] = -U^{(bac)} \\ &[K_{ab}, U^c] = U_{(acb)} - U_{(bca)} \\ &[K^{ab}, U_c] = -U^{(bca)} + U^{(acb)} \\ &[S_a^b, S_c^d] = S_{(abc)}^d - S_c^{(bad)} \\ &[S_a^b, K_{cd}] = K_{(abc)d} + K_{c(abd)} \\ &[S_a^b, K^{cd}] = -K^{(bac)d} - K^{c(bad)} \\ &[K_{ab}, K^{cd}] = S_{(acb)}^d - S_{(bca)}^d - S_{(adb)}^c + S_{(bda)}^c \end{aligned} \tag{2.7}$$

The Jacobi identities of g follow from the defining identities of a Kantor triple system (KTS) [21, 22]

$$(ab(cdx)) - (cd(abx)) - (a(dcb)x) + ((cda)bx) = 0 \tag{2.8}$$

$$\{(ax(cbd)) - ((cbd)xa) + (ab(cxd)) + (c(bax)d)\} - \{c \leftrightarrow d\} = 0 \tag{2.9}$$

In general a given simple Lie algebra can be constructed in several different ways by the above method corresponding to different choices of the subalgebra g^0 and different KTS's. Note that the second defining identity of a KTS is trivially satisfied by a JTS for which the grade ± 2 subspaces vanish.

Consider now the affine Lie algebra $\hat{g}$ defined by the Lie algebra g constructed over a KTS V. The commutation relations $\hat{g}$ can be written in the form of operator products as follows:

$$\begin{aligned}
U_a(z)U^b(w) &= \frac{k\delta_a^b}{(z-w)^2} + \frac{S_a^b(w)}{(z-w)} + \cdots \\
U^a(z)U_b(w) &= \frac{k\delta_a^b}{(z-w)^2} - \frac{S_a^b(w)}{(z-w)} + \cdots \\
U_a(z)U_b(w) &= \frac{K_{ab}(w)}{(z-w)} + \cdots \\
S_a^b(z)U_c(w) &= \frac{U_{(abc)}(w)}{(z-w)} + \cdots \\
S_a^b(z)U^c(w) &= \frac{-U^{(bac)}(w)}{(z-w)} + \cdots \\
S_a^b(z)S_c^d(w) &= \frac{k\Sigma_{ac}^{bd}}{(z-w)^2} - \frac{1}{(z-w)}(S_{(abc)}^d - S_c^{(bad)})(w) + \cdots \\
S_a^b(z)K_{cd}(w) &= \frac{1}{(z-w)}(K_{(abc)d} + K_{c(abd)})(w) + \cdots \\
S_a^b(z)K^{cd}(w) &= \frac{-1}{(z-w)}(K^{(bac)d} + K^{c(bad)})(w) + \cdots \\
K_{ab}(z)K^{cd}(w) &= \frac{k\Omega_{ab}^{cd}}{(z-w)^2} + \frac{1}{(z-w)}(S_{(acb)}^d - S_{(bca)}^d - S_{(adb)}^c + S_{(bda)}^c)(w) + \cdots \\
K_{ab}(z)U^c(w) &= \frac{1}{(z-w)}(U_{(acb)} - U_{(bca)})(w) + \cdots \\
K^{ab}(z)U_c(w) &= \frac{1}{(z-w)}(U^{(acb)} - U^{(bca)})(w) + \cdots
\end{aligned} \tag{2.10}$$

where Σ_{ab}^{cd} are the structure constants of the underlying KTS V:

$$U_{(abc)} = \Sigma_{ac}^{bd} U_d \tag{2.11}$$

and the tensor Ω is defined as

$$\Omega_{ab}^{cd} = \Sigma_{ab}^{cd} - \Sigma_{ab}^{dc} \tag{2.12}$$

We choose a basis V such that

$$\Omega_{ab}^{ca} = \frac{2d}{D}(\check{g} - \check{s})\delta_b^c \tag{2.13}$$

where $\check{g}$ and $\check{s}$ denote the dual coxeter numbers of the Lie algebra g and its subalgebra s generated by the elements of the grade ± 2 subspaces and their commutant $[g^{-2}, g^{+2}]$, respectively. D is the dimension of the triple system V and d is the dimension of the grade $+2$ subspace.

Furthermore one introduces fermion fields $\psi_a(\psi^a)$ and $\psi_{ab}(\psi^{ab})$ corresponding to the grade $+1(-1)$ and $+2(-2)$ subspaces of g, respectively. They are normalized such that

$$\begin{aligned}
\psi_a(z)\psi^b(w) &= \frac{\delta_a^b}{(z-w)} + \cdots \\
\psi_{ab}(z)\psi^{cd}(w) &= \frac{\Omega_{ab}^{dc}}{(z-w)} + \cdots
\end{aligned} \tag{2.14}$$

The operators of conformal dimension 3/2 defined by the following expressions:

$$\begin{aligned}
G(z) &= \sqrt{\frac{2}{k+\check{g}}}\{U_a\psi^a + \frac{1}{2(\check{g}-\check{s})}K_{ab}\psi^{ab} - \tfrac{1}{2}\psi^a\psi^b\psi_{ab}\}(z) \\
\bar{G}(z) &= \sqrt{\frac{2}{k+\check{g}}}\{U^a\psi_a - \frac{1}{2(\check{g}-\check{s})}K^{ab}\psi_{ab} + \tfrac{1}{2}\psi^{ab}\psi_a\psi_b\}(z)
\end{aligned} \tag{2.15}$$

generate an N = 2 superconformal algebra [14, 15] with the Virasoro generator

$$\begin{aligned} T(z) = \tfrac{1}{k+\check{g}}\{ & \tfrac{1}{2}(U_a U^a + U^a U_a) + \tfrac{1}{4(\check{g}-\check{s})}(K_{ab}K^{ba} + K^{ab}K_{ba}) \\ & + \tfrac{k}{2}(\partial\psi^a \psi_a + \partial\psi_a \psi^a) + \tfrac{1}{2}\Omega^{cb}_{ac}(\partial\psi^a\psi_b + \partial\psi_b\psi^a) \\ & + \tfrac{(k+\check{g}-\check{s})}{4(\check{g}-\check{s})}(\partial\psi^{ab}\psi_{ab} + \partial\psi_{ab}\psi^{ab}) + S^b_a\psi^a\psi_b \\ & + \tfrac{1}{(\check{g}-\check{s})}S^b_a\psi^{ac}\psi_{bc} + \psi_{ab}\psi^{ac}\psi^b\psi_c + \tfrac{1}{4}\Omega^{cd}_{ab}\psi^a\psi^b\psi_c\psi_d\}(z) \end{aligned} \tag{2.16}$$

and the $U(1)$ current

$$\begin{aligned} J(z) = \tfrac{1}{k+\check{g}}\{ & S^a_a - \tfrac{1}{(\check{g}-\check{s})}\Omega^{cb}_{ca}S^a_b + k\psi^a\psi_a \\ & + \Omega^{bc}_{ab}\psi^a\psi_c + \tfrac{(k-\check{g}+\check{s})}{2(\check{g}-\check{s})}\psi^{ab}\psi_{ab}\}(z) \end{aligned} \tag{2.17}$$

Its central charge turns out to be :

$$c = \frac{3}{(k+\check{g})}\{kD + \frac{k}{4(\check{g}-\check{s})^2}\Omega^{cd}_{ab}\Omega^{ab}_{cd} + \frac{1}{2}\Omega^{ba}_{ab}\} \tag{2.18}$$

This realization of N=2 SCA's is equivalent to their realization over the coset spaces G/H where G and H are the groups generated by g and g^0, respectively. These spaces are, in general, not symmetric spaces. For further details and a complete classification of the $N = 2$ SCA's constructed over KTS's we refer to [14, 15].

3 Construction of $N = 2$ Subalgebras of Maximal $N = 4$ Superconformal Algebras over Freudenthal Triple Systems

For a very special subclass of Kantor triple systems, namely the Freudenthal triple systems (FTS), the $N = 2$ superconformal algebras as constructed in the previous section can be extended to the maximal $N = 4$ superconformal algebras [14, 15]. Freudenthal introduced these triple systems in his study of the geometries associated with exceptional groups [23]. Kantor and Skopets classified FTS's and showed that there is a one-to-one correspondence between simple Lie algebras and simple FTS's with a non-degenerate bilinear form [24]. The Freudenthal triple product (abc) can be written in the form:

$$(abc) = \{abc\} - (c,b)a - (c,a)b - (a,b)c \tag{3.1}$$

where $\{abc\}$ is completely symmetric in its arguments and $(,)$ is a skew-symmetric bilinear form defined over the FTS. For a simple Lie algebra g constructed over a FTS the grade ± 2 subspaces become one dimensional as a consequence of the identity

$$(abc) - (cba) = 2(a,c)b \tag{3.2}$$

The realization of $N = 2$ superconformal algebra over it corresponds to the coset space $G/H_0 \times U(1)$ where H_0 is such that $G/H_0 \times SU(2)$ is the unique quaternionic symmetric space of G .

Let g, h_0 be the Lie algebras of G and H_0 listed above, respectively. The 5-graded structure of g

$$g = g^{-2} \oplus g^{-1} \oplus g^{0} \oplus g^{+1} \oplus g^{+2} \tag{3.3}$$

is such that $g^0 = h_0 \oplus K_3$ where K_3 is the generator of the $U(1)$ factor and the elements K_{ab} and K^{ab} of grade ± 2 subspaces can be written in the form

$$\begin{aligned} K_{ab} &= \Omega_{ab} K_{+} \\ K^{ab} &= \Omega^{ab} K^{+} \end{aligned}$$

where Ω_{ab} is a symplectic form defined over the FTS. [1] The tensor Ω^{ab} is the inverse of Ω_{ab} and

$$\begin{aligned} \Omega_{ab}\Omega^{bc} &= \delta_a^c \\ \Omega_{ab}^{\dagger} &= \Omega^{ba} = -\Omega^{ab} \end{aligned} \tag{3.4}$$

The elements K_{+} and K^{+} are hermitian conjugates of each other and can be written as

$$\begin{aligned} K_{+} &= K_1 + iK_2 \in g^{+2} \\ K^{+} \equiv K_{-} &= K_1 - iK_2 \in g^{-2} \end{aligned} \tag{3.5}$$

There is a universal relation between the dual Coxeter number $\check{g}$ of G and the dimension D of the underlying FTS :

$$D = 2(\check{g} - 2) \tag{3.6}$$

The generators K_{-}, K_{+} and K_3 form an $SU(2)$ subalgebra of g.

$$\begin{aligned} [K_{+}, K_{-}] &= 2K_3 \\ [K_3, K_{\pm}] &= \pm K_{\pm} \end{aligned} \tag{3.7}$$

The commutation relations of the U's can now be written in the form

$$\begin{aligned} [U_a, U_b] &= \Omega_{ab} K_{+} \\ [U^a, U^b] &= \Omega^{ab} K_{-} \\ [U_a, U^b] &= S_a^b \end{aligned} \tag{3.8}$$

Ω_{ab} is an invariant tensor of H_0 and S_a^b are the generators of the subgroup $H_0 \times U(1)$. The trace component of S_a^b gives the $U(1)$ generator

$$K_3 = \frac{1}{2(\check{g} - 2)} S_a^a \tag{3.9}$$

Hence we have the decomposition

$$S_a^b = H_a^b + \delta_a^b K_3 \tag{3.10}$$

[1] In a given basis e_a of the FTS one can set $\Omega_{ab} = (e_a, e_b)$.

where $H_a^b = S_a^b - \frac{1}{D}\delta_a^b S_c^c$ are the generators of the subgroup H_0. Note that H_a^b commutes with K_3, K_+ and K_-. The other non-vanishing commutators of g are

$$
\begin{aligned}
[K_+, U^a] &= \Omega^{ab} U_b \\
[K_-, U_a] &= \Omega_{ab} U^b \\
[K_3, U^a] &= -\tfrac{1}{2} U^a \\
[K_3, U_a] &= \tfrac{1}{2} U_a \\
[S_a^b, U_c] &= \Sigma_{ac}^{bd} U_d \\
[S_a^b, U^c] &= -\Sigma_{ad}^{bc} U^d \\
[S_a^b, S_c^d] &= \Sigma_{ac}^{be} S_e^d - \Sigma_{ae}^{bd} S_c^e
\end{aligned}
\tag{3.11}
$$

where Σ_{ab}^{cd} are the structure constants of the FTS which in our normalization satisfy

$$
\begin{aligned}
\Sigma_{ab}^{ac} &= (\breve{g} - 2)\delta_b^c \\
\Sigma_{ab}^{bc} &= (\breve{g} - 1)\delta_a^c \\
\Sigma_{ab}^{cd} - \Sigma_{ab}^{dc} &= \Omega_{ab}\Omega^{cd}
\end{aligned}
\tag{3.12}
$$

The complex Fermi fields associated with the grade ± 2 subspaces can be represented as

$$
\begin{aligned}
\psi_{ab}(z) &= \Omega_{ab}\psi_+(z) \\
\psi^{ab}(z) &= \Omega^{ab}\psi^+(z)
\end{aligned}
\tag{3.13}
$$

where $\psi_+(z)$ and $\psi^+(z)$ satisfy [14, 15]:

$$
\psi_+(z)\psi^+(w) = \frac{1}{(z-w)} + \cdots \tag{3.14}
$$

The supersymmetry generators of the $N = 2$ superconformal algebra simplify

$$
G(z) = \sqrt{\frac{2}{k+\breve{g}}}\{U_a\psi^a + K_+\psi^+ - \tfrac{1}{2}\Omega_{ab}\psi^a\psi^b\psi_+\}(z) \tag{3.15}
$$

$$
\bar{G}(z) = \sqrt{\frac{2}{k+\breve{g}}}\{U^a\psi_a + K^+\psi_+ - \tfrac{1}{2}\Omega^{ab}\psi_a\psi_b\psi^+\}(z) \tag{3.16}
$$

The Virasoro generator takes the form

$$
\begin{aligned}
T(z) = \frac{1}{k+\breve{g}}\{&\tfrac{1}{2}(U_aU^a + U^aU_a) + \tfrac{1}{2}(K_+K^+ + K^+K_+) \\
&-\tfrac{k+1}{2}(\psi_a\partial\psi^a + \psi^a\partial\psi_a) - \tfrac{1}{2}(k + \breve{g} - 2)(\psi_+\partial\psi^+ + \psi^+\partial\psi_+) \\
&+H_a^b\psi^a\psi_b + K_3(\psi^a\psi_a + 2\psi^+\psi_+) + \psi_+\psi^+\psi^a\psi_a + \tfrac{1}{4}\Omega_{ab}\psi^a\psi^b\Omega^{cd}\psi_c\psi_d\}(z)
\end{aligned}
\tag{3.17}
$$

and the $U(1)$ current is given by

$$
J(z) = \frac{1}{k+\breve{g}}\{2(\breve{g} - 1)K_3 + (k+1)\psi^a\psi_a + (k - \breve{g} + 2)\psi^+\psi_+\}(z) \tag{3.18}
$$

The central charge of the N = 2 SCA defined by a FTS is

$$c = \frac{6(k+1)(\breve{g}-1)}{(k+\breve{g})} - 3 \tag{3.19}$$

As stated earlier the above realization of $N = 2$ SCA's corresponds to the coset $G/H_0 \times U(1)$ where the $U(1)$ generator K_3 determines the 5-graded structure of g

$$[2K_3, g^m] = mg^m \tag{3.20}$$

where g^m denotes the subspace of grade m with $m = 0, \pm 1, \pm 2$. The Lie algebra g can also be given a 5-graded structure with respect to K_1 as well as K_2. Therefore, one can realize the $N = 2$ SCA equivalently over the coset $G/H_0 \times U(1)'$ or the coset $G/H_0 \times U(1)''$, where the generators of $U(1)'$ and $U(1)''$ are K_1 and K_2, respectively. The generators belonging to grade ± 1 and ± 2 subspaces with respect to K_1 are

$$\begin{aligned} U_a' &= \tfrac{1}{\sqrt{2}}(U_a + \Omega_{ab}U^b) \\ U^{a'} &= \tfrac{1}{\sqrt{2}}(U^a - \Omega^{ab}U_b) \\ K_+' &= i(K_2 + iK_3) \\ K_-' &= -i(K_2 - iK_3) \end{aligned} \tag{3.21}$$

They satisfy

$$\begin{aligned} [U_a', U_b'] &= \Omega_{ab}K_+' \\ [U^{a'}, U^{b'}] &= \Omega^{ab}K_-' \\ [K_+', U^{a'}] &= \Omega^{ab}U_b' \\ [K_-', U_a'] &= \Omega_{ab}U^{b'} \end{aligned} \tag{3.22}$$

Whereas the grade ± 1 and ± 2 subspaces with respect to K_2 are

$$\begin{aligned} U_a'' &= \tfrac{1}{\sqrt{2}}(U_a + i\Omega_{ab}U^b) \\ U^{a''} &= \tfrac{1}{\sqrt{2}}(U^a + i\Omega^{ab}U_b) \\ K_+'' &= -i(K_3 + iK_1) \\ K_-'' &= i(K_3 - iK_1) \end{aligned} \tag{3.23}$$

with analogous commutation relations to (3.22).

For every simple Lie group G (except for $SU(2)$) there exists a subgroup $H_0 \times U(1)$, unique up to automorphisms corresponding to $SU(2)$ rotations, such that its Lie algebra g has a 5-graded structure with respect to the subalgebra $h_0 \oplus K$, where K is the generator of the $U(1)$ subgroup that determines the 5-grading. This follows from the fact that there is a one-to-one correspondence between simple Lie algebras and simple FTS's with a non-degenerate bilinear form [24].

4 The Construction of Maximal $N = 4$ Superconformal Algebras

The $N = 2$ superconformal algebras constructed over FTS's admit extensions to maximal $N = 4$ superconformal algebras of references [16]. To achieve this one needs to introduce a $N = 2$ "matter multiplet" and define two additional supersymmetry generators as well as adding the matter contributions to the first two supersymmetry generators. The required currents of the matter multiplet turn out to be the $U(1)$ current generated by K_3 that gives the 5-graded structure of the Lie algebra g, and an additional $U(1)$ current whose generator K_0 commutes with g together with the associated fermions which we denote as a complex fermion χ_+ and its conjugate χ^+. Then the four supersymmetry generators of the $N = 4$ superconformal algebra can be written as

$$
\begin{aligned}
G^+ \equiv \tfrac{1}{\sqrt{2}}(G_1 + iG_2) &= \sqrt{\tfrac{2}{k+\breve{g}}}\{U_a\psi^a + K_+\psi^+ + K_3\chi_+ \\
&\quad -\tfrac{1}{2}\Omega_{ab}\psi^a\psi^b\psi_+ - \tfrac{1}{2}\psi^a\psi_a\chi_+ - \psi^+\psi_+\chi_+\} + iZ\chi_+ \\
G^- \equiv \tfrac{1}{\sqrt{2}}(G_1 - iG_2) &= \sqrt{\tfrac{2}{k+\breve{g}}}\{U^a\psi_a + K^+\psi_+ + K_3\chi_+ \\
&\quad -\tfrac{1}{2}\Omega^{ab}\psi_a\psi_b\psi^+ - \tfrac{1}{2}\psi^a\psi_a\chi^+ - \psi^+\psi_+\chi^+\} - iZ\chi^+ \\
G^{+K} \equiv \tfrac{1}{\sqrt{2}}(G_3 + iG_4) &= \sqrt{\tfrac{2}{k+\breve{g}}}\{\Omega^{ab}U_a\psi_b + K_+\chi^+ + K_3\psi_+ \\
&\quad +\tfrac{1}{2}\Omega^{ab}\psi_a\psi_b\chi_+ + \tfrac{1}{2}\psi^a\psi_a\psi_+ - \chi^+\chi_+\psi_+\} + iZ\psi_+ \\
G^{-K} \equiv \tfrac{1}{\sqrt{2}}(G_3 - iG_4) &= \sqrt{\tfrac{2}{k+\breve{g}}}\{\Omega_{ab}U^b\psi^a + K^+\chi_+ + K_3\psi^+ \\
&\quad +\tfrac{1}{2}\Omega_{ab}\psi^a\psi^b\chi^+ + \tfrac{1}{2}\psi^a\psi_a\psi^+ - \chi^+\chi_+\psi^+\} - iZ\psi^+
\end{aligned}
\tag{4.24}
$$

The Virasoro generator of the maximal $N = 4$ superconformal algebra $\mathcal{A}_\gamma$ is given by

$$
\begin{aligned}
T(z) = \tfrac{1}{2}\left[Z^2 - (\chi_+\partial\chi^+ + \chi^+\partial\chi_+) - (\psi_+\partial\psi^+ + \psi^+\partial\psi_+)\right](z) \\
+\tfrac{1}{k+\breve{g}}\{\tfrac{1}{2}(U_aU^a + U^aU_a) + \tfrac{1}{2}(K_+K^+ + K^+K_+) + K_3^2 \\
-\tfrac{k+1}{2}(\psi_a\partial\psi^a + \psi^a\partial\psi_a) + H_a^b\psi^a\psi_b + \tfrac{1}{4}\Omega_{ab}\psi^a\psi^b\Omega^{cd}\psi_c\psi_d\}(z)
\end{aligned}
\tag{4.25}
$$

The generators of the two $SU(2)$ currents take the form

$$
\begin{aligned}
V_3^+(z) &= K_3(z) + \tfrac{1}{2}(\psi_+\psi^+ + \chi_+\chi^+)(z) \\
V_+^+(z) &= (V_1^+ + iV_2^+)(z) = (K_+ - \psi_+\chi_+)(z) \\
V_-^+(z) &= (V_1^+ - iV_2^+)(z) = (K^+ + \psi^+\chi^+)(z) \\
V_3^-(z) &= \tfrac{1}{2}(\psi^a\psi_a + \psi^+\psi_+ + \chi_+\chi^+)(z) \\
V_+^-(z) &= (V_1^- + iV_2^-)(z) = (\psi^+\chi_+ - \tfrac{1}{2}\Omega_{ab}\psi^a\psi^b)(z) \\
V_-^-(z) &= (V_1^- - iV_2^-)(z) = (\chi^+\psi_+ - \tfrac{1}{2}\Omega^{ab}\psi_a\psi_b)(z)
\end{aligned}
\tag{4.26}
$$

The U(1) current of the $N = 4$ SCA is $Z(z)$ and the four dimension $\frac{1}{2}$ generators are simply the fermion fields $\psi_+(z), \psi^+(z), \chi_+(z)$ and $\chi^+(z)$. One finds that the levels of the two $SU(2)$ currents are $k^+ = k + 1$ and $k^- = \breve{g} - 1$. The central charge of the $N=4$ SCA turns out to be $c=\frac{6k^+k^-}{k^++k^-}$. The above realization of the $N=4$ SCA corresponds to the coset space $G\times U(1)/H$.

By decoupling the four dimension $\frac{1}{2}$ generators and the $U(1)$ current $Z(z)$ one obtains a non-linear $N=4$ SCA [25, 26] a la Bershadsky and Knizhnik [27, 28]. The realization given above for the $N=4$ SCA leads to very simple expressions for the supersymmetry generators of the non-linear $N=4$ algebra [19]

$$\begin{aligned}
\tilde{G}^{+} &\equiv \frac{1}{\sqrt{2}}(\tilde{G}_1+i\tilde{G}_2)=\sqrt{\frac{2}{k+\breve{g}}}U_a\psi^a \\
\tilde{G}^{-} &\equiv \frac{1}{\sqrt{2}}(\tilde{G}_1-i\tilde{G}_2)=\sqrt{\frac{2}{k+\breve{g}}}U^a\psi_a \\
\tilde{G}^{+K} &\equiv \frac{1}{\sqrt{2}}(\tilde{G}_3+i\tilde{G}_4)=\sqrt{\frac{2}{k+\breve{g}}}\Omega^{ab}U_a\psi_b \\
\tilde{G}^{-K} &\equiv \frac{1}{\sqrt{2}}(\tilde{G}_3-i\tilde{G}_4)=\sqrt{\frac{2}{k+\breve{g}}}\Omega_{ba}U^a\psi^b
\end{aligned} \tag{4.27}$$

with the central charge $\tilde{c}=c-3$. It is clear from the above expressions for the generators that the non-linear $N=4$ SCA is realized over the symmetric space

$$G/H\times SU(2) \tag{4.28}$$

which is the unique quaternionic symmetric space associated with G [29] .

5 $N=4$ Supersymmetric Gauged WZW Models

So far we have been discussing the construction of extended superconformal algebras and study of their chiral rings using algebraic methods. In this section we shall study a certain class of Lagrangian field theories , namely supersymmetric gauged WZW models, that are invariant under extended superconformal groups. Gauged WZW models were studied in [30, 31]. The $N=1$ supersymmetric ordinary WZW models were studied in [32, 33, 34] and their gauged versions in [35]. $N=2$ supersymmetric gauged WZW models were studied by Witten [36]. Witten's results were extended to $N=4$ gauged WZW models in [19].

The WZW action at level k is given by $kI(g)$ where

$$I(g)=-\frac{1}{8\pi}\int_{\Sigma}d^2\sigma\sqrt{h}h^{ij}Tr(g^{-1}\partial_i g\cdot g^{-1}\partial_j g)-i\Gamma \tag{5.1}$$

with the WZ functional [37] given by [38]

$$\Gamma=\frac{1}{12\pi}\int_M d^3\sigma\epsilon^{ijk}Tr(g^{-1}\partial_i g\cdot g^{-1}\partial_j g\cdot g^{-1}\partial_k g) \tag{5.2}$$

M is a three manifold whose boundary is the Riemann surface Σ with metric h. We choose the metric $h_{z\bar{z}}=h^{z\bar{z}}=1$ and work with complex coordinates z and $\bar{z}$. g represents the group element that maps Σ into the group G. The supersymmetric WZW action $I(g,\Psi)$ is obtained by adding to $I(g)$ the free action of Weyl fermions Ψ_L and Ψ_R in the complexification of the adjoint representation of G [36]:[2]

$$I(g,\Psi) = I(g) + \frac{i}{4\pi}\int d^2zTr(\Psi_L\partial_{\bar{z}}\Psi_L + \Psi_R\partial_z\Psi_R) \tag{5.3}$$

It is invariant under the supersymmetry transformations

$$\begin{aligned} \delta g &= i\epsilon_- g\Psi_L + i\epsilon_+\Psi_R g \\ \delta\Psi_L &= \epsilon_-(g^{-1}\partial_z g - i\Psi_L^2) \\ \delta\Psi_R &= \epsilon_+(\partial_{\bar{z}}gg^{-1} + i\Psi_R^2) \end{aligned} \tag{5.4}$$

Gauging any diagonal subgroup H of the $G_L \times G_R$ symmetry of the WZW model leads to an anomaly free theory. The gauge invariant action, which does not involve any kinetic energy term for the gauge fields, can be written as:

$$I(g,A) = I(g) + \frac{1}{2\pi}\int_\Sigma d^2zTr(A_{\bar{z}}g^{-1}\partial_z g - A_z\partial_{\bar{z}}gg^{-1} + A_{\bar{z}}g^{-1}A_z g - A_{\bar{z}}A_z) \tag{5.5}$$

where $A_z, A_{\bar{z}}$ are the matrix valued gauge fields belonging to the subgroup H. It is invariant under the gauge transformations [36]:

$$\begin{aligned} \delta g &= [u,g] \\ \delta A_i &= -D_i u = -\partial u - [A_i,u] \end{aligned} \tag{5.6}$$

Denoting as $\mathcal{G}$ and $\mathcal{H}$ the complexifications the Lie algebras of G and H one has the orthogonal decomposition

$$\mathcal{G} = \mathcal{H} \oplus \mathcal{T} \tag{5.7}$$

where $\mathcal{T}$ is the orthocomplement of $\mathcal{H}$. To supersymmetrize the WZW model with the gauged subgroup H one introduces Weyl fermions with values in $\mathcal{T}$ minimally coupled to the gauge fields and otherwise free

$$I(g,A,\Psi) = I(g,A) + \frac{i}{4\pi}\int d^2zTr(\Psi_L D_{\bar{z}}\Psi_L + \Psi_R D_z\Psi_R) \tag{5.8}$$

This action is invariant under the supersymmetry transformation laws:

$$\begin{aligned} \delta g &= i\epsilon_- g\Psi_L + i\epsilon_+\Psi_R g \\ \delta\Psi_L &= \epsilon_-(1-\Pi)(g^{-1}D_z g - i\Psi_L^2) \\ \delta\Psi_R &= \epsilon_+(1-\Pi)(D_{\bar{z}}gg^{-1} + i\Psi_R^2) \\ \delta A &= 0 \end{aligned} \tag{5.9}$$

where Π is the orthogonal projection of $\mathcal{G}$ onto $\mathcal{H}$. As shown by Witten if the gauge group H is such that the coset space G/H is Kahler the above action has

[2] We shall consider only models that have equal number of supersymmetries in both the left and the right moving sectors.

$N=2$ supersymmetry[36]. For G/H Kahler the subspace $\mathcal{T}$ can be decomposed as

$$\mathcal{T}=\mathcal{T}_+\oplus\mathcal{T}_- \tag{5.10}$$

where $\mathcal{T}_+$ and $\mathcal{T}_-$ are in complex conjugate representations of H. The action can then be written in the form [36]

$$I(g,\Psi,A)=I(g,A)+\frac{i}{2\pi}\int d^2zTr(\beta_L D_{\bar z}\alpha_L+\beta_R D_z\alpha_R) \tag{5.11}$$

where

$$\begin{aligned}\alpha_L&=\Pi_+\Psi_L\\ \beta_L&=\Pi_-\Psi_L\\ \alpha_R&=\Pi_+\Psi_R\\ \beta_R&=\Pi_-\Psi_R\end{aligned} \tag{5.12}$$

with Π_+ and Π_- representing the projectors onto the subspaces $\mathcal{T}_+$ and $\mathcal{T}_-$. Denoting the chiral and anti-chiral supersymmetry generators in left and right moving sectors as $G_L,\bar G_L$ and $G_R,\bar G_R$, respectively, one finds [36] that they satisfy the $N=2$ supersymmetry algebra

$$\begin{aligned}\{G_L,\bar G_L\}&=-iD_z\\ \{G_R,\bar G_R\}&=-iD_{\bar z}\end{aligned} \tag{5.13}$$

The gauged WZW models are known to be conformally invariant. Hence the $N=2$ global supersymmetry of these theories implies $N=2$ superconformal invariance.

Since the existence of a second supersymmetry in a supersymmetric gauged WZW model is guaranteed by the Kählerian property of the coset space G/H , i.e when it admits a complex structure, to have $N=4$ supersymmetry one needs coset spaces with three complex structures which anti-commute with each other and form a closed algebra. Therefore one expects supersymmetric WZW models based on the groups $G\times U(1)$ of the previous section with a gauged subgroup H such that $G/H\times SU(2)$ is a quaternionic symmetric space to actually have $N=4$ supersymmetry. Let us show that this is indeed the case [19].

We shall designate the generators of $G\times U(1)$ as we did in previous sections ,i.e K_0,K_1,K_2,K_3,U_a,U^a and H_a^b , where K_0 is the generator of the additional $U(1)$ factor normalized such that

$$TrK_0^2=TrK_1^2=TrK_2^2=TrK_3^2 \tag{5.14}$$

The fermions associated with the grade ±1 subspaces of the Lie algebra of G will be denoted as ψ_a,ψ^a as before and the fermions associated with K_0 and K_i will be denoted as $\xi^0,\xi^i(i=1,2,3)$.Thus the fermions in the coset $G\times U(1)/H$ can be represented as

$$\Psi=2K_0\xi^0+2K_i\xi^i+U_a\psi^a+U^a\psi_a \tag{5.15}$$

for both the left and the right moving sectors. The coset $G \times U(1)/H$ can be given a Kahler decomposition such that

$$\Psi = \alpha + \beta \tag{5.16}$$

where

$$\begin{aligned}\alpha &= U_a\psi^a + K_+(\xi^1 - i\xi^2) + (K_3 + iK_0)(\xi^3 - i\xi^0)\\ \beta &= U^a\psi_a + K_-(\xi^1 + i\xi^2) + (K_3 - iK_0)(\xi^3 + i\xi^0)\end{aligned} \tag{5.17}$$

The complex structure C_3 corresponding to this Kahler decomposition acts on Ψ as

$$C_3\Psi = -i\alpha + i\beta \tag{5.18}$$

where the index 3 in C_3 signifies the fact in the subspace $G/H \times SU(2)$ its action corresponds to commutation with the generator $-iK_3$. One can similarly give a Kahler decomposition of the coset space $G \times U(1)/H$ which selects out K_1 or K_2. For the decomposition with respect to K_1 we have:

$$\begin{aligned}\Psi &= \alpha' + \beta'\\ \alpha' &= U'_a\psi^{a'} + (K_2 + iK_3)(\xi^2 - i\xi^3) + (K_1 + iK_0)(\xi^1 - i\xi^0)\\ \beta' &= U^{a'}\psi'_a + (K_2 - iK_3)(\xi^2 + i\xi^3) + (K_1 - iK_0)(\xi^1 + i\xi^0)\end{aligned} \tag{5.19}$$

Under the action of the corresponding complex structure C_1 we have

$$C_1\Psi = -i\alpha' + i\beta' \tag{5.20}$$

In the case of K_2 one finds

$$\begin{aligned}\Psi &= \alpha'' + \beta''\\ \alpha'' &= U''_a\psi^{a''} + (K_3 + iK_1)(\xi^3 - i\xi^1) + (K_2 + iK_0)(\xi^2 - i\xi^0)\\ \beta'' &= U^{a''}\psi''_a + (K_3 - iK_1)(\xi^3 + i\xi^1) + (K_2 - iK_0)(\xi^2 + i\xi^0)\end{aligned} \tag{5.21}$$

with the complex structure action

$$C_2\Psi = -i\alpha'' + i\beta'' \tag{5.22}$$

The fermionic part of the action $I(g, \Psi, A)$ can then be written in three different ways involving the pairs $(\alpha, \beta), (\alpha', \beta')$ and (α'', β'').

$$\begin{aligned}I(\Psi, A) &= \frac{i}{4\pi}\int d^2z Tr(\Psi_L D_{\bar z}\Psi_L + \Psi_R D_z\Psi_R)\\ &= \frac{i}{2\pi}\int d^2z Tr(\beta_L D_{\bar z}\alpha_L + \beta_R D_z\alpha_R)\\ &= \frac{i}{2\pi}\int d^2z Tr(\beta'_L D_{\bar z}\alpha'_L + \beta'_R D_z\alpha'_R)\\ &= \frac{i}{2\pi}\int d^2z Tr(\beta''_L D_{\bar z}\alpha''_L + \beta''_R D_z\alpha''_R)\end{aligned} \tag{5.23}$$

For each form of the action in terms of an (α, β) pair one can define a pair of supersymmetry transformations in each sector. Let us denote them as $(G, \bar{G})$, $(G'.\bar{G}')$ and $(G'', \bar{G}'')$:

$$
\begin{aligned}
(G, \bar{G}) &\leftrightarrow (\alpha, \beta) \\
(G', \bar{G}') &\leftrightarrow (\alpha', \beta') \\
(G'', \bar{G}'') &\leftrightarrow (\alpha'', \beta'')
\end{aligned}
\tag{5.24}
$$

Each pair of these operators in both sectors satisfy the $N = 2$ supersymmetry algebra given by the equations 5.13. However they are not all independent. The sum of each pair gives the manifest $N = 1$ supersymmetry generator of the model in both sectors, which we shall denote as G^0:

$$
G^0 = \frac{1}{\sqrt{2}}(G + \bar{G}) = \frac{1}{\sqrt{2}}(G' + \bar{G}') = \frac{1}{\sqrt{2}}(G'' + \bar{G}'') \tag{5.25}
$$

They satisfy

$$
\begin{aligned}
\{G_L^0, G_L^0\} &= -iD_z \\
\{G_R^0, G_R^0\} &= -iD_{\bar{z}}
\end{aligned}
\tag{5.26}
$$

One then has three additional supersymmetry generators (in each sector)

$$
\begin{aligned}
G^3 &= \frac{1}{i\sqrt{2}}(G - \bar{G}) \\
G^1 &= \frac{1}{i\sqrt{2}}(G' - \bar{G}') \\
G^2 &= \frac{1}{i\sqrt{2}}(G'' - \bar{G}'')
\end{aligned}
\tag{5.27}
$$

Each one of these three supersymmetry generators anticommute with G^0. To prove that $G^\mu (\mu = 0, 1, 2, 3)$ form an $N = 4$ superalgebra we need to further show that the $G^i (i = 1, 2, 3)$ anticommute with each other. To prove this we first note that the complex structures C_i obey the relation

$$
C_i C_j = C_k \tag{5.28}
$$

where i, j, k are in cyclic permutations of $(1, 2, 3)$ and the action is invariant under the replacement of Ψ by $C_i \Psi$. If we start from an action with Ψ replaced by ,say, $C_1 \Psi$ then the manifest $N = 1$ supersymmetry will be generated by G^1 and the second supersymmetry generated by the Kahler decomposition with respect to the complex structure C_3 will be G^2 since $C_3 C_1 = C_2$. Hence by the results of Witten on $N = 2$ supersymmetric gauged WZW models we have

$$
\{G^1, G^2\} = 0 \tag{5.29}
$$

and by cyclic permutation we find

$$
\{G^2, G^3\} = \{G^3, G^1\} = 0 \tag{5.30}
$$

Thus the four supersymmetry generators G^μ satisfy the $N = 4$ supersymmetry algebra:

$$\begin{aligned} \{G_L^\mu, G_L^\nu\} &= -i\delta^{\mu\nu} D_z \\ \{G_R^\mu, G_R^\nu\} &= -i\delta^{\mu\nu} D_{\bar{z}} \\ \mu, \nu, \ldots &= 0, 1, 2, 3 \end{aligned} \tag{5.31}$$

Since the gauged WZW models considered above are known to be conformally invariant we have thus proven that they are invariant under $N = 4$ superconformal transformations.

Acknowledgements: This talk was written up during my stay at the Institute for Theoretical Physics of the University of Helsinki. I would like to thank Antti Niemi and other members of the Institute for their kind hospitality.

References

1. M. Green, J. H. Schwarz and E. Witten, "Superstring Theory" , Cambridge Univ. Press, (1987, Cambridge).
2. T. Banks, L. Dixon, D. Friedan and E. Martinec, *Nucl. Phys.* **B299** (1988) 613.
3. W. Boucher, D. Friedan and A. Kent, *Phys. Lett.* **172B** (1986) 316 ; P. DiVecchia, J.L. Petersen and M. Yu, *Phys. Lett.* **172B** (1986) 211 ; A.B. Zamolodchikov and V.A. Fateev, *Zh. Eksp. Theor. Fiz.* **90** (1986) 1553 and *Sov. Phys. JETP* **6** (1985) 215 ; Z. Qiu, *Phys. Lett.* **188B** (1987) 207; D. Gepner and Z. Qiu, *Nucl. Phys.* **B285** (1987) 423 ;
4. W. Lerche, C. Vafa and N.P. Warner, *Nucl. Phys.* **B324** (1989) 427.
5. Y. Kazama and H. Suzuki, *Nucl. Phys.* bf B321 (1989) 232; *Phys. Lett.* **B216** (1989) 112;
6. L.J. Dixon, J. Lykken and M.E. Peskin, *Nucl. Phys.* **B325** (1989) 329.
7. I. Bars, *Nucl. Phys.* **B334** (1990) 125;
8. N. Seiberg,*Nucl. Phys.* **B303** (1988) 286.
9. See P. Fendley, W. Lerche, S. D. Mathur and N.P. Warner, "N=2 Supersymmetric Integrable Models from Affine Toda Theories", Preprint CTP1865 (CALT-68-1631; HUTP-90/A036) and the references therein.
10. E. Witten, *Nucl. Phys.* **B340** (1990) 281 ; E. Witten, *Comm. Math. Phys.* **118** (1988) 411 ; C. Vafa, Harvard Preprint HUTP-90/A064 ; J. Distler and P. Nelson, *Phys. Rev. Lett.* **66** (1991) 1955 ; R. Dijkgraaf, H. Verlinde and E. Verlinde, Preprint PUPT-1217 (IASSNS-HEP-90/80).
11. P. Goddard, A. Kent and D. Olive, *Phys. Lett.* **152B** (1985) 88 ; *Comm. Math. Phys.* **103** (1986)105 ; P. Goddard, W. Nahm and D. Olive, *Phys. Lett.* **160B** (1985) 111.
12. H. Sugawara, *Phys. Rev.* **170** (1968) 1659 ; C. Sommerfield, *Phys. Rev.* **176** (1968) 2019.
13. M. Günaydin, *Phys. Lett.* **B255** (1991) 46.
14. M. Günaydin and S. Hyun, *Mod. Phys. Lett.* **A6** (1991) 1733.
15. M. Günaydin and S. Hyun, *Nucl. Phys.* **B373** (1992) 688.

16. K. Schoutens, *Nucl. Phys.* **B295** (1988) 634 ; A. Sevrin, W. Troost and A. Van Proeyen, *Phys. Lett.* **B208** (1988) 447.
17. M. Günaydin, "Ternary Algebraic Approach to Extended Superconformal Symmetry", invited talk to appear in the proceedings of the *V*th Regional Conference on Mathematical Physics, Trakya University, Edirne, Türkiye, Dec. 1991.
18. M. Günaydin, *Int. Jour. Mod. Phys.* **A8** (1993) 301.
19. M. Günaydin, *Phys. Rev.* **D47** (1993) 3600.
20. J. Tits, *Nederl. Akad. van Wetens.* **65** (1962) 530 ; I. L. Kantor, *Sov. Math. Dok.* **5** (1964) 1404 ; M. Koecher, *Amer. J. Math.* **89** (1967) 787.
21. I.L. Kantor, *Trudy Sem. Vektor. Anal.* **16** (1972) 407 ; *Sov. Math. Dokl.* **14** (1973) 254.
22. I. Bars and M. Günaydin, *J. Math. Phys.* **20** (1979)1977.
23. H. Freudenthal, *Proc. Konikl.Nederl. Akad. Wet. Ser.* **A57** (1954) 218-230 and 363-408 ; K. Meyberg, ibid **A71** (1968) 162-174 and 175-190.
24. I.L. Kantor and I.M. Skopets,*Sel. Math. Sov.* **2** (1982)293.
25. P. Goddard and A. Schwimmer, *Phys. Lett.* **B214** (1988) 209.
26. M. Günaydin, J.L. Petersen, A. Taormina and A. Van Proeyen, *Nucl. Phys.* **B322** (1989)402.
27. V.G. Knizhnik, *Theor. Math. Phys.* **66** (1986) 68.
28. M. Bershadsky, *Phys.Lett.*,**174B** (1986) 285.
29. A. Van Proeyen, *Class. Quantum Gravity*, **6** (1989) 1501; A. Sevrin and G. Theodoridis, *Nucl. Phys.*,**B332** (1990) 380.
30. E. Guadagnini, M. Martellini and M. Minchev, *Phys.Lett.*, **B191** (1987) 69; W. Nahm, "Gauging Symmetries of Two Dimensional Conformally Invariant Models", Davis Preprint, UCD-88-02 (unpublished); M. R. Douglas, "G/H Conformal Theory", Ph.D. Thesis, Caltech (1988), unpublished; A. Altschuler, K. Bardakci and E. Rabinovici, *Comm. Math. Phys.*,**118**(1988) 157; D. Karabali, Q.H. Park , H. J. Schnitzer and Z. Yang, *Phys.Lett.*,**B216** (1989) 307; D. Karabali and H. J. Schnitzer, *Nucl. Phys.*, **B329**(1990) 649; K. Gawedzki and A. Kupiainen , *Phys.Lett.*,**B215**(1988) 119; *Nucl. Phys.*,**B320**(1989) 649 ; P. Bowcock, *Nucl. Phys.*,**B316** (1989) 80.
31. E. Witten, *Comm. Math. Phys.*,**144**(1992) 189.
32. R. Rohm, *Phys. Rev.* **D32** (1984) 2849.
33. P. Di Vecchia, V.G. Knizhnik, J.L. Petersen and P. Rossi, *Nucl. Phys.* **B253** (1985) 701.
34. A. Sevrin, W. Troost, A. Van Proeyen and Ph. Spindel, *Nucl. Phys.*,**B308** (1988) 662.
35. H. J. Schnitzer, *Nucl. Phys.*,**B324** (1989) 412.
36. E. Witten, *Nucl. Phys.*, **B371** (1992) 191.
37. J. Wess and B. Zumino, *Phys.Lett.*, **B37** (1971) 95.
38. E. Witten, *Comm. Math. Phys.*, **92** (1984) 455.

A Three-Dimensional Fractional String

Philip C. Argyres

School of Natural Sciences, Institute for Advanced Study
Princeton, NJ 08540, USA

A natural and important question in string theory is whether heterotic and superstrings are the only kinds of string theories that admit realistic low-energy physics—having a flat Lorentz space-time interpretation with gravity, Yang-Mills gauge theories, chiral fermions and no tachyons in the low-energy physics. Though so far only the superstring and heterotic string fit this bill, I argue in this talk that there is a promising candidate class of string theories that may also have these properties, namely, fractional strings.

String theories are characterized by the local symmetries of two-dimensional field theories on the string world-sheet. For example, the superstring and the heterotic string are invariant, respectively, under (1,1) and (1,0) superconformal super-reparametrizations on the world-sheet. It is natural to ask whether other symmetries on the world-sheet can give rise to consistent string theories. This idea is the basis for "W-strings," a proposed class of strings based on W-symmetries of the string world-sheet which involve integer and half-integer spin currents greater than 2. The large class of *fractional strings* [1] are based on a similar idea: since fractional-spin fields exist in two-dimensional theories, one can imagine new local symmetries on the world-sheet involving fractional-spin currents (replacing the spin-3/2 supercurrent of the superstring). Evidence has been presented [2-6] that fractional superstrings with spin 4/3, 6/5, and 10/9 currents on the world-sheet have potentially interesting phenomenologies in 6, 4 and 3 space-time dimensions, respectively.

To make these ideas clearer, let us first briefly review the conceptual formulation of the superstring. A string propagating in a D-dimensional space-time sweeps out a two-dimensional world-sheet, which can be (locally) parameterized by two intrinsic coordinates σ and τ. Then the motion of the world-sheet in space-time is described by the D coordinate functions $X^\mu(\sigma,\tau)$. From the point of view of a description intrinsic to the world-sheet, these coordinate functions are just a collection of D scalar fields on the string world-sheet. In the Ramond-Neveu-Schwarz description of the superstring, there are additional degrees of freedom on the world-sheet which do not have a simple space-time geometric interpretation, namely the D world-sheet spinor fields $\psi^\mu(\sigma,\tau)$. Upon gauge-fixing the local symmetries on the superstring world-sheet, one obtains an algebra of the constraint operators which determine the physical states in the string spectrum. For the superstring this algebra is just the superVirasoro algebra, generated by the world-sheet spin-2 energy-momentum tensor T and the spin-3/2 superconformal current S. The superVirasoro algebra is an infinite dimensional algebra of the modes of these operators which, very heuristically, has the form

$$TT \sim T\ , \quad TS \sim S\ , \quad SS \sim T\ . \tag{1}$$

In terms of the X^μ and ψ^μ fields on the world-sheet T and S have the familiar expressions $T \sim \partial X^\mu \partial X_\mu + \psi^\mu \partial \psi_\mu$ and $S \sim \partial X^\mu \psi_\mu$. However, the important idea for fundamental strings is that it is the algebra (1) and not its specific representation in terms of fields such as X^μ and ψ^μ on the world-sheet that characterizes the string. In this way string theory is not tied to any particular space-time interpretation—"space-time" itself becomes a derived notion of fundamental strings.

One can describe the fractional string idea in analogy to the forgoing discussion of the superstring. Consider again a string in D-dimensional space-time, *i.e.* with D scalar fields X^μ on its world-sheet. Now, instead of adding the spin-1/2 ψ^μ "internal" degrees of freedom, say we add D spin-$1/N$ world-sheet fields ϵ^μ, where N is some integer greater than 3. (Such fractional-spin fields abound in two-dimensional field theories; the $\mathbf{Z}_N$ parafermions [7] are familiar examples. I will describe a simple example of a two-dimensional field theory with such fields in more detail below.) Then the constraint algebra has the form

$$TT \sim T\ , \quad TG \sim G\ , \quad GG \sim G + T\ . \tag{2}$$

Note that GG can close back on G as well as T. We expect G to be a spin-$(1 + 1/N)$ current of the form $G \sim \partial X^\mu \epsilon_\mu + \ldots$. But again, it is the fractional conformal algebra (2) that defines the fractional string, not any specific conformal field theory representation in terms of fields like X^μ and ϵ^μ.

This talk focuses on the fractional string with N=3, that is to say, with spin-4/3 fractional currents G. This fractional string is disscussed extensively in Ref. 8. The main questions we need to answer concerning this string are the following: What, explicitly, is the world-sheet spin-4/3 fractional algebra? Is the resulting spin-4/3 string consistent (does it have unitary scattering amplitudes)? Does the spin-4/3 string have sensible low-energy physics?

We can answer the first question by taking the spin-4/3 fractional algebra to be the spin-4/3 generalized parafermion algebra introduced by Zamolodchikov and Fateev [7]. This algebra is generated by two chiral currents $G^\pm$ of conformal dimension (world-sheet spin) 4/3 in addition to the energy-momentum tensor $T(z)$, and is encoded in in the singular terms of the operator product expansions:

$$\begin{aligned}
T(z)T(w) &= \frac{1}{(z-w)^4}\left\{\frac{c}{2} + 2(z-w)^2 T(w) + (z-w)^3 \partial T(w)\right\}\ , \\
T(z)G^\pm(w) &= \frac{1}{(z-w)^2}\left\{\frac{4}{3}G^\pm(w) + (z-w)\partial G(w)\right\}\ , \\
G^\pm(z)G^\pm(w) &= \frac{\lambda}{(z-w)^{4/3}}\left\{G^\mp(w) + \frac{1}{2}(z-w)\partial G^\mp(w)\right\},
\end{aligned}$$

$$G^{\pm}(z)G^{\mp}(w) = \frac{1}{(z-w)^{8/3}}\left\{\frac{3c}{8}+(z-w)^2T(w)\right\}. \quad (3)$$

The z and w coordinates one can take as right-moving light-cone coordinates on the string world-sheet. By only showing the algebra for the right-movers, we are effectively describing open strings. To describe closed strings (at tree level) we would simply have to include a second copy of the algebra for the left-movers. The first operator product just states that $T(z)$ obeys the Virasoro algebra, while the second implies that $G^{\pm}(z)$ are dimension-4/3 chiral primary fields. Although one can imagine other chiral algebras involving dimension-4/3 currents, we have chosen the above algebra to define the spin-4/3 fractional string because this algebra is known to have a sensible representation theory [7,8]. In particular, it is known that associativity fixes λ as a function of the central charge c: $\lambda^2 = (8-c)/6$.

One can derive the algebra of constraints from the above operator product algebra as a set of generalized commutation relations satisfied by the modes of the currents. The physical state conditions are imposed by the requirement that the positive (annihilation) modes of the currents vanish when acting on physical states.

We now turn to the second, harder, question of the construction of unitary scattering amplitudes for the physical states of this string. I will just state that tree-level amplitudes for the spin-4/3 string have been constructed, and that spurious state decoupling (a non-trivial preprequisite for unitarity) has been shown for them based on the properties of the spin-4/3 algebra [8]. To prove unitarity depends on the specific CFT representation (choice of space-time background), and has to be done on a case by case basis. Tree unitarity has been shown [9] for the specific 3-dimensional example to be described below. One-loop consistency has not been shown, however. Some heuristic arguments [1-6] suggest that the spin-4/3 string may be one-loop consistent at critical central charge $c = 10$, corresponding to flat space-time dimension $D = 6$. If this is the case, then the $c = 5$, $D = 3$ example to be described below does not correspond to a consistent fractional string vacuum.

This brings us to the final question of whether the spin-4/3 string can describe sensible physics. I will try to answer this question in the affirmative by constructing a conformal field theory (CFT) representation of the spin-4/3 algebra which gives an interpretation of the fractional string as propagating in a flat Minkowski space-time (albiet of dimension 3 only) with low energy physics describing gravity, non-abelian gauge fields, chiral fermions and no tachyons. As mentioned above, this $D = 3$ fractional string vacuum is not consistent at the string loop level; it is analogous to formulating the bosonic string in $D < 26$ flat dimensions—tree scattering amplitudes are well-behaved, but the theory is known to be inconsistent at the loop level, where $D = 26$ is required. Thus, this three-dimensional example should only be taken as evidence that realistic low-energy physics is possible for the spin-4/3 string.

The CFT in question is easy to construct: it consists of five free (massless) scalar fields on the world-sheet with two of them compactified on a specific lattice. The three uncompactified fields, $X^\mu(z)$, $\mu = 0,1,2$, are interpreted as the string coordinate fields, and obey the standard operator product $X^\mu(z)X^\nu(w) = -\eta^{\mu\nu}\ln(z-w)$, where $\eta^{\mu\nu}$ is the three-dimensional Minkowski metric with signature $(-++)$. This X^μ CFT has a global $SO(2,1)$ Lorentz symmetry.

The remaining two fields, $\varphi^i(z)$, $i = 1,2$, are compactified on a triangular lattice—the $SU(3)$ root lattice. In a basis in which the φ^i boundary conditions are diagonalized, $\varphi^i = \varphi^i + 2\pi$, their operator products read

$$\varphi^i(z)\varphi^j(w) = -g^{ij}\ln(z-w)\,, \qquad g^{ij} = \frac{1}{3}\begin{pmatrix} 2 & -1 \\ -1 & 2 \end{pmatrix}\,. \tag{4}$$

This CFT is well-known to be a realization of the $SU(3)_1$ Wess-Zumino-Witten model, (a $\mathbf{Z}_2$ orbifold of) which is conformally equivalent to the $SO(2,1)_2$ model. So, all the fields in the φ^i CFT can be organized in $SO(2,1)$ representations, corresponding to the space-time Lorentz properties of states in the string theory.

Furthermore, currents obeying the spin-4/3 algebra Eq. (3) can be constructed in this $c = 5$ CFT.

Using the prescription for fractional string tree scattering amplitudes alluded to above, one can calculate the scattering amplitudes for the states in the spectrum of this theory. One finds that the spectrum starts with massless vector particles (and a graviton in the closed string case). The potential tachyon states can be consistently eliminated by an analog of the GSO projection of the superstring. The massless vector particle three-point coupling is precisely the Yang-Mills coupling plus a non-linear term which is higher-order in the string tension and therefore is suppressed at energies far below the Planck scale.

I conclude by pointing out two fundamental questions which arise upon considering fractional strings.

What is the local world-sheet symmetry underlying the spin-4/3 constraint algebra? Though the form of the spin-4/3 algebra provides a rigid guide to such a symmetry, its identification remains an open question. Trying to formulate such a gauge symmetry involving spin-4/3 currents gives rise to the interesting question of whether there is some (presumably non-commutative) generalization of the notion of a two-manifold with transition functions which admit globally defined spin-1/3 fields. This would be a fractional analog of superRiemann surfaces.

What are the space-time features of critical fractional strings? This, also, is an open question, since its answer depends largely on the construction of new CFT representations of the spin-4/3 algebra. So far, only hints of possibly new space-time structures have been gleaned from the fractional superstring partition function [2-6].

This work was supported by NSF grant PHY-92-45317 and by the Ambrose Monell Foundation.

References

1. P.C. Argyres, A. LeClair and S.-H.H. Tye, *Phys. Lett.* **253B** (1991) 306.
2. P.C. Argyres and S.-H.H. Tye, *Phys. Rev. Lett.* **67** (1991) 3339.
3. K.R. Dienes and S.-H.H. Tye, *Nucl. Phys.* **B376** (1992) 297.
4. P.C. Argyres, K.R. Dienes and S.-H.H. Tye, *Commun. Math. Phys.* **154** (1993) 471.
5. P.C. Argyres and K.R. Dienes, *Phys. Rev. Lett.* **71** (1993) 819.
6. K.R. Dienes, *Nucl. Phys.* **B413** (1994) 103.
7. A.B. Zamolodchikov and V.A. Fateev, *Sov. Phys. J.E.T.P.* **62** (1985) 215; **63** (1986) 913; *Theor. Math. Phys.* **71** (1987) 451.
8. P.C. Argyres and S.-H.H. Tye, *Phys. Rev.* **D49** (1994) 5326, 5349.
9. Z. Kakushadze and S.-H.H. Tye, *Phys. Rev.* **D49** (1994) 4122.

Why Don't We Have a Covariant Superstring Field Theory?

Martin Cederwall

Institute for Theoretical Physics
Chalmers University of Technology and Göteborg University
S-412 96 Göteborg, Sweden
email: *tfemc@fy.chalmers.se*

Abstract. This talk deals with the old problem of formulating a covariant quantum theory of superstrings, "covariant" here meaning having manifest Lorentz symmetry and supersymmetry. The advantages and disadvantages of several quantization methods are reviewed. Special emphasis is put on the approaches using twistorial variables, and the algebraic structures of these. Some unsolved problems are identified.

Before going into supersymmetric strings, let me examine the situation in bosonic string theory.

The bosonic string has a well defined (Lorentz) covariant $1^{\underline{st}}$ quantized formulation as a gauge theory, preferably through BRST [1]. This formulation is the starting point and an absolute prerequisite for the $2^{\underline{nd}}$ quantization, the *field theory* [2]. It is probable that field theory can provide a framework for posing questions about the big symmetries of string theory, including general coordinate invariance. Some aspects of background invariance have already been addressed in bosonic string theory [3].

In this perspective, what is the corresponding status of superstring theory? It is not so good. Why is this so?

To be clear about the ambitions, one would like a covariant quantum superstring theory to fulfill the following requirements:

1. It should have manifest space-time symmetry, including supersymmetry.
2. It should contain $1^{\underline{st}}$ class constraints only.

Concerning the second of these points, there may well be $2^{\underline{nd}}$ class constraints present, but they have to be dealt with in a covariant manner before $1^{\underline{st}}$ quantization. There are different methods for doing this – in Dirac's original treatment of constraints [4] $2^{\underline{nd}}$ class constraints are eliminated consistently by letting remaining variables (parametrizing the $2^{\underline{nd}}$ class constraint surface) obey Dirac brackets; in the method by Batalin and Fradkin [5] additional constraints are added to turn the constraints into 1st class ones.

There are a number of quite different formulations of superstring theory. Let us examine them with respect to the requirements! The three main classes of models are

1. Spinning string
2. Green-Schwarz superstring
3. Twistor superstrings (main subject of this talk).

In each of these approaches, there may be alternative formulations or modifications. I will briefly review the different models, and make some indications on to what extent and on which points the requirements we have set up fail to be fulfilled.

The spinning/fermionic/NSR string [6] has an $N=1$ world-sheet supersymmetry. The action is the $N=1$ generalization of

$$S = \int d^2\sigma \sqrt{-g} g^{\alpha\beta} \partial_\alpha X^\mu \partial_\beta X_\mu$$

where a world-sheet gravitino and a fermionic space-time vector have been included [7]. The *space-time* supersymmetry of the spinning string is not present until the GSO projection [8] is performed on the spectrum and it is highly non-manifest. A functioning field theory exists for the spinning string [9], but it has of course the same drawbacks as the 1st quantized theory. The only thing that this formulation does not give us is manifest supersymmetry. The calculational power that two-dimensional conformal field theory comprises makes this approach to superstring field theory the best one so far.

Green and Schwarz found a space-time supersymmetric action for the superstring [10],

$$\begin{aligned} S = \int d^2\sigma \{ & \sqrt{-g} g^{\alpha\beta} \Pi^\mu_\alpha \Pi_{\mu\beta} - i\epsilon^{\alpha\beta} \partial_\alpha X^\mu (\overline{\theta}^1 \gamma_\mu \partial_\beta \theta^1 - \overline{\theta}^2 \gamma_\mu \partial_\beta \theta^2) \\ & + i\epsilon^{\alpha\beta} \overline{\theta}^1 \gamma^\mu \partial_\alpha \theta^1 \overline{\theta}^2 \gamma_\mu \partial_\beta \theta^2 \} \end{aligned}$$

where

$$\Pi^\mu_\alpha = \partial_\alpha X^\mu - i\overline{\theta}^A \gamma^\mu \partial_\alpha \theta^A$$

This action has some interesting properties, that can as well be analyzed in the simpler superparticle case, containing only the first term in the action. Its constraint structure is given by

$$\begin{aligned} L &\equiv P^2 \approx 0 \\ \Phi &\equiv p_\theta - iP_\mu \gamma^\mu \theta \approx 0 \end{aligned}$$

There is one bosonic constraint, generating translations along the world-line, and a fermionic spinor of local supersymmetry generators. Due to $P^2 = 0$, the rank of

$$\{\Phi_a, \Phi_b\} = -2iP_\mu \gamma^\mu_{ab}$$

is 8 (out of 16). The chiral spinor Φ thus contains eight $1^{\underline{st}}$ class constraints ("κ-symmetry" [11]) and eight $2^{\underline{nd}}$ class constraints. There is no way (naïvely) of eliminating covariantly the second class constraints before quantization [12]. Φ is a chiral spinor, transforming in the *irreducible* representation 16 of $Spin(1,9)$, and cannot be decomposed without giving up Lorentz covariance. Exactly the same is true for the superstring.

The difficulty of getting rid of the second class constraints in a covariant manner is closely connected to the problems with finding covariant supersymmetric field theories in ten dimensions.

One can always choose a light-front gauge, where only the physical degrees of freedom remain (no gauge invariance). Then the step to field theory is straightforward [13].

A couple of approaches exist, where one tries to deal with the constraint structure of the Green-Schwarz string by modifying it. Siegel proposed [14] that only the first class constraints should be kept, by only demanding (in the superparticle version) $\Psi \equiv P_\mu(\gamma^\mu \Phi) \approx 0$, and then introducing additional bosonic constraints removing fermionic degrees of freedom. The problem here is that the spinor Ψ has an infinite level of reducibility. It is not clear how to treat the infinite tower of ghosts that arise. Similar approaches are advocated in [15]. Berkovits [16] has shown how the Green-Schwarz superstring is obtained as an $N = 2$ supersymmetric string in four dimensions, with the extra six dimension compactified on a Calabi-Yau manifold. It is unclear if a similar construction can be made in six or ten dimensions.

Much of the original work in supertwistors was motivated by the observation that they have the potential of solving the problem with separation of the fermionic constraints in $1^{\underline{st}}$ and $2^{\underline{nd}}$ class parts. Let me therefore briefly describe the fundamental ideas of division algebra twistors for massless bosonic particles and superparticles, and then discuss application to string theory.

The classical dimensionalities of the superstring are $D = 3, 4, 6, 10$. The gamma matrix identities $(\lambda_1\gamma_\mu\lambda_2)\gamma^\mu\lambda_3 + \text{cycl.} = 0$ needed are directly related to the existence of the (alternative) division algebras $\mathbb{K}_\nu = \mathbb{R}$, $\mathbb{C}$, $\mathbb{H}$ and $\mathbb{O}$. More specifically:

Existence of Clifford algebra $\leftrightarrow$ alternativity,

$v^\mu = \lambda\gamma^\mu\lambda$ lightlike $\leftrightarrow$ division property $|ab| = |a||b|$.

This opens the way to twistor transformations of the lightlikeness constraints in these dimensionalities:

$$P^\mu = \frac{1}{2}\lambda\gamma^\mu\lambda \quad \Longleftrightarrow \quad P^{a\dot{a}} = \lambda^a\lambda^{\dagger\dot{a}}$$

The lightlike directions form the sphere S^ν. The spinor, modulo $\mathbb{R}_+$, lies on $S^{2\nu-1}$, where $\nu = D - 2$. The spinor λ is a two-component object $\lambda = \left[\lambda^1\ \lambda^2\right]^t$ with entries in $\mathbb{K}_\nu$, transforming under $SL(2;\mathbb{K}_\nu) \approx Spin(1,\nu+1)$. A vector is a hermitean matrix.

$$v = \begin{bmatrix} v^+ & v^* \\ v & v^- \end{bmatrix}$$

The twistor transform [17] from λ to P is the *Hopf map* $S^{2\nu-1} \to S^{\nu}$ with fiber $S^{\nu-1}$. The realization of the last (octonionic) Hopf map relies on the understanding of S^7 as "almost a Lie group" [18].

The twistor transform can be extended to superparticles [19], and it *solves the* $2^{\underline{nd}}$ *class constraint problem!* The fermionic variables become a Lorentz scalar element of the division algebra, except in D=10, where such things do not exist, and the fermion carries a vector representation, which actually can be identified as the fermionic varibles of the spinning particle [20]. All phase space variables sit in a representation of $OSp(1|4;\mathbb{K}_\nu)$ $(\nu \neq 8)$, which is the superconformal group [21] in D=3,4,6.

Strings are not (space-time) conformally invariant, except in the zero tension limit. A twistor transformation of the lightlikeness condition $(\partial X)^2 = 0$ as $\partial X = \lambda\lambda^\dagger$ introduces $2^{\underline{nd}}$ class constraints between λ and its canonical momentum ω, since X already spans the entire phase space for the left-(right-)moving sector.

Simple counting of the number of degrees of freedom in D=10 gives

$$8\,(\text{phys.}) = 2 \times 16 - 2 \times 1\,(\text{Vir.}) - 2 \times 7\,(\text{affine } S^7) - n \quad \Longrightarrow \quad n = 8\;,$$

so there are must be 8 $2^{\underline{nd}}$ class constraints for the bosonic twistor string in D=10. These constraints are quite analogous to the fermionic ones in the space-time picture of the superstring. The problem with fermionic constraints can be solved, but it reappears in the bosonic sector! This is quite general for twistor formulation of strings. The problem is not universally recognized, but seems to be generic in the sense that it appears as soon as chiral spinors form part of phase space. It is not at all clear whether the problem can be circumvented. We are not in the position that we dare to formulate a no-go theorem.

Even though the problem I have pointed out seems to be a very severe one concerning the prospect of finding a covariant quantization scheme for superstrings, there is a lot of very interesting structure in twistor superstrings. Different versions exist, each with its own advantages.

1. N=8 superconformal algebra (based on S^7) as a gauge group [22]. This formulation is manifestly "octonionic", which I consider as fundamental.
2. N=8 superfield formulation where the κ symmetry is identified as a local world-sheet supersymmetry [23]. The rôle of superconformal symmetry here is less clear.

It is likely that there exists an N=8 superfield formulation of 1., but the theory of N=8 superconformal field theory is unexplored.

A very interesting and intriguing observation is that the Green-Schwarz superstring gauge-fixed to the light-cone exhibits an N=8 superconformal symmetry [24]. The spin 2 and spin 3/2 generators are remnants of the Virasoro and local fermionic constraints, but the S^7 generators can not be traced back to a symmetry in this way. It is tempting to think that it is actually a sign of a gauge symmetry in an action where the $2^{\underline{nd}}$ class constraints reflect a partial choice of gauge. This still remains a speculation – we have not been able to find such a formulation.

References.

1. S. Hwang, *Phys.Rev.* **D28**(1983)2614.
2. W. Siegel and B. Zwiebach, *Nucl.Phys.* **B263**(1986)105,
   T. Banks and M.F. Peskin, in "Unified String Theories", World Scientific (1986),
   A. Neveu and P. West, *Nucl.Phys.* **B268**(1986)125.
   E. Witten, *Nucl.Phys.* **B268**(1986)253
3. B. Zwiebach, Lectures at Les Houches Summer School 1992,
   `hep-th/9305026`,
   E. Witten, *Phys.Rev.* **D46**(1992)5467,
   A. Sen and B. Zwiebach, *Nucl.Phys.* **414**(1994)649.
4. P.A.M. Dirac, *Rev.Mod.Phys.* **21**(1949)392.
5. I.A. Batalin and E.S. Fradkin, *Nucl.Phys.* **B279**(1987)514.
6. P. Ramond, *Phys.Rev.* **D3**(1971)2415,
   A. Neveu and J.H. Schwarz, *Nucl.Phys.* **B31**(1971)1109.
7. L. Brink, P. DiVecchia and P. Howe, *Phys.Lett.* **65B**(1976)471.
8. F. Gliozzi, J. Scherk and D. Olive, *Nucl.Phys.* **B122**(1977)253.
9. D. Friedan, E. Martinec and S. Shenker, *Phys.Lett.* **160B**(1985)55,
   E. Witten, *Nucl.Phys.* **B276**(1986)291
   C.R. Preitschopf, C.B. Thorn and S.A. Yost, *Nucl.Phys.* **B337**(1990)363.
10. M.B. Green and J.H. Schwarz, *Phys.Lett.* **136B**(1984)367.
11. W. Siegel, *Phys.Lett.* **128B**(1983)397.
12. I. Bengtsson and M. Cederwall, Göteborg-ITP-84-21.
13. M.B. Green, J.H. Schwarz and L. Brink, *Nucl.Phys.* **B219**(1983)437.
14. W. Siegel, *Nucl.Phys.* **B263**(1986)93.
15. R. Kallosh, *Phys.Lett.* **225B**(1989)49.
16. N. Berkovits, `hep-th/9404162`.
17. R. Penrose and M.A.H. McCallum, *Phys.Rep.* **6**(1972)241,
    I. Bengtsson and M. Cederwall, *Nucl.Phys.* **B302**(1988)81,
    N. Berkovits, *Phys.Lett.* **247B**(1990)45,
    M. Cederwall, *J.Math.Phys.* **33**(1992)388.
18. M. Cederwall and C.R. Preitshopf, `hep-th/9309030`, *Commun.Math.Phys.* in press.
19. A. Ferber, *Nucl.Phys.* **B132**(1978)55,
    T. Shirafuji, *Progr.Theor.Phys.* **70**(1983)18,
    I. Bengtsson and M. Cederwall, *Nucl.Phys.* **B302**(1988)81,
    N. Berkovits, *Phys.Lett.* **247B**(1990)45,
    M. Cederwall, *J.Math.Phys.* **33**(1992)388.
20. M. Cederwall, *Mod.Phys.Lett.* **A9**(1994)967.
21. A. Sudbery, *J.Phys.* **A17**(1984)939.
22. N. Berkovits, *Nucl.Phys.* **B358**(1991)169,
    M. Cederwall and C.R. Preitshopf, `hep-th/9309030`,
    Commun.Math.Phys. in press.

23. D.P. Sorokin, V.I. Tkach and D.V. Volkov, *Mod.Phys.Lett.* **A4**(1989)901,
M. Tonin, *Int.J.Mod.Phys.* **A7**(1992)6013,
F. Delduc, A. Galperin, P. Howe and E. Sokatchev,
Phys.Rev. **D47**(1992)578,
I.A. Bandos, M. Cederwall, D.P. Sorokin and D.V. Volkov,
hep-th/9403181, *Mod.Phys.Lett.* **A** in press.
24. L. Brink, M. Cederwall and C. R. Preitschopf, *Phys.Lett.* **311B**(1993)76.

The reference list, and also the text, gives a very fragmented rendering of the contributions to a vast subject. The author sincerely apologizes for that.

Search for Duality Symmetries in p-Branes

E. Sezgin

Center for Theoretical Physics, Texas A&M University, College Station, Texas 77843, U.S.A.

Abstract. The requirement of an SL(2) duality symmetry, mixing the worldvolume field equations with Bianchi identities, leads to a highly nonlinear equation involving the transformation parameters and certain worldvolume currents. In general, this equation seems to admit a solution only for a two parameter subgroup of the seeked SL(2). These transformations also leave invariant the first class constraints generating the worldvolume reparametrizations. In the special case of p-branes in p+1 dimensions, the full SL(2) is realized

Keywords. Duality, membranes, p-branes.

I met Feza Gürsey a number of times, but unfortunately I never had a chance to collaborate with him. I have always been inspired by his beautiful work and his kind and gentle personality. I have a great respect and admiration for him. He is greatly missed and he will always be remembered.

Feza Gürsey had a keen sense for beauty of symmetries in physics and he made profound contributions in search of them. I believe that the topic of this talk, which is dedicated to his memory, is very much in the spirit of his research philosophy that emphasizes symmetries. More specifically, I will talk about a search for duality symmetries in theories of extended objects, known an p-branes.

It is well known that the ten dimensional heterotic string theory compactified on a six torus has target space duality symmetry group O(6,22; Z) which mixes momentum modes with winding modes [1,2]. From the world-sheet field theory point of view, this group mixes the world-sheet field equations with Bianchi identities [2]. It is natural to search for similar duality symmetries for higher p-branes.

In view of the possibility of a connection between the strongly coupled heterotic string and a weakly coupled fivebrane theory in ten dimensions [3], the issue of duality symmetry in fivebrane theory especially seems interesting. The expected target space duality group in this case is an SL(2,Z) group [4] (For a review, see [5]). Although it has been shown that an SL(2,Z) duality group indeed mixes the momentum modes with the winding modes of the fivebrane theory [6], so far this symmetry group has not been understood as a worldvolume duality group transforming the worldvolume equations of motion into Bianchi identities. In a recent paper [7], we studied the problem from this point of view. We found that, with a reasonable set of assumptions about the form of the duality transformations, the existence of an SL(2) symmetry requires a solution to a highly nonlinear equation involving the transformation parameters and the worldvolume currents that transform into each other under duality transformations. In general, this equation seems to admit only a two parameter solution which forms a subgroup of the seeked SL(2).In the special case of p-branes in p+1 dimensions, the full SL(2) is realized. In this note, we will describe the main result of [7], and we shall furthermore show that the two parameter duality group also leaves invariant the first class constraints which generate the worldvolume reparametrizations. An attempt to find duality symmetries in p-branes has been made before in somewhat different setting [8]. Further comments on this paper can be found in [7].

While the case of most interest is the fivebrane, we shall consider all p-branes, without much more effort. The dynamical variables describing the p-brane are scalar fields $x^\mu(\sigma)$, $y^\alpha(\sigma)$ and a worldvolume metric $\gamma_{ij}(\sigma)$. Here σ^i (i=0,...,p) are the worldvolume coordinates, x^μ, $\mu = 0,...,m-1$ are coordinates on m dimensional space M and y^α, $\alpha = 1,...,p+1$ are coordinates on a compact (p+1)-dimensional manifold N. We shall take the background fields to be the metrics $g_{\mu\nu}(x)$ anrd $g_{\alpha\beta}(x)$ on M and N respectively and an antisymmetric tensor field $b_{\alpha_1...\alpha_{p+1}}(x) = \lambda_1(x)\varepsilon_{\alpha_1...\alpha_{p+1}}$. In this background, the usual Polyakov type action for the p-brane is

$$S = \int d^{p+1}\sigma\ \mathcal{L} = \int d^{p+1}\sigma\ [-\frac{1}{2}\sqrt{-\gamma}(\gamma^{ij}\partial_i x^\mu \partial_j x^\nu g_{\mu\nu} + \gamma^{ij}\partial_i y^\alpha \partial_j y^\beta g_{\alpha\beta}) + \frac{p-1}{2}\sqrt{-\gamma} + \frac{1}{(p+1)!}\varepsilon^{i_1...i_{p+1}}\partial_{i_1}y^{\alpha_1}...\partial_{i_{p+1}}y^{\alpha_{p+1}}\lambda_1\varepsilon_{\alpha_1...\alpha_{p+1}}] . \tag{1}$$

Let us concentrate on the field equation for the internal coordinates y^α. It reads $\partial_i P^i{}_\alpha=0$, where

$$P^i{}_\alpha = \frac{\partial \mathcal{L}}{\partial \partial_i y^\alpha} = -\sqrt{-\gamma}\,\gamma^{ij}\partial_j y^\beta g_{\beta\alpha} + \lambda_1 J^i{}_\alpha, \tag{2}$$

$$J^i{}_\alpha = \frac{1}{p!}\ \varepsilon^{ij_1...j_p}\partial_{j_1}y^{\beta_1}...\partial_{j_p}y^{\beta_p}\,\varepsilon_{\alpha\beta_1...\beta_p} . \tag{3}$$

In searching for a duality symmetry in p-brane theories, we also need to know how the induced metric on the worldvolume transforms. We know from ten dimensional supergravity compactified on a six-torus that under SL(2) the metrics $g_{a\beta}$ and $g_{\mu\nu}$ rescale. Therefore let us define

$$g_{\mu\nu} = \lambda_2{}^K \bar{g}_{\mu\nu}, \qquad g_{\alpha\beta} = \lambda_2{}^L \bar{g}_{\alpha\beta}, \tag{4}$$

where $\bar{g}_{\alpha\beta}$ and $\bar{g}_{\mu\nu}$ are assumed to be inert under SL(2), and det $\bar{g}_{\alpha\beta} = 1$. Thus $\lambda_2(x) = (\det g_{\alpha\beta})^{1/(p+1)L})$. In the case p = 5 it is known from the SL(2) duality symmetry of the effective field theory limit that K = –1 and L = 1/3 [5].

The equation of motion for the worldvolume metric γ gives

$$\gamma_{ij} = \lambda_2{}^K \partial_i x^\mu \partial j x^\upsilon \bar{g}_{\mu\nu} + \lambda_2{}^L \partial_i y^\alpha \partial_j y^\beta \bar{g}_{\alpha\beta} \,. \tag{5}$$

Thus, the duality transformation rule for the worldvolume metric follows from the transformation rules assigned to quantities occuring in this equation.

We now look for transformation rules that mix the field equation $\partial_i P^i{}_\alpha = 0$, with the Bianchi identity $\partial_i J^i{}_\alpha = 0$. The most natural way to do this is to consider the transformations of the currents $P^i{}_\alpha$ and $J^i{}_\alpha$ into each other as

$$\delta P^i{}_\alpha = aP^i{}_\alpha + bJ^i{}_\alpha, \tag{6a}$$

$$\delta P^i{}_\alpha = cP^i{}_\alpha + bJ^i{}_\alpha, \tag{6b}$$

with a, b, c, d being constants. It is important to take the parameters to be constant, so that these transformations indeed map the field equations and Bianchi identities into a combination of each other. The key point in establishing the duality symmetry is to show that (2) is invariant under the transformations (6), combined with appropriate transformation rules for the background fields λ_1 and λ_2. In this regard, we are following the approach of Gaillard and Zumino [9].

Since all relevant quantities have two indices, it is convenient to use matrix notation. We define matrices P and J with components $P^i{}_\alpha$ and $J^i{}_\alpha$, matrices ∂y with components $(\partial y)^\alpha{}_i = \partial_i y^\alpha$ and $\bar{g}^{(p+1)}$ with components $\bar{g}_{\alpha\beta}$. From the definition (3) we find that $\partial y = J^{-1}(\det J)^{1/p}$. This equation allows us to calculate the variation of ∂y under the duality transformations.

The central result of [7] is that, taking into account the variation of the worldvolume metric γ, and allowing any transformation rules for the scalar fields λ_1, λ_2, the invariance of (2) under the duality transformations (6) implies requires that the following highly nonlinear equation be satisfied:

$$\begin{aligned} & cX^2 + [a - 2c\lambda_1 - \frac{1}{p}(d + c\operatorname{tr}X) - (\frac{p-1}{2}K + L)\lambda_2{}^{-1}\delta\lambda_2]X \\ & + b - \frac{p-1}{p}d\lambda_1 + \frac{1}{p}c\lambda_1\operatorname{tr}X + (\frac{p-1}{2}K + L)\lambda_2\lambda_2{}^{-1}\delta\lambda_2 - \delta\lambda_1 = \\ & = \lambda_2{}^L \gamma^{-1} V \{ 2cX^2 - [\frac{2}{p}(d + c\operatorname{tr}X) + 2c\lambda_1 + (L - K)\lambda_2{}^{-1}\delta\lambda_2)]X \\ & + \frac{1}{p}c(\operatorname{tr}X)^2 - c\operatorname{tr}(X^2) + (\frac{1}{p}c\lambda_1 + \frac{1}{p}d + \frac{1}{2}(L - K)\lambda_2{}^{-1}\delta\lambda_2)\operatorname{tr}X \\ & \qquad - \frac{p-1}{2}d\lambda_1 - \frac{1}{2}(p-1)(L-K)\lambda_1\lambda_2{}^{-1}\delta\lambda_2 \}, \end{aligned} \tag{7}$$

where

$$X = P \,.\, J^{-1} = \lambda_1 + \lambda_2{}^{L} \; \frac{\gamma^{-1}V}{\sqrt{-\det(\gamma^{-1}V)}} \;, \quad V = (\partial y)^{T}\bar{g}^{(p+1)}\partial y. \tag{8}$$

Since $\gamma^{-1}V$ can be reexpressed in terms of X via (8), this is an infinite polynominal equation in the (p+1) x (p+1) matrix X. One is free to determine $\delta\lambda_1$ and $\delta\lambda_2$ as functions of a, b, c, d, λ_1, λ_2 to satisfy this equation, and also if necessary to put restrictions on the transformation parameters a, b, c, d. It is important to realize that these transformations have to be the same for all X. In order to prove that this equation has no solution it would therefore be sufficient to find two particular matrices X which give incompatible values for the variations $\delta\lambda_1$ and $\delta\lambda_2$.

In [7], we implemented this idea by choosing a particular matrix X and expanding equation (7) around it. For our background we choose the fields x^μ and y^α such that $\gamma^{-1}V = \lambda_2{}^{-L}\eta$, where η is the Minkowski metric, and then write $\gamma^{-1}V = \lambda_2{}^{-L}(\eta+Y)$. Expanding (7) in powers of Y, we then find that the solution of (7) is given by

$$\begin{aligned} c &= 0, \\ d &= p\,\frac{K-L}{pK+L}\,a, \\ \delta\lambda_1 &= b + \frac{(p+1)L}{pK+L}\,a\lambda_1, \\ \delta\lambda_2 &= \frac{2}{pK+L}\,a\lambda_2 . \end{aligned} \tag{9}$$

In fact, one can check this is the solution of the full equation (7). Furthermore, one can show that the x^μ-equation of motion is also invariant under these transformations. Therefore, we have a two parameter group of duality transformations of the p-brane. It is easy to check that the transformation rules (6) and (9) yield the same commutator algebra. Denoting the transformations by a and b, the only nonvanishing commutator is [a,b] = b. One can have d = –a by choosing $K = \frac{p-1}{2p}\,L$. In this way the two parameter group appears to be a subgroup of the expected group SL(2). Notice that, from a solution with magnetic charge, one can obtain a solution with magnetic and electric charge. However, the two parameter group does not contain the important $R \rightarrow 1/R$ transformations.

In the special case of a p-brane propagating in p+1 dimensions, it is straightforward to show that the full SL(2) duality symmetry group is realized..

Finally, let us consider the action of the two parameter duality group described above on the reparametrization generating constraints. These constraints are [10].

$$H = g^{\mu\nu} p_\mu p_\nu + g^{\alpha\beta}(p_\alpha - \lambda_1 j_\alpha)(p_\beta - \lambda_1 j_\beta)$$
$$+ \det(\partial_a x^\mu \partial_b x^\nu g_{\mu\nu} + \partial_a y^\alpha \partial_b y^\beta g_{\alpha\beta}), \quad (10)$$

$$H_a = \partial_a x^\mu p_\mu + \partial_a y^\alpha p_\alpha, \quad (11)$$

where the index $a = 1,\ldots, p$ labels the spatial directions, p_μ and p_α are the momentum variables associated with x_μ and y^α, respectively, and $j_\alpha = J^0{}_\alpha$, which can be read off from $J^i{}_\alpha$ given in (3).

Consistent with the definition (10) and the transformations rules (6) and (9) one finds $\delta\partial_a y^\alpha = \frac{d}{p}\,\partial_a y^\alpha$. Using this variation, and taking x^μ, and therefore $\partial_a x^\mu$ to be inert, we find that under the duality transformations (6) and (9) the constraint H_a is preserved: $\delta H_a = \frac{(p+1)K}{pK+L} aH_a$, provided that we assign the transformation rule $\delta p_\mu = \frac{(p+1)K}{pK+L}\, a p_\mu$. Finally, one finds that the constraint H is also preserved under the duality transformation: $\delta H = \frac{(p+1)K}{pK+L}\, aH$.

In the special case of p-branes in (p+1)-dimensions, the Hamiltonian simplifies drastically to

$$H = g^{\alpha\beta}(p_\alpha - \lambda_1 j_\alpha)(p_\beta - \lambda_1 j_\beta) + \lambda_2{}^{(p+1)L} g^{\alpha\beta} J_\alpha J_\beta \,. \quad (12)$$

One can show that the Hamiltonian constraint and the space reparametrization constraint $H_a = \partial_a x^\mu p_\mu + \partial_a y^\alpha p_\alpha$ are preserved under the full SL(2) duality transformations.

Acknowledgments. I would like to thank to organizers of the Gürsey Memorial Conference I on String and Symmetries for their kind invitation. I also thank Roberto Percacci, with whom we collaborated on the material summarized here (see ref. [7]). This work has been supported in part by the U.S. National Science Foundation, under grant PHY-9106593.

References

[1] K.S. Narain, M.H. Sarmadi and E. Witten, Nucl. Phys. **B279** (1987) 367;
A. Shapere and F. Wilczek, Nucl. Phys. **B320** (1989) 669;
A. Giveon, E. Rabinovici and G. Veneziano, Nucl. Phys. **B322** (1989) 167;
A. Giveon, N. Malkin and E. Rabinovici, Phys. Lett. **220** (1989) 551;
W. Lerche, D. Lüst and N.P. Warner, Phys. Lett. **B231** (1989) 417.

[2] M.J. Duff, Phys. Lett. **B173** (1986) 289;
S. Cecotti, S. Ferrara and L. Girardello, Nucl. Phys. **B308** (1988) 436;
J. Molera and B. Ovrut, Phys. Rev. **D40** (1989) 1146;
M.J. Duff, Nucl. Phys. **B335** (1990) 610;
J. Maharana and J.H. Schwarz, Nucl. Phys. **B390** (1993) 3.

[3] M.J. Duff, Class. Quant. Grav. **5** (1988) 189;
A. Strominger, Nucl. Phys. **B343** (1990) 167;
M.J. Duff and J.X. Lu, Nucl. Phys. **B354** (1991) 141; Phys. Rev. Lett. **66** (1991) 1402; Class. Quant. Grav. **9** (1991) 1;
C.G. Callan, J.A. Harvey and A. Strominger, Nucl. Phys. **B359** (1991) 611; Nucl. Phys. **B367** (1991) 60;
M.J. Duff, R. Khuri and J.X. Lu, Nucl. Phys. **B377** (1992) 281;
J.Dixon, M.J. Duff and J. Plefka, Phys. Rev. Lett. **69** (1992) 3009.

[4] J.H. Schwarz and A. Sen, Phys. Lett. **B312** (1993) 105;
J.H. Schwarz and A. Sen, Phys. Lett. **B312** (1993) 105;
P. Binetrúy, Phys. Lett. **B315** (1993) 80.

[5] A. Sen, preprint, TIFR/TH/94-03 (hep-th/9402002).

[6] A. Sen, Nucl. Phys. **B388** (/1992) 457; Phys. Lett. **B303** 22; Mod. Phys. Lett. **A8** (1993) 2023;
J.H. Schwarz and A. Sen, Phys. Lett. **312** (1993) 105;
T. Ortin, Phys. Rev. **D47** (1993) 3136;
M.J. Duff and R. Khuri, preprint, CTP-TAMU-17/93.

[7] R. Percacci and E. Sezgin, preprint, CTP TAMU-15/94, SISSA 44/94/EP (hep-th/9407021).

[8] M.J. Duff and J.X. Lu, Nucl. Phys. **B347** (1990) 394.

[9] M.K. Gaillard and B.Zumino, Nucl. Phys. **B193** (1981) 221.

[10] E. Bergshoeff, R.Percacci, E. Sezgin, K.S. Stelle and P.K. Townsend, Nucl. Phys. **B398** (1993) 343.

Non Linear Identities Between Unitary Minimal Virasoro Characters

Anne Taormina[1]

[1] Department of Mathematical Sciences, University of Durham, England

Abstract. Non linear identities between unitary minimal Virasoro characters at low levels ($m = 3$, 4, 5) are presented as well as a sketch of some proofs. The first identity gives the Ising model characters ($m = 3$) as bilinears in tricritical Ising model characters ($m = 4$), while the second one gives the tricritical Ising model characters as bilinears in the Ising model characters and the six combinations of $m = 5$ Virasoro characters which do not appear in the spectrum of the three state Potts model.

1 Notations and conventions

The algebra underlying the two-dimensional conformal symmetry is called the Virasoro algebra. It has an infinity of generators L_a, $a \in Z$ which obey the following commutation relations,

$$[L_a, L_b] = (a-b)L_{a+b} + \frac{c}{12}a(a^2-1)\delta_{a+b,0} \tag{1}$$

where c is a c-number called the central charge. The representation theory of this infinite-dimensional algebra has been studied in great detail, in particular the unitary highest weight state representations [1, 2, 3, 4, 5]. When the central charge takes one of the following fractional values,

$$c = 1 - \frac{6}{m(m+1)}, \qquad m = 2, 3, ..., \tag{2}$$

it is known that only a finite number a primary fields with conformal dimensions

$$h_{r,s} = \frac{[(m+1)r - ms]^2 - 1}{4m(m+1)}, \qquad r = 1, 2, ...m-1; \qquad s = 1, ..., r \tag{3}$$

are highest weight states of irreducible unitary representations. The corresponding characters at level m are defined as,

$$\chi_{r,s}^{Vir\ (m)}(q) = \eta^{-1}(q)\left[\theta_{r(m+1)-sm,m(m+1)}(q) - \theta_{r(m+1)+sm,m(m+1)}(q)\right]. \tag{4}$$

The Dedekind function, $\eta(q)$, is defined by,

$$\eta(q) = q^{\frac{1}{24}} \prod_{n=0}^{\infty} \left(1 - q^{n+1}\right), \tag{5}$$

while the generalised level k theta functions, $\theta_{m,k}(q)$, are given by,

$$\theta_{m,k}(q) = \sum_{n \in Z} q^{k\left(n+\frac{m}{2k}\right)^2}, \tag{6}$$

with the properties,

$$\theta_{m,k}(q) = \theta_{-m,k}(q) = \theta_{2k-m,k}(q). \tag{7}$$

This definition, (4), of the Virasoro characters coincides with the definition given in [5] up to a factor $q^{-\frac{c}{24}}$. For completeness, we reproduce here the Kac tables giving the conformal dimensions (3) of the primary fields for the first three theories,

	$r=1$	$r=2$
$s=2$		1/16
$s=1$	0	1/2

$c = \frac{1}{2}, m = 3$

	$r=1$	$r=2$	$r=3$
$s=3$			1/10
$s=2$		3/80	3/5
$s=1$	0	7/16	3/2

$c = \frac{7}{10}, m = 4$

	$r=1$	$r=2$	$r=3$	$r=4$
$s=4$				1/8
$s=3$			1/15	2/3
$s=2$		1/40	21/40	13/8
$s=1$	0	2/5	7/5	3

$c = \frac{4}{5}, m = 5$

2 The identities

The first set of identities relates the unitary Virasoro characters at levels $m = 3$ and $m = 4$ in such a way that the three Ising model characters are given by the vector product of two 3-vectors whose components are the six tricritical Ising model characters,

$$\chi_{1,i}^{Vir\ (3)}(q) = \epsilon_{ijk}(-1)^{j+k}\chi_{j,4}^{Vir\ (4)}(q)\ \chi_{k,2}^{Vir\ (4)}(q). \tag{8}$$

It is interesting to note that the vector $\chi_{1,i}^{Vir\ (3)}(q)$ $(i = 1,2,3)$ being orthogonal to the vectors $(-1)^j\chi_{j,4}^{Vir\ (4)}(q)$ and $(-1)^k\chi_{k,2}^{Vir\ (4)}(q)$ (no summation on j and k) is a trivial consequence of the identity (8). It produces relations which can be derived from repeated use of the Goddard-Kent-Olive (GKO) sum-rules [3],

$$\chi_{2\ell}^{k}(q,z)\chi_{2\ell'}^{1}(q,z) = \sum_{2\ell''=0}^{k+1} \chi_{2\ell''}^{k+1}(q,z)\ \chi_{2\ell+1,2\ell''+1}^{Vir\ (k+2)}(q), \tag{9}$$

where $2\ell'' \equiv 2\ell + 2\ell' \pmod 2$ and where the $SU(2)_k$ characters for isospin ℓ ($2\ell = 0, 1, \ldots, k$) are defined by,

$$\begin{aligned}\chi^k_{2\ell}(q,z) &= q^{-1/8} z^{-1} \prod_{n=1}^{\infty} (1-q^n)^{-1}(1-q^n z^2)^{-1}(1-q^{n-1}z^{-2})^{-1} \\ &\times \sum_{m \in Z + \frac{\ell+1/2}{k+2}} q^{(k+2)m^2} \left[z^{2(k+2)m} - z^{-2(k+2)m} \right] . \end{aligned} \tag{10}$$

In this instance, one considers the coset $[SU(2)_1 \times SU(2)_1 \times SU(2)_1]/SU(2)_3$ and applies the GKO sumrules twice on the following trilinear in $\widehat{SU(2)}_1$ characters,

$$\left[\chi^1_0(q,z)\,\chi^1_0(q,z)\right]\chi^1_1(q,z) = \chi^1_0(q,z)\left[\chi^1_0(q,z)\,\chi^1_1(q,z)\right] . \tag{11}$$

In the second set of identities, the tricritical Ising model characters are obtained as the product of unitary Virasoro characters at levels $m = 3$ and $m = 5$ in the following way (implicit dependence in the variable q),

$$\begin{aligned}
\chi^{Vir\ (4)}_{2,1} &= \chi^{Vir\ (3)}_{2,2} \left[\chi^{Vir\ (5)}_{2,1} - \chi^{Vir\ (5)}_{3,1}\right] , \\
\chi^{Vir\ (4)}_{2,2} &= \chi^{Vir\ (3)}_{2,2} \left[\chi^{Vir\ (5)}_{1,1} - \chi^{Vir\ (5)}_{4,1}\right] , \\
\chi^{Vir\ (4)}_{1,1} \pm \chi^{Vir\ (4)}_{3,1} &= \left[\chi^{Vir\ (3)}_{1,1} \pm \chi^{Vir\ (3)}_{2,1}\right]\left[\chi^{Vir\ (5)}_{2,2} \mp \chi^{Vir\ (5)}_{3,2}\right] , \\
\chi^{Vir\ (4)}_{1,2} \pm \chi^{Vir\ (4)}_{3,2} &= \left[\chi^{Vir\ (3)}_{1,1} \pm \chi^{Vir\ (3)}_{2,1}\right]\left[\chi^{Vir\ (5)}_{1,2} \mp \chi^{Vir\ (5)}_{4,2}\right] .
\end{aligned} \tag{12}$$

It is remarkable that the six combinations of level $m = 5$ Virasoro characters involved are precisely those which do *not* appear in the spectrum of the three state Potts model. The identities (12) are consistent with the weaker identities,

$$\chi^{Vir\ (4)}_{2,2}(q) \left[\chi^{Vir\ (5)}_{2,1}(q) - \chi^{Vir\ (5)}_{3,1}(q)\right] = \chi^{Vir\ (4)}_{2,1}(q) \left[\chi^{Vir\ (5)}_{1,1}(q) - \chi^{Vir\ (5)}_{4,1}(q)\right] , \tag{13}$$

and,

$$\begin{aligned}
&\left[\chi^{Vir\ (4)}_{1,1}(q) \pm \chi^{Vir\ (4)}_{3,1}(q)\right] \left[\chi^{Vir\ (5)}_{1,2}(q) \mp \chi^{Vir\ (5)}_{4,2}(q)\right] \\
&= \left[\chi^{Vir\ (4)}_{1,2}(q) \pm \chi^{Vir\ (4)}_{3,2}(q)\right] \left[\chi^{Vir\ (5)}_{2,2}(q) \mp \chi^{Vir\ (5)}_{3,2}(q)\right] ,
\end{aligned} \tag{14}$$

which can be obtained from the GKO character sumrules for the coset $[SU(2)_1 \times SU(2)_2 \times SU(2)_1]/SU(2)_4$ when considering the following trilinears in $\widehat{SU(2)}$ characters,

$$\left[\chi^1_0(q,z)\,\chi^2_0(q,z)\right]\chi^1_1(q,z) = \chi^1_0(q,z)\left[\chi^2_0(q,z)\,\chi^1_1(q,z)\right] , \tag{15}$$

and,

$$\left[\chi^1_0(q,z)\,\chi^2_1(q,z)\right]\chi^1_1(q,z) = \chi^1_0(q,z)\left[\chi^2_1(q,z)\,\chi^1_1(q,z)\right] . \tag{16}$$

3 Proof of the identities

A proof of the identities (8), and one of the identities (12) involving the Jacobi triple product identity and standard properties of the generalised theta functions (6) has been explicitly given in [6]. We write here the proof of one of the remaining identities in (12), the others being discussed along similar lines.

The identity we prove here is,

$$\chi_{2,1}^{Vir\ (4)}(q) = \chi_{2,2}^{Vir\ (3)}(q)\ \left(\chi_{2,1}^{Vir\ (5)}(q) - \chi_{3,1}^{Vir\ (5)}(q)\right), \tag{17}$$

which we rewrite

$$\begin{aligned}\prod_{n=0}^{\infty}\left(1-q^{n+1}\right) &= \prod_{n=0}^{\infty}\left(1+q^{5n+2}\right)\ \left(1+q^{5n+3}\right)\ \left(1-q^{5n+2}\right)\ \left(1-q^{5n+3}\right) \\ &\times\ q^{-\frac{49}{120}}\ \left[\theta_{7,30}(q)+\theta_{23,30}(q)-\theta_{13,30}(q)-\theta_{17,30}(q)\right].\end{aligned} \tag{18}$$

We have used the infinite product representation of $m=3$ and $m=4$ characters (see for instance [5])

$$\chi_{1,2}^{Vir\ (3)}(q) = \chi_{2,2}^{Vir\ (3)}(q) = q^{\frac{1}{24}}\prod_{n=0}^{\infty}\left(1+q^{n+1}\right), \tag{19}$$

and,

$$\chi_{2,1}^{Vir\ (4)}(q) = q^{-\frac{7}{240}+\frac{7}{16}}\prod_{n=0}^{\infty}\frac{\left(1+q^{5n+1}\right)\ \left(1+q^{5n+4}\right)\ \left(1+q^{5n+5}\right)}{\left(1-q^{5n+2}\right)\ \left(1-q^{5n+3}\right)}, \tag{20}$$

as well as the definition of $m=5$ Virasoro characters in terms of theta functions (4). The infinite product on the RHS can be expressed as a product of level $k=30$ theta functions. Indeed, one has,

$$\begin{aligned}\prod_{n=0}^{\infty}(1\pm q^{5n+2})(1\pm q^{5n+3}) &= \prod_{n=0}^{\infty}\prod_{m=0}^{m=2}(1\pm q^{15n+2+5m})(1\pm q^{15n+15-(2+5m)}) \\ &= \prod_{n=0}^{\infty}(1-q^{15n+15})^{-3}\times \\ &\quad [\sum_{n=-\infty}^{\infty}(\pm1)^n q^{\frac{15}{2}n^2+\frac{11}{2}n}]\,[\sum_{n=-\infty}^{\infty}(\pm1)^n q^{\frac{15}{2}n^2+\frac{9}{2}n}]\,[\sum_{n=-\infty}^{\infty}(\pm1)^n q^{\frac{15}{2}n^2+\frac{1}{2}n}] \\ &= \prod_{n=0}^{\infty}(1-q^{15n+15})^{-3}\times \\ &\quad q^{-1-\frac{1}{120}}[\theta_{11,30}\pm\theta_{19,30}]\ q^{-\frac{81}{120}}[\theta_{9,30}\pm\theta_{21,30}]\ q^{-\frac{1}{120}}[\theta_{1,30}\pm\theta_{29,30}],\end{aligned} \tag{21}$$

where we used the Jacobi triple identity,

$$\sum_{n=-\infty}^{+\infty} z^n q^{n^2} = \prod_{n=0}^{\infty} \left(1-q^{2n+2}\right) \left(1+zq^{2n+1}\right) \left(1+z^{-1}q^{2n+1}\right). \qquad (22)$$

The infinite product on the RHS of (18) then becomes

$$q^{-\frac{49}{120}} \prod_{n=0}^{\infty} \left(1+q^{5n+2}\right) \left(1+q^{5n+3}\right) \left(1-q^{5n+2}\right) \left(1-q^{5n+3}\right)$$
$$= q^{-4+\frac{5}{24}} \prod_{n=0}^{\infty} \left(1-q^{15n+15}\right)^{-6} \prod_{r=0}^{2} \left[\theta^2_{9+10r,30}(q) - \theta^2_{21-10r,30}(q)\right], \qquad (23)$$

while the LHS is given by,

$$\prod_{n=0}^{\infty} \left(1-q^{n+1}\right) = q^{-4+\frac{5}{24}} \prod_{n=0}^{\infty} \left(1-q^{15n+15}\right)^{-6} \prod_{r=0}^{6} \left[\theta_{2r+1,30}(q) - \theta_{29-2r,30}(q)\right]. \qquad (24)$$

The identity to prove then reduces to,

$$\prod_{r=0}^{2} \left[\theta_{7+10r,30}(q) - \theta_{23-10r,30}(q)\right] \left[\theta_{5,30}(q) - \theta_{25,30}(q)\right] =$$
$$\prod_{r=0}^{2} \left[\theta_{9+10r,30}(q) + \theta_{21-10r,30}(q)\right] \left[\theta_{7,30}(q) + \theta_{23,30}(q) - \theta_{13,30}(q) - \theta_{17,30}(q)\right], \qquad (25)$$

which is equivalent to,

$$\left[\theta_{1,15}(q)\,\theta_{10,15}(q) + \theta_{5,15}(q)\,\theta_{14,15}(q)\right] \left[\theta_{3,15}(q)\,\theta_{4,15}(q) + \theta_{11,15}(q)\,\theta_{12,15}(q)\right] =$$
$$\left[\theta_{1,15}(q)\,\theta_{8,15}(q) + \theta_{7,15}(q)\,\theta_{14,15}(q)\right] \left[\theta_{5,15}(q)\,\theta_{6,15}(q) + \theta_{9,15}(q)\,\theta_{10,15}(q)\right], \qquad (26)$$

as can be seen by using the product of two theta functions at level 30, according to the rule,

$$\theta_{m,k}(q)\,\theta_{m',k'}(q) = \sum_{\ell=1}^{k+k'} \theta_{mk'-m'k+2\ell kk',kk'(k+k')}(q)\,\theta_{m+m'+2\ell k,k+k'}(q), \qquad (27)$$

as well as the standard properties,

$$\theta_{m,60}(q) = \sum_{\ell=0}^{29} \theta_{30m+3600\ell,54000}(q), \qquad (28)$$

and,

$$\theta_{m,15}(q) = \theta_{2m,60}(q) + \theta_{60-2m,60}(q). \qquad (29)$$

This last equality (26) can be derived by considering the product,

$$[\theta_{9,30} + \theta_{21,30}]\ [\theta_{11,30} + \theta_{19,30}]\ [\theta_{7,30} + \theta_{23,30}]\ [\theta_{1,30} + \theta_{29,30}]\,, \tag{30}$$

and using (27), (28) and (29) by pairing the first and second factors together (and the third and fourth factors) and then by pairing the first and third factors together (and the second and fourth factors). This completes the proof of (17).

4 Conclusion

Two sets of identities between unitary minimal Virasoro characters at level $m = 3, 4, 5$ have been presented and the proof of one of them explicitly given. The existence of these non linear identities for low levels is quite remarkable. Whether these identities are exceptional or not remains to be investigated.

References

[1] Cappelli A., Itzykson C., Zuber J.-B., Nucl. Phys. **B280** 445 (1987).

[2] Friedan D., Qiu Z., Shenker S., Phys. Rev. Lett. **52** 1575 (1984).

[3] Goddard P., Kent A., Olive D., Commun. Math. Phys. **103** 105 (1986).

[4] Nam S., Phys. Lett. **B172** 323 (1986).

[5] Rocha-Caridi, A., In: Vertex operators in mathematics and physics. Lepowsky, J. et al.(eds.). MSRI publications No. 3, p. 451. Berlin, Heidelberg, New York: Springer 1984.

[6] Taormina A., Commun. Math. Phys. **165** 69 (1994).

Finite Deformations of CFT and Space-Time Geometry

Gregory Pelts*

Department of Physics, The Rockefeller University, New York, NY 10021-6399

Memory of Victor Nikolaevich Popov

Abstract. We demonstrate in detail how the space of 2D quantum field theories can be parameterized by off-shell states of free closed string. The dynamic equation corresponding to the condition of conformal invariance includes an infinite number of higher order terms, and we give an explicit procedure for their calculation. The symmetries corresponding to equivalence relations of CFT are described. In this framework we show how to perform nonperturbative analysis in the low-energy limit and prove that it corresponds to Brans-Dicke theory of gravity interacting with a skewsymmetric tensor field.

Key words: string, CFT, deformation, Brans-Dicke

1. Introduction

Classical closed string states are believed to be associated with quantum conformal field theories in two dimensions (CFT), which are usually defined as theories of the single string moving in some nontrivial space-time background. The condition of anomaly cancellation leads to the so-called β-function equation on the background fields. The main advantage of this approach is its more or less explicit connection to space-time geometry, and the main drawback is that it usually focuses only on massless fields. Treatment of massive fields is problematic and, therefore, characterization of dynamical degrees of freedom is obscure. Symmetries also are not explicit because classically equivalent CFT may correspond to inequivalent quantum theories. Approaches [1-5] based on the operator formalism [6] encounter problems dealing with ambiguity of the vertex operator commutator due to contact singularity of their T-product. As we will see, this ambiguity is principal as it actually makes the string theory nonlinear.

2. Vertex operators

We will consider CFT as a family of amplitudes $\langle 0\rangle_\Sigma$ assigned to Riemann surfaces Σ and obeying sewing property (see details in [11]). It is similar to the Segal's definition [7]. In this formalism vertex operators $\Psi(z_0)$ inserted at the point z_0 can be defined as a family of states $\langle\Psi(z_0)\rangle_D \in \mathcal{H}^{\partial D}$ associated with disc like environments D of z_0 obeying the condition

$$\langle\Psi(z_0)\rangle_{D_2} = Sp_{\partial D_1}\langle 0\rangle_{D_2\setminus D_1} \otimes \langle\Psi(z_0)\rangle_{D_1} \quad (D_1 \subset D_2).$$

Here Sp_Γ denotes contraction of amplitudes corresponding to sewing of Riemann surfaces along the contour Γ. The T-product of such vertex operators can be defines as

$$\langle\Psi_0(z_0)\cdots\Psi_n(z_n)\rangle_\Sigma = Sp_{\partial\Sigma_{ext}} \bigotimes_{i=0}^{n} \langle\Psi_i(z_i)\rangle_{D_i} \otimes \langle 0\rangle_{\Sigma_{ext}}. \tag{1}$$

Here D_i are nonintersecting environments of z_i and Σ_{ext} is their complement in Σ.

The Virasoro algebra does not have a bounded natural representation in $\mathcal{H}^{\partial D}$. The conformal transformations deform the boundary of the disk and, therefore, corresponding to them linear operators are not automorphisms. However, we can define such a representation in the space of vertex operators, which is independent of the position of the boundary.

3. Infinitesimal deformation of CFT

Infinitesimal deformations of amplitudes can be parameterized by (1,1)-primary fields

$$\langle 0\rangle_\Sigma \longrightarrow \langle 0\rangle_\Sigma + \delta\langle 0\rangle_\Sigma, \quad \delta\langle 0\rangle_\Sigma = \frac{1}{\pi}\int_\Sigma \langle\Psi(z)\rangle_\Sigma\, d^2z.$$

Formally we can parameterize vertex operators of the deformed theory by vertex operators of the initial theory as follows:

$$\langle\Upsilon\rangle_D \longrightarrow \langle\Upsilon\rangle_D + \delta\langle\Upsilon\rangle_D, \; \delta\langle\Upsilon(z_0)\rangle_D = \frac{1}{\pi}\int_D \langle\Psi(z)\Upsilon(z_0)\rangle_D\, d^2z.$$

However, in general, the integral here may be divergent because of the contact singularity of the T-product. The simple cutoff regularization

$$\delta_R\langle\Upsilon(z_0)\rangle_\Sigma = \frac{1}{\pi}\int_{\Sigma\setminus D_{z_0,R}} \langle\Psi(z)\Upsilon(z_0)\rangle_\Sigma\, d^2z, \;\; D_{z_0,R} = \{z\in\Sigma,\, |z-z_0|\le R\}$$

will violate sewing properties for small disks. Instead we will make an overage of such regularizations over infinitely small cutoff radiuses

$$\delta\langle\Upsilon(z_0)\rangle_\Sigma = \int_0^\infty \delta_r\langle\Upsilon(z_0)\rangle_\Sigma d\mu(r). \tag{2}$$

Here $d\mu$ is a generalized measure in $\mathbf{R}_+$ having support in 0 and integrable in a product with all the functions having a finite degree singularity at $r = 0$. Such a measure exists and is fully described by the function

$$\Lambda(\alpha) = \int_0^\infty r^{2\alpha}\mu(r)\, dr \quad (\alpha\in\mathbf{R}),$$

satisfying

$$\Lambda(0) = 1, \quad \Lambda(\alpha) = 0 \quad (\alpha > A)$$

for some positive A.

4. Residue-like operations

We will call local linear operators from the space of functions on Σ^{N+1} having diagonal singularities to the space of functions on Σ residue-like operators of rank N and denote them as

$$G(z_0) = \mathcal{R}_{z_N=\cdots=z_0} F(z_1, \ldots, z_0) \quad \text{or} \quad G(z_0) = \mathcal{R}_{\bar{z}_N=\cdots=\bar{z}_0} F(z_1, \ldots, z_0).$$

Using T-product (1) we can define representation of residue-like operations by multilinear products in the space of vertex operator functions:

$$\begin{aligned} \{\Upsilon_i\}_{i=1}^N \longrightarrow \Upsilon &= \mathcal{R}_{z_N=\cdots=z_0} \Upsilon_0 \ldots \Upsilon_N, \\ \langle \Upsilon(z_0) \rangle_\Sigma &\stackrel{\text{def}}{=} \mathcal{R}_{z_N=\cdots=z_0} \langle \Upsilon_0 \ldots \Upsilon_N \rangle_\Sigma. \end{aligned}$$

The deformation (2) induce the following deformation of this representation:

$$\delta \mathcal{R}_{z_n=\cdots=z_0} \Upsilon_0(z_0) \cdots \Upsilon_n(z_n) = \mathcal{R}_{z \dot{=} z_n=\cdots=z_0} \Upsilon_0(z_0) \cdots \Upsilon_n(z_n) \Psi(z). \qquad (3)$$

Here $\mathcal{R}_{z \dot{=} z_n=\cdots=z_0}$ is a next rank residue-like operation defined as follows:

$$\mathcal{R}_{z \dot{=} z_n=\cdots=z_0} F = \mathcal{R}_{z_n=\cdots=z_0} \frac{1}{\pi} \int_\Sigma F \, d^2 z - \frac{1}{\pi} \int_\Sigma \mathcal{R}_{z_n=\cdots=z_0} F \, d^2 z. \qquad (4)$$

It is, indeed, a residue-like operation, because the right part of (4) does not depend on the area of integration as far as it includes z_0. We will call this operation a successor of $\mathcal{R}_{z_n=\cdots=z_0}$. A successor of antiholomorphic derivative can be shown to be

$$\partial_{\bar{z} \dot{=} \bar{z}_0} F = -\mathrm{Res}_{z=z_0} F.$$

Here $\mathrm{Res}_{z=z_i}$ is a generalized residue operation defined for nonholomorphic functions as follows:

$$\mathrm{Res}_{z=z_0} \frac{(z-z_0)^k}{|z-z_0|^{2\alpha}} = \frac{\Lambda(k-\alpha)}{k!(k+1)!} \delta_{k,-1} \quad (\alpha \in \mathbf{R}, k \in \mathbf{Z}).$$

5. Finite deformations

Let deformed amplitudes be defined by the formula

$$\langle 0 \rangle_\Sigma^\Psi = \left\langle \exp \frac{1}{\pi} \int_\Sigma \Psi \, d^2 z \right\rangle_\Sigma .$$

Here Ψ is some vertex operator function (not necessarily primary), and the contact divergences are regularized by the method (2). Then the sewing property is automatically satisfied, and only the condition of conformal invariance remains to be implemented. Hereafter we mark all the deformed objects with the superscript symbol of the vertex operator function parameterizing the deformation. We will use the following formula to parameterize vertex operators of the deformed theory:

$$\langle \Upsilon(z_0) \rangle_\Sigma^\Psi = \left\langle \Upsilon(z_0) \exp \frac{1}{\pi} \int_\Sigma \Psi(z) \, d^2 z \right\rangle_\Sigma .$$

The energy-momentum tensors for the family of deformed theories parameterize by scaled vertex operator function $\tau\Psi$ ($\tau \in \mathbf{R}$) can be shown to obey the following differential equation

$$\begin{aligned}\frac{\partial}{\partial\tau}T^{\tau\Psi}_{zz}(z) &= \mathcal{B}^{\tau\Psi}_{z_1=z}\Psi(z_1)T^{\tau\Psi}_{zz}(z) + \mathcal{A}^{\tau\Psi}_{\bar z_1=\bar z}\Psi(z_1)T^{\tau\Psi}_{z\bar z}(z)\\ \frac{\partial}{\partial\tau}T^{\tau\Psi}_{z\bar z}(z) &= \mathcal{A}^{\tau\Psi}_{z_1=z}\Psi(z_1)T^{\tau\Psi}_{zz}(z) + \mathcal{B}^{\tau\Psi}_{\bar z_1=\bar z}\Psi(z_1)T^{\tau\Psi}_{z\bar z}(z) + \Psi(z).\end{aligned} \tag{5}$$

Here $\mathcal{A}_{z_1=z}$, $\mathcal{B}_{z_1=z}$ are residue-like operations satisfying

$$\partial_z\mathcal{A}_{z_1=z} + \partial_{\bar z}\mathcal{B}_{z_1=z} = \mathrm{Res}_{z_1=z} + \mathrm{Res}_{z=z_1}.$$

Differentiating (5) with respect to τ and applying (3) we can recurrently calculate all the higher derivatives of the energy-momentum tensor and then substitute them to the Taylor expansion. Thus we have a perturbative formula for T^{Ψ} with higher order terms expressed through residue-like operations $\mathrm{Res}_{z=z_0}$, $\mathcal{A}_{z=z_0}$, $\mathcal{B}_{z=z_0}$ and their successors (4). Details on calculation of these operations can be found in [11]. The following transformation of energy-momentum tensor does not effect translation operators:

$$\begin{aligned} T^{\Psi,\Phi}_{zz} &= T^{\Psi}_{zz} - \partial^{\Psi}_{z}\Phi_z, & T^{\Psi,\Phi}_{z\bar z} &= T^{\Psi}_{z\bar z} + \partial^{\Psi}_{\bar z}\Phi_z,\\ T^{\Psi,\Phi}_{\bar z\bar z} &= T^{\Psi}_{\bar z\bar z} - \partial^{\Psi}_{\bar z}\Phi_{\bar z}, & T^{\Psi,\Phi}_{\bar z z} &= T^{\Psi}_{\bar z z} + \partial^{\Psi}_{z}\Phi_{\bar z}.\end{aligned}$$

If the theory is conformally symmetrical, there exists Φ_z, $\Phi_{\bar z}$ trivializing the contradiagonal components of $T^{\Psi,\Phi}$. Then the diagonal components of $T^{\Psi,\Phi}$, will be (anti)holomorphic. They can be used for calculation of the deformed Virasoro representation.

6. Symmetries

The following transformations of Ψ can be shown not to effect equivalence classes of theories:

$$\delta_\xi\Psi(z) = \partial_{\bar z}\xi_z(z) + \mathrm{Res}_{z_1=z}\underbrace{\xi_z(z_1)\Psi(z)}_{asym} + z \leftrightarrow \bar z + O(\Psi^2). \tag{6}$$

They are parameterized by the pair of vertex operator functions $\xi = (\xi_z, \xi_{\bar z})$. The corresponding *covariant* transformation of vertex operators are

$$\hat\xi\Upsilon(z_0) = \mathrm{Res}_{\bar z=\bar z_0}\xi_z(z)\Upsilon(z_0) + \frac{1}{2}\mathrm{Res}_{z_2\dot= z_1=z_0}\Psi(z_2)\xi(z_1)\Upsilon(z_0) + z \leftrightarrow \bar z + O(\Psi^2).$$

The symmetry transformation of Φ is defined by the requirement for the energy-momentum tensor $T^{\Psi,\Phi}$ to transform covariantly. Note that the commutator of such symmetries depend on Ψ and regularization parameters.

7. Linear approximation

In the linear approximations the energy-momentum tensor and symmetries can be shown to be

$$T^{\Psi,\Phi}_{z\bar{z}} = \mathcal{O}\Psi - \bar{L}_{-1}\Phi_z, \quad T^{\Psi,\Phi}_{\bar{z}z} = \bar{\mathcal{O}}\Psi - L_{-1}\Phi_{\bar{z}},$$
$$T^{\Psi,\Phi}_{zz} = T_{zz} + L_{-1}\Phi_z, \quad T^{\Psi,\Phi}_{\bar{z}\bar{z}} = T_{\bar{z}\bar{z}} + \bar{L}_{-1}\Phi_{\bar{z}},$$
$$\delta_\xi\Psi = -\bar{L}_1\xi_z - L_1\xi_{\bar{z}}, \quad \delta_\xi\Phi_z = \mathcal{O}\xi_z, \quad \delta_\xi\Phi_{\bar{z}} = \bar{\mathcal{O}}\xi_{\bar{z}}.$$

Here

$$\mathcal{O} = \mathbf{1} + \sum_{j=0}^{\infty}\frac{(L_{-1})^j L_j}{(j+1)!}, \quad \bar{\mathcal{O}} = \mathbf{1} + \sum_{j=0}^{\infty}\frac{(\bar{L}_{-1})^j \bar{L}_j}{(j+1)!}.$$

Then the equations

$$T^{\Psi,\Phi}_{z\bar{z}} = T^{\Psi,\Phi}_{\bar{z}z} = 0$$

are satisfied if Φ is trivial and Ψ is a primary field. It corresponds to the following deformations of the Virasoro operators

$$L^{\Psi}_k = L_k + \frac{1}{2\pi i}\oint_\Gamma \Psi(z-z_0)^k d\bar{z}, \quad \bar{L}^{\Psi}_k = \bar{L}_k + \frac{1}{2\pi i}\oint_\Gamma \Psi(\bar{z}-\bar{z}_0)^k dz,$$

which is equivalent to the deformations proposed in [1]. Some deformations corresponding to nonprimary fields were first discribed in [8-10] in the low-energy limit. This relaxation of the equations of motion is compensated by the symmetries and does not create additional physical degrees of freedom.

8. Low-energy limit

Let us consider deformations corresponding to the vertex operator function $\Psi = H_{\nu\mu}\partial X^\nu\bar{\partial}X^\mu(z')$, where $H_{\nu\mu}$ is a Hermitian matrix with slowly-varying coefficients. Then the coordinate vertex operators can be shown to have the following properties:

$$\langle X^\nu(z')X^\mu(z)\rangle^{\Psi}_{\Sigma} \approx -2\langle :g^{\nu\mu}(z):\rangle^{\Psi}_{\Sigma}\ln|z'-z|,$$
$$\bar{\partial}^\Psi\partial^\Psi :\psi: \approx :(\psi_{;\nu\mu} + i\psi^{;\eta}d\omega_{\eta\nu\mu})\,\partial^\Psi X^\nu\bar{\partial}^\Psi X^\mu:.$$

The contravariant metric $g^{\nu\mu}$ and skewsymmetric tensor $\omega_{\nu\mu}$ here are equal to

$$g = U(1)U^T(1), \quad \omega = \Im\left(\int_0^1 (U^{-1})^T\frac{\partial}{\partial\tau}U^{-1}\,d\tau\right),$$

where

$$U(\tau) = \cosh\left(\tau\sqrt{HH^T}\right) + H^T\frac{\sinh\left(\tau\sqrt{HH^T}\right)}{\sqrt{HH^T}}.$$

Local space-time symmetries

$$\delta g_{\nu\mu} = \varepsilon_{\nu;\mu} + \varepsilon_{\mu;\nu}, \quad \delta\omega_{\nu\mu} = \varepsilon^\eta\omega_{\nu\mu;\eta} + \varepsilon^{\nu;\eta}\omega_{\eta\mu} + \varepsilon^{\mu;\eta}\omega_{\nu\eta} + d\varsigma_{\nu\mu}$$

are a particular case of more general symmetries (6) with the following choice of parameters:

$$\xi_z =:(\varepsilon_\nu + i\varsigma_\nu)\partial X^\nu :, \quad \xi_{\bar z} =:(\varepsilon_\mu - i\varsigma_\mu)\bar\partial X^\mu : .$$

The contradiagonal components of the energy-momentum tensor can be shown to be

$$T^{\Psi}_{z\bar z} \approx T^{\Psi}_{\bar z z} \approx -\frac{1}{2} :(R_{\nu\mu} - d\omega_\nu{}^{\sigma\rho} d\omega_{\mu\sigma\rho} + i d\omega_{\eta\nu\mu}{}^{;\eta})\, \partial^\Psi X^\nu \bar\partial^\Psi X^\mu : .$$

Putting here $\Phi_z = \partial^\Psi_z :\phi:$, $\Phi_{\bar z} = \partial^\Psi_{\bar z} :\phi:$ we will come to the following equations of motion:

$$R_{\nu\mu} = d\omega_\nu{}^{\sigma\rho} d\omega_{\mu\sigma\rho} + 2\phi_{;\nu\mu}, \quad d\omega_{\eta\nu\mu}{}^{;\eta} = 2\phi^{;\eta} d\omega_{\eta\nu\mu}.$$

As a consequence of this equations it can be shown that

$$2\,\Box\,\phi + 4\phi^{;\nu}\phi_{;\nu} = m^2 - \frac{2}{3} d\omega^{\nu\sigma\rho} d\omega_{\nu\sigma\rho}.$$

The parameter m here is the topological constant of the theory which can be interpreted as a dilaton mass It is responsible for deformation of central charge

$$c = D + \frac{1}{2}m^2.$$

Acknowledgement. I am very grateful to Mark Evans for inspiring me to develop the *deformation* approach, for discussions and for many important suggestions. It is also a pleasure to thank Toni Weil for significant editing help.

References

1. M. Evans and B. Ovrut, Phys. Rev. D **41**, 3149 (1990).
2. A. Sen, Nucl. Phys. B **345**, 551 (1990).
3. M. Campbell, P. Nelson and E. Wong, Int. J. Mod. Phys. A **6**, 4909 (1991).
4. B. Zwiebach, Nucl. Phys. B **390**, 33 (1993), hep-th/9206084.
5. K. Ranganathan, H. Sonoda and B. Zwiebach, Report FERMILAB-PUB-MIT-CTP-2193, April 1993, hep-th/9304053.
6. Alvarez-Gaume, C. Gomez, G. Moore and C.Vafa, Nucl. Phys B **303**, 445 (1988).
7. G.B. Segal, in *Proceedings of the XVI International Conference on Differential Geometrical Methods in Theoretical Physics*, Como, Italy, 1987, edited by K. Bleuler and M. Werner, (Univ. di Como, Como, Italy, 1987), pp. 165-171.
8. M. Evans and I. Giannakis, Phys. Rev. D **44**, 2467 (1991).
9. M. Evans and I. Giannakis, in *Proceedings of the XX International Conference on Differential Geometric Methods in Theoretical Physics*, edited by S. Catto and A. Rocha, New York, 1991 (City College, New York, 1991), pp. 759-768, hep-th/9109055.
10. Mark Evans, Ioannis Giannakis and D.V. Nanopoulos, Report RU93-8, CTP-TAMU-2/94, CERN-TH.7022/93, hep-th/9401075.
11. G. Pelts, *Finite deformations of CFT beyond the first order and space time geometry*, Phys. Rev. D **3** (to be published), hep-th/9406022.

Non–Standard Fermion Propagators from Conformal Field Theory

Rainer Dick

Department of Physics, University of Munich
Theresienstr. 37, 80333 Munich, Germany

and

School of Natural Sciences, Institute for Advanced Study
Olden Lane, Princeton, NJ 08540, USA

1 Introduction

The quest for a four–dimensional notion of analyticity and the related problem to define four–dimensional analogues of two–dimensional conformal field theories is a subject of much interest and under intense study since many years. Already in 1956 Feza Gürsey employed the quaternionic formulation of the Dirac equation to derive a conformally invariant nonlinear spinor equation [1]. He later returned to the topic of quaternionic analyticity several times yielding important insights and results, see e.g. [2]. Finally, his life–long fascination in this topic and the harmonic space approach developed by the Dubna group merged in a remarkable and beautiful recent paper with Mark Evans and Victor Ogievetsky [3]. In this paper the relevance and applicability of quaternionic analyticity in the framework of self–dual theories is worked out very clearly.

Notions of quaternionic analyticity arise in a similar manner also in the twistor approach to space–time [4].

Here I would like to introduce another approach to analyticity in 3+1 dimensions: I would like to point out that *left or right handed massless spinors in 3+1 dimensions can be interpreted as half–differentials on spheres in momentum space*. This implies the possibility to formulate covariant phase space constraints on spinors of definite helicity in terms of (anti–)meromorphy constraints. More specifically, the entries of a spinor of negative helicity are shown to yield local representations $\psi(z,\bar{z},E)$ of a primary field of weight $(\frac{1}{2},0)$, where $z(p)$ denotes a stereographic coordinate in momentum space, see Eq. (6) below.

The Weyl equation then appears as a particular consequence of the transformation behavior under holomorphic reparametrizations:

$$\psi'(z',\bar{z}',E') = \psi(z,\bar{z},E)\left(\frac{\partial z'}{\partial z}\right)^{-\frac{1}{2}} \tag{1}$$

Covariance of the construction follows because Lorentz transformations induce via $SL(2,\mathbf{C})$ holomorphic transformations of spheres in momentum space, and the resulting transformation behavior of left handed spinors complies with the corresponding transformations of half–differentials. The construction implies in particular, that left handed spinors can be subjected to covariant constraints

$$\frac{\partial\psi}{\partial\bar{z}} = 0 \tag{2}$$

Another source of motivation for the present work besides the formulation of covariant analyticity constraints is due to applications to low energy QCD: For an explanation of this note that the isomorphy between chiral Weyl spinors on the one hand and half–differentials on the other hand offers the possibility to write correlators of massless fermions as a sum of correlators of primary fields with a factorized transformation behavior under Lorentz transformations. Exploiting this observation, we find that the correlator is determined up to 2 functions f_1 and f_2 which depend on single, but different arguments [5]:

$$\langle\Psi(\vec{p})\overline{\Psi}(\vec{p}')\rangle = \tag{3}$$

$$\begin{pmatrix} 0 & 1 \\ 0 & 0 \end{pmatrix} \otimes \begin{pmatrix} \bar{z}z' & \bar{z} \\ z' & 1 \end{pmatrix} \langle\phi(\vec{p})\phi^+(\vec{p}')\rangle + \begin{pmatrix} 0 & 0 \\ 1 & 0 \end{pmatrix} \otimes \begin{pmatrix} 1 & -\bar{z}' \\ -z & z\bar{z}' \end{pmatrix} \langle\psi(\vec{p})\psi^+(\vec{p}')\rangle$$

$$+ \begin{pmatrix} 1 & 0 \\ 0 & 0 \end{pmatrix} \otimes \begin{pmatrix} \bar{z} & -\bar{z}\bar{z}' \\ 1 & -\bar{z}' \end{pmatrix} \langle\phi(\vec{p})\psi^+(\vec{p}')\rangle + \begin{pmatrix} 0 & 0 \\ 0 & 1 \end{pmatrix} \otimes \begin{pmatrix} z' & 1 \\ -zz' & -z \end{pmatrix} \langle\psi(\vec{p})\phi^+(\vec{p}')\rangle$$

$$\langle\psi(\vec{p}_1)\psi^+(\vec{p}_2)\rangle = \langle\phi(\vec{p}_2)\phi^+(\vec{p}_1)\rangle = f_1\left(\frac{|\vec{p}_1|}{|\vec{p}_2|}\right) \frac{1 + z_1\bar{z}_2}{\sqrt{|\vec{p}_1||\vec{p}_2|}} \delta_{z\bar{z}}(z_1 - z_2) \tag{4}$$

$$\langle\psi(\vec{p}_1)\phi^+(\vec{p}_2)\rangle = \overline{\langle\phi(\vec{p}_2)\psi^+(\vec{p}_1)\rangle} = \frac{1}{z_1 - z_2} f_2\left(|\vec{p}_1||\vec{p}_2| \frac{(z_1 - z_2)(\bar{z}_1 - \bar{z}_2)}{(1 + z_1\bar{z}_1)(1 + z_2\bar{z}_2)}\right) \tag{5}$$

Applications of this result to low energy QCD arise from expectations that chiral symmetry is broken even in the massless limit of QCD. Note the consistency of Eq. (3–5): Since our result provides a Lorentz covariant *massless* propagator, those parts of it which break chiral symmetry necessarily must also break translational invariance. This is in agreement with Eq. (5), since the right hand side of this equation cannot accomodate for a δ–function in external momenta, and complies with the onset of confinement in the low energy domain. The f_1–terms in turn preserve chiral symmetry: They do not contribute to a chiral condensate and anticommute with γ_5. Consistency of the result in this sector is expressed by the fact that these terms contain a δ–function which restricts the correlator to parallel momenta.

A short outline on the derivation of these results is given in the next section.

2 Massless Fermions and Half–Differentials

For the present purposes, it is sufficient to define half–differentials by their transformation behavior under holomorphic transformations. A half–differential of weight $(\frac{1}{2}, 0)$ then transforms according to (1), while a half–differential of weight $(0, \frac{1}{2})$ transforms with the complex conjugate factor. A fully covariant definition of primary fields including half-differentials and a discussion of topological restrictions on fractional weights was given in Chapter 2 of [6], while their exact relations to 2D spinors have been clarified in [7].

For convenience, I employ the Weyl representation for Dirac matrices. To clarify the relation between spinors in 3+1 dimensions and half-differentials, we introduce stereographic coordinates in momentum space:

$$z = \frac{p_+}{|\vec{p}| - p_3} \tag{6}$$

$$\tilde{z} = -\frac{1}{z}$$

According to Eq. (1) the relation between local representations $\psi(z, \bar{z}, |\vec{p}|)$ and $\psi(\tilde{z}, \bar{\tilde{z}}, |\vec{p}|)$ of a half-differential of weight $(\frac{1}{2}, 0)$ is

$$\psi(\tilde{z}, \bar{\tilde{z}}, |\vec{p}|) = -z\psi(z, \bar{z}, |\vec{p}|) \tag{7}$$

where the sign ambiguity has been resolved in such a way to avoid minus signs in the construction of the Weyl spinors below. Insertion of (6) demonstrates that (7) is equivalent to the Weyl equation for a massless spinor of negative helicity:

$$(|\vec{p}| + \vec{p}\cdot\vec{\sigma}) \begin{pmatrix} \psi(z, \bar{z}, |\vec{p}|) \\ \psi(\tilde{z}, \bar{\tilde{z}}, |\vec{p}|) \end{pmatrix} = 0 \tag{8}$$

Similarly, the relation between local representations of a primary field of weight $(0, \frac{1}{2})$

$$\phi(\tilde{z}, \bar{\tilde{z}}, |\vec{p}|) = \bar{z}\phi(z, \bar{z}, |\vec{p}|) \tag{9}$$

is the Weyl equation for a massless spinor of positive helicity:

$$(|\vec{p}| - \vec{p}\cdot\vec{\sigma}) \begin{pmatrix} \phi(\tilde{z}, \bar{\tilde{z}}, |\vec{p}|) \\ \phi(z, \bar{z}, |\vec{p}|) \end{pmatrix} = 0 \tag{10}$$

To complete the proof that local representations of half-differentials create Weyl spinors as indicated in (8, 10), it remains to demonstrate that these objects exhibit a spinorial transformation behavior under the full Lorentz group: Under parity or time reversal $z(\vec{p})$ goes to $-\bar{z}(\vec{p})^{-1}$ and thus half-differentials of weight $(\frac{1}{2}, 0)$ become half-differentials of weight $(0, \frac{1}{2})$ and vice versa. Under proper orthochronous Lorentz transformations $\Lambda(\omega) = \exp(\frac{1}{2}\omega^{\mu\nu}L_{\mu\nu})$, with ω the usual set of rotation and boost parameters, $z(\vec{p})$ goes to

$$z'(\vec{p}\,') = \overline{U} \circ z(\vec{p}) = \frac{\bar{a}z + \bar{b}}{\bar{c}z + \bar{d}} \tag{11}$$

if $E = |\vec{p}|$. For the corresponding results in the negative energy case, see [5]. Here U is the positive chirality spin representation of Λ:

$$U(\omega) = \exp(\frac{1}{2}\omega^{\mu\nu}\sigma_{\mu\nu}) = \begin{pmatrix} a & b \\ c & d \end{pmatrix} \in SL(2, C)$$

A detailed proof of the transformation law (11) can be found in Ref. [5][1].

A half-differential ϕ of weight $(0, \frac{1}{2})$ then transforms according to (1) into

$$\phi'(z', \bar{z}', |\vec{p}'|) = (c\bar{z} + d)\phi(z, \bar{z}, |\vec{p}|)$$

implying

$$\begin{pmatrix} \phi'(\bar{z}', \bar{\bar{z}}', |\vec{p}'|) \\ \phi'(z', \bar{z}', |\vec{p}'|) \end{pmatrix} = U \cdot \begin{pmatrix} \phi(\bar{z}, \bar{\bar{z}}, |\vec{p}|) \\ \phi(z, \bar{z}, |\vec{p}|) \end{pmatrix}$$

Thus a half-differential of weight $(0, \frac{1}{2})$ is equivalent to a spin-$\langle \frac{1}{2}, 0\rangle$-representation of the proper orthochronous Lorentz group $\mathcal{L}_+^\uparrow$. Similarly, it is proved that a half-differential of weight $(\frac{1}{2}, 0)$ is equivalent to a spin-$\langle 0, \frac{1}{2}\rangle$-representation of the proper orthochronous Lorentz group:

$$\begin{pmatrix} \psi'(z', \bar{z}', |\vec{p}'|) \\ \psi'(\bar{z}', \bar{\bar{z}}', |\vec{p}'|) \end{pmatrix} = U^{-1\dagger} \cdot \begin{pmatrix} \psi(z, \bar{z}, |\vec{p}|) \\ \psi(\bar{z}, \bar{\bar{z}}, |\vec{p}|) \end{pmatrix}$$

For the case $E = -|\vec{p}|$, see again [5]. Half-differentials thus yield Weyl spinors in Minkowski space and vice versa.

As a consequence, the expansion of a massless spinor in terms of helicity states can be written in terms of half-differentials:

$$\Psi(p) = \begin{pmatrix} 1 \\ 0 \end{pmatrix} \otimes \begin{pmatrix} \bar{z} \\ 1 \end{pmatrix} \phi(z, \bar{z}, |\vec{p}|) + \begin{pmatrix} 0 \\ 1 \end{pmatrix} \otimes \begin{pmatrix} 1 \\ -z \end{pmatrix} \psi(z, \bar{z}, |\vec{p}|) \qquad (12)$$

Eq. (12) yields a representation of the corresponding correlation functions in terms of primary fields, which was displayed already in Eq. (3). Therefore, the 2-point functions on the right hand side of (3) transform under a factorized representation of the Lorentz group. This makes this representation very convenient for the investigation of all correlations $\langle \Psi(p)\overline{\Psi}(p')\rangle$ which comply with Lorentz covariance.

While the behavior of the parameters $(z, \bar{z}, |\vec{p}|)$ under rotations is completely specified by (11), for boosts we also have to specify the behavior of $|\vec{p}|$. For a boost $\exp(-uL_{03})$ we have e.g.

$$\begin{aligned} z' &= \exp(-u)z \\ |\vec{p}'| &= \frac{|\vec{p}|}{z\bar{z}+1}(\exp(-u)z\bar{z} + \exp(u)) \end{aligned}$$

We may now determine the 3-dimensional correlators of primary fields appearing on the right hand side of (3) by methods similar to the methods employed to fix 2- and 3-point functions of primary fields in 2D conformal field theory: Choose a generating set of the symmetry group and construct the general solutions of the corresponding covariance conditions [5]. Lorentz covariance then

[1] The analog of Eq. (11) in configuration space is known since a long time. However, it seems to have escaped attention that this implies a covariant notion of analyticity in momentum space, and that expressing z as a ratio actually means to introduce half-differentials.

fixes the $(\frac{1}{2},0)\otimes(0,\frac{1}{2})$–differential $\langle\psi(\vec{p}_1)\psi^+(\vec{p}_2)\rangle$ up to a function $f_1(|\vec{p}_1|/|\vec{p}_2|)$:

$$\langle\psi(\vec{p}_1)\psi^+(\vec{p}_2)\rangle = f_1\left(\frac{|\vec{p}_1|}{|\vec{p}_2|}\right)\frac{1+z_1\bar{z}_2}{\sqrt{|\vec{p}_1||\vec{p}_2|}}\delta_{z\bar{z}}(z_1-z_2) \tag{13}$$

Similarly, the $(\frac{1}{2},0)\otimes(\frac{1}{2},0)$–differential $\langle\psi(\vec{p}_1)\phi^+(\vec{p}_2)\rangle$ takes the form

$$\langle\psi(\vec{p}_1)\phi^+(\vec{p}_2)\rangle = \frac{1}{z_1-z_2}f_2\left(|\vec{p}_1||\vec{p}_2|\frac{(z_1-z_2)(\bar{z}_1-\bar{z}_2)}{(1+z_1\bar{z}_1)(1+z_2\bar{z}_2)}\right) \tag{14}$$

while invariance under **C**, **P** or **T** implies

$$\langle\psi(\vec{p}_1)\phi^+(\vec{p}_2)\rangle = \overline{\langle\phi(\vec{p}_2)\psi^+(\vec{p}_1)\rangle}$$
$$\langle\psi(\vec{p}_1)\psi^+(\vec{p}_2)\rangle = \langle\phi(\vec{p}_2)\phi^+(\vec{p}_1)\rangle$$

thus establishing the result we were seeking.

The unperturbed result for the on–shell correlation

$$\langle\psi(\vec{p})\overline{\psi}(\vec{p}')\rangle = -2p\cdot\gamma|\vec{p}|\delta(\vec{p}-\vec{p}')$$

is recovered from Eqs. (3–5) for $f_1(x)=\delta(x-1)$, $f_2=0$.

Off–shell extensions of (3) can be inferred from the requirement to yield the same propagator in configuration space:

$$\begin{aligned} S(x,x') &= \frac{\Theta(t-t')}{(2\pi)^3}\int\frac{d^3\vec{p}}{2|\vec{p}|}\int\frac{d^3\vec{p}'}{2|\vec{p}'|}\exp(\mathrm{i}p\cdot x)\mathrm{i}\langle\Psi(\vec{p})\overline{\Psi}(\vec{p}')\rangle\exp(-\mathrm{i}p'\cdot x') \\ &- \frac{\Theta(t'-t)}{(2\pi)^3}\int\frac{d^3\vec{p}}{2|\vec{p}|}\int\frac{d^3\vec{p}'}{2|\vec{p}'|}\exp(-\mathrm{i}p\cdot x)\mathrm{i}\langle\Psi(\vec{p})\overline{\Psi}(\vec{p}')\rangle\exp(\mathrm{i}p'\cdot x') \\ &= \frac{1}{(2\pi)^4}\int d^4p\int d^4p'\exp(\mathrm{i}p\cdot x)S(p,p')\exp(-\mathrm{i}p'\cdot x') \end{aligned} \tag{15}$$

thus fixing the structure up to the 2 functions f_1, f_2.

References

[1] F. Gürsey, *Nuovo Cimento* **3** (1956) 988.

[2] F. Gürsey and W.X. Jiang, *J. Math. Phys.* **33** (1992) 682.

[3] M.J. Evans, F. Gürsey and V. Ogievetsky, *Phys. Rev.* **D47** (1993) 3496.

[4] R. Penrose and W. Rindler, *Spinors and Space–Time*, Vol. 2, Cambridge University Press, Cambridge 1986.

[5] R. Dick, *Half–Differentials and Fermion Propagators*, IASSNS–HEP–94/83, hep–th/9410099, to appear in *Rev. Math. Phys.*

[6] R. Dick, Thesis, Universität Hamburg 1990; *Fortschr. Phys.* **40** (1992) 519.

[7] H. Nicolai, *Nucl. Phys.* **B414** (1994) 299.

Manifestly Space-Time Conformally Invariant Null Strings

Jan Isberg

Department of Mathematics, King's College London,
Strand, London WC2R 2LS, UK

Abstract

We propose an action for the null string, where space-time conformal invariance is manifest, discuss it's gauge symmetries, and study the quantization of this model using covariant techniques. The presence of a BRST anomaly for every space-time dimensionality confirms previously noted problems of preserving space-time conformal symmetry of the null string in the quantum theory.

Lie Algebras and Exactly Solvable Problems in Quantum Mechanics

F. Iachello[1]

[1] Center for Theoretical Physics, Sloane Laboratory,
Yale University, New Haven, CT 06520-8120, USA

Abstract. A formulation of quantum mechanics in terms of Lie algebras is introduced. Exactly solvable problems in an arbitrary number of dimensions are classified and discussed briefly.

1. Introduction

Exactly solvable problems play an important role in quantum mechanics for two reasons: (i) in many situations the underlying dynamics is such that it can be well described by exactly solvable interactions; (ii) when this is not the case, exactly solvable problems provide a basis in which calculations can be done. The problem arises then of classifying all exactly solvable problems in an arbitrary number of dimensions, ν. Exactly solvable problems are particularly useful when the number of dimensions ν is large. An exactly solvable problem can be defined as that situation in which all observables of the quantum system (in particular its energy levels) are given explicitly in terms of the quantum numbers that characterize the states uniquely. Several examples (for ν small) are quoted in basic textbooks[1]. It turns out that all exactly solvable quantum mechanical problems can be found by studying the algebraic structure of certain Lie algebras. In this article, dedicated to the memory of Feza Gürsey, who, among many other things, contributed to the development of this subject, I will briefly discuss the relation between Lie algebras and exactly solvable problems.

In quantum mechanics, there are three types of spectra: discrete (bound-state), continuous (scattering) and band-like (band-structure). All three types will be discussed briefly, although most of the emphasis will be on discrete (finite and infinite dimensional) spectra. It is for this type of spectra that exactly solvable problems, or more generally Lie algebraic methods, have found their most useful application, especially in the study of nuclear[2] and molecular[3] spectra.

2. Algebraic Theory

Exactly solvable problems can be studied by making use of algebraic theory. Algebraic theory[4] is a map of a quantum mechanical problem onto an algebraic structure, $\mathcal{G}$. The logic scheme of algebraic theory is given in the following diagram:

$$\begin{array}{c}\text{Quantum mechanical system}\\ \Downarrow \\ \text{Algebraic Structure}\end{array}$$

$$\left\{\begin{array}{l}\text{Lie Algebras}\\ \text{Graded Lie Algebras}\\ \text{Infinite Dimensional (Kac-Moody) Algebras}\\ \text{q-deformed (Hopf) Algebras}\\ \cdots\end{array}\right.$$

$$\begin{array}{c}\Downarrow \\ \text{Observables}\end{array}$$

$$\left\{\begin{array}{l}\text{Energy Spectra}\\ \text{Transition Rates}\\ \cdots\end{array}\right.$$

$$\begin{array}{c}\Downarrow \\ \text{Experiment}\end{array}$$

All physical operators are written in terms of the elements of an algebra, $G_\alpha \epsilon \mathcal{G}$. Although, in general, one can take any arbitrary functional of the G_α's, I will consider in detail in this article only the case in which the quantum operators are quadratic polynomials in the G_α's. For example, the Hamiltonian operator is

$$H = E_o + \sum_\alpha \epsilon_\alpha G_\alpha + \sum_{\alpha\beta} u_{\alpha\beta} G_\alpha G_\beta \quad ; \qquad G_\alpha \epsilon \mathcal{G} \quad . \tag{1}$$

Thus H is in the envelopping algebra of $\mathcal{G}$.[1] The algebra $\mathcal{G}$ is called the spectrum generating algebra (SGA). In addition to specifying the algebra $\mathcal{G}$, one must also specify the space on which $\mathcal{G}$ acts. It has been suggested[4] that, for non-relativistic quantum mechanical problems with no spin in ν space dimensions, one can always take the Lie algebra $gl(\nu+1, C)$ as the spectrum generating algebra, and the totally symmetric representations $[k, \dot{0}]$ as the corresponding quantum mechanical space of states on which $gl(\nu+1, C)$ acts.[2] The introduction of a (complex) algebra related to the number of dimensions ν is obvious, since there are ν coordinates and momenta. The novel aspect is the embedding onto an algebra with one additional (complex) degree of freedom, which allows one to describe all states simultaneously. Discrete and band-structure states are described by the totally symmetric irreducible representation $[N, \dot{0}]$ of the real form $u(\nu+1)$ of $gl(\nu+1, C)$[4] while scattering states are described by the continuous representation $[k, \dot{0}]$ of the real non-compact form $u(\nu, 1)$[5]. Other embeddings have been suggested in the past for specific problems[6], such as $sp(2\nu, R)$ for the harmonic oscillator and $so(\nu+1, 2)$ for the Coulomb problem. The algebra $gl(\nu+1, C)$ has the advantage of covering all cases.

In a generic problem, in order to find the eigenvalues and eigenstates of H, one must diagonalize it in the appropriate basis. However, in some special cases, the Hamiltonian

[1] Another interesting class of problems is that in which the inverse of the Hamiltonian, H^{-1}, is a polynomial in the G_α's.

[2] In this article, lowercase letters will be used to denote algebras. The algebra $gl(\nu+1, C)$ is isomorphic to the complex extension $u(\nu+1, C)$, with real forms $u(\nu+1)$ and $u(\nu, 1)$. In the following sections, the letters C and R will be used only when needed. The totally symmetric representations of $gl(\nu+1, C)$ will be denoted by $[k, \dot{0}] \equiv [k, 0, 0, \cdots, 0]$ or simply by $[k]$.

H may not contain all elements of $\mathcal{G}$, but only those combinations which form invariant (Casimir) operators, $\mathcal{C}$, of $\mathcal{G}$ and of a subalgebra chain $\mathcal{G} \supset \mathcal{G}' \supset \mathcal{G}'' \supset \cdots$,

$$H = \alpha \mathcal{C}(\mathcal{G}) + \alpha' \mathcal{C}(\mathcal{G}') + \alpha'' \mathcal{C}(\mathcal{G}'') + \cdots \quad . \tag{2}$$

In these cases, the eigenvalue problem for H can be solved in explicit form,

$$E = \alpha < \mathcal{C}(\mathcal{G}) > + \alpha' < \mathcal{C}(\mathcal{G}') > + \alpha'' < \mathcal{C}(\mathcal{G}'') > + \cdots \quad , \tag{3}$$

where $< \mathcal{C}(\mathcal{G}) >$ denotes the expectation value of $\mathcal{C}(\mathcal{G})$ in the appropriate representation $\mathcal{G}$. Hence, these situations correspond to exactly solvable problems. (These situations are also called dynamic symmetries (DS).) A classification of all exactly solvable problems in ν dimensions can then be obtained from the classification of all subalgebra chains of $gl(\nu+1, C)$.

3. Structure of the Solutions. Discrete Spectra.

In this section, I will discuss the structure of solutions for discrete spectra in $\nu = 1, 3$ and 5 dimensions. The appropriate algebras here are $u(\nu+1)$.

$\underline{\nu = 1}$

The algebra $u(2)$ has 4 elements and two (trivial) Abelian subalgebras

$$u(2) \begin{array}{l} \nearrow \quad u(1) \qquad (I) \quad , \\ \\ \searrow \quad o(2) \qquad (II) \quad . \end{array} \tag{4}$$

The two subalgebras are obviously isomorphic: $u(1) \approx o(2)$.

Chain I

States in this chain are characterized by the quantum numbers

$$\left| \begin{array}{ccc} u(2) & \supset & u(1) \\ \downarrow & & \downarrow \\ N & & n \end{array} \right\rangle \quad . \tag{5}$$

where $[N] \equiv [N, 0]$ and $n = N, N-1, \cdots, 0$. The exactly solvable Hamiltonian for this chain is the trivial Hamiltonian

$$H^{(I)} = E_o + \epsilon C_1(u1) + \alpha C_2(u1) \quad , \tag{6}$$

where $C_1(\mathcal{G})$ and $C_2(\mathcal{G})$ denote linear and quadratic invariants of $\mathcal{G}$. The eigenvalues are

$$E^{(I)}(N, n) = E_o + \epsilon n + \alpha n^2 \quad . \tag{7}$$

Chain II

States in this chain are characterized by the quantum numbers

$$\left| \begin{array}{ccc} u(2) & \supset & o(2) \\ \downarrow & & \downarrow \\ N & & m \end{array} \right\rangle \quad . \tag{8}$$

where $[N] \equiv [N,0]$ and $m = \pm N, \pm(N-2), \cdots, \pm 1$ or 0 (N odd or even). The exactly solvable Hamiltonian for chain II is again trivially given by

$$H^{(II)} = E_o + AC_1(o2) + BC_2(o2) \quad , \tag{9}$$

with eigenvalues

$$E^{(II)}(N,m) = E_o + Am + Bm^2 \quad . \tag{10}$$

Both chains lead to the same type of spectra. In one dimension, because of the isomorphism $u(1) \approx o(2)$, all exactly solvable problems with quadratic Hamiltonians are isospectral, i.e. they have the same functional form for the bound state spectrum.

$\underline{\nu = 3}$

The algebra $u(4)$ has 16 elements. If no condition is imposed on the form of the Hamiltonian, $u(4)$ can be split into abelian $u(1)$'s and the solution is as above for each dimension. A more interesting (and important) situation is that in which H has rotational invariance. This implies that one should solve the branching problem with the condition that $o(3)$ be contained in $\mathcal{G}$. There are then only two solutions[3]:

$$u(4) \begin{array}{l} \nearrow \; u(3) \supset o(3) \supset o(2) \qquad (I) \quad , \\ \\ \searrow \; o(4) \supset o(3) \supset o(2) \qquad (II) \quad . \end{array} \tag{11}$$

Chain I

States in this chain are characterized by the quantum numbers

$$\left| \begin{array}{ccccccc} u(4) & \supset & u(3) & \supset & o(3) & \supset & o(2) \\ \downarrow & & \downarrow & & \downarrow & & \downarrow \\ N & & n & & L & & M \end{array} \right\rangle \quad . \tag{12}$$

where $[N] \equiv [N,0,0,0]$; $n \equiv [n,0,0] = N, N-1, \cdots, 0$; $L = n, n-2, \cdots, 1$ or 0 (n=odd or even), $-L \leq M \leq +L$. The exactly solvable problem for chain I is

$$H^{(I)} = E_o + \epsilon C_1(u3) + \alpha C_2(u3) + \beta C_2(o3) \quad . \tag{13}$$

[The Casimir operator of $o(2)$ is usually not added since it amounts to placing the system in an external electric or magnetic field.] The eigenvalues of (13) are

$$E^{(I)}(N,n,L,M) = E_o + \epsilon n + \alpha n(n+2) + \beta L(L+1) \quad . \tag{14}$$

Chain II

States in this chain are characterized by the quantum numbers

$$\left| \begin{array}{ccccccc} u(4) & \supset & o(4) & \supset & o(3) & \supset & o(2) \\ \downarrow & & \downarrow & & \downarrow & & \downarrow \\ N & & \omega & & L & & M \end{array} \right\rangle \quad . \tag{15}$$

where $[N] \equiv [N,0,0,0]$; $\omega \equiv (\omega,0) = N, N-2, \cdots, 1$ or 0 ($N =$ odd or even); $L = \omega, \omega-1, \cdots, 0$; $-L \leq M \leq +L$. The exactly solvable problem for chain II is

$$H^{(II)} = E_o + AC_2(o4) + BC_2(o3) \quad , \tag{16}$$

with eigenvalues

$$E^{(II)}(N,\omega,L,M) = E_o + A\omega(\omega+2) + BL(L+1) \quad . \tag{17}$$

$\nu = 5$.

The study of exactly solvable problems can be repeated in any number of dimensions. In $\nu = 5$ dimensions the spectrum generating algebra is $u(6)$ with 36 elements. If one imposes the condition of rotational invariance there are three solutions to the branching problem[2]:

$$u(6) \begin{array}{ll} \nearrow & u(5) \supset o(5) \supset o(3) \supset o(2) \quad , \quad (I) \quad , \\ \longrightarrow & su(3) \supset o(3) \supset o(2) \quad , \quad (II) \quad , \\ \searrow & o(6) \supset o(5) \supset o(3) \supset o(2) \quad , \quad (III) \quad . \end{array} \tag{18}$$

The corresponding exactly solvable problems can be written down as above and will not be quoted here. One can observe that the complexity of the problem of classifying all exactly solvable problems grows obviously with the number of dimensions from being trivial in $\nu = 1$ dimension, (the subalgebras of $u(2)$ are abelian and there is only one class of exactly solvable problems, up to equivalences), to being of moderate difficulty in $\nu = 5$ dimensions. The same comment applies to the generic algebraic theory which is of little interest when $\nu = 1$ but acquires more and more importance, as a tool to solve problems, as the number of dimensions increases.

4. More on the Structure of Solutions. Contractions.

In the preceding section, the general structure of algebraic theory and of exactly solvable problems in ν dimensions has been introduced. It is of interest to elaborate further on this structure, because it may appear, at first glance, that some problems known to be exactly solvable, are missing from the classification of Sect.3. I return therefore to the case of $\nu = 1$ and write the algebra $gl(2,C)$ in a slightly different way[7]. Consider the four dimensional Lie algebra $\mathcal{G}(a,b)$, composed of elements F_+, F_-, F_3 and F, with commutation relations [3]

$$\left\{ \begin{array}{lcl} [F_3, F_\pm] & = & \pm F_\pm \\ [F_+, F_-] & = & 2aF_3 - bF \\ [F_\pm, F] & = & 0 \\ [F_3, F] & = & 0 \end{array} \right. \tag{19}$$

and Casimir (invariant) operators

[3]In Ref.[7], a is replaced by a^2.

$$\begin{cases} \mathcal{C}_1 &= F \\ \mathcal{C}_2 &= F_+F_- + a\,F_3^2 - a\,F_3 - b\,F_3F \quad . \end{cases} \tag{20}$$

One can see that, as a, b take the values 0 or 1, one has the following algebras:

$$\begin{aligned} \mathcal{G}(1,0) &\approx so(3) \oplus u(1) \approx su(2) \oplus u(1) \quad , \\ \mathcal{G}(0,1) &\approx h(2) \quad , \\ \mathcal{G}(0,0) &\approx e(2) \oplus u(1) \quad . \end{aligned} \tag{21}$$

[Also $\mathcal{G}(-1,0) \approx so(2,1) \oplus u(1) \approx su(1,1) \oplus u(1)$.] In Eq.(21), $h(2)$ denotes the Heisenberg algebra, while $e(2)$ denotes the Euclidean algebra. These algebras can be obtained from $u(2)$ by the process of contraction. We thus see that, starting from $u(2)$, we can form the following diagram:

$$u(2) \begin{array}{l} \nearrow \; su(2) \oplus u(1) \\ \underset{c}{\rightarrow} \; h(2) \\ \underset{c}{\searrow} \; e(2) \oplus u(1) \end{array} \quad . \tag{22}$$

The arrow with a c denotes contractions. It turns out that the contraction to $e(2)$ leads to quantum mechanics for free particles, which is not of interest in the present context. However, the contraction to $h(2)$ leads to an interesting situation, because it can be achieved by letting the dimension of the representation N of $u(2)$ go to ∞. Thus $h(2)$ (and its representations) describes a situation with an infinite number of bound states, such as the harmonic oscillator. [Indeed a simple realization of $h(2)$ in terms of boson operators, $\tau^\dagger, \tau$, is:

$$h(2): \; \tau^\dagger, \; \tau, \; \tau^\dagger\tau, \; 1 \quad , \tag{23}$$

where 1 is a c-number. One can see that the four-dimensional Lie algebra (23) satisfies the commutation relations (19) with $a = 0$, $b = 1$.]

Thus, $u(\nu+1)$ describes both the usual situation in which the number of bound states is finite and, also, by the process of contraction (or $N \to \infty$), the situation in which the number of bound states is infinite. In ν dimensions this can be written as

$$u(\nu+1) \begin{array}{l} \nearrow \; su(\nu+1) \oplus u(1) \\ \\ \underset{c}{\searrow} \; h(\nu+1) \end{array} \quad . \tag{24}$$

Here $su(\nu+1)$ describes systems with a finite number of bound states while $h(\nu+1)$ describes the case of systems with an infinite number of bound states.[4] Exactly solvable problems for these cases can be found by considering the branching problem for $h(\nu+1)$.

[4]For scattering problems, $su(\nu+1)$ is replaced by $su(\nu,1)$.

$\nu = 1$

The algebra $h(2)$ has a trivial Abelian subalgebra $u(1)$ [the operator $\tau^\dagger\tau$ in (23)]

$$h(2) \supset u(1) \qquad (I) \quad . \tag{25}$$

States in this chain are characterized by the quantum numbers[5]

$$\left| \begin{array}{ccc} h(2) & \supset & u(1) \\ \downarrow & & \downarrow \\ \infty & & n \end{array} \right\rangle \quad . \tag{26}$$

where $n = 0, 1, \cdots, \infty$. The exactly solvable problem has still the structure of (6) and (7) but now $n = 0, 1, \cdots, \infty$. This chain describes the harmonic oscillator in one dimension.

$\nu = 3$

If rotational invariance is imposed on the problem, there is only one solution to the branching problem of $h(4)$:

$$h(4) \supset u(3) \supset o(3) \supset o(2) \qquad (I) \quad . \tag{27}$$

States in this chain are characterized by the quantum numbers

$$\left| \begin{array}{ccccccc} h(4) & \supset & u(3) & \supset & o(3) & \supset & o(2) \\ \downarrow & & \downarrow & & \downarrow & & \downarrow \\ \infty & & n & & L & & M \end{array} \right\rangle \quad , \tag{28}$$

with $n = 0, 1, \cdots, \infty$; $L = n, n-2, \cdots 1$ or 0 ($n =$ odd or even); $-L \le M \le +L$. This is the chain of the 3 dimensional harmonic oscillator. For any number of dimensions $\nu, h(\nu+1)$ contains the algebra $u(\nu)$. Chains originating from $h(\nu+1) \supset u(\nu)$ form the class of exactly solvable problems with an infinite number of bound states for Hamiltonians of the type (1). This completes the classification of exactly solvable problems in ν dimensions for Hamiltonians of the type (1).

A similar classification can be done for problems in ν dimensions with Hamiltonian

$$H^{-1} = E_o^{-1} + \sum_\alpha \eta_\alpha \mathcal{G}_\alpha + \sum_{\alpha\beta} w_{\alpha\beta} \mathcal{G}_\alpha \mathcal{G}_\beta \; ; \quad \mathcal{G}_\alpha \epsilon gl(\nu+1, C) \quad . \tag{29}$$

(See footnote 1.) The possible dynamical symmetries of systems with Hamiltonian (29) are identical to those of Sects. 3 and 4. However, for these problems, the inverse of the energy, E^{-1}, rather than the energy, E, has eigenvalue expressions linear and quadratic in the quantum numbers. Although this class of exactly solvable problems includes the Coulomb potential, which is of great interest, it will not be discussed here, since the presence of the elements of $\mathcal{G}$ in the denominator poses some problems which require a full (and separate) discussion. For the purposes of this article it suffices to say that the Coulomb problem in three dimensions belongs to the type II of exactly solvable problems discussed in Sect. 3. It has $o(4)$ dynamic symmetry, and its Hamiltonian can be written as[6]

[5] The representations of $\mathcal{G}(0,1) \approx h(2)$ are discussed in Ref.[7], p.41. The representation of interest here, not bounded from above, is denoted by $\uparrow \omega, \mu$. In particular, $\omega = 0, \mu = 1$. The representation $\uparrow 0, 1$ is denoted by ∞ in (26), in order to make the comparison with the corresponding representation of $su(2)$ transparent.

$$H = -\frac{A}{2(C_2(o4)+1)} \quad , \tag{30}$$

i.e. as in Eq. (29). The eigenvalues of (30) are

$$E(\omega, L, M) = -\frac{A}{2[\omega(\omega+2)+1]} \quad , \tag{31}$$

and, with the substitution $n = \omega + 1$, can be brought to the usual Bohr form

$$E(n) = -\frac{A}{2n^2} \quad . \tag{32}$$

5. Differential Realizations. Schrödinger Form.

In view of the fact that quantum mechanics is usually formulated in differential form (Schrödinger equation) it is of interest to realize the abstract Lie algebras of Sects. 3 and 4 in terms of differential operators. There are several ways in which a Lie algebra can be realized in terms of differential operators. I begin with the realization of $gl(\nu+1, C)$ in terms of differential operators in $\nu+1$ variables. In particular, I consider the one-dimensional case, $\nu = 1$, and realize the algebra $gl(2, C)$ in two variables.

5.1 Realizations of $gl(\nu+1, C)$ in $\nu+1$ variables.

These realizations have been studied by Miller[8], who considered the class of realizations of the form

$$\begin{aligned} F_3 &= \frac{\partial}{\partial y} \quad , \\ F_\pm &= e^{\pm y}\left(\pm\frac{\partial}{\partial x} - k(x)\frac{\partial}{\partial y} + j(x)\right) \quad , \\ F &= \mu \quad . \end{aligned} \tag{33}$$

The commutation relations (19) are satisfied if the functions $k(x)$ and $j(x)$ satisfy the equations

$$\begin{cases} \frac{dk(x)}{dx} + k^2(x) &= -a \\ \frac{dj(x)}{dx} + k(x)\, j(x) &= -\frac{b\mu}{2}. \end{cases} \tag{34}$$

Eqs. (34) admit several solutions, listed by Miller. When, $\nu = 1$, the algebraic structure of the bound state problem is

$$u(2) \supset o(2)[\approx u(1)] \quad . \tag{35}$$

The basis states are given by (8). They are functions of two variables, $f_m(x, y)$. They satisfy the equation

$$F f_m(x,y) = \mu f_m(x,y) \quad , \tag{36}$$

trivially, and the equations

$$\begin{cases} F_3 f_m(x,y) & = & m f_m(x,y) \\ C_2(a,b) f_m(x,y) & = & \lambda f_m(x,y), \end{cases} \tag{37}$$

i.e. they are simultaneous eigenfunctions of the Casimir operators of (20). The first equation is solved by writing

$$f_m(x,y) = e^{my}\, g_m(x) \quad . \tag{38}$$

The dependence on y can be factored out leading to a second order differential equation for $g_m(x)$. Thus, although one starts with two variables, one ends with a differential equation in one variable, since the additional variable is effectively eliminated by the condition of sitting on a given representation of $u(2)$, Eqs. (36) and (37). Starting from $u(2)$ and the realizations (33), one obtains most of the equations of mathematical physics: hypergeometric functions, confluent hypergeometric functions, parabolic cylinder functions, Bessel functions. Furthermore, since for each algebra in the diagram (22) the eigenvalues of the Casimir operators are known, for example for $su(2)$ the eigenvalue λ can be written as $\lambda = \frac{N}{2}(\frac{N}{2}+1), N =$ integer, this procedure leads to the class of one-dimensional second order differential equations which are exactly solvable. The differential equations can be brought into the Schrödinger form

$$-\frac{d^2}{dx^2}\psi(x) + V(x)\psi(x) = E\psi(x) \quad , \tag{39}$$

and give rise to a class of exactly solvable Schrödinger problems, listed in Table I. All potentials in Table I have spectra which are of the general form given in (7). Those in (a),(b),(c) are quadratic in the quantum number n, while the harmonic oscillator, (d), is linear in n. [The class of exactly solvable problems in one dimension is actually slightly wider than that given in Table I, since some variations of the potentials in (a),(b),(c),(d) are also exactly solvable. For example, the potential $V_0/\sin^2\alpha x + V_1/\cos^2\alpha x$ is also exactly solvable. All these potentials originate however from the procedure described above starting with $u(2)$.]

Table I. Exactly solvable Schrödinger problems in one dimension.

(a)	$V(x) = \left(\frac{\hbar^2\alpha^2}{2m}\right)\frac{\kappa(\kappa-1)}{\sin^2\alpha x}$	$E(n) = \left(\frac{\hbar^2\alpha^2}{2m}\right)(\kappa+2n)^2$
(b)	$V(x) = -\left(\frac{\hbar^2\alpha^2}{2m}\right)\frac{\kappa(\kappa-1)}{ch^2\alpha x}$	$E(n) = -\left(\frac{\hbar^2\alpha^2}{2m}\right)(\kappa-1-n)^2$
(c)	$V(x) = D\left(e^{-2\alpha x} - 2e^{-\alpha x}\right)$	$E(n) = -D + \left(\frac{\hbar^2\alpha^2}{2m}\right)\left[\frac{2\sqrt{2mD}}{\alpha\hbar}\left(n+\frac{1}{2}\right) - \left(n+\frac{1}{2}\right)^2\right]$
(d)	$V(x) = \frac{1}{2}m\omega^2x^2$	$E(n) = \hbar\omega\left(n+\frac{1}{2}\right)$

The procedure described above can be generalized to any number of dimensions by starting from a differential realization of $u(\nu+1)$ in terms of differential operators in

$\nu+1$ variables, and requiring that the basis states be simultaneous eigenfunctions of all linear and quadratic Casimir operators of a chain of algebras starting with $u(\nu+1)$ and ending with $o(3)$ (if rotational invariance is required). This leads to the theory of special functions in ν complex variables, since the additional variables can again be eliminated by the condition similar to (36)

$$\hat{N} f_{m_1,\cdots,m_\nu}(x_1,x_2,\cdots,x_\nu,x_{\nu+1}) = N\, f_{m_1,\cdots,m_\nu}(x_1,x_2,\cdots,x_\nu,x_{\nu+1}) \quad , \tag{40}$$

which states that we are in a given irreducible representation $[N,\dot{0}]$ of $u(\nu+1)$.

A new problem arises here if one attempts to bring the resulting differential equation in ν variables to its Schrödinger form:

$$\nabla^2\psi(x_1,x_2,\cdots,x_\nu) + V(r)\psi(x_1,x_2,\cdots,x_\nu) = E\psi(x_1,x_2,\cdots,x_\nu) \quad , \tag{41}$$

where ∇^2 is the Laplace operator in ν variables and

$$r = (x_1^2 + x_2^2 + \cdots + x_\nu^2)^{1/2} \quad . \tag{42}$$

If one insists that the Schrödinger equation be of the form (41) (i.e. rotational invariance is imposed on the problem), only type (d) will survive in $\nu > 1$ dimensions. It is for these situation that Lie algebras give something new. In fact, as one can see from (11) and (18), the abstract formulation of the problem gives additional exactly solvable problems with rotational invariance. These problems, however do not correspond to a Schrödinger equation with a local interaction, as in (41), but to more complicated forms with non-local interactions. Nonetheless they are exactly solvable and therefore provide a very useful tool to analyze experimental data.

5.2. Realizations of $gl(\nu+1,C)$ in ν variables.

The algebra $gl(\nu+1,C)$ can also be realized in ν variables. In particular, the algebra $gl(2,C)$ can be realized in one variable. These realizations have been studied by Kamran and Olver[8] who considered the problem of classifying all the second order differential equations originating from the Hamiltonian (1), up to equivalences. In their analysis, Kamran and Olver considered not only differential equations originating from the semisimple Lie algebra $sl(2,C)$ in

$$gl(2,C) \approx u(2) \supset su(2) \oplus u(1) \approx sl(2,C) \oplus u(1) \quad , \tag{43}$$

but also those originating from solvable algebras and contractions of $u(2)$. Their result is summarized in Table II. In this table, $P(z)$ denotes an arbitrary polynomial $Q(z)$ an arbitrary quadratic polynomial, u and v arbitrary constants and $\mathcal{P}(x)$ is the Weierstrass elliptic function.

The classification of Kamran and Olver includes, in addition to the four classes (b), (c), (d), (e), which, with particular choices of the coefficients v and u, lead to the four classes of exactly solvable problems of Table I, also two new classes (a) and (f). Class (a) is not interesting for applications in physics, while the non-exactly solvable class (f) was discovered in 1983[5].

Table II. Lie algebraic potentials in one dimension[8].

(a)	$V(x) =$ solution of a linear, homogeneous, constant coefficients, ordinary differential equation.
(b)	$V(x) = v\ x^{-2} + P(x^2)$
(c)	$V(x) = v\ e^{-2x} + u\ e^{-x} + P\ (e^x)$
(d)	$V(x) = v\ \sec^2 x + u\ \sec x\ \tan x + P(\cos x)$
(e)	$V(x) = v\ \mathrm{sech}^2 x + u\ \mathrm{sech} x\ \ \tanh x + P(\ \cosh x)$
(f)	$V(x) = v\ \mathcal{P}(x) + u + Q[\mathcal{P}(x)][\mathcal{P}'(x)]^{-2}$

This class will be discussed in the following section. Here it is of interest to explicitly display some of the realizations of Ref.[8], in particular those corresponding to class (c) and (e) of Table II which have considerable application in physics. Let $\mathcal{D}_1, \mathcal{D}_2, \mathcal{D}_3$ be the three elements of the Lie algebra $sl(2, R)$.
Useful realizations are:

Type (c) :

$$\left\{ \begin{array}{lcl} \mathcal{D}_1 & = & D - \frac{\alpha-1}{2} - \frac{\beta}{2}e^{-x} \\ \mathcal{D}_2 & = & e^{-x}D - \frac{\alpha-1}{2}e^{-x} - \frac{\beta}{2}e^{-2x} \\ \mathcal{D}_3 & = & e^{x}D - \frac{4\kappa-\alpha+1}{2}e^{x} - \frac{\beta}{2} \end{array} \right. . \tag{44}$$

Type (e) :

$$\left\{ \begin{array}{lcl} \mathcal{D}_1 & = & \sinh x\ \mathcal{D}_2 \\ \mathcal{D}_2 & = & \mathrm{sech} x\ \left(D - \frac{1}{2}\alpha\ \sinh x\right) - \frac{1}{4} \\ \mathcal{D}_3 & = & \sinh^2 x\ \mathcal{D}_2 + 2\kappa\ \sinh x \end{array} \right. . \tag{45}$$

In these expressions, $D = \frac{d}{dx}$, and α, β, κ are parameters.

6. Structure of the Solutions. Continuous Spectra.

Although most of the emphasis is applications of Lie algebraic methods has been on discrete spectra, it is of interest to discuss briefly the situation for continuous spectra. Continuous spectra can be described algebraically by using the real form $u(\nu, 1)$ of $gl(\nu + 1, C)$. The only case that has been discussed extensively so far is that of $\nu = 1$[5, 9].

$\underline{\nu = 1}$
The algebraic structure of scattering in one dimension is

$$u(1,1) \supset o(1,1) \quad , \tag{46}$$

where $o(1,1)$ is the Abelian algebra formed by the non-compact element of $u(1,1)$. The Hamiltonian is

$$H = BC_2(o(1,1)) \quad , \tag{47}$$

with eigenvalues

$$E(k) = Bk^2 \quad , \tag{48}$$

Table III. S-matrices for exactly solvable problems[9].

(b) $V(x) = -\frac{j(j+1)}{\mathrm{ch}^2 \alpha x}$

$$\begin{cases} R = \frac{\Gamma(ik)\Gamma(-ik-j)\Gamma(ik+j+1)}{\Gamma(-ik)\Gamma(-j)\Gamma(j+1)} \\ \\ T = \frac{\Gamma(1-ik+j)\Gamma(-ik-j)}{\Gamma(1-ik)\Gamma(-ik)} \end{cases}$$

(c) $V(x) = (2j+1)^2(\mathrm{e}^{-2\alpha x} - 2\mathrm{e}^{-\alpha x})$

$$\begin{cases} R = (2j+1)^{-2ik}\frac{\Gamma(2ik+1)\Gamma(ik-j)}{\Gamma(-2ik+1)\Gamma(-ik-j)} \\ \\ T = 0 \end{cases}$$

where now k is continuous $0 \leq k \leq \infty$. In the case of scattering, an exactly solvable problem is defined as that problem for which the S-matrix can be obtained in explicit form in terms of the quantum numbers that characterize the state uniquely. It has been possible to prove that problems which are exactly solvable for bound states are also exactly solvable for scattering. In particular, returning to Table I, one can see that only types (b) and (c) have continuous spectra. Using algebraic theory it has been possible to determine the S-matrices corresponding to these two types, given in Table III. In one dimension, the S-matrix reduces to the reflection and transmission coefficients, R and T. One may note that, apart from some overall phases, the S-matrices of exactly solvable problems appear always to be ratios of Γ functions.

$\nu = 3$

The algebraic structure of scattering in three dimensions, for problems with rotational invariance, is

$$u(3,1) \supset o(3,1) \supset o(3) \supset o(2) \quad . \tag{49}$$

This structure has not been investigated in detail, apart from the case of the Coulomb interaction which, however, does not belong to the class of problems with Hamiltonian (1). [6]

It is important to note that here too the abstract formulation of scattering in terms of Lie algebras has brought something new, in the sense that it has been possible to

[6] The Coulomb problem in three dimensions has S-matrix given by $S_\ell(k) = \Gamma(\ell+1+i\eta)/\Gamma(\ell+1-i\eta)$, where $\eta = mZ_1Z_2e^2/k$, i.e. it is still a ratio of Γ-functions.

generate exactly solvable problems with rotational invariance which produce S-matrices in explicit form. These S-matrices are quite useful in the analysis of experimental data.

7. Structure of the Solutions. Band Spectra.

Lie algebras are also relevant for band spectra. Consider in fact the class (f) of Lie algebraic potentials in Table II. A special case of (f) leads to the Schrödinger equation[5]

$$\left[-\frac{d^2}{dx^2}+j(j+1)\kappa^2\mathrm{sn}^2(x,\kappa)\right]\psi(x)=E\psi(x) \quad , \tag{50}$$

where $\mathrm{sn}(x,\kappa)$ is a Jacobi function with modulus κ. For $0 \leq \kappa^2 < 1, \mathrm{sn}^2(x,k)$ is periodic and real. Eq.(50) describes therefore a one-dimensional crystal. The Lie algebraic Hamiltonian which corresponds to (50) is of the generic form (1) and, in particular, it can be written in terms of the elements of $su(2)$, introduced in Sect.5, as

$$H = F_x^2 + \kappa^2 F_y^2 \quad . \tag{51}$$

This Hamiltonian does not have dynamic symmetry, except in the case $\kappa^2 = 1$, when it reduces to

$$H = F^2 - F_z^2 \quad . \tag{52}$$

When $\kappa^2 = 1$, the Schrödinger equation (50) reduces to the equation with the potential $\mathrm{sech}^2 x$ (Pöschl-Teller potential) which is exactly solvable. However, the eigenvalues of the algebraic Hamiltonian (51) can be found easily by diagonalizing a matrix which is $(2F+1)\times(2F+1)$. One can show that these eigenvalues correspond to the edges of the band spectra. Other properties can be calculated by writing down explicitly the algebraic form of the eigenfunctions $\Theta_{n,k}(x) = e^{ikx}u_{n,k}(x)$ in the potential (50), where k is the crystal momentum and n labels the bands; $u_{n,k}(x)$ is periodic. Thus, once more, all properties of the system can be obtained algebraically, although in this case there is no dynamic symmetry.

The method outlined above can also be extended to band structure in a larger number of dimensions. However, as in the case of bound state problems, a complication arises since the algebraic structure must respect the point symmetry of the crystal. Thus, while in Sect.3 the branching problem was solved by insisting that $o(3)$ be contained in the chain, in the case of band-structures one must insist that the discrete group $\mathcal{A}$ be contained in the chain. The group $\mathcal{A}$ is that describing the geometric symmetry of the crystal (cubic, monoclinic, $\cdots$).

8. Conclusions

Algebraic theory (i.e. the expansion of quantum mechanical operators onto elements of a Lie algebra) is a powerful technique which allows one to attack problems in physics and chemistry. This theory associates a quantum mechanical problem (non-relativistic and with no spin) in ν dimensions to a Lie algebra $gl(\nu+1,C)$ in $\nu+1$ dimensions, and a representation $[k,\dot{0}]$. All types of spectra, (discrete, continuous and band) can be obtained within this algebraic structure

$$
gl(\nu+1,C) \begin{array}{lll} \nearrow & su(\nu+1)\oplus u(1) \quad , & \text{discrete spectra} \quad , \\ \longrightarrow & su(\nu,1)\oplus u(1) \quad , & \text{continuous spectra} \quad , \\ \searrow & su(\nu+1)\oplus u(1) \quad , & \text{band spectra} \quad . \end{array} \tag{53}
$$

By considering all possible subalgebra chains of $gl(\nu+1, C)$ subject to some conditions if any, and all contractions of $gl(\nu+1, C)$, and imposing the condition that the Hamiltonian be a combination of invariant operators, one obtains all exactly solvable problems in ν dimensions. Exactly solvable problems are defined as those problems for which the discrete spectrum is given explicitly in terms of quantum numbers that characterize the states uniquely and for which the S-matrix is also given explicitly. However, the most important aspect of algebraic theory is not so much the possibility that it provides to classify all exactly solvable problems, but rather the fact that it provides a framework in which realistic calculations can be done. The method is the more useful the more complex is the situation one is analyzing, i.e. large number of dimensions, ν, and/or large number of degrees of freedom.

Acknowledgements

This paper is dedicated to the memory of Feza Gürsey, one of the greatest physicists and scholars of this half-century. His insight and knowledge have contributed to the development of the subject discussed here as well as to many others.

This work has been supported in part by D.O.E. Grant DE-FG02-91ER40608.

References

[1] S. Flügge, "Practical Quantum Mechanics", *Springer-Verlag*, Berlin (1974).

[2] A. Arima and F. Iachello, *Phys. Rev. Lett.* **35**, 1069 (1975); F. Iachello and A. Arima, "The Interacting Boson Model", *Cambridge University Press*, Cambridge (1987).

[3] F. Iachello, *Chem. Phys. Lett.* **78**, 581 (1981); F. Iachello and R.D. Levine, *J. Chem. Phys.* **77**, 3046 (1982); F. Iachello and R.D. Levine, "Algebraic Theory of Molecules", *Oxford University Press*, Oxford (1994).

[4] F. Iachello, *Nucl. Phys.* **A560**, 23 (1993); F. Iachello, in "Lie Algebras, Cohomologies and New Applications of Quantum Mechanics", *Contemporary Mathematics*, AMS, **Vol. 160** (1994), p. 151-171.

[5] Y. Alhassid, F. Gürsey and F. Iachello, *Phys. Rev. Lett.* **50**, 873 (1983).

[6] See, B.W. Wybourne, "Classical Groups for Physicists", *J. Wiley and Sons*, N.Y. (1976), *Chapts. 20 and 21.*

[7] W. Miller, Jr.," Lie Theory and Special Functions", *Academic Press*, New York (1968), *Chapt. 2.*

[8] N. Kamran and P.J. Olver, *J. Math. Anal. Appl.* **145**, 342 (1990).

[9] Y. Alhassid, F. Gürsey and F. Iachello, *Ann. Phys.* (N.Y.) **148**, 346 (1983).

Geometry and Arithmetic of Factorized S-Matrices

Peter G. O. Freund[1] and Anton V. Zabrodin[2]

1 Enrico Fermi Institute, Department of Physics and Mathematical Disciplines Center, University of Chicago, Chicago, IL 60637, USA
2 Institute of Chemical Physics, Kosygina Str. 4, SU-117334, Moscow, Russia

Dedicated to the memory of Feza Gürsey, whose work so deeply influenced the development of Particle Physics.

In realistic four-dimensional quantum field theories integrability is elusive. Relativity, when combined with quantum theory does not permit an infinity of local conservation laws except for free fields, for which the S-matrix is trivial $S = 1$. In two space-time dimensions, where forward and backward scattering are the *only* possibilities, nontrivial S-matrices are possible even in integrable theories. Such S-matrices are known to factorize [1]. This means that there is no particle production, so that the 4-point amplitudes determine all higher n-point amplitudes. In our recent work [2, 3, 4, 5, 6] we found that in such integrable two-dimensional theories, even the input 4-point amplitudes are determined by a simple principle. Roughly speaking these amplitudes describe the S-wave scattering which one associates with free motion on certain quantum-symmetric spaces. The trivial S-matrix of free field theory describes the absence of scattering which one associates with free motion on a euclidean space, itself a symmetric space. As is well known [7, 8, 9], for curved symmetric spaces the S-matrices for S-wave scattering are no longer trivial, but rather they are determined by the Harish-Chandra c-functions of these spaces [10]. The quantum deformation of this situation is what appears when one considers excitation scattering in two-dimensional integrable models.

All this is intimately connected with the algebraic approach of Alhassid, Gürsey and Iachello [9] to certain bound state and scattering problems in quantum mechanics.

The wave function of a free particle moving on a symmetric space is an eigenfunction of the laplacian on this space. Take as the simplest example, the 2-sphere $S^2 = SO(3)/SO(2)$. Functions on S^2 depend on two angular variables θ and ϕ. Functions invariant under the $SO(2)$ group of rotations around the z-axis depend only on the *radial* variable θ (this variable is called radial because upon stereographic projection from the North pole, it becomes the radial variable in the plane). The θ part of the laplacian on the sphere is called the radial laplacian. Its eigenfunctions are the $SO(2)$-invariant "S- waves" on the two-dimensional sphere. They are, of course, the Legendre polynomials $P_n(\theta)$, corresponding to eigenvalue $E_n = -n(n+1)$. The Legendre polynomials are precisely the zonal spherical functions on the 2-sphere. Notice that the radial coordinate θ on the sphere is continuous, the sphere being a continuous manifold. The spectral variable E_n on the other hand, is discrete, the sphere being compact.

Let us now consider the non-compact real hyperbolic plane [10] $H_\infty = SL(2,\mathbf{R})/SO(2,\mathbf{R})$ instead. Then both the radial coordinate and the spectral variable are continuous. The $SO(2)$-invariant wave functions are now Legendre functions of complex index. This time, as was to be expected, these wave functions are the zonal spherical functions (zsf's) of the real hyperbolic plane. In this non-compact case it is meaningful to inquire about the asymptotic behavior of the wave functions, or in other words the zsf's, for large values of the radial coordinate. This asymptotic behavior, as in any scattering problem, involves a superposition of an incoming and of an outgoing wave. The corresponding Jost function can be expressed explicitly in terms of gamma functions, it is precisely the Harish-Chandra c-function of the real hyperbolic plane, given [10] by

$$c_\infty(k) = \frac{\zeta_\infty(ik)}{\zeta_\infty(ik+1)}, \quad \zeta_\infty(s) = \pi^{-\frac{s}{2}}\Gamma(\frac{s}{2}) \tag{1}$$

and leading to the S-matrix

$$S_\infty(k) = \frac{c_\infty(k)}{c_\infty(-k)} \,. \tag{2}$$

In all this one can readily replace [11] the real hyperbolic plane H_∞ by the p-adic hyperbolic plane $H_p = SL(2,\mathbf{Q}_p)/SL(2,\mathbf{Z}_p)$, which is no longer a continuous manifold, but rather a *discrete* one. Specifically it is an infinite homogeneous tree, with p+1 edges incident at each vertex, in other words an infinite Bethe lattice. One can repeat the whole reasoning we just sketched, using the Cartier [12] laplacian on the tree. The c-function is then found to be

$$c_p(k) = \frac{\zeta_p(ik)}{\zeta_p(ik+1)}, \quad \zeta_p(s) = \frac{1}{1-p^{-s}} \,. \tag{3}$$

The corresponding S-matrix is

$$S_p(k) = \frac{c_p(k)}{c_p(-k)} \,. \tag{4}$$

The zsf's themselves in these cases are polynomials in the variable $x=p^{ik/2}+p^{-ik/2}$, the Mautner-Cartier polynomials [12, 13]. Remarkably, as discovered by Macdonald [14], there exists a quantum deformed symmetric space with two deformation parameters, which, as one varies the deformation parameters, interpolates, as it were, between the real and all the (infinitely many) p-adic symmetric spaces we just considered. The scattering amplitudes obtained from the quantum mechanics on this interpolating deformed symmetric space, correspond to excitation scattering in a certain limit of a class of XYZ-like models. This interpolation is a special case of a general *p-adics-quantum-group connection* [14, 15, 16]

Before we go into all this, let us point out that by taking the infinite product over all primes (i.e. Euler product) of all S-matrices (4) and then multiplying it with the S-matrix (2) we get an *adelic* S-matrix [11] which remarkably describes scattering on the non-compact (though finite area) fundamental domain of $SL(2,Z)$ on the real hyperbolic plane [17]. This adelic S-matrix involves the Riemann zeta function. The absence of bound states in the continuum required by unitarity, or equivalently by the probabilistic interpretation of quantum mechanics, then translates for this S-matrix into a mathematical statement about the zeta function. This statement, it turns out [18], is equivalent to the prime number theorem of Hadamard and de la Vallée Poussin [19]. A genuine number-theoretic theorem thus acquires a simple physical meaning.

Now let us return to the excitation scattering in the XYZ-like models, as it relates to the interpolation between all the S-matrices mentioned in the beginning. The XYZ model is a deformation of the $SU(2)$-invariant Heisenberg spin chain with the individual spins in the fundamental two-dimensional representation. The generalization we have in mind involves deformed $SU(n)$ spin chains again with the spins in the fundamental representation. There are n -1 types of excitations in such a model and for the time being we concern ourselves with the first level excitations. The S-matrix for the scattering of two such excitations is known and it depends on one variable, the rapidity u, and on three parameters: the elliptic modulus q, the anisotropy parameter γ, and n the integer which identifies the $SU(n)$ whose deformation we are considering. In the limit $n \to \infty$ the S-matrix for the scattering of two first level excitations can be written in the very simple form [2, 3]

$$S(u) = \frac{c(u;l|q)}{c(-u;l|q)}, \quad c(u;l|q) = \frac{\Gamma_q(iu)}{\Gamma_q(iu+l)}. \tag{5}$$

Here $\Gamma_q(iu)$ is the q-analogue gamma function [20] and

$$l = \frac{-2\gamma}{log q}. \tag{6}$$

Since for $q \to 1$, the q-gamma function approaches the ordinary gamma function, we see that for $l = \frac{1}{2}$, the limit $q \to 1$ reproduces the real hyperbolic plane result (2). For q=0, $t = q^l = \frac{1}{p}$ the function $c(u;l\,|\,q)$ reduces to the p-adic hyperbolic plane result (4).

In the real case the solution of the Schrödinger equation was given by Legendre polynomials and by Legendre functions in the compact and noncompact cases respectively. In the p-adic case it was given by the Mautner-Cartier polynomials. In the deformed case the Jost-Harish-Chandra function (5) corresponds to the Macdonald polynomials of type A_l [14], which coincide with the classical Rogers-Askey-Ismail polynomials [21, 22]. These polynomials reduce to the Legendre and Mautner-Cartier polynomials in the appropriate limits, and thus interpolate between them. Since in all these limits one obtains zsf's, it stands to reason to expect that even the full Rogers-Askey-Ismail (RAI) polynomials should correspond to zsf's of a suitable quantum-symmetric space. This expectation is buttressed by the fact that these RAI polynomials are known [20] to be eigenfunctions of a second order difference operator with all the earmarks of a laplacian. The appearance of two deformation parameters q and l, or q and t, indicates a connection with an elliptic Sklyanin-type deformation [23, 24, 25]

We had to take the particular limit in which the n of $SU(n)$ approached infinity, even though the RAI polynomials are connected with n=2. The need to do this is essentially connected with the fact that the elliptic algebras of the Sklyanin type have no coproduct except in the limit $n \rightarrow \infty$. It is as if to get to the representation theory in the n=2 case, we first have to embed in the "simpler" $n = \infty$ case.

So far we only discussed the scattering of first level excitations in the generalized XYZ models. Including all higher excitations introduces further parameters into the problem, namely the levels of the two excitations. This leads [5] to the Hall-Littlewood-Kerov [26, 27] polynomials. These polynomials come with an infinite supply of (pairs of) parameters. These extra parameters have to do with more general XYZ type models in which not only is the underlying group $SU(n)$, $n \geq 2$, but the individual spins of the chain are no longer in the fundamental representation, but in an arbitrary representation. This indeed needs an infinite supply of parameters, as the specification of the representation for $SU(n)$ involves n—1 parameters and this increases indefinitely with n.

All this is related to work on the q-Knizhnik-Zamolodchikov equations [28,29] specifically to the spectral problem [5,6] for these equations, rather than the monodromy problem.

The main lesson to be learned from this work is that in a very large class of integrable quantum field theories in 1+1 dimensions, the S-matrix is not only factorized, but even the input four-point amplitude which determines the whole S-matrix, is in some sense trivial. It does not outright vanish as would be the case for free motion in euclidean space, but it has the Harish-Chandra-like form for "free" motion on some, in general, quantum-symmetric space. It is natural to conjecture that this is the case quite generally for two-dimensional integrable quantum field theories. It is also fair to say that this goes a long way towards extending to integrable quantum field theory, the algebraic approach developed by Alhassid, Gürsey and Iachello [9] in ordinary quantum mechanics. A byproduct of this work is the unexpected p-adics-quantum-group connection which we discussed above.

This work was supported in part by the grant NSF: PHY-91-23780

References

[1] Zamolodchikov, A.B., Zamolodchikov, Al.B.: Ann. Phys. (N.Y.) **120**, 253 (1979)

[2] Freund, P.G.O., Zabrodin, A.V.: Phys. Lett. **B284**, 283 (1992)

[3] Freund, P.G.O., Zabrodin, A.V.: Comm. Math. Phys. **147**, 277 (1992)

[4] Freund, P. G. O., Zabrodin, A.V.: Phys. Lett. **B294**, 347 (1992)

[5] Freund, P.G.O., Zabrodin, A.V.: Phys. Lett. **B311**, 103 (1993)

[6] Freund, P.G.O., Zabrodin, A.V.: preprint EFI-93-44, to be published

[7] Olshanetsky, M.A., Perelomov, A.M.: Phys. Rep **94**, 313 (1983)

[8] Wehrhahn, R.F.: Phys. Rev. Lett. **65**, 1294 (1990)

[9] Alhassid, Y., Gürsey, F., Iachello, F.: Phys. Rev. Lett. **50**, 873 (1983)

[10] Helgason, S.: Topics in Harmonic Analysis on Homogeneous Spaces. Basel: Birkhäuser 1981

[11] Freund, P.G.O.: Phys. Lett. **257B**, 119 (1991)

[12] Cartier, P.: Proc. Symp. Pure Math., vol 26, Providence: A.M.S. 1973.

[13] Mautner, F.: Am. J. Math. **80**, 441(1958)

[14] Macdonald, I.G.: In: Orthogonal Polynomials: Theory and Practice, Nevai P. (ed.) Dordrecht: Kluwer Academic 1990;
Macdonald I.G.: Queen Mary College preprint 1989.

[15] Freund, P.G.O.: In: Superstrings and Particle Theory. Clavelli, L., Harms, B. (eds.) . Singapore: World Scientific 1990

[16] Zabrodin, A.V.: Mod. Phys. Lett. **A7**, 441(1992)

[17] Faddeev, L.D., Pavlov, B.S.: Sem. Stekl. Math. Inst. Leningrad **27**, 161 (1972)

[18] Brekke, L., Freund, P.G.O.: Phys. Rep. **233**,1(1993)

[19] Titchmarsh, E.C.: The Theory of the Riemann Zeta Function. Oxford: Clarendon 1986

[20] Gasper, G., Rahman, M.: Basic hypergeometric series. Cambridge: Cambridge University Press 1990.

[21] Rogers, L.J.: Proc. Lond. Math. Soc. **26**, 318 (1895)

[22] Askey, R., Ismail, M.E.H. In: Studies in Pure Math. Erdös, P. (ed.) Basel: Birkhäuser 1983.

[23] Sklyanin, E.K.: Func Anal. Appl. **16**, 263 (1986); ibid. **17**, 273 (1983)

[24] Odeskii, A.V., Feigin, B.L.: Func. Anal. Appl. **23**, 207 (1989)

[25] Cherednik, I.V.: Func. Anal. Appl. **19**, 77 (1985)

[26] Macdonald, I.G.: Symmetric Functions and Hall Polynomials. Oxford: Clarendon 1979

[27] Kerov, S.V.: Func. Anal. Appl. **25**, 65 (1991)

[28] Knizhnik, V.G., Zamolodchikov, A.B.: Nucl. Phys. **B247**, 83 (1984).

[29] Frenkel, I., Reshetikhin, N.Yu.: Comm. Math. Phys. **146** ,1 (1992).

Duality Principle and Braided Geometry

S. Majid[1]

DAMTP, University of Cambridge, Cambridge CB3 9EW, UK

Abstract We give an overview of a new kind symmetry in physics which exists between observables and states and which is made possible by the language of Hopf algebras and quantum geometry. It has been proposed by the author as a feature of Planck scale physics. More recent work includes corresponding results at the semiclassical level of Poisson-Lie groups and at the level of braided groups and braided geometry.

Keywords: quantum groups – noncommutative geometry – braided geometry – integrable systems – Mach principle – κ-deformation – q-deformation

1 Introduction

I did not know Feza Gürsey personally, but I would guess that he understood very well one of the principles close to my own heart: the deep interdependence of theoretical physics and pure mathematics. A fairly conventional view, expressed neatly by Pierre Ramond in his talk, is that we theoretical physicists are probing into higher and higher energy scales building finer and more accurate theories in the process. Some say that we can learn in this task from pure mathematicians to supply us with the right conceptual structures, others say that they can learn from us to find examples. My own view, however, is more radical but I will try nevertheless to outline it here and to reflect it in my talk. The starting point is that after all Nature does not care what mathematics has been discovered until now. So it would be rather naive to think that the maths that we know now is going to be enough to reach higher and higher levels until we reach the Planck scale or beyond to the ultimate theory of physics. Most likely if we did meet someone who knew this ultimate theory of physics and asked them to tell us, we would not understand a word of it because of *completely new* mathematical concepts that are centuries from being invented. Put another way, we physicists should not only try to apply known or fashionable mathematics, we should be ready and willing to invent whatever we think Nature uses as we go along in our search: we should be equally good *pure* mathematicians (in the creative sense, not necessarily in terms of rigour) as we are calculators and applied ones. Sometimes one can forget that mathematics is about creating natural ideas and concepts, and not about theorems and lemmas *per se*, though these are how a mathematician knows that the concepts are worthwhile. Physicists have perhaps other intuitive ways of knowing what is worthwhile.

[1] Royal Society University Research Fellow and Fellow of Pembroke College, Cambridge

Put another way, the naive view is that Nature exists and maths exists, and we just have to apply the latter to describe the former. By contrast, I would like to argue that the word 'exists' is questionable in both cases and that in fact each creates and justifies the other as we climb to higher and higher energy scales in our understanding. I mean that Nature and Language (which for me means mathematics) define each other and are thereby interdependent.

I would like to describe here how this philosophy works in my own approach to unifying physics, which includes, as for many physicists, the goal of unifying quantum mechanics and gravity. It concerns a new kind of geometry more general than Riemannian and powerful enough to not break down in the quantum domain. Planck scale physics is the reason that I first became interested in Hopf algebras (quantum groups) and quantum or non-commutative geometry. It has nothing to do with the reason that quantum groups subsequently became very fashionable and important, which has more to do with work on inverse scattering and applications to knot and three-manifold invariants. It is gratifying that the same kind of new mathematics is emerging from all these different directions.

My goal in this lecture is to convince you of two things:

a) There already *is* today a fairly complete theory of quantum geometry, at least as a new mathematical language. There are several approaches and the one I describe here and which I consider fairly complete is my own one which I called the *braided approach* (or *braided geometry*[1][2]). There are braided lines[3], planes[4], matrices[5][6], differentials and plane waves (exponentials)[7], Gaussians and integration[8], forms, epsilon tensors[9] and more known in this approach in association which a general R-matrix, i.e. a general solution of the celebrated quantum Yang-Baxter equations (QYBE). These equations are usually associated with the quantum groups of Faddeev, Jimbo and Drinfeld[10][11][12] and these do indeed play a role in the braided approach as a background covariance of our constructions, but not as the geometry itself. I will begin with the more conventional (but not so complete) approach in which quantum groups are geometrical objects too, and then move on the systematic braided theory. This approach is quite different also from the approach to non-commutative geometry comming out of the work of A. Connes and others.

b) This new language of quantum or braided geometry does indeed make it possible to formulate new and previously unsayable concepts in theoretical physics. The duality principle which is the title of my lecture will be such a new concept. This in turn allows the possibility to realise the new concept experimentally and thereby to discover new and previously inconceivable physical phenomena. This is what I mean by the deep interdependence of theoretical physics and pure mathematics. One can develop this point of view into a Kantian or Hegelian perspective on the concept of reality in physics[13].

The new and previously unsayable physical phenomenon which I want to concentrate on here is a generalisation of wave-particle duality or position-momentum symmetry. I would like to formulate this mathematically as a consequence of something a bit deeper, which I call a *quantum-geometry transforma-*

tion. This is the assertion that the symmetry algebra (or generalised momentum) of a system could also be isomorphic to the coordinate algebra of some geometry. That geometry would be momentum space as a geometrical object. This is certainly possible in flat space where

$$U(\mathbb{R}^n) = \mathbb{C}[\mathbf{p}] = \mathbb{C}(\mathbb{R}^n). \tag{1}$$

Here $U(\mathbb{R}^n)$ is the enveloping algebra of the momentum generators $\mathbf{p} = \{p^i\}$ which is the symmetry point of view, while $\mathbb{C}(\mathbb{R}^n)$ is the algebra of coordinate functions $\{p^i\}$ on momentum space as a geometrical object. They are obviously isomorphic since both are polynomials in n-variables. In fact, this isomorphism is as *Hopf algebras*. The axioms of a Hopf algebra will be given next in Section 2 but they have a coproduct map Δ which on the left is trivial (the Lie algebra structure of $\mathbb{R}^n$ is reflected in the product of the enveloping algebra). On the right the same generators are viewed as coordinates and have a coproduct which corresponds to addition in momentum space, while the algebra is trivial. So the isomorphism is perhaps remarkable. The right hand side is the way that we will deal with geometry in this lecture, in terms of the algebra of coordinates. It is a geometrical picture of momentum against a picture as operators $\frac{\partial}{\partial x_i}$.

The algebra $\mathbb{C}(\mathbb{R}^n) = \mathbb{C}[\mathbf{x}]$ as generated by position coordinates $\{x_i\}$ is how we think of position space. So (1) allows us to think of the momentum in the same language as we think of position, but with $\mathbf{p}$ in the role of $\mathbf{x}$. Likewise

$$C(\mathbb{R}^n) = \mathbb{C}[\mathbf{x}] = U(\mathbb{R}^n) \tag{2}$$

whereby we think of $x_i = \frac{\partial}{\partial p^i}$ as differential operators on momentum space.

As well as being isomorphic, the position and momentum are connected in a symmetrical way by this differentiation. This can be formulated as the assertion that the two Hopf algebras $\mathbb{C}[\mathbf{x}]$, $\mathbb{C}[\mathbf{p}]$ are dually paired by

$$\langle \phi(\mathbf{p}), \psi(\mathbf{x}) \rangle = (\phi(\partial)\psi)(0) \tag{3}$$

making them dual as Hopf algebras. An element of one defines a linear functional on the other. We will see the precise definition in Section 2 but it should be clear that the dual Hopf algebra to $\mathbb{R}^n$ is $\hat{\mathbb{R}}^n \cong \mathbb{R}^n$ again because $\mathbb{R}^n$ is a self-dual Abelian group. A basis and dual basis are

$$\phi^{m_1 \cdots m_n} = (p^1)^{m_1} \cdots (p^n)^{m_n}, \quad \psi_{m_1 \cdots m_n} = \frac{x_1^{m_1} \cdots x_n^{m_n}}{m_1! m_2! \cdots m_n!}.$$

The map (3) is the evaluation $\mathbb{C}[\mathbf{p}] \otimes \mathbb{C}[\mathbf{x}] \to \mathbb{C}$ as dual linear spaces. But there is also a canonical element called the *coevaluation* and defined simply as the element in $\mathbb{C}[\mathbf{x}] \otimes \mathbb{C}[\mathbf{p}]$ corresponding to the identity map $\mathbb{C}[\mathbf{x}] \to \mathbb{C}[\mathbf{x}]$ in the usual way since $\mathbb{C}[\mathbf{p}]$ is dual to $\mathbb{C}[\mathbf{x}]$. In our case it is a formal powerseries (rather than living in the algebraic tensor product) and comes out as

$$\mathrm{coev} = \sum_{m_i} \psi_{m_1 \cdots m_n} \otimes \phi^{m_1 \cdots m_n} = e^{\mathbf{p} \cdot \mathbf{x}}. \tag{4}$$

This is a rather modern way of thinking about exponentials and Fourier theory, but one that generalises well to quantum geometry[8].

So conceptually, $\mathbb{C}[\mathbf{p}]$ and $\mathbb{C}[\mathbf{x}]$ are dual to each other, but they are also isomorphic (i.e. self-dual). This is our formulation of wave-particle duality. It works just as well for any self-dual Abelian group (such as $\mathbb{Z}_n$). It also works when the group is not self-dual but still Abelian. Then the Fourier conjugate space is the dual group, e.g. $\hat{\mathbb{Z}} = S^1$. The dual point of view with position and momentum interchanged is no longer isomorphic but both are still possible.

On the other hand, this duality phenomenon completely fails to get off the ground when the symmetry or generalised momentum group is non-Abelian. If g is a non-Abelian Lie algebra and we ask for a geometrical picture

$$U(g) = \mathbb{C}(?) \tag{5}$$

as functions on some ordinary manifold, we can never succeed simply because the left hand side is non-commutative, while the pointwise multiplication of functions on a manifold on the right hand side is always commutative. So for a geometrical picture we need some more general non-commutative or quantum geometry. To give an elementary example, consider the sphere S^2 with a standard Poisson bracket. Its Kirillov-Kostant quantisation is

$$[x_i, x_i] = \lambda \epsilon_{ij}{}^k x_k, \quad \sum_i x_i^2 = 1.$$

This is just $U(su_2)$ modulo a 'constant distance' relation. So at least in this case we can think geometrically of $U(su_2)$ as a quantum space even though the x_i are non-commuting coordinates. Moreover, with the greater generality of non-commutative or quantum geometry we will be able to pursue our position-momentum symmetry or duality phenomenon.

Acknowledgements

I would like to thank the Erwin Schrödinger Institute in Vienna for support during preparation, and the organisers here in Istanbul for inviting me.

2 What IS a Hopf algebra or quantum group?

Probably the reader has seen the definition of a Hopf algebra or quantum group elsewhere. There are several points of view which, remarkably, all lead to the same set of axioms. Here I want to give the axioms from the duality point of view, which is the one that interests us.

In this case a Hopf algebra A is just an algebra equipped with further structure such that the dual linear space A^* of functionals on A is also an algebra in a compatible way. In terms of A this additional structure is a *coproduct* $\Delta : A \to A \otimes A$. It looks like a product but goes in the opposite direction.

While an algebra is of course associative, the coproduct is *coassociative*. Thus

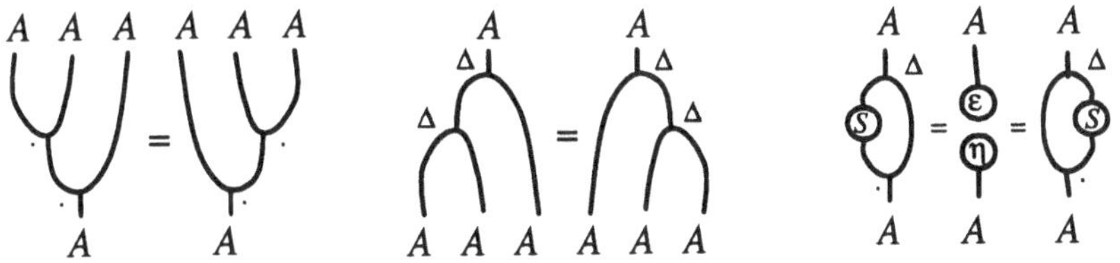

where we also show axioms for an *antipode* or 'linearised inverse' map $S : A \to A$. In addition, we require that Δ is an algebra homomorphism. We also require a *counit* map $\epsilon : A \to \mathbb{C}$ obeying $(\epsilon \otimes \mathrm{id}) \circ \Delta = \mathrm{id} = (\mathrm{id} \otimes \epsilon) \circ \Delta$.

The nice thing about these axioms is their input-output symmetry: Turn them up-side-down and you get the same axioms with the roles of $\cdot$ and Δ interchanged! The counit axioms become the ones for the inclusion of the identity element $\mathbb{C} \to A$ which we usually take for granted. This input-output symmetry is a deep feature of quantum geometry in the form that comes out of Hopf algebras. For our present purposes we realise it as follows: *If A is a Hopf algebra* (say finite-dimensional) *then so is A^**. The structures correspond according to

$$\langle \phi\psi, a \rangle = \langle \phi \otimes \psi, \Delta a \rangle, \quad \langle \phi, ab \rangle = \langle \Delta\phi, a \otimes b \rangle, \quad \langle \phi, Sa \rangle = \langle S\phi, a \rangle$$

for all $\phi, \psi \in A^*$ and $a, b \in A$. In the non-finite dimensional case we can still look for such a *duality pairing* between two Hopf algebras.

Thus $\mathbb{C}[\mathbf{x}]$ and $\mathbb{C}[\mathbf{p}]$ are Hopf algebras which are dually paired in this way. The Hopf algebra structure is

$$\Delta x_i = x_i \otimes 1 + 1 \otimes x_i, \quad \epsilon x_i = 0, \quad S x_i = -x_i$$

and similarly for $\mathbf{p}$. For a general function $a(\mathbf{x})$ the coproduct $\Delta a(\mathbf{x})$ is a function of two variables with value $(\Delta a)(x, y) = a(x + y)$. So the coproduct expresses the addition law on $\mathbb{R}^n$ in terms of the co-ordinate functions. The addition law is of course the basis for the geometry of $\mathbb{R}^n$.

More generally, if G is a group there will typically be a Hopf algebra $\mathbb{C}(G)$ generated by coordinates on G and again with a coproduct corresponding to the group law. If the group is non-Abelian the geometry here has curvature. This corresponds to the output of Δ not being symmetric or cocommutative. We also have another Hopf algebra $U(g)$ generated by the Lie algebra of G. It is dual

$$U(g) = \mathbb{C}(G)^*$$

in the loose sense that we have a duality pairing between these Hopf algebras. Since enveloping algebras are typically part of the algebra of quantum observables of a quantum system, we see that Hopf algebras suggest a general principle that *quantum theory and geometry are in a dual relationship*, implemented in the simplest case by Hopf algebra duality.

When we look away from commutative examples such as $\mathbb{C}(G)$ we can still think of a Hopf algebra as a generalisation of the functions on a group manifold, even when there is no manifold in the usual sense. Likewise, when we look away from cocommutative examples such $U(g)$ we can still think of a Hopf algebra as like an enveloping algebra or a Lie algebra. Because the axioms of a Hopf algebra are self dual, one algebraic concept contains both points of view. This means that the category of Hopf algebras unifies two concepts in physics: one linked to geometry and the other to quantum mechanics. This is why the language of Hopf algebras opens up the possibility to solve (5) by generalising both sides slightly, and why we call it a ***quantum-geometry transformation***.

This also suggests that Hopf algebras are a natural category in which to develop some simple models in which quantum theory and gravity are unified. This is the approach to Planck scale physics introduced in the author's thesis[14][15][16][13] on the basis of the above duality considerations. Obviously a natural source of non-commutative algebras A is as quantum algebras of observables, which are non-commutative versions of the classical algebra of functions on phase space. So we ask:

- When is a quantum algebra of observables a Hopf algebra?
- If so, what does it mean physically?

The answer to the first question is that it happens quite often; demanding it is a non-trivial but interesting constraint to put on a physical system. I will describe a large class of such physical systems associated to certain homogeneous spaces[14][16]. The answer to the second question is two fold:

(a) We now have geometry of phase space (in the crude form of a group law) even in the quantum setting, i.e. a unification of quantum theory and geometry. A general quantum system would include in its algebra of observables the enveloping algebra $U(g)$ for any generalised momentum generators; so to render such an algebra as like functions on a manifold involves for the momentum part exactly the quantum-geometry transformation (5). We also have non-commutativity coming from the momentum-position cross relations which we likewise have to deal with via quantum geometry.

(b) We have the possibility of a new kind of symmetry in physics in which we reinterpret A^* as the algebra of observables of a dual system, and $A \subset A^{**}$ as containing the states of the dual system. Thus we think of

$$\langle \phi, a \rangle = \phi(a) = \sum_i \rho_i \langle \phi_i | a | \phi_i \rangle$$

as the expectation value of observable a in mixed state $\phi = \sum \rho_i \langle \phi_i | \, | \phi_i \rangle$ while the dual system considers the same number as $\langle \phi, a \rangle = a(\phi)$, the expectation value of the dual-observable ϕ in the dual-state a. Note that we work with states in the mathematical way as (typically positive) linear functionals on the algebra of observables. These are typically convex linear combinations of expectations against Hilbert-space states, as indicated here. Thus we have the possibility of a second physical system with observables A^* dual to our original one. This is a rather radical phenomenon in the quantum world which we propose to call *observable-state* or *micro-macro* duality[16][15]. It is also connected as we have seen with input-output symmetry, a point which is developed further in [17].

The simplest example in 1-dimension is the Hopf algebra $\mathbb{C}[x]\blacktriangleright\!\!\triangleleft\mathbb{C}[p]$ generated by x, p with the relations and Hopf structure[15][18]

$$[p,x] = -\tfrac{A}{B}(1-e^{-Bx}),\quad \Delta x = x\otimes 1 + 1\otimes x,\quad \Delta p = p\otimes e^{-Bx} + 1\otimes p$$

$$\epsilon x = \epsilon p = 0,\quad Sx = -x,\quad Sp = -pe^{Bx}$$

depending on two real parameters A, B. It is the general solution of the problem in 1-dimension within the bicrossproduct construction in the next section. We identify $\hbar = -\frac{A}{B}$ and have a deformation of usual quantum mechanics. We keep p as a conserved momentum and Hamiltonian $-p^2/2m$ so that the particle is in free-fall: the modified commutation relation then results in modified dynamics. We use conventions in which p is antihermitian, i.e. $\frac{-\imath p}{m}$ is the velocity. Then

$$\frac{dx}{dt} = \frac{\imath}{\hbar}[-\frac{p^2}{2m}, x] = (\frac{-\imath p}{m})(1-e^{-Bx}) + O(\hbar),\quad \frac{dp}{dt} = 0.$$

Hence if we consider a particle falling in from ∞ we see that as the particle approaches the origin $x = 0$ it goes more and more slowly. In fact, it takes an infinite amount of time to reach the origin, which therefore behaves in some ways like a black-hole event horizon. This analogy should not be pushed too far since our present treatment is in nonrelativistic quantum mechanics, but it gives us an estimate $B = \frac{c^2}{MG}$ as introducing the distortion in the geometry comparable to a gravitational mass M. Here G is Newton's constant.

On the other hand we get a commutative algebra if we take the limit $\hbar \to 0$. In this case $\mathbb{C}[x]\blacktriangleright\!\!\triangleleft\mathbb{C}[p] \cong \mathbb{C}(\mathbb{R}\rtimes\!\!\!\ltimes\mathbb{R})$ where $\mathbb{R}\rtimes\!\!\!\ltimes\mathbb{R}$ has a non-Abelian group law corresponding to the coproduct above. It has curvature related to B. Combining with the above we see that our Hopf algebra has two limits[15][18]:

$$\begin{array}{lll} & mM \ll m_P^2 & \nearrow\ \mathbb{C}[x]\rtimes\!\!\!\triangleleft\mathbb{C}[p] \text{ usual quantum mechanics} \\ \mathbb{C}[x]\blacktriangleright\!\!\triangleleft\mathbb{C}[p] & & \\ & mM \gg m_P^2 & \searrow\ \mathbb{C}(X) \text{ usual curved geometry} \end{array}$$

where m_P is the Planck mass. In the first limit the particle motion is not detectably different from usual quantum mechanics outside the Compton wavelength from the origin. Of course, there is still a singularity in our model right at the origin. In the second limit the non-commutativity would not show up for length scales larger than the background gravitational scale set by M. This demonstrates our goal of unifying quantum mechanical and gravitational effects.

The dual Hopf algebra is generated by linear functionals ϕ, ψ defined by

$$\langle\phi, :a(x,p):\rangle = (\frac{\partial a}{\partial x})(0,0),\quad \langle\psi, :a(x,p):\rangle = (\frac{\partial a}{\partial p})(0,0)$$

where : $a(x,p) := \sum a_{n,m} x^n p^m$ is a normal-ordered function. Then

$$[\psi,\phi] = \hbar^{-1}(1-e^{-A\psi}),\quad \Delta\phi = \phi\otimes 1 + e^{-A\psi}\otimes\phi,\quad \Delta\psi = \psi\otimes 1 + 1\otimes\psi$$

$$\epsilon\phi = \epsilon\psi = 0,\quad S\phi = -e^{A\psi}\phi,\quad S\psi = -\psi$$

which is of just the same type as the above:

$$(\mathbb{C}[x]\blacktriangleright\!\!\triangleleft\mathbb{C}[p])^* = \mathbb{C}[\phi]\triangleright\!\!\blacktriangleleft\mathbb{C}[\psi]\cong\mathbb{C}[p]\triangleright\!\!\blacktriangleleft\mathbb{C}[x].$$

So this particular Hopf algebra is self-dual. It means that the observable-state duality is indeed realised in this model, the dual quantum system being of a similar form. This is a new kind of symmetry principle which one can propose as a speculative idea for the structure of Planck scale physics[16].

3 Bicrossproduct Hopf algebras

The example at the end of the last section is one of a large class of bicrossproduct models[16] which we describe now. We consider first the finite group case and then move on to the Lie version.

The data for the bicrossproduct construction is a *group factorisation*. This means a group X with subgroups $G, M \subset X$ such that $G\times M \to X$ given by multiplying within X is a bijection. In this case multiplying in the group and projecting down to M and G gives mutual actions $\triangleleft, \triangleright$ obeying

$$\triangleleft : M\times G \to M,\qquad \triangleright : M\times G\to G$$

$$s\triangleleft e = s,\ (s\triangleleft u)\triangleleft v = s\triangleleft(uv);\quad e\triangleleft u = e,\ (st)\triangleleft u = (s\triangleleft(t\triangleright u))\,(t\triangleleft u) \tag{6}$$

$$e\triangleright u = u,\ s\triangleright(t\triangleright u) = (st)\triangleright u;\quad s\triangleright e = e,\ s\triangleright(uv) = (s\triangleright u)\,((s\triangleleft u)\triangleright v)$$

for all $u, v\in G$ and $s, t\in M$. The first two in each line say that we have an action, and the second say that the action is almost by automorphisms, but twisted by the other action.

This data is just what it takes to obtain a Hopf algebra $\mathbb{C}(M)\blacktriangleright\!\!\triangleleft\mathbb{C}G$ built on the vector space with basis $G\times M$. We write the basis elements as labelled squares $s\underset{u}{\square} = s\overset{s\triangleright u}{\underset{u}{\square}}s\triangleleft u$ where the convention is that if we label the left and lower edges then the other two are labelled by the values transformed by the actions $\triangleright, \triangleleft$. In this notation, the Hopf algebra structure is[18]

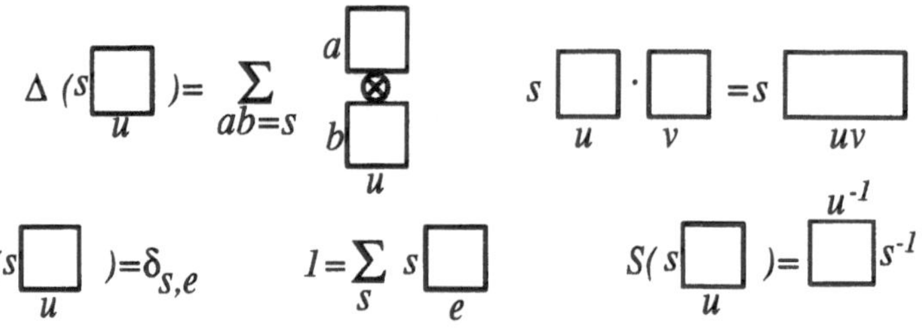

The product consists of gluing the squares horizontally whenever the edges are suitably matched. The coproduct by contrast consist of *ungluing* vertically, i.e. it is the sum of all pairs of squares which when glued vertically would give the square we began with. The dual Hopf algebra is $(\mathbb{C}(M){\blacktriangleright\!\!\triangleleft}\mathbb{C}G)^* = \mathbb{C}M{\triangleright\!\!\blacktriangleleft}\mathbb{C}(G)$ and has just the same form with vertical gluing and horizontal ungluing. The roles of $\triangleright, \triangleleft$ and G, M are interchanged.

In mathematical terms the algebra here is a *cross product algebra*. This is a more or less standard way to quantise particles moving under the action of a group[19][20]. In nice cases the action $\triangleleft$ of G on M induces a metric with the particle then moving on geodesics. The Lie algebra g of G plays the role of momentum since its elements generate the geodesics or 1-parameter flows. This works for any action $\triangleleft$. The main result of [14] is that if in this setting of particles on homogeneous spaces there is a *backreaction* $\triangleright$ obeying (6) then the resulting quantum algebra of observables is a Hopf algebra. Not every action $\triangleleft$, i.e. not every homogeneous space, admits such a backreaction, i.e. it is a strong constraint on the system. We have seen the flavour of the constraint in the example $\mathbb{C}[x]{\blacktriangleright\!\!\triangleleft}\mathbb{C}[p]$: it forces non-linear dynamics. One can think of it within this class of models as some kind of 'Einstein's equation'[16] since it is an (integrated) second order constraint on $\triangleleft$. It is forced by the duality principle which leads us to look for a Hopf algebra structure of this type.

Note that we gave the Hopf algebras above for finite groups, while for physics we need to work with Lie groups and topological spaces. This can be done with Hopf-von Neumann algebras[21] as well as algebraically in the form $\mathbb{C}(M){\blacktriangleright\!\!\triangleleft}U(g)$ with dual $U(m){\triangleright\!\!\blacktriangleleft}\mathbb{C}(G)$ where g, m are the respective Lie algebras. There are examples for all g compact semisimple, with $m = g^{\star\mathrm{op}}$ a suitable solvable Lie algebra coming from the Iwasawa decomposition[22].

For the simplest 3-dimensional example we take $g = su_2$ and $M = SU(2)^{\star\mathrm{op}}$. Then $\mathbb{C}(SU(2)^{\star\mathrm{op}}){\blacktriangleright\!\!\triangleleft}U(su_2)$ is generated by position co-ordinates $\{x_1, x_2, x_3, (x_3+1)^{-1}\}$ and su_2 generators $\{e_1, e_2, e_3\}$ with

$$[x_i, x_j] = 0, \quad \Delta x_i = x_i \otimes 1 + (x_3+1) \otimes x_i, \quad \epsilon x_i = 0, \quad S x_i = -\tfrac{x_i}{x_3+1}$$

$$[e_i, e_j] = \epsilon_{ij}{}^k e_k, \quad [e_i, x_j] = \epsilon_{ij}{}^k x_k - \tfrac{\epsilon_{ij3}}{2}\tfrac{x^2}{x_3+1}, \quad \epsilon e_i = 0$$

$$\Delta e_i = e_i \otimes \tfrac{1}{x_3+1} + e_3 \otimes \tfrac{x_i}{x_3+1} + 1 \otimes e_i, \quad S e_i = e_3 x_i - e_i(x_3+1).$$

That this is a Hopf algebra is rather hard to check directly, and is best done via the theory above. We start with the action

$$e_i{\triangleright}x_j = \epsilon_{ij}{}^k\Big(x_k - \frac{\delta_{k3}}{2}\frac{x^2}{x_3+1}\Big)$$

of su_2 on $\mathbb{C}(SU(2)^{\star\mathrm{op}})$ induced by $\triangleleft$ above, and then make a cross product $\mathbb{C}(SU(2)^{\star\mathrm{op}}){>\!\!\triangleleft}U(su_2)$ to obtain the algebra. For the coalgebra we follow a dual construction: we construct from $\triangleright$ in (6) a *coaction*

$$\beta(e_i) = e_i \otimes \frac{1}{x_3+1} + e_3 \otimes \frac{x_i}{x_3+1}$$

and make a *cross coproduct* construction $\mathbb{C}(SU(2)^{\star\mathrm{op}}) \blacktriangleright\!\!< U(su_2)$ to obtain Δ. The axioms for a coaction β here are like the axioms for an action but with the arrows reversed. There is also a $*$-structure $e_i^* = -e_i$ and $x_i^* = x_i$ making the above into a Hopf $*$-algebra. The Hopf-von Neumann algebra version is in [21].

The manifold $SU(2)^{\star\mathrm{op}}$ is the region $\{s \in \mathbb{R}^3, \quad s_3 > -1\}$ equipped with a deformed (non-Abelian) group law[16][22]. The action $\triangleleft$ is a nonlinear deformation of the usual action of su_2 by rotations. Its orbits are non-concentrically nested spheres in this region, accumulating as $s_3 = -1$. Particles quantised on these orbits are described by the above Hopf algebra. It is a nonlinear quantum top. The dual system consists of the Lie algebra $su_2^{\star\mathrm{op}}$ acting on the group manifold $SU(2)$ with orbits which are in fact the symplectic leaves in $SU(2)$ for the Sklyanin bracket, i.e. an equally interesting system.

Recently, a four dimensional example similar to this one has been found in [23] with M a non-Abelian version of $\mathbb{R}^{1,3}$ and $g = so(1,3)$. Writing $\{x_\mu\}$ for the position-coordinates and $\{N_i, M_i\}$ for the Lorentz generators we have the Hopf algebra $U(so(1,3)) \triangleright\!\!\blacktriangleleft \mathbb{C}(M)$ as [23]

$$[x_\mu, x_\nu] = 0, \quad [M_i, M_j] = \epsilon_{ij}{}^k M_k, \quad [N_i, N_j] = \epsilon_{ij}{}^k N_k, \quad [M_i, N_j] = 0$$

$$[x_0, M_i] = 0, \quad [x_i, M_j] = \epsilon_{ij}{}^k x_k$$

$$[x_0, N_i] = -x_i, \quad [x_i, N_j] = -\delta_{ij}\left(\tfrac{\kappa}{2}(1 - e^{-\frac{2x_0}{\kappa}}) + \tfrac{x^2}{2\kappa}\right) + \tfrac{x_i x_j}{\kappa}$$

$$\Delta x_0 = x_0 \otimes 1 + 1 \otimes x_0, \quad \Delta x_i = x_i \otimes 1 + e^{-\frac{x_0}{\kappa}} \otimes x_i$$

$$\Delta M_i = M_i \otimes 1 + 1 \otimes M_i, \quad \Delta N_i = N_i \otimes 1 + e^{-\frac{x_0}{\kappa}} \otimes N_i + \tfrac{\epsilon_i{}^{jk}}{\kappa} x_j \otimes M_k$$

along with a suitable counit, antipode and $*$-structure. According to the above, we consider this the quantum algebra of observables of a system consisting of particles on orbits in our curved $\mathbb{R}^{1,3}$ under a nonlinear Lorentz transformation.

On the other hand, it is obvious that both this example and the preceding $\mathbb{R}^3$ example from [16][22][21] could be considered instead as deformed enveloping algebras of 3-dimensional or 4-dimensional Poincaré Lie algebras

$$\mathbb{C}(SU(2)^{\star\mathrm{op}}) \blacktriangleright\!\!\triangleleft U(su_2) \equiv U_\kappa(\mathrm{p}_3), \quad U(so(1,3)) \triangleright\!\!\blacktriangleleft \mathbb{C}(M) \cong U_\kappa(\mathrm{p}(1,3)).$$

In the first case we introduce a parameter κ in the formulae above. In the second case we recover a Hopf algebra isomorphic[23] to the κ-deformed Poincaré Hopf algebra introduced by other means in [24]. We regard the x_i or x_μ generators now as momentum generators. This demonstrates once again the quantum geometry transformation (5), this time dualising the interpretation of the position coordinates to view them as momentum.

4 Bicrossproduct Poisson structures and Lie bialgebras

This section announces new material in which we look at the above ideas at the semiclassical level. The semiclassical notion of a Hopf algebra is a *Poisson-Lie group*. At the Lie algebra level it is a *Lie bialgebra*[25]. These notions are useful in the theory of classical inverse scattering where Lie algebra splittings lead to solutions of the Classical Yang-Baxter equations. The examples we give in this section are however, quite different from this standard theory: they are by contrast the actual Poisson brackets whose quantisation is the bicrossproduct quantum algebras of observables described in the last section. The latter are likewise far removed from the usual quantum groups $U_q(g)$ of Drinfeld and Jimbo, being a different origin of quantum groups and Hopf algebras as coming out of ideas for Planck scale physics in [16]. Further details are in [26, Chapter 8].

Briefly, a Lie bialgebra is a Lie algebra g such that g^* is also a Lie algebra in a compatible way. In terms of g it means that there is an additional structure $\delta : g \to g \otimes g$ called the *Lie cobracket* and required to obey

$$\delta = -\tau \circ \delta, \quad (\mathrm{id} \otimes \delta) \circ \delta\xi + \mathrm{cyclic} = 0, \quad \delta([\xi, \eta]) = \mathrm{ad}_\xi(\eta) - \mathrm{ad}_\eta(\xi) \tag{7}$$

for all $\xi, \eta \in g$. Here τ is the usual transposition and ad is the adjoint action extended to $g \otimes g$ as a derivation. The dual of a finite-dimensional Lie bialgebra is also a Lie bialgebra with the role of [,] and δ interchanged.

The last condition in (7) can be understood as a Lie algebra cocycle. Hence if G is a connected simply connected Lie group with Lie algebra g then δ exponentiates to a group cocycle $D : G \to g \otimes g$. This defines a Poisson bracket

$$\{f, g\}(u) = (R_* D)(f \otimes g)(u), \quad \forall f, g \in C^\infty(G)$$

where R_{*u} is the right translation $g = T_e G \to T_u G$ applied to the tensor square $g \otimes g$ to give the 2-tensor field $R_* D$. We evaluate this tensor field as a differential operator on the functions f and g. This is the geometrical meaning of a Lie bialgebra[25]. It corresponds to a class of Poisson brackets on the group manifold. The Poisson brackets in this class behave well with respect to the group product.

We find such a structure now in the semiclassical part of our bicrossproduct Hopf algebras. These are quantisations of $C^\infty(X)$ where X is the classical phase space and in our case is a Lie group corresponding to the coproduct structure. The commutator in our quantum algebra of observables to lowest order in $\hbar$ gives us the Poisson bracket. To describe the result we start with the bicrossproduct data (6) in Lie algebra form. This is a pair g, m of Lie algebras acting on each other in a matching way as [15]

$$\triangleleft : m \otimes g \to m, \quad \triangleright : m \otimes g \to g$$

$$\begin{gathered}
[\phi, \psi] \triangleright \xi = \phi \triangleright (\psi \triangleright \xi) - \psi \triangleright (\phi \triangleright \xi), \quad \phi \triangleleft [\xi, \eta] = (\phi \triangleleft \xi) \triangleleft \eta - (\phi \triangleleft \eta) \triangleleft \xi \\
\phi \triangleright [\xi, \eta] = [\phi \triangleright \xi, \eta] + [\xi, \phi \triangleright \eta] + (\phi \triangleleft \xi) \triangleright \eta - (\phi \triangleleft \eta) \triangleright \xi \\
[\phi, \psi] \triangleleft \xi = [\phi \triangleleft \xi, \psi] + [\phi, \psi \triangleleft \xi] + \phi \triangleleft (\psi \triangleright \xi) - \psi \triangleleft (\phi \triangleright \xi)
\end{gathered} \tag{8}$$

for all $\xi, \eta \in g$ and $\phi, \psi \in m$.

It was shown in [22] that subject to a completeness condition for some resulting vector fields, such Lie algebras acting on each other exponentiate to Lie groups as in (6). Moreover, we have such a *matched pair* whenever a Lie algebra splits as a vector space into sub-Lie algebras g, m. The bigger Lie algebra is reconstructed as a *double cross sum* $g\bowtie m$ built on $g \oplus m$ with the Lie bracket in [15]. It is generated by g, m as sub-Lie algebras and the cross-bracket

$$[\phi, \xi] = \phi \triangleright \xi + \phi \triangleleft \xi.$$

This construction can be useful for generating more and more solutions of (8):

Proposition 1 *Let $(g, m, \triangleleft, \triangleright)$ be a Lie algebra matched pair as in (8). Then $(g^{\mathrm{op}} \oplus m, g\bowtie m)$ is also a Lie algebra matched pair, where $g\bowtie m$ acts by the left adjoint action and $g^{\mathrm{op}} \oplus m$ acts by $-\triangleleft, \triangleright$ for the action of g^{op}, m.*

Explicitly, g^{op} is g with the negated Lie bracket. The action of $g\bowtie m$ and matching 'backreaction' of $g^{\mathrm{op}} \oplus m$ are

$$(\xi \oplus \phi)\triangleright(\eta \oplus \psi) = ([\xi, \eta] + \phi\triangleright\eta - \psi\triangleright\xi) \oplus ([\phi, \psi] + \phi\triangleleft\eta - \psi\triangleleft\xi)$$

$$(\xi \oplus \phi)\triangleleft(\eta \oplus \psi) = (-\psi\triangleright\xi) \oplus (-\phi\triangleleft\eta)$$

when both are built on the vector space $g \oplus m$, using the Lie brackets of g, m and the original actions. The new actions obey (8) for our bigger system. We obtain a new Lie algebra $(g^{\mathrm{op}} \oplus m)\bowtie(g\bowtie m)$, and so on by iteration.

By contrast to these double cross sum Lie algebras, we now obtain from the same data (8) a Lie bialgebra.

Proposition 2 *Let $(g, m, \triangleleft, \triangleright)$ be a Lie algebra matched pair as in (8). There is a* bicross-sum *Lie bialgebra $m^* \blacktriangleright\!\!\triangleleft g$ generated by m^* with zero Lie bracket, g and the cross relations and Lie coalgebra*

$$[\xi, f] = \xi\triangleright f, \quad \delta f = \langle f, [e_a, e_b]\rangle e^a \otimes e^b, \quad \delta\xi = e_a\triangleright\xi \otimes e^a - e^a \otimes e_a\triangleright\xi$$

for all $f \in m^*$ and $\xi \in g$. Here $\{e_a\}$ is a basis of m and $\{e^a\}$ a dual basis.

This is built as a vector space on $m^* \oplus g$ with the Lie bracket etc., defined by the above on the two components. The proof is a matter of some detailed calculations to check the axioms (7) for a Lie bialgebra. To explain this construction in detail one has to introduce the notion of Lie bialgebra coactions β dual to the notion of a Lie algebra action. Then the Lie bracket is the usual semidirect sum Lie bracket for the action of g on m^* induced by $\triangleleft$, while the cobracket δ is a co-semidirect sum by a Lie coaction β induced by $\triangleright$. The dual Lie bialgebra is

$$(m^* \blacktriangleright\!\!\triangleleft g)^* = m \triangleright\!\!\blacktriangleleft g^*$$

built in an analogous way as another bicross-sum with m, g interchanged.

These constructions, like $g\bowtie m$, all have generalisations to the case when g, m are Lie bialgebras to begin with. This more general theory is the semiclassical part of the double cross product and bicrossproduct of general Hopf algebras in [15, Sec. 3]. We do not try to cover it here, but see [26]. Instead, we give some examples of Proposition 2 and its dual.

For the first bicrossproduct example in Section 3, the Lie algebra data is as follows. We have $g = su_2 = \{e_i\}$ and $m = su_2^{\star\mathrm{op}} = \{x^i\}$ say, with

$$[x^1, x^2] = 0, \quad [x^3, x^1] = x^1, \quad [x^3, x^2] = x^2 \tag{9}$$

$$x^i \triangleleft e_j = \epsilon^i{}_{jk} x^k, \quad x^i \triangleright e_j = e_3 \delta^i{}_j - \delta^i{}_3 e_j \tag{10}$$

which solve (8). Both Lie algebras here are 3-dimensional and $m \blacktriangleright\!\!\triangleleft g^*$ is six-dimensional. It has the structure (9) and

$$[e^i, e^j] = 0, \quad [x^i, e^j] = \delta^i{}_3 e^j - e^i \delta^j{}_3$$

$$\delta e^i = \epsilon^i{}_{jk} e^j \otimes e^k, \quad \delta x^i = \epsilon^i{}_{jk}(e^j \otimes x^k - x^k \otimes e^j)$$

where $\{e^i\}$ is a dual basis to the su_2 basis. Its associated Lie group is a six-dimensional manifold X on which the cobracket δ defines a natural Poisson structure as explained above. Its quantisation is the bicrossproduct Hopf algebra in Section 3. From another point of view, the same Hopf algebra is a deformation of the enveloping algebra of a Poincaré Lie bialgebra $m^* \blacktriangleright\!\!\triangleleft g$ with structure

$$[x_i, x_j] = 0, \quad [e_i, e_j] = \epsilon_{ij}{}^k e_k, \quad [e_i, x_j] = \epsilon_{ij}{}^k x_k$$

$$\delta x_i = x_3 \otimes x_i - x_i \otimes x_3, \quad \delta e_i = e_3 \otimes x_i - x_i \otimes e_3 + x_3 \otimes e_i - e_i \otimes x_3$$

from Proposition 2, where $\{x_i\}$ is a dual basis to the $\{x^i\}$.

The same theory applies to the example at the end of Section 3. We have $g = so(1,3) = \{M_i, N_j\}$ and $m = \{x^\mu\}$ say, with[23]

$$[x^i, x^j] = 0, \quad [x^i, x^0] = x^i \tag{11}$$

$$\begin{gathered} M_i \triangleleft x^0 = M_i \triangleleft x^j = 0, \quad N_i \triangleleft x^0 = -N_i, \quad N_i \triangleleft x^j = \epsilon_i{}^{jk} M_k \\ M_i \triangleright x^0 = 0, \quad M_i \triangleright x^j = \epsilon_i{}^j{}_k x^k, \quad N_i \triangleright x^0 = -\delta_{ij} x^j, \quad N_i \triangleright x^j = -\delta_i{}^j x^0. \end{gathered} \tag{12}$$

One can check that $(m, g, \triangleleft, \triangleright)$ is again a Lie algebra matched pair. The Lie bialgebra $g^* \blacktriangleright\!\!\triangleleft m$ is ten-dimensional and comes from Proposition 2 as (11) and

$$[M^i, M^j] = [N^i, N^j] = [M^i, N^j] = [x^i, N^j] = [x^0, M^i] = 0$$

$$[x^0, N^i] = -N^i, \quad [x^i, M^j] = \epsilon^{ij}{}_k N^k, \quad \delta N^i = \epsilon^i{}_{jk} N^j \otimes N^k$$

$$\delta M^i = \epsilon^i{}_{jk} M^j \otimes M^k, \quad \delta x^0 = (N^i \otimes x^j - x^j \otimes N^i)\delta_{ij}$$

$$\delta x^i = N^i \otimes x^0 - x^0 \otimes N^i + \epsilon^i{}_{jk}(M^j \otimes x^k - x^k \otimes M^j)$$

where $\{N^i, M^i\}$ are a dual basis to the Lorentz basis. Its associated Lie group is a ten-dimensional manifold X on which δ defines a natural Poisson structure as explained. Its quantisation is the bicrossproduct Hopf algebra at the end of Section 3. From another point of view, the same Hopf algebra is a deformation of the enveloping algebra of a Poincaré Lie bialgebra $g\bowtie m^*$ with structure

$$[x_\mu, x_\nu] = [x_0, M_i] = [M_i, N_j] = 0, \quad [M_i, M_j] = \epsilon_{ij}{}^k M_k, \ [N_i, N_j] = \epsilon_{ij}{}^k N_k$$

$$[x_i, M_j] = \epsilon_{ij}{}^k x_k, \quad [x_i, N_j] = -\delta_{ij} x_0, \quad [x_0, N_i] = -x_i$$

$$\delta x_0 = 0, \quad \delta x_i = x_i \otimes x_0 - x_0 \otimes x_i$$

$$\delta M_i = 0, \quad \delta N_i = N_i \otimes x_0 - x_0 \otimes N_i + \epsilon_i{}^{jk}(x_j \otimes M_k - M_k \otimes x_j)$$

where $\{x_\mu\}$ is a dual basis to the $\{x^\mu\}$.

5 Braided geometry and q-Minkowski space

Now we outline another and more complete approach to quantum geometry which has emerged recently in [1][5] and subsequent works by the author and collaborators. In this approach the deformation of usual geometry is not connected conceptually with quantisation, but with *braid statistics.* We want to explore our Hopf algebra duality ideas in this context as well.

The idea of this *braided geometry* is to start with super-geometry and replace its usual Bose-Fermi statistics ± 1 by something more general. This can be an arbitrary factor q [27] or more generally it can be a matrix solution R of the quantum Yang-Baxter equation (QYBE). This equation is connected historically with quantum groups[10] and quantum inverse scattering, but we will not use it in this historical way. I.e., this is a new approach to q-deformation. There are already two big reviews by the author[2][28] so we shall be brief.

The starting point for braided geometry is the concept of a *braided group*[1]. In mathematical terms a braided group B is defined like a Hopf algebra or quantum group as in Section 2, but the coproduct $\underline{\Delta} : B \to B \otimes B$ is no longer multiplicative as it was there. Instead, it is *braided-multiplicative* in the sense

$$\underline{\Delta}(ab) = (\underline{\Delta}a)(\underline{\Delta}b), \quad (a \otimes c)(b \otimes d) = a\Psi(c \otimes b)d \tag{13}$$

for all $a, b, c, d \in B$. In diagrammatic form we use an operator $\Psi = \times : B \otimes B \to B \otimes B$ called the braiding in those places where diagrammatically we would need to write a braid crossing, i.e. in those places where there is a transposition in our constructions. We are working in fact in a kind of braided mathematics. The familiar case on which we model everything is the super case where

$$\Psi(c \otimes b) = b \otimes c(-1)^{|c||b|}$$

on homogeneous elements of degree $|\ |$.

The simplest braided group is the braided line $B = \mathbb{C}[x]$ with structure

$$\underline{\Delta}x = x\otimes 1 + 1\otimes x, \quad \underline{\epsilon}x = 0, \quad \underline{S}x = -x, \quad \Psi(x^m\otimes x^n) = q^{mn}x^n\otimes x^m$$

The braiding Ψ means for example that

$$\underline{\Delta}x^m = \sum_{r=0}^{m}\left[{m \atop r};q\right]x^r\otimes x^{m-r}; \quad \left[{m \atop r};q\right] = \frac{[m;q]!}{[r;q]![m-r;q]!}, \quad [m;q] = \frac{1-q^m}{1-q}$$

which is the origin in braided geometry of the well-known q-integers $[m,q]$.

The next simplest example is the braided plane B generated by x, y with

$$yx = qxy, \quad \underline{\Delta}x = x\otimes 1 + 1\otimes x, \quad \underline{\Delta}y = y\otimes 1 + 1\otimes y$$

$$\underline{\epsilon}x = \underline{\epsilon}y = 0, \quad \underline{S}x = -x, \quad \underline{S}y = -y$$

$$\Psi(x\otimes x) = q^2 x\otimes x, \quad \Psi(x\otimes y) = qy\otimes x, \quad \Psi(y\otimes y) = q^2 y\otimes y$$

$$\Psi(y\otimes x) = qx\otimes y + (q^2-1)y\otimes x$$

The algebra here is sometimes called the 'quantum plane'; the new part is the coproduct $\underline{\Delta}$ and the braiding Ψ. The latter is the same one that leads to the Jones knot polynomial in a more standard context.

The general construction of which these are examples is associated to a pair of invertible matrices $R, R' \in M_n\otimes M_n$ obeying

$$R_{12}R_{13}R_{23} = R_{23}R_{13}R_{12}, \quad R_{21}R' = R'_{21}R$$

$$R_{12}R_{13}R'_{23} = R'_{23}R_{13}R_{12}, \; R'_{12}R_{13}R_{23} = R_{23}R_{13}R'_{12} \quad (PR+1)(PR'-1) = 0$$

where P is the permutation matrix and $R_{12} = R\otimes\mathrm{id}$ in $M_n^{\otimes 3}$, etc. Then we define the associated braided covector algebra B with generators $\{x_i\}$ as[4]

$$x_ix_j = x_bx_aR'^a{}_i{}^b{}_j, \quad \text{i.e.,} \quad \mathbf{x}_1\mathbf{x}_2 = \mathbf{x}_2\mathbf{x}_1R'$$

$$\underline{\Delta}x_i = x_i\otimes 1 + 1\otimes x_i, \quad \underline{\epsilon}x_i = 0 \quad \underline{S}x_i = -x_i$$

$$\Psi(x_i\otimes x_j) = x_b\otimes x_aR^a{}_i{}^b{}_j, \quad \text{i.e.,} \quad \Psi(\mathbf{x}_1\otimes\mathbf{x}_2) = \mathbf{x}_2\otimes\mathbf{x}_1R.$$

We use here and below a shorthand notation where the suffices on bold-face vectors etc., refer to the position of the indices in a tensor product of matrices.

We can proceed to develop geometry in this general setting, starting with this $\underline{\Delta}$, which we call 'coaddition' since it has the additive form. We can bring this out by using a notation $\mathbf{x}\equiv\mathbf{x}\otimes 1$, $\mathbf{x}'\equiv 1\otimes\mathbf{x}$. In our braided tensor product (13) these two do not commute but rather obey the braid-statistics

$$\mathbf{x}'_1\mathbf{x}_2 = \mathbf{x}_2\mathbf{x}'_1R$$

corresponding to Ψ. The braided-homomorphism property of $\underline{\Delta}$ is then just the statement that $\mathbf{x}'' = \mathbf{x} + \mathbf{x}'$ obey the same relations of B provided we use these braid statistics between our two independent copies.

In this notation, we define braided-differentiation as[7]

$$\partial^i f(\mathbf{x}) = \left(a_i^{-1}(f(\mathbf{a}+\mathbf{x}) - f(\mathbf{x}))\right)_{\mathbf{a}=0} \equiv \text{coeff of } a_i \text{ in } f(\mathbf{a}+\mathbf{x})$$

and find that these operators obey the relations $\partial_1\partial_2 = R'\partial_2\partial_1$ of a *braided vector algebra*. This is defined like our covectors B but with upper indices. We have a braided-Leibniz rule

$$\partial^i(bc) = (\partial^i b)c + \cdot\Psi^{-1}(\partial^i \otimes b)c, \quad \forall b, c \in B$$

where there is a natural braiding between vectors and covectors defined also by R. In braided geometry all independent objects enjoy braid statistics with respect to each other.

On monomials the braided differentiation comes out as

$$\partial^i(\mathbf{x}_1 \cdots \mathbf{x}_m) = \mathbf{e}^i{}_1\mathbf{x}_2 \cdots \mathbf{x}_m \, [m; R]_{1\cdots m}$$

where $\mathbf{e}^i$ is a basis covector $(\mathbf{e}^i)_j = \delta^i{}_j$ and

$$[m; R] = 1 + (PR)_{12} + (PR)_{12}(PR)_{23} + \cdots + (PR)_{12} \cdots (PR)_{m-1,m}$$

is a certain *braided integer matrix*[7]. In the 1-dimensional case the latter become the usual $[m, q]$ and ∂^i becomes the celebrated Jackson q-derivative. Using the braided integers we can define braided-binomial coefficient matrices[7] and recover the coproduct of $x_{i_1} \cdots x_{i_m}$ much as in the 1-dimensional case above.

It is also possible to define a braided-exponential or 'plane wave' following the same lines as in (4): we define a pairing between the braided covectors and vectors of the form (3) with ∂ the braided one, and define exp to be the corresponding coevaluation as a formal powerseries. It is an eigenfunction of the ∂^i as well as an eigenfunction with respect to braided differentiation $\frac{\partial}{\partial p^i}$. We have a braided wave-particle duality. Note that if the braided integer matrices are all invertible then exp takes a simple form using $([m; R]!)^{-1}$ in the powerseries. In the one-dimensional case we recover the celebrated q-exponential $\sum_{m=0}^{\infty} \frac{x^m p^m}{[m,q]!}$.

We can also define a braided Gaussian[8] as the solution g_η of the equation

$$\partial^i g_\eta = -x_a \eta^{ai} g_\eta, \quad \underline{\epsilon}(g_\eta) = 1$$

again as a formal powerseries. Here η^{ij} is a *braided metric* defined in such a way that $x_a\eta^{ai}$ behaves like a braided vector. It obeys a number of identities with R, R'. If η obeys some further identities, the Gaussian takes a nice form involving the 1-dimensional q-exponential of $x_b x_a \eta^{ab}$.

One can also define translation-invariant integration $\int$. More precisely, it turns out to be more natural to define a linear functional $\mathcal{Z} : B \to \mathbb{C}$ where[8]

$$\mathcal{Z}[f(\mathbf{x})] = \left(\int f(\mathbf{x}) g_\eta\right)\left(\int g_\eta\right)^{-1}; \quad \mathcal{Z}[x_i] = 0, \quad \mathcal{Z}[x_i x_j] = \eta_{ba} R^a{}_i{}^b{}_j \lambda^2$$

etc. Here λ is the quantum group normalisation constant[29], depending on R.

Finally, we add to our assumptions on R the additional equations

$$R_{12}R'_{13}R'_{23} = R'_{23}R'_{13}R_{12}, \quad R'_{12}R'_{13}R_{23} = R_{23}R'_{13}R'_{12}$$
$$R'_{12}R'_{13}R'_{23} = R'_{23}R'_{13}R'_{12}$$

which gives a kind of symmetry $R \leftrightarrow -R'$. Using it, we can define forms $\theta_i = dx_i$ as like the above braided covectors but with $-R$ in place of R'. This leads to

$$\mathrm{d}\mathbf{x}_1\mathrm{d}\mathbf{x}_2 = -\mathrm{d}\mathbf{x}_2\mathrm{d}\mathbf{x}_1 R, \quad \mathbf{x}_1\mathrm{d}\mathbf{x}_2 = \mathrm{d}\mathbf{x}_2\mathbf{x}_1 R$$
$$\mathrm{d}(\mathrm{d}x_{i_1}\cdots \mathrm{d}x_{i_p} f(\mathbf{x})) = \mathrm{d}x_{i_1}\cdots \mathrm{d}x_{i_p}\mathrm{d}x_a \tfrac{\partial}{\partial x_a} f(\mathbf{x})$$

for exterior differentials in the braided approach. As this point we make contact with similar formulae for $\mathrm{d}x_i$ imposed in other approaches[30] by consistency arguments. We also define the epsilon tensor as[9]

$$\epsilon^{i_1 i_2 \cdots i_n} = \frac{\partial}{\partial\theta_{i_1}} \cdots \frac{\partial}{\partial\theta_{i_n}} \theta_1 \cdots \theta_n = ([n; -R']!)^{i_n \cdots i_1}_{12 \cdots n}.$$

This completes our survey of braided geometry. We also have a natural candidate in this approach for braided-Minkowski space. We just use the braided matrices $B(R)$ introduced in [5] with generators $\{u^i{}_j\}$ and relations $R_{21}\mathbf{u}_1 R\mathbf{u}_2 = \mathbf{u}_2 R_{21}\mathbf{u}_1 R$. These relations can be put in the form of a braided covector space with $u_I = u^{i_0}{}_{i_1}$ in a multi-index notation and $\mathbf{R}'$ a suitable matrix built from R. We did this in [5] while the correct $\mathbf{R}$ for the additive braiding Ψ was found by U. Meyer in [31] and corresponds to the braid statistics $R^{-1}\mathbf{u}'_1 R\mathbf{u}_2 = \mathbf{u}_2 R_{21}\mathbf{u}'_1 R$. It assumes that R obeys a certain Hecke condition. There is also a third matrix $\mathbf{R}_{\cdot}$ corresponding to a different braiding Ψ needed for a braided group with multiplicative $\underline{\Delta}\mathbf{u} = \mathbf{u} \otimes \mathbf{u}$ as introduced in [5]. In short, we have a natural matrix-like braided group with both coaddition and comultiplication.

Moreover, for R obeying a certain reality condition, there is a natural $*$-structure defined by $u^i{}_j{}^* = u^j{}_i$, i.e. our braided matrices are hermitian[32]. Then it is obvious that 2×2 braided hermitian matrices are a good definition of q-Minkowski space. The standard Jones R-matrix for example gives the braided-matrices $BM_q(2)$ from [5] with four generators $\mathbf{u} = \begin{pmatrix} a & b \\ c & d \end{pmatrix}$ and relations

$$ba = q^2 ab, \quad ca = q^{-2}ac, \quad da = ad, \qquad bc = cb + (1 - q^{-2})a(d - a)$$
$$db = bd + (1 - q^{-2})ab, \quad cd = dc + (1 - q^{-2})ca.$$

Previously [33][34] proposed a similar algebra as q-Minkowski space on the basis of tensoring two copies of the quantum plane as spinors. The braided approach on the other hand means that we get all the structure above: braid statistics, coaddition, differentiation, exponentiation, Gaussians, integration, forms and the epsilon tensor. See [35][8][36][9] and the review [28] for details.

Finally, we return to the braided theory and ask about braided-Lie algebras and braided-enveloping algebras. There is such a theory introduced by the author and based on the axioms[37]

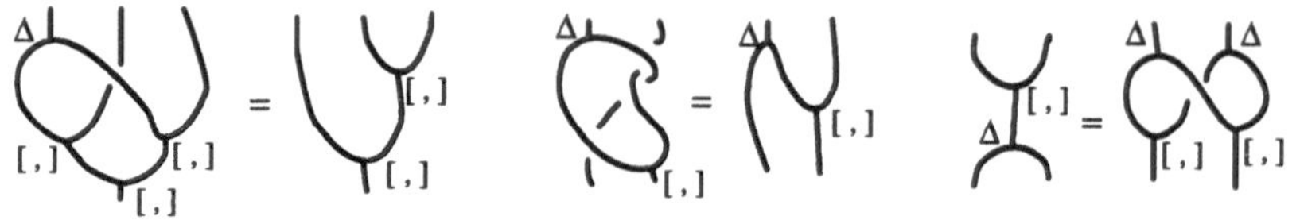

where our braided Lie-algebra $\mathcal{L}$ has a bracket [,] and an additional 'sharing out' map Δ which we usually take for granted when working with ordinary Lie algebras. There is also a map ϵ. We refer to [37] for details and for the theorem that every braided-Lie algebra has a universal enveloping braided group $U(\mathcal{L})$. Also in [37] is a general class of examples of the form

$$\mathcal{L} = \mathrm{span}\{u^i{}_j\}, \quad \Delta\mathbf{u} = \mathbf{u}\otimes\mathbf{u}, \quad [\mathbf{u}_1, R\mathbf{u}_2] = R_{21}^{-1}\mathbf{u}_2 R_{21} R$$
$$\Psi(R^{-1}\mathbf{u}_1 \otimes R\mathbf{u}_2) = \mathbf{u}_2 R^{-1} \otimes \mathbf{u}_1 R.$$

One has $U(\mathcal{L}) = B(R)$ for this braided-Lie algebra. We can also generate it by

$$\chi^i{}_j = u^i{}_j - \delta^i{}_j, \quad R_{21}\chi_1 R\chi_2 - \chi_2 R_{21}\chi_1 R = R_{21}R\chi_2 - \chi_2 R_{21} R.$$

Again the Jones R-matrix can be fed into this construction and gives us the braided-Lie algebra $gl_{2,q}$. It has generators h, x_+, x_-, γ with braided-Lie bracket

$$\begin{aligned}
&[h, x_+] = (q^{-2}+1)q^{-2}x_+ = -q^{-2}[x_+, h]\\
&[h, x_-] = -(q^{-2}+1)x_- = -q^{2}[x_-, h], \quad [x_+, x_-] = q^{-2}h = -[x_-, x_+]\\
&[h, h] = (q^{-4}-1)h, \quad [\gamma, \begin{cases} h \\ x_+ \\ x_- \end{cases}] = (1-q^{-4}) \begin{cases} h \\ x_+ \\ x_- \end{cases}
\end{aligned}$$

and zero for the others. We see that as $q \to 1$ the γ mode decouples and we have the Lie algebra $su_2 \oplus u(1)$, but for $q \neq 1$ these are unified. On the other hand, the enveloping bialgebra $U(\mathcal{L})$ has an isomorphism[37]

$$U(gl_{2,q}) \cong \mathbb{R}_q^{1,3}, \quad \begin{pmatrix} h \\ x_+ \\ x_- \\ \gamma \end{pmatrix} = (q^2-1)^{-1} \begin{pmatrix} a-d \\ c \\ b \\ q^{-2}a + d - (q^{-2}+1) \end{pmatrix}$$

which is another example of a quantum-geometry transformation, now in our braided setting! What is remarkable about this is that when $q \to 1$ the left hand side is the enveloping algebra of $su(2) \oplus u(1)$ and is non-commutative, but the right hand side is the coordinate algebra of Minkowski space and becomes commutative, i.e. the isomorphism is only possible in the braided $q \neq 1$ world! Such an isomorphism between two of the key ingredients of the standard model in particle physics suggests a deep application of q-deformation in this context.

References

[1] S. Majid. Braided groups and algebraic quantum field theories. *Lett. Math. Phys.*, 22:167–176, 1991.

[2] S. Majid. Beyond supersymmetry and quantum symmetry (an introduction to braided groups and braided matrices). In M-L. Ge and H.J. de Vega, editors, *Quantum Groups, Integrable Statistical Models and Knot Theory*, pages 231–282. World Sci., 1993.

[3] S. Majid. Anyonic quantum groups. In Z. Oziewicz et al, editor, *Spinors, Twistors, Clifford Algebras and Quantum Deformations (Proc. of 2nd Max Born Symposium, Wroclaw, Poland, 1992)*, pages 327–336. Kluwer.

[4] S. Majid. Braided momentum in the q-Poincaré group. *J. Math. Phys.*, 34:2045–2058, 1993.

[5] S. Majid. Examples of braided groups and braided matrices. *J. Math. Phys.*, 32:3246–3253, 1991.

[6] S. Majid. On the addition of quantum matrices. *J. Math. Phys.*, 35:2617–2633, 1994.

[7] S. Majid. Free braided differential calculus, braided binomial theorem and the braided exponential map. *J. Math. Phys.*, 34:4843–4856, 1993.

[8] A. Kempf and S. Majid. Algebraic q-integration and Fourier theory on quantum and braided spaces, 1994. In press *J. Math. Phys.*

[9] S. Majid. q-epsilon tensor for quantum and braided spaces, 1994.

[10] L.D. Faddeev, N.Yu. Reshetikhin, and L.A. Takhtajan. Quantization of Lie groups and Lie algebras. *Leningrad Math J.*, 1:193–225, 1990.

[11] V.G. Drinfeld. Quantum groups. In A. Gleason, editor, *Proceedings of the ICM*, pages 798–820, Rhode Island, 1987. AMS.

[12] M. Jimbo. A q-difference analog of $U(g)$ and the Yang-Baxter equation. *Lett. Math. Phys.*, 10:63–69, 1985.

[13] S. Majid. Principle of representation-theoretic self-duality. *Phys. Essays*, 4(3):395–405, 1991.

[14] S. Majid. *Non-commutative-geometric Groups by a Bicrossproduct Construction.* PhD thesis, Harvard mathematical physics, 1988.

[15] S. Majid. Physics for algebraists. *J. Algebra*, 130:17–64, 1990.

[16] S. Majid. Hopf algebras for physics at the Planck scale. *J. Classical and Quantum Gravity*, 5:1587–1606, 1988.

[17] S. Majid. Quantum random walks and time reversal. *Int. J. Mod. Phys.*, 8:4521–4545, 1993.

[18] S. Majid. Cross product quantization, nonabelian cohomology and twisting of Hopf algebras. In press *Proc. Generalised Symmetries, Clausthal, Germany, July 1993*. World Sci.

[19] G.W. Mackey. *Induced Representations*. Benjamin, New York, 1968.

[20] H.D. Doebner and J. Tolar. Quantum mechanics on homogeneous spaces. *J. Math. Phys.*, 16:975–984, 1975.

[21] S. Majid. Hopf-von Neumann algebra bicrossproducts, Kac algebra bicrossproducts, and the classical Yang-Baxter equations. *J. Funct. Analysis*, 95:291–319, 1991.

[22] S. Majid. Matched pairs of Lie groups associated to solutions of the Yang-Baxter equations. *Pac. J. Math.*, 141:311–332, 1990.

[23] S. Majid and H. Ruegg. Bicrossproduct structure of the κ-Poincaré group and non-commutative geometry. *Phys. Lett. B*, 1994.

[24] J. Lukierski, A. Nowicki, H. Ruegg, and V.N. Tolstoy. q-Deformation of Poincaré algebra. *Phys. Lett. B*, 264,271:331,321, 1991.

[25] V.G. Drinfeld. Hamiltonian structures on Lie groups, Lie bialgebras and the geometric meaning of the classical Yang-Baxter equations. *Sov. Math. Dokl.*, 27:68, 1983.

[26] S. Majid. *Foundations of Quantum Group Theory*, 1994. In press CUP.

[27] S. Majid. $\mathbb{C}$-statistical quantum groups and Weyl algebras. *J. Math. Phys.*, 33:3431–3444, 1992.

[28] S. Majid. Introduction to braided geometry and q-Minkowski space. In press *Proc. School on Quantum Groups, Varenna, Italy, June 1994*.

[29] S. Majid. Quantum and braided linear algebra. *J. Math. Phys.*, 34:1176–1196, 1993.

[30] J. Wess and B. Zumino. Covariant differential calculus on the quantum hyperplane. *Proc. Supl. Nucl. Phys. B*, 18B:302, 1990.

[31] U. Meyer. q-Lorentz group and q-Minkowski space with both braided coaddition and q-spinor decomposition, 1993. In press *Comm. Math. Phys.*

[32] S. Majid. The quantum double as quantum mechanics. *J. Geom. Phys.*, 13:169–202, 1994.

[33] U. Carow-Watamura, M. Schlieker, M. Scholl, and S. Watamura. A quantum Lorentz group. *Int. J. Mod. Phys.*, 6:3081–3108, 1991.

[34] O. Ogievetsky, W.B. Schmidke, J. Wess, and B. Zumino. q-Deformed Poincaré algebra. *Comm. Math. Phys.*, 150:495–518, 1992.

[35] S. Majid and U. Meyer. Braided matrix structure of q-Minkowski space and q-Poincaré group. *Z. Phys. C*, 1994.

[36] U. Meyer. Wave equations on q-Minkowski space, 1994.

[37] S. Majid. Quantum and braided Lie algebras. *J. Geom. Phys.*, 13:307–356, 1994.

HyperKähler Quotients and N=4 Gauge Theories in D=2 [1]

Marco Billó and Pietro Fré

SISSA - International School for Advanced Studies, via Beirut 2-4, 34140 Trieste, Italy
and I.N.F.N. - Sezione di Trieste - Trieste, Italy

1 Introduction

In this contribution we review some results we have recently obtained on the relationship between certain geometrical constructions of HyperKähler manifolds and N=4 supersymmetry in D=2 space-time. HyperKähler manifolds are of interest in connection with several issues in contemporary theorethical physics:

(i) non-compact HyperKähler manifolds in 4 dimensions can be interpreted as gravitational instantons;

(ii) 4-dimensional σ-models with compact or non-compact HyperKähler $4n$-dimensional target manifolds describe the interaction of hypermultiplets in the N=2, D=4 supersymetric theories.

In particular very recently the role *(ii)* played by HyperKähler manifolds has come under attention in relation with monopole theory, non-perturbative calculations in N=2 gauge theories and topological field theories in D=4 [1, 2, 3, 4, 5].

From the point of view *(i)* an important question not yet fully answered concerns gravitational instanton effects in string theory. Some progress has been made in constructing (4,4), c=6 superconformal models describing string propagation on such backgrounds [6, 7]; yet one has still to go a long way in order to master this kind of problems. Since string propagation on any manifold $\mathcal{M}$ is described by a $2D$ σ-model with $\mathcal{M}$ as target space it is clear that any $2D$ field theory whose effective theory is a σ-model on a HyperKähler $\mathcal{M}$ has a deep meaning in the above context.

Here we shall discuss precisely this kind of theories. They are N=4 gauge theories coupled to hypermultiplets with canonical kinetic terms, the only interaction being induced by the minimal gauge coupling. Yet the structure of N=4 supersymmetry provides the link with the geometrical set-up of HyperKähler quotients. The basic result [8] is that the auxiliary fields of the gauge multiplet realize in a lagrangian way the triholomorphic momentum map and

[1] *Talk given by P. Fré at the F. Gursey Memorial Conference, Instanbul, June 1994*

lead to an effective theory that has the HyperKähler quotient as target space. What we have found is the N=4 generalization of a pattern discovered by Witten in N=2,D=2 gauge theories. The essential difference is that while the N=2 case displays a two-phase structure, a σ-model phase and a Landau-Ginzburg phase, the N=4 case admits only the σ-model phase.

In this framework and using the geometrical results of Kronheimer [9, 10] (reviewed in this note) we show how to construct a *microscopic field theory* that admits as effective theory the N=4 σ-model on all Asymptotically Locally Euclidean (ALE) self-dual spaces. The physical data contained in the microscopic lagrangian (the Fayet-Iliopoulos parameters [11]) are put by our construction into *explicit* correspondence with the geometrical data associated with the ALE manifold, in particular the moduli of the complex structures. This correspondence is believed to be of relevance in the topological theories obtained by twisting either our microscopic two-dimensional theory [8] or N=2 supergravity in four dimensions [12, 13, 1].

2 HyperKähler quotients

HyperKähler manifolds

On a HyperKähler manifold $\mathcal{M}$, which is necessarily $4n$-dimensional, there exist three covariantly constant complex structures $\mathcal{J}^i : T\mathcal{M} \to T\mathcal{M}$, $i = 1,2,3$; the metric is hermitean with respect to all of them and they satisfy the quaternionic algebra: $\mathcal{J}^i \mathcal{J}^j = -\delta^{ij} + \varepsilon^{ijk}\mathcal{J}^k$.

In a vierbein basis $\{V^A\}$, hermiticity of the metric is equivalent to the statement that the matrices $\mathcal{J}^i_{AB}$ are antisymmetric. By covariant constancy, the three HyperKähler two-forms $\Omega^i = \mathcal{J}^i_{AB} V^A \wedge V^B$ are closed: $d\Omega^i = 0$. In the four-dimensional case, because of the quaternionic algebra constraint, the $\mathcal{J}^i_{AB}$ can be either selfdual or antiselfdual; if we take them to be antiselfdual: $\mathcal{J}^i_{AB} = -\frac{1}{2}\epsilon_{ABCD}\mathcal{J}^i_{CD}$, then the integrability condition for the covariant constancy of $\mathcal{J}^i$ forces the curvature two-form R^{AB} to be selfdual: thus, in the four-dimensional case, HyperKähler manifolds are particular instances of gravitational instantons.

A HyperKähler manifold is a Kähler manifold with respect to each of its complex structures.

Momentum map

Consider a compact Lie group G acting on a HyperKähler manifold $\mathcal{S}$ of real dimension $4n$ by means of Killing vector fields $\mathbf{X}$ that are holomorphic with respect to the three complex structures of $\mathcal{S}$; then these vector fields preserve also the HyperKähler forms:

$$\left.\begin{array}{l} \mathcal{L}_{\mathbf{X}} g = 0 \leftrightarrow \nabla_{(\mu} X_{\nu)} = 0 \\ \mathcal{L}_{\mathbf{X}} \mathcal{J}^i = 0 \ , \ i = 1,2,3 \end{array}\right\} \Rightarrow 0 = \mathcal{L}_{\mathbf{X}}\Omega^i = i_{\mathbf{X}} d\Omega^i + d(i_{\mathbf{X}}\Omega^i) = d(i_{\mathbf{X}}\Omega^i) . \tag{1}$$

Here $\mathcal{L}_{\mathbf{X}}$ and $i_{\mathbf{X}}$ denote respectively the Lie derivative along the vector field $\mathbf{X}$ and the contraction (of forms) with it.

If $\mathcal{S}$ is simply connected, $d(i_{\mathbf{X}}\Omega^i) = 0$ implies the existence of three functions $\mathcal{D}_i^{\mathbf{X}}$ such that $d\mathcal{D}_i^{\mathbf{X}} = i_{\mathbf{X}}\Omega^i$. The functions $\mathcal{D}_i^{\mathbf{X}}$ are defined up to a constant, which can be arranged so to make them equivariant: $\mathbf{X}\mathcal{D}_i^{\mathbf{Y}} = \mathcal{D}_i^{[\mathbf{X},\mathbf{Y}]}$.

The $\{\mathcal{D}_i^{\mathbf{X}}\}$ constitute then a *momentum map*. This can be regarded as a map $\mathcal{D} : \mathcal{S} \longrightarrow \mathbf{R}^3 \otimes \mathcal{G}^*$, where $\mathcal{G}^*$ denotes the dual of the Lie algebra $\mathcal{G}$ of the group G. Indeed let $x \in \mathcal{G}$ be the Lie algebra element corresponding to the Killing vector $\mathbf{X}$; then, for a given $m \in \mathcal{S}$, $\mathcal{D}_i(m) : x \longmapsto \mathcal{D}_i^{\mathbf{X}}(m) \in \mathbf{C}$ is a linear functional on $\mathcal{G}$. In practice, expanding $\mathbf{X} = X_a\mathbf{k}^a$ in a basis of Killing vectors $\mathbf{k}^a$ such that $[\mathbf{k}^a, \mathbf{k}^b] = f^{abc}\mathbf{k}^c$, where f^{abc} are the structure constants of $\mathcal{G}$, we also have $\mathcal{D}_i^{\mathbf{X}} = X_a\,\mathcal{D}_i^a$, $i = 1,2,3$; the $\{\mathcal{D}_i^a\}$ are the components of the momentum map.

HyperKähler quotient

It is a procedure that provides a way to construct from $\mathcal{S}$ a lower-dimensional HyperKähler manifold $\mathcal{M}$, as follows. Let $\mathcal{Z}^* \subset \mathcal{G}^*$ be the dual of the centre of $\mathcal{G}$. For each $\zeta \in \mathbf{R}^3 \otimes \mathcal{Z}^*$ the level set of the momentum map

$$\mathcal{N} \equiv \bigcap_i \mathcal{D}_i^{-1}(\zeta^i) \subset \mathcal{S}, \tag{2}$$

which has dimension $\dim\mathcal{N} = \dim\mathcal{S} - 3\,\dim G$, is invariant under the action of G, due to the equivariance of $\mathcal{D}$. It is thus possible to take the quotient

$$\mathcal{M} = \mathcal{N}/G.$$

$\mathcal{M}$ is a smooth manifold of dimension $\dim\mathcal{M} = \dim\mathcal{S} - 4\dim G$ as long as the action of G on $\mathcal{N}$ has no fixed points. The three two-forms ω^i on $\mathcal{M}$, defined via the restriction to $\mathcal{N} \subset \mathcal{S}$ of the Ω^i and the quotient projection from $\mathcal{N}$ to $\mathcal{M}$, are closed and satisfy the quaternionic algebra thus providing $\mathcal{M}$ with a HyperKähler structure.

Once $\mathcal{J}^3$ is chosen as the preferred complex structure, the momentum maps $\mathcal{D}_\pm = \mathcal{D}_1 \pm i\mathcal{D}_2$ are holomorphic (resp. antiholomorphic) functions.

The standard use of the HyperKähler quotient is that of obtaining non trivial HyperKähler manifolds starting from a flat $4n$ real-dimensional manifold $\mathbf{R}^{4n}$ acted on by a suitable group G generating triholomorphic isometries [14, 15].

This is the way it was utilized by Kronheimer [9, 16] in its exhaustive construction of all self-dual asymptotically locally Euclidean four-spaces (ALE manifolds) [17, 10, 18, 19], that we consider later.

Indeed the manifold $\mathbf{R}^{4n}$ can be given a quaternionic structure, and the corresponding quaternionic notation is sometimes convenient. For $n = 1$ one has the flat quaternionic space $\mathbf{H} \stackrel{\text{def}}{=} \left(\mathbf{R}^4, \{J^i\}\right)$. We represent its elements

$$q \in \mathbf{H} = x + iy + jz + kt = x^0 + x^i J^i, \qquad x, y, z, t \in \mathbf{R}$$

realizing the quaternionic structures J^i by means of Pauli matrices: $J^i = i\left(\sigma^i\right)^T$. Thus

$$q = \begin{pmatrix} u & iv^* \\ iv & u^* \end{pmatrix} \qquad \longrightarrow \qquad q^\dagger = \begin{pmatrix} u^* & -iv^* \\ -iv & u \end{pmatrix} \tag{3}$$

where $u = x^0 + ix^3$ and $v = x^1 + ix^2$. The euclidean metric on $\mathbf{R}^4$ is retrieved as $dq^\dagger \otimes dq = ds^2\mathbf{1}$. The HyperKähler forms are grouped into a quaternionic two-form

$$\Theta = dq^\dagger \wedge dq \stackrel{def}{=} \Omega^i J^i = \begin{pmatrix} i\Omega^3 & i\Omega^+ \\ i\Omega^- & -i\Omega^3 \end{pmatrix} . \tag{4}$$

For generic n, we have the space $\mathbf{H}^n$, of elements

$$q = \begin{pmatrix} u^A & iv^{A^*} \\ iv^A & u^{A^*} \end{pmatrix} \qquad \longrightarrow \qquad q^\dagger = \begin{pmatrix} u^{A^*} & -iv^{A^*} \\ -iv^A & u^A \end{pmatrix} \qquad \begin{matrix} u^A, v^A \in \mathbf{C}^n \\ A = 1, \dots n \end{matrix} \tag{5}$$

Thus $dq^\dagger \otimes dq = ds^2\mathbf{1}$ gives $ds^2 = du^{A^*} \otimes du^A + dv^{A^*} \otimes dv^A$ and the HyperKähler forms are grouped into the obvious generalization of the quaternionic two-form in eq.(4): $\Theta = \sum_{A=1}^n dq^{A\dagger} \wedge dq^A = \Omega^i J^i$, leading to $\Omega^3 = 2i\partial\bar{\partial}K$ where the Kähler potential K is $K = \frac{1}{2}\left(u^{A^*}u^A + v^{A^*}v^A\right)$, and to $\Omega^+ = 2idu^A \wedge dv^A$, $\Omega^- = (\Omega^+)^*$.

Let $(T_a)^A_B$ be the antihermitean generators of a compact Lie group G in its $n \times n$ representation. A triholomorphic action of G on $\mathbf{H}^n$ is realized by the Killing vectors of components

$$X_a = \left(\hat{T}_a\right)^A_B q^B \frac{\partial}{\partial q^A} + q^{B\dagger} \left(\hat{T}_a\right)^B_A \frac{\partial}{\partial q^{A\dagger}} \qquad ; \qquad \left(\hat{T}_a\right)^A_B = \begin{pmatrix} (T_a)^A_B & 0 \\ 0 & (T_a^*)^A_B \end{pmatrix} . \tag{6}$$

Indeed one has $\mathcal{L}_{\mathbf{X}}\Theta = 0$. The corresponding components of the momentum map are:

$$\mathcal{D}^a = q^{A\dagger} \begin{pmatrix} (T_a)^A_B & 0 \\ 0 & (T_a^*)^A_B \end{pmatrix} q^B + \begin{pmatrix} c & \bar{b} \\ b & -ic \end{pmatrix} \tag{7}$$

where $c \in \mathbf{R}$, $b \in \mathbf{C}$ are constants.

3 HyperKähler quotients in D=2, N=4 theories

Consider a supersymmetric σ-model from a 2-dimensional N=4 super-world sheet to a target space $\mathcal{S}$. It is well known that supersymmetry requires $\mathcal{S}$ to be HyperKähler . Introduce also the supersymmetric gauge multiplet for a group G acting triholomorpiycally on $\mathcal{S}$. It turns out that the auxiliary fields of the gauge multiplet $\{\mathcal{P}, \mathcal{Q}\}$, $(\mathcal{P} \in \mathbf{R}, \mathcal{Q} \in \mathbf{C})$ are identified with the momentum functions $\{\mathcal{D}^3, \mathcal{D}^\pm\}$for the G-action on $\mathcal{S}$.

In view of this fundamental property, the HyperKähler quotient offers a natural way to construct a N=4, D=2 σ-model on a non-trivial manifold $\mathcal{M}$ starting from a free σ-model on a flat-manifold $\mathcal{S} = \mathbf{H}^n$. It suffices to gauge appropriate triholomorphic isometries by means of non-propagating gauge multiplets. Omitting the kinetic term of these gauge multiplets and performing the gaussian integration of the corresponding fields one realizes the HyperKähler quotient in a Lagrangian way.

Actually the HyperKähler quotient is a generalization of a similar Kähler quotient procedure, where the momentum map $\mathcal{D} : \mathcal{S} \rightarrow \mathbf{R} \otimes \mathcal{G}^*$ consists just of one hamiltonian function, rather than three. The Kähler quotient is related with N=2,D=2 supersymmetry, the reason being that in this case the vector multiplet contains just one real auxiliary field $\mathcal{P}$.

Recently, Witten has reconsidered the Kähler quotient construction of an N=2 two-dimensional σ-model in [20]. We constructed in [8] the analogue N=4 model, and compared it with the N=2 model. We review here some fundamental points of this work.

Results for the N=2 theory

In the N=2 case a vector multiplet is composed of a gauge boson $\mathcal{A}$, namely a world-sheet 1-form, two spin 1/2 gauginos, whose four components we denote by λ^+,λ^-, $\tilde{\lambda}^+,\tilde{\lambda}^-$, a complex physical scalar M and a real auxiliary scalar $\mathcal{P}$. For simplicity we write here the formulæ for a U(1) gauge group [2] and we consider a linear superpotential [20]. $i\frac{t}{4}M$, with the coupling t is $t = ir + \frac{\theta}{2\pi}$. Denote $\partial_\pm = \partial/\partial z^o \pm \partial/\partial z^1$, z^α being the world-sheet coordinates. Then we have:

$$\begin{aligned} \mathcal{L}^{(2)}_{\text{gauge}} &= \frac{1}{2}\mathcal{F}^2 - i\left(\tilde{\lambda}^+\partial_+\tilde{\lambda}^- + \lambda^+\partial_-\lambda^-\right) - 4(\partial_+ M^*\partial_- M + \partial_- M^*\partial_+ M) \\ &\quad + 2\mathcal{P}^2 - 2r\mathcal{P} + \frac{\theta}{2\pi}\mathcal{F} \end{aligned} \tag{8}$$

$r\mathcal{P}$ is the Fayet-Iliopoulos term [11], $\theta\mathcal{F}/2\pi$ a topological term. We consider then n chiral multiplets $X^i, \psi^i, \tilde{\psi}^i, \mathcal{H}_i$ (plus their complex conjugates) of charge q^i with respect to the above $U(1)$ group. The complex scalars X^i span $\mathbf{C}^n$. The complex auxiliary field $\mathcal{H}^i$ is identified with the derivative of a holomorphic superpotential $W(X)$: $\mathcal{H}^i = \partial_{i^*}W^*$. The supersymmetric lagrangian is

$$\begin{aligned} \mathcal{L}^{(2)}_{\text{chiral}} &= -(\nabla_+ X^{i^*}\nabla_- X^i + \nabla_- X^{i^*}\nabla_+ X^i) + 4i(\psi^i\nabla_-\psi^{i^*} + \tilde{\psi}^i\nabla_+\tilde{\psi}^{i^*}) \\ &\quad + 8\Big(\left(\psi^i\tilde{\psi}^j\partial_i\partial_j W + \text{c.c.}\right) + \partial_i W\partial_{i^*}W^*\Big) + 2i\sum_i q^i\left(\psi^i\tilde{\lambda}^- X^{i^*} - \tilde{\psi}^i\lambda^- X^{i^*} - \text{c.c.}\right) \\ &\quad + 8i\left(M^*\sum_i q^i\psi^i\tilde{\psi}^{i^*} - \text{c.c.}\right) + 8M^*M\sum_i (q^i)^2 X^{i^*}X^i - 2\mathcal{P}\sum_i q^i X^{i^*}X^i \end{aligned} \tag{9}$$

Here and henceforth ∇ is the covariant derivative constructed by means of the gauge connection $\mathcal{A}$.

The total lagrangian is $\mathcal{L}^{(2)} = \mathcal{L}^{(2)}_{\text{gauge}} - \mathcal{L}^{(2)}_{\text{chiral}}$, the relative sign being fixed by the requirement of positivity of the energy. Note that $\mathcal{D}^{\mathbf{X}}(X, X^*) = \sum_i q^i|X^i|^2$ is the momentum map function for the holomorphic action of the gauge group on the matter multiplets. Eliminating $\mathcal{P}$ through its own equation of motion we get: $\mathcal{P} = -\frac{1}{2}(D^{\mathbf{X}}(X, X^*) - r)$ and the bosonic potential reduces to:

$$U = \frac{1}{2}\left[r - \sum_i q^i|X^i|^2\right]^2 + 8|\partial_i W|^2 + 8|M|^2\sum_i (q^i)^2|X^i|^2 \tag{10}$$

[2] The extension to several abelian groups is trivial and for a generic group it is contained in [8] in the same setting as here.

Let us focus now on the low energy effective theory for this model. First of all consider the structure of the classical vacua. We must extremize the potential (10). Witten considered [20] the case with L.G. potential $W = X^0 \mathcal{W}(X^i)$ where $\mathcal{W}(X^i)$ is quasi homogeneous of degree d if the X^i's are assigned as homogeinity weigths their charges q^i, and is transverse: $\partial_i \mathcal{W} = 0 \; \forall i$ iff $X^i = 0 \; \forall i$. X^0 has charge $-d$. Then two phases appear:

- $r > 0$: *σ-model phase.* The space of classical vacua is a transverse hypersurface embedded in $\mathbf{WCP}^N_{q^1 \ldots q^N}$: $X^0 = M = 0$, $\sum_i q^i |X^i|^2 = r$, $\mathcal{W}(X^i) = 0$.

- $r < 0$: *Landau-Ginzburg phase.* The space of vacua is a point: $X^0 = \sqrt{\frac{-r}{d}}$, $X^i = M = 0$. The low-energy theory describes massles fields governed by a Landau-Ginzburg potential $\mathcal{W}(X^i)$

Our interest is in the low-energy theory around a vacuum of the first type. This theory turns out to be correctly described by a N=2 σ-model [8]. It emerges via the lagrangian realization of a Kähler quotient. Here we look just at the bosonic fields. To be simple and definite, consider the $\mathbf{CP}^n$ model, that is the case $q^A = 1$, $A = 0, 1, \ldots, n$ and $W(X^A) = 0$. In this case there is only the σ-model phase.

Reinstall the gauge coupling constant g in the lagrangian $\mathcal{L}^{(2)}$. Then let $g \to \infty$. We are left with a gauge invariant lagrangian describing matter coupled to gauge fields that have no kinetic terms. Then we vary the action in these fields. The resulting equations of motion express the gauge fields in terms of the matter fields. This procedure is nothing else, from the functional integral viewpoint, but the gaussian integration over the gauge multiplet in the limit $g \longrightarrow \infty$. It amounts to deriving the low-energy effective action around the classical vacua of the complete, gauge plus matter system. Indeed we have seen that around these vacua the oscillations of the gauge fields are massive, and thus decouple from the low-energy point of view.

In our example, the variation in the gauge connection components $\mathcal{A}_\pm$ and in M identifies them in terms of the matter fields. In particular $\mathcal{A}_\pm = -\mathrm{i}(X\partial_\pm X^* - X^*\partial_\pm X)/2X^*X$. This has to be substituted into the lagrangian. This latter is by construction invariant under the U(1) transformation $X^A \to e^{i\Phi} X^A$, $\Phi \in \mathbf{R}$. Allow now $\Phi \in \mathbf{C}$, thus complexifying U(1) to $\mathbf{C}^* = \mathbf{C} - \{0\}$. Introduce an extra field v, transforming under $\mathbf{C}^*$ as $v \to v + \frac{i}{2}(\Phi - \Phi^*)$. Then the combinations $e^{-v} X^A$ undergo just a $U(1)$ transformation: $e^{-v} X^A \to e^{i\mathrm{Re}\Phi} e^{-v} X^A$.

By substituting in the lagrangian $X^A \to e^{-v} X^A$ it becomes $\mathbf{C}^*$-invariant. In particular the term involving $\mathcal{P}$ becomes $-2\mathcal{P}(r - e^{-2v} X^* X)$. Performing the variation with respect to the auxiliary field $\mathcal{P}$, the resulting equation of motion identifies the extra scalar field v in terms of the matter fields. Introducing $\rho^2 \equiv r$ the result is that $e^{-v} = \frac{\rho}{\sqrt{X^* X}}$.

What is the geometrical meaning of the above "tricks" (introduction of the extra field v, consideration of the complexified gauge group)? The answer relies on the properties of the Kähler quotient construction [14]; Let us recall a few concepts, using notions and notations introduced in section 1

Let $\mathbf{Y}(s) = Y^a \mathbf{k}_a(s)$ be a Killing vector on $\mathcal{S}$ (in our case $\mathbf{C}^{N+1}$), belonging to $\mathcal{G}$ (in our case $\mathbf{R}$), the algebra of the gauge group. In our case $\mathbf{Y}$ has a single component: $\mathbf{Y} = i\Phi(X^A \frac{\partial}{\partial X^A} - X^{A^*} \frac{\partial}{\partial X^{A^*}})$ ($\Phi \in \mathbf{R}$). The X^A's are the coordinates on $\mathcal{S}$. Consider the vector field $I\mathbf{Y} \in \mathcal{G}^c$ (the complexified algebra), I being the complex structure acting on $T\mathcal{S}$. In our case $I\mathbf{Y} = \Phi(X^A \frac{\partial}{\partial X^A} + X^{A^*} \frac{\partial}{\partial X^{A^*}})$. This vector field is orthogonal to the hypersurface $\mathcal{D}^{-1}(\zeta)$, for any level ζ; that is, it generates transformations that change the level of the surface. In our case the surface $\mathcal{D}^{-1}(\rho^2) \in \mathbf{C}^{N+1}$ is defined by the equation $X^{A^*}X^A = \rho^2$. The infinitesimal transformation generated by $I\mathbf{Y}$ is $X^A \rightarrow (1+\Phi)X^A$, $X^{A^*} \rightarrow (1+\Phi)X^{A^*}$ so that the transformed X^A's satisfy $X^{A^*}X^A = (1+2\Phi)\rho^2$. As recalled in section I, the Kähler quotient consists in starting from $\mathcal{S}$, restricting to $\mathcal{N} = \mathcal{D}^{-1}(\zeta)$ and taking the quotient $\mathcal{M} = \mathcal{N}/G$. The above remarks about the action of the complexified gauge group suggest that this is equivalent (at least if we skip the problems due to the non-compactness of G^c) to simply taking the quotient $\mathcal{S}/G^c$, the so-called "algebro-geometric" quotient [14, 21].

The Kähler quotient allows in principle to determine the expression of the Kähler form on $\mathcal{M}$ in terms of the original one on $\mathcal{S}$. Schematically, let j be the inclusion map of $\mathcal{N}$ into $\mathcal{S}$, p the projection from $\mathcal{N}$ to the quotient $\mathcal{M} = \mathcal{N}/G$, Ω the Kähler form on $\mathcal{S}$ and ω the Kähler form on $\mathcal{M}$. It can be shown [14] that

$$\begin{array}{ccccc} \mathcal{S} & \stackrel{j}{\longleftarrow} & \mathcal{N} = \mathcal{D}^{-1}(\zeta) & \stackrel{p}{\longrightarrow} & \mathcal{M} = \mathcal{N}/G \\ \Omega & \longrightarrow & j^*\Omega = p^*\omega & \longleftarrow & \omega \end{array} \tag{11}$$

In the algebro-geometric setting, the holomorphic map that associates to a point $s \in \mathcal{S}$ (for us, $\{X^A\} \in \mathbf{C}^{N+1}$) its image $m \in \mathcal{M}$ is obtained as follows:
i) Bringing s to $\mathcal{N}$ by means of the finite action infinitesimally generated by a vector field of the form $\mathbf{V} = I\mathbf{Y} = V^a \mathbf{k}_a$

$$\pi: \quad s \in \mathcal{S} \longrightarrow e^{-V} s \in \mathcal{D}^{-1}(\zeta) \tag{12}$$

ii) Projecting e^{-V} to its image in the quotient $\mathcal{M} = \mathcal{N}/G$.

Thus we can consider the pullback of the Kähler form ω through the map $p \cdot \pi$:

$$\begin{array}{ccccc} \mathcal{S} & \stackrel{\pi}{\longrightarrow} & \mathcal{N} = \mathcal{D}^{-1}(\zeta) & \stackrel{p}{\longrightarrow} & \mathcal{N}/G \\ \pi^* p^* \omega & \longleftarrow & p^*\omega & \longleftarrow & \omega \end{array} \tag{13}$$

Looking at (11) we see that $\pi^* p^* \omega = \pi^* j^* \Omega$ so that at the end of the day, in order to recover the pullback of ω to $\mathcal{S}$ it is sufficient:
i) to restrict Ω to $\mathcal{N}$
ii) to pull back this restriction to $\mathcal{M}$ with respect to the map $\pi = e^{-V}$.

We see from (12) that the components of the vector field $\mathbf{V}$ must be determined by requiring

$$\mathcal{D}(e^{-V}s) = \zeta \tag{14}$$

In the lagrangian context,after having introduced the extra field v (which is now interpreted as the unique component of the vector field $\mathbf{V}$) to make the lagrangian $\mathbf{C}^*$-invariant, eq. (14) is retrieved as the equation of motion for $\mathcal{P}$. We have determined the form of the map π : it corresponds on the bosonic fields to $X^A \to e^{-v}X^A$.

The remaining steps in treating the lagrangian just consist in implementing the Kähler quotient as in (13). At the end we obtain the σ-model on the target space $\mathcal{M}$ (in our case $\mathbf{CP}^N$) endowed with the Kähler metric corresponding to the Kähler form ω. In our example such metric is the Fubini-Study metric. Indeed in full generality one can show [14] that the Kähler potential $\hat{K}$ for the manifold $\mathcal{M}$, such that $\omega = 2i\partial\bar{\partial}\hat{K}$ is given by

$$\hat{K} = K|_{\mathcal{N}} + V^a\zeta_a \tag{15}$$

Here K is the Kähler potential on $\mathcal{S}$; $K|_{\mathcal{N}}$ is its restriction to $\mathcal{N}$, that is, it is computed after acting on the point $s \in \mathcal{S}$ with the transformation e^{-V} determined by eq. (14); V^a are the components of the vector field $\mathbf{V}$ along the ath generator of the gauge group, and ζ_a those of the level ζ of the momentum map; recall that the ζ belong to the dual of the center, $\mathcal{Z}^*$ and therefore only the components of $\mathbf{V}$ along the center actually contribute to eq. (15). In our case we have the single component v given by $e^{-v} = \rho^2/\sqrt{X^*X}$; ρ^2 is the single component of the level. The original Kähler potential on $\mathcal{S} = \mathbf{C}^{N+1}$ is $K = \frac{1}{2}X^{A^*}X^A$ so that when restricted to $\mathcal{D}^{-1}(\rho^2)$ it takes an irrelevant constant value $\frac{\rho^2}{2}$. Thus we deduce from (15) that the Kähler potential for $\mathcal{M} = \mathbf{CP}^N$ that we obtain is $\hat{K} = \frac{1}{2}\rho^2\log(X^*X)$. Fixing a particular gauge to perform the quotient with respect to $\mathbf{C}^*$ (see later), this potential can be rewritten as $\hat{K} = \frac{1}{2}\rho^2\log(1+x^*x)$, namely the Fubini-Study potential.

Indeed it is trivial to rewrite the lagrangian after the substitution $X^A \to (\rho^2/\sqrt{X^*X})X^A$. We can then utilize the gauge invariance to fix for instance, in the coordinate patch where $X^0 \neq 0$, $X^0 = 1$. That is, we fix completely the gauge going from the homogeneous coordinates (X^0, X^i) to the inhomogeneous coordinates $(1, x^i = X^i/X^0)$ on $\mathbf{CP}^N$.

Having chosen our gauge, we rewrite the lagrangian in terms of the fields x^i Thus the bosonic lagrangian reduces finally to that of a σ-model on $\mathbf{CP}^n$:

$$\mathcal{L} = g_{ij^*}(\partial_+ x^i\partial_- x^{j^*} + \partial_- x^i\partial_+ x^{j^*}) \tag{16}$$

where g_{ij^*} is the Fubiny-Study metric, $g_{ij^*} = \frac{\rho^2}{1+x^*x}\left(\delta_{ij} - \frac{x^{i^*}x^j}{1+x^*x}\right)$.

The N=4 theory

The N=4 vector multiplet [3], in addition to the gauge boson, namely the 1-form $\mathcal{A}$, contains four spin 1/2 gauginos whose eight components are denoted by λ^+,λ^-, $\tilde{\lambda}^+$,$\tilde{\lambda}^-$,μ^+,μ^-, $\tilde{\mu}^+$,$\tilde{\mu}^-$, two complex physical scalars $M \neq M^*$, $N \neq N^*$,

[3] As for N=2 we write the formulae for a U(1) gauge group and we refer for extension to a group G to [8].

and three auxiliary fields arranged into a real scalar $\mathcal{P} = \mathcal{P}^*$ and a complex scalar $\mathcal{Q} \neq \mathcal{Q}^*$.

$$\begin{aligned}
\mathcal{L}^{(4)}_{\text{gauge}} &= \frac{1}{2}\mathcal{F}^2 - i(\tilde{\lambda}^+\partial_+\tilde{\lambda}^- + \tilde{\mu}^+\partial_+\mu^- + \lambda^+\partial_-\lambda^- + \mu^+\partial_-\tilde{\mu}^-) \\
&+ 4(\partial_+ M^*\,\partial_- M + \partial_- M^*\,\partial_+ M + \partial_+ N^*\,\partial_- N + \partial_- N^*\,\partial_+ N) \\
&+ \frac{\theta}{2\pi}\mathcal{F} + 2\mathcal{P}^2 + 2\mathcal{Q}^*\mathcal{Q} - 2r\mathcal{P} - (s\mathcal{Q}^* + s^*\mathcal{Q}) \qquad (17)
\end{aligned}$$

Note that in addition to the Fayet-Iliopoulos and the topological term present in the N=2 case we have term linear in the complex auxiliary field $\mathcal{Q}$, involving a new complex parameter s.

The quaternionic hypermultiplets are the N=4 analogues of the N=2 chiral multiplets. They are described by a set of bosonic complex fields u^i, v^i, that can be organized in quaternions

$$Y^i = \begin{pmatrix} u^i & iv^{i^*} \\ iv^i & u^{i^*} \end{pmatrix} \qquad (18)$$

spanning $\mathbf{H}^n$. Their supersymmetric partners are four spin 1/2 fermions, whose eight components we denote by $\psi^i_u, \tilde{\psi}^i_u, \psi^i_v, \tilde{\psi}^i_v$ together with their complex conjugates $\psi^{i^*}_u, \tilde{\psi}^{i^*}_u, \psi^{i^*}_v, \tilde{\psi}^{i^*}_v$. On these matter fields the abelian gauge group acts in a *triholomorphic* fashion, which in our setting means that $\{u^i, v^i\}$ have charges $\{q^i, -q^i\}$. The lagrangian is:

$$\begin{aligned}
\mathcal{L}^{(4)}_{\text{quatern}} &= -(\nabla_+ u^{i^*}\,\nabla_- u^i + \nabla_- u^{i^*}\,\nabla_+ u^i + \nabla_+ v^{i^*}\,\nabla_- v^i + \nabla_+ v^{i^*}\,\nabla_- v^i) \\
&+ 4i(\psi^i_u\,\nabla_-\psi^{i^*}_u + \psi^i_v\,\nabla_-\psi^{i^*}_v + \tilde{\psi}^i_u\nabla_+\tilde{\psi}^{i^*}_u + \tilde{\psi}^i_v\,\nabla_+\tilde{\psi}^{i^*}_v) \\
&+ 2i\sum_i q^i\Big\{\Big[\psi^i_u\,(\tilde{\lambda}^- u^{i^*} + \tilde{\mu}^+ v^i) - \text{c.c.}\Big] - \Big[\psi^i_v\,(\tilde{\lambda}^- v^{i^*} - \tilde{\mu}^+ u^i) - \text{c.c.}\Big] \\
&- \Big[\tilde{\psi}^i_u\,(\lambda^- u^{i^*} + \mu^+ v^i) - \text{c.c.}\Big] + \Big[\tilde{\psi}^i_v\,(\lambda^- v^{i^*} - \mu^+ u^i) - \text{c.c.}\Big]\Big\} \\
&+ 8i\Big[M^*\sum_i q^i(\psi^i_u\tilde{\psi}^{i^*}_u - \psi^i_v\,\tilde{\psi}^{i^*}_v) - \text{c.c.}\Big] - 8i\Big[N\sum_i q^i(\psi^i_u\,\tilde{\psi}^i_v + \psi^i_v\,\tilde{\psi}^i_u) - \text{c.c.}\Big] \\
&+ 8(|M|^2 + |N|^2)\sum_i (q^i)^2(|u^i|^2 + |v^i|^2) - 2\mathcal{P}\sum_i q^i(|u^i|^2 - |v^i|^2) \\
&+ 2i(\mathcal{Q}\sum_i q^i u^i v^i - \text{c.c.}) \qquad (19)
\end{aligned}$$

Comparing with formulae (7) we see that the auxiliary field $\mathcal{P}$ multiplies the real component $\mathcal{D}^3(u^i, v^i) = \sum_i q^i\,(|u^i|^2 - |v^i|^2)$, while $\mathcal{Q}$ multiplies the holomorphic component $\mathcal{D}^-(u^i, v^i) = -2i\sum_i q^i\,u^i\,v^i$ of the momentum map.

Consider $\mathcal{L}^{(4)} = \mathcal{L}^{(4)}_{\text{gauge}} - \mathcal{L}^{(4)}_{\text{quatern}}$. Varying in $\mathcal{P}$ and $\mathcal{Q}$ we obtain :

$$\mathcal{P} = \frac{1}{2}\Big[r - \sum_i q^i(|u^i|^2 - |v^i|^2)\Big] = \frac{1}{2}\Big[r - \mathcal{D}^3(u, v)\Big]$$

$$\mathcal{Q} = \frac{1}{2}\Big[s - 2i\sum_i q^i u^i v^i\Big] = \frac{1}{2}\Big[s - \mathcal{D}^+(u,v)\Big] \tag{20}$$

Using eq.s (20) from the lagrangian we can extract the final form of the N=4 bosonic potential. We make an exception to our choice of focusing on U(1) gauge group and write this potential in the case where the gauge group is the direct sum of several U(1)'s. This is done in order to make contact with the HyperKähler quotients of the next section. Use an index a to enumerate such U(1) factors; the triholomorphic action is described by the matrices $(F^a)^i_j$ acting on the u^j and $-F^a$ on the v^j fields. In practice one has to replace $q^i \longrightarrow (F^a)^i_j$ and sum over a. Then we get

$$U = \sum_a \Big(\frac{1}{2}\big(r_a - \mathcal{D}^3_a\big)^2 + \frac{1}{2}|s_a - \mathcal{D}^+_a|^2 + 8\big(|M_a|^2 + |N_a|^2\big) \sum_{i,j} (F^a F^a)_{ij}\,(u^{i^*} u^j + v^{i^*} v^j)\Big) \tag{21}$$

As we see, the parameters r, s of the N=4 Fayct-Iliopoulos term are identified with the levels of the triholomorphic momentum-map.

Minimizing the potential (21) we find only a N=4 σ-model phase; the reason is the absence of an N=4 analogue of the Landau-Ginzburg potential. Besides $M = N = 0$, we must impose $\mathcal{D}^3(u,v) = r$ and $\mathcal{D}^+ = s$. Taking into account the gauge invariance of the Lagrangian, this means that the classical vacua are characterized by having $M = N = 0$ and the matter fields u, v lying on the HyperKähler quotient

$$\mathcal{M} = \mathcal{D}_3^{-1}(r) \cap \mathcal{D}_+^{-1}(s)/U(1) \tag{22}$$

of the quaternionic space $\mathbf{H}^n$ spanned by the fields u^i, v^i with respect to the triholomorphic action of the $U(1)$ gauge group. Considering the fluctuations around this vacuum, we can see that the fields of the gauge multiplet are massive, together with the modes of the matter fields not tangent to $\mathcal{M}$. The low-energy theory will turn out to be the N=4 σ-model on $\mathcal{M}$.

It is useful at this point, in order to compare with the N=2 case, to note that N=4 theories are nothing else but particular N=2 theories whose structure allows the existence of additional supersymmetries.

Thus if we look at the N=4 theory described above from an N=2 point of view, it contains one gauge multiplet and $2n + 1$ chiral multiplets, whose bosonic components we denote as $\{X^A\}$, $A = 0, \ldots n$. They are explicitely $\{X^0 = 2N, u^i, v^i\}$, with charges $\{0, q^i, -q^i\}$. The Landau Ginzburg potential is

$$W\left(X^A\right) = -\frac{1}{4}X^0\left(s^* - \mathcal{D}^-(u,v)\right) = -\frac{1}{4}X^0\left(s^* + 2i\sum_i q^i u^i v^i\right) \tag{23}$$

where $\mathcal{D}^-(u,v) = -2i\sum_i q^i u^i v^i$ is the holomorphic part of the momentum map for the triholomorphic action of the gauge group on $\mathbf{H}^n \equiv \mathbf{C}^{2n}$. The Landau-Ginzburg potential being given by eq. (23), the form (10) of the N=2 bosonic potential reduces exactly to the potential of eq. (21) (for U(1) gauge group):

$$\begin{aligned} U &= \frac{1}{2}\left(r-\mathcal{D}^3\right)^2+\sum_A |\partial_A W|^2+8|M|^2\sum_i (q^i)^2\left(|u^i|^2+|v^i|^2\right) \\ &= \frac{1}{2}\left(r-\mathcal{D}^3\right)^2+\frac{1}{2}|s-\mathcal{D}^+|^2+\left(8|M|^2+2|X^0|^2\right)\sum_i (q^i)^2\left(|u^i|^2+|v^i|^2\right) \end{aligned} \tag{24}$$

From this N=2 point of view we do not see two different phases in the structure of the classical vacuum because of the expression of $\mathcal{D}^3(u,v)$, see eq. (20). It is indeed clear that by setting

$$r-\sum_i q^i\left(|u^i|^2-|v^i|^2\right)=0 \tag{25}$$

the exchange of $r>0$ with $r<0$ just corresponds to the exchange of the u's with the v's. Everywhere else the u's and the v's appear symmetrically, hence the two phases $r>0$ and $r<0$ are actually the same thing. This is far from being accidental. The charge of v^i is opposite to the one of u^i because of the triholomorphicity of the action of the gauge group, which is essential in a N=4 theory; thus the indistinguishability of the two phases is intrinsic to any N=4 theory of the type we are considering in this paper.

To complete the definition of the vacuum, we must set $M=X^0=0$ and require $\mathcal{D}^+(u,v)=s$.

We now examine the reconstruction of the low-energy theory. We again focus on the bosonic sector. We utilize the above N=2 point of view, and proceed as before. We just insist now on the peculiarities due to the form of the potential (24).

By letting the gauge coupling constant go to infinity, we eliminate the gauge kinetic terms and also that for X^0, since it originates from the N=4 gauge multiplet.

Note that the holomorphic contraint $\mathcal{D}^+=s$ is not implemented in the N=2 lagrangian we are starting from through a Lagrange multiplier. This would be the case (by means of the auxiliary field $\mathcal{Q}$) had we chosen to utilize the N=4 formalism, see eq.(19), and this is the case for the real constraint $\mathcal{D}^3=r$, through the auxiliary field $\mathcal{P}$. This fact causes no problem, as it is perfectely consistent with what happens, from the geometrical point of view, taking the HyperKähler quotient. Indeed the HyperKähler quotient procedure is schematically represented by

$$\mathcal{S}\ \overset{j^+}{\longleftarrow}\ \mathcal{D}_+^{-1}(s)\ \overset{j^3}{\longleftarrow}\ \mathcal{N}\equiv\mathcal{D}_3^{-1}(r)\cap\mathcal{D}_+^{-1}(s)\ \overset{p}{\longrightarrow}\ \mathcal{M}\equiv\mathcal{N}/G \tag{26}$$

where we have extended in an obvious way the notation of eq. (11): j^+ and j^3 are the inclusion maps and p the projection on the quotient.

We already remarked that the surface $\mathcal{D}_3^{-1}(r)$ is not invariant under the action of the *complexified* gauge group G^c. Instead it is easy to verify that the holomorphic surface $\mathcal{D}_+^{-1}(s)$ *is* invariant under the action of G^c. Just as in the Kähler quotient procedure we can therefore replace the restriction to $\mathcal{D}_3^{-1}(r)$ and the G quotient with a G^c quotient, without modifying the need of taking the restriction to $\mathcal{D}_+^{-1}(s)$. The HyperKähler quotient can be realized as follows:

$$\mathcal{S} \stackrel{j^+}{\longleftarrow} \mathcal{D}_+^{-1}(s) \stackrel{p^c}{\longrightarrow} \mathcal{M} \equiv \mathcal{D}_+^{-1}(s)/G^c \tag{27}$$

We see that, in any case, we have to implement the constraint $\mathcal{D}^+ = s$. This does not affect the procedure of extending the action of the gauge group to its complexification. Setting now, to be simple and definite, $q^i = 1 \; \forall i$ [4], the complexified group acts as:

$$u^i \longrightarrow e^{i\Phi}u^i \qquad ; \qquad v^i \longrightarrow e^{-i\Phi}v^i$$
$$v \longrightarrow v + \frac{i}{2}(\Phi - \Phi^*) \tag{28}$$

One obtains the invariance of the lagrangian under this action by means of the substitutions $u^i \to e^{-v}u^i$, $v^i \to e^{-v}v^i$. The variation, after these replacements, in the auxiliary field $\mathcal{P}$ gives the equation $\mathcal{D}^3(e^{-v}u, e^v v) = r$, that is

$$r - e^{-2v}\sum_i |u^i|^2 + e^{2v}\sum_i |v^i|^2 = 0 \tag{29}$$

This is easily solved for v. We have still to implement the holomorphic constraint $\mathcal{D}_+ = s$. This task is simplified by the $\mathbf{C}^*$ gauge invariance of our lagrangian. As it is clear from the form of the $\mathbf{C}^*$-transformations one can for instance choose the gauge $u^n = v^n$. Solving explicitely the constraint gives the $\{u^i, v^i\}$, $i = 1, \dots n$ in terms of some irreducible coordinates $\{\hat{u}^I, \hat{v}^I\}$,$I = 1, \dots n-1$. The final result of the appropriate manipulations that should be made on the lagrangian will be the reconstruction of the action for the N=2 σ-model having as target space the HyperKähler quotient $\mathbf{H}^n/U(1)$, endowed with the Kähler metric which is naturally provided by this onstruction, exactly ias it happened in the Kähler quotient case. The Kähler quotient is again obtained through eq. (15). It is convenient to call $\beta = \sum_i |u^i|^2$ and $\gamma = \sum_i |v^i|^2$ (both to be considered as functions of $\hat{u}^I, \hat{v}^I$). Differently from the $\mathbf{CP}^N$ case, the part of the target space Kähler potential coming from the restriction of the Kähler potential for $\mathbf{H}^n$ to the surface $\mathcal{D}_3^{-1}(r) \cap \mathcal{D}_+^{-1}(s)$ is not an irrelevant constant. Indeed it is given (see section 1) by:

$$K|_{\mathcal{N}} = \frac{1}{2}(e^{-2v}\sum_i |u^i|^2 + e^{2v}\sum_i |v^i|^2) = \frac{1}{2}\sqrt{\rho^4 + 4\beta\gamma} \tag{30}$$

The final expression of the Kähler potential for the Calabi metric is:

$$\hat{K} = \frac{1}{2}\sqrt{\rho^4 + 4\beta\gamma} + \frac{\rho^2}{2}\log\frac{-\rho^2 + \sqrt{\rho^4 + 4\beta\gamma}}{2\gamma} \tag{31}$$

[4] This corresponds to the obvious N=4 generalization of $\mathbf{CP}^N$. The spaces obtained by means of the hyperKähler quotient procedure of $\mathbf{H}^n$ with respect to this $U(1)$ action have real dimension $4(n-1)$; the Kähler metric metric they inherit from the quotient construction are called Calabi metrics [22]

In the case $n = 2$, the target space has 4 real dimensions and the Calabi metric is nothing else that the Eguchi-Hanson metric, i.e. the simplest Asymptotically Locally Euclidean (ALE) gravitational instanton [17, 14, 9].

4 ALE spaces as hyperKähler quotients

The most natural gravitational analogues of the Yang-Mills instantons are geodesically complete Riemannian four-manifolds with (anti)selfdual curvature 2-form [5]. One would like their metric to approach the euclidean metric at infinity, in agreement with the "intuitive" picture of instantons as being localized in finite regions of space-time. This behaviour is however possible only modulo an additional subtlety: the base manifold has a boundary at infinity S^3/Γ, Γ being a finite group of identifications. "Outside the core of the instanton" the manifold looks like $\mathbf{R}^4/\Gamma$ instead of $\mathbf{R}^4$. This is the reason of the name given to these spaces: the asymptotic behaviour is only *locally* euclidean. The unique *globally* euclidean gravitational instanton is euclidean four-space itself, wich has boundary S^3.

The simplest of the ALE metrics is the Eguchi-Hanson metric [17], which corresponds to the case where the boundary infinity is $S^3/\mathbf{Z}_2$. The so-called Multi-center metrics [10, 19] correspond to the cases $S^3/\mathbf{Z}_n$. The general picture [19, 9] is as follows: every ALE space is determined by its group of identifications Γ, which must be a finite Kleinian subgroup of SU(2). Kronheimer described indeed manifolds having such a boundary; he showed that in principle a unique selfdual metric can be obtained for each of these manifolds [9] and, moreover, that every selfdual metric approaching asymptotically the euclidean one can be recovered in such a manner [16].

The Kleinian subgroups of SU(2)

Choosing complex coordinates $z_1 = x - iy, z_2 = t + iz$ on $\mathbf{R}^4 \sim \mathbf{C}^2$, and representing a point (z_1, z_2) by a quaternion, the group $\mathrm{SO}(4) \sim \mathrm{SU}(2)_L \times \mathrm{SU}(2)_R$, which is the isometry group of the sphere at infinity, acts on the quaternion by matrix multiplication:

$$\begin{pmatrix} z_1 & i\bar{z}_2 \\ iz_2 & \bar{z}_1 \end{pmatrix} \longrightarrow M_1 \cdot \begin{pmatrix} z_1 & i\bar{z}_2 \\ iz_2 & \bar{z}_1 \end{pmatrix} \cdot M_2 \tag{32}$$

the element $M \in \mathrm{SO}(4)$ being represented as $(M_1 \in \mathrm{SU}(2)_L, M_2 \in \mathrm{SU}(2)_R)$. The group Γ can be seen as a finite subgroup of $\mathrm{SU}(2)_L$, acting on $\mathbf{C}^2$ in the natural way by its two-dimensional representation:

$$\forall U \in \Gamma \subset \mathrm{SU}(2), \qquad U : \mathbf{v} = \begin{pmatrix} z_1 \\ z_2 \end{pmatrix} \longrightarrow U\mathbf{v} = \begin{pmatrix} \alpha & i\beta \\ i\bar{\beta} & \bar{\alpha} \end{pmatrix} \begin{pmatrix} z_1 \\ z_2 \end{pmatrix}. \tag{33}$$

It is a classic result that the possible finite subgroups of SU(2) are organized in two infinite series and three exceptional cases; each subgroup Γ is in correspon-

[5] In 4 dimensions (anti)selfduality of the curvature implies that the spaces are HyperKähler and that their metric automatically satisfies the vacuum Einstein equations

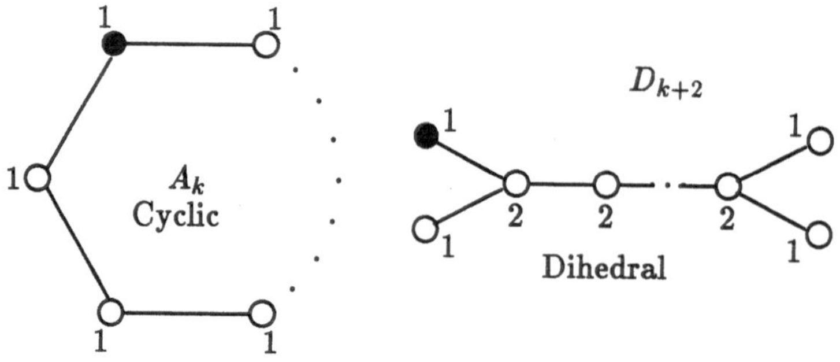

Figure 1: *Extended Dynkin diagrams of the infinite series*

dence with a simply laced Lie algebra $\mathcal{G}$, and we write $\Gamma(\mathcal{G})$ for it. See fig.s 1,2 for the explicit correspondence.

The 2-dimensional defining representation $\mathcal{Q}$ is obtained by regarding the group Γ as an SU(2) subgroup [that is, $\mathcal{Q}$ is the representation which acts in eq.(33)].

For the Kleinian groups Γ it is particularly important the decomposition of the tensor product of an irreducible representation D_μ with the defining 2-dimensional representation $\mathcal{Q}$. It is indeed at the level of this decomposition that the relation between these groups and the simply laced Dynkin diagrams is more explicit [23]. Furthermore this decomposition plays a crucial role in the explicit construction of the ALE manifolds [9]. Setting

$$\mathcal{Q} \otimes D_\mu = \bigoplus_{\nu=0}^{r} A_{\mu\nu} \, D_\nu \tag{34}$$

where D_0 denotes the identity representation, one finds that the matrix $\bar{c}_{\mu\nu} = 2\delta_{\mu\nu} - A_{\mu\nu}$ is the *extended Cartan matrix* encoded in the *extended Dynkin diagram* corresponding to the given group. Recall that the extended Dynkin diagram contains in addition to the *dots* representing the *simple roots* $\{\alpha_1 \ldots\ldots \alpha_r\}$ an *extra dot* (marked black in Fig.s 4, 4) representing the negative of the highest root $\alpha_0 = \sum_{i=1}^{r} n_i \, \alpha_i$ (n_i are the Coxeter numbers). There is thus a correspondence between the non-trivial conjugacy classes or equivalently the non-trivial irrepses of the group $\Gamma(\mathcal{G})$ and the simple roots of $\mathcal{G}$; in this correspondence, the extended Cartan matrix provides us with the Clebsch-Gordan coefficients (34), while the Coxeter numbers n_i express the dimensions of the irreducible representations. All these informations are summarized in Fig.s 2,3 where the numbers n_i are attached to each of the dots; the extra dot stands for the identity representation.

•Example *Consider the cyclic subgroups of $SU(2)$, that is the A_k-series. The defining*

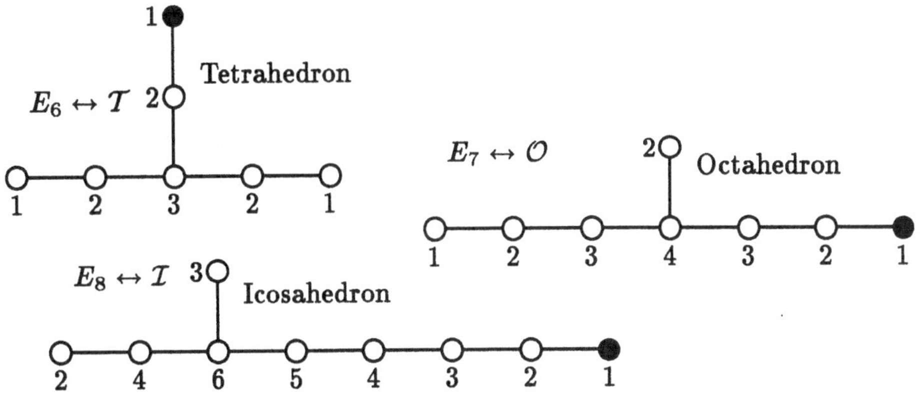

Figure 2: *Exceptional extended Dynkin diagrams*

2-dimensional representation $\mathcal{Q}$ is given by the matrices

$$\gamma_l \in \Gamma(A_k) \quad ; \quad \gamma_l = \mathcal{Q}_l \stackrel{\text{def}}{=} \begin{pmatrix} e^{2\pi i l/(k+1)} & 0 \\ 0 & e^{-2\pi i l/(k+1)} \end{pmatrix} \{l = 1, \ldots\ldots, k\}. \tag{35}$$

It is not irreducible since all irreducible representations are one-dimensional as one sees from Fig. 2. In the j-th irreducible representation the l-th element of the group is represented by $D^{(j)}(e_l) = \nu^{jl}$, where $\nu = \exp 2\pi/k$, $j = 1, \ldots k$. The $(k+1) \times (k+1)$ array of phases ν^{jl} appearing in the above equation is the character table. Given the $\mathbf{C}^2$ carrier space of the defining representation [see eq.s (33)] one can construct three algebraic invariants, namely

$$z = z_1\, z_2 \quad ; \quad x = (z_1)^{k+1} \quad ; \quad y = (z_2)^{k+1} \tag{36}$$

that satisfy the polynomial relation

$$W_{A_k}\ (x,\, y,\, z) \stackrel{\text{def}}{=} x\, y - z^{k+1} = 0\,. \qquad \square \tag{37}$$

Analogous invariants and polynomials (see table 1) can be constructed for the other kleinian subgroups.

ALE manifolds and resolution of simple singularities

The polynomial constraint $W_\Gamma(x, y, z) = 0$ plays an important role in the construction of the ALE manifolds. Indeed, as we are going to see, the vanishing locus in $\mathbf{C}^3$ of the *potential* $W_\Gamma(x, y, z)$ coincides with the space of equivalence classes $\mathbf{C}^2/\Gamma$, that is with the singular orbifold limit of the self-dual manifold $\mathcal{M}_\Gamma$. According to the standard procedure of deforming singularities [24, 25, 26, 27] there is a corresponding family of smooth manifolds $\mathcal{M}_\Gamma(t_1, t_2, \ldots\ldots, t_r)$ obtained as the vanishing locus $\mathbf{C}^3$ of a *deformed potential*:

$$\tilde{W}_\Gamma(x, y, z; t_1, t_2, \ldots\ldots, t_r) \;=\; W_\Gamma(x, y, z) + \sum_{\alpha=1}^{r} t_\alpha\, \mathcal{P}^{(\alpha}(x, y, z) \tag{38}$$

where t_α are complex numbers (the moduli of the complex structure of $\mathcal{M}_\Gamma$) and $\mathcal{P}^{(\alpha)}(x,y,z)$ is a basis spanning the chiral ring

$$\mathcal{R} = \frac{\mathbf{C}[x,y,z]}{\partial W} \tag{39}$$

of polynomials in x,y,z that do not vanish upon use of the vanishing relations $\partial_x W = \partial_y W = \partial_z W = 0$. The dimension of this chiral ring $|\mathcal{R}|$ is precisely equal to the number of non-trivial conjugacy classes (or of non trivial irreducible representations) of the finite group Γ. From the geometrical point of view this implies an identification between the number of complex structure deformations of the ALE manifold and the number r of non-trivial conjugacy classes discussed above.

In this framework one can describe the homology of the manifolds $\mathcal{M}_\Gamma(t_\alpha)$ with $t \neq 0$. The non-contractible two-cycles $c_\alpha, \alpha = 1,\dots r$ (each isomorphic to a copy of $\mathbf{CP}^1$) can be put into correspondence with the vertices of the non-extended Dynkin diagram for Γ. The intersection matrix of the c_α is the negative of the Cartan matrix:

$$c_\alpha \cdot c_\beta = \bar{c}_{\alpha\beta} \,. \tag{40}$$

The Kronheimer construction, that we shortly describe, shows that the base manifold (simply denoted as $\mathcal{M}$) of an ALE space is diffeomorphic to the space $\mathcal{M}_\Gamma(t_\alpha)$ supporting the resolution of the orbifold $\mathbf{C}^2/\Gamma$. Therefore the equation (40) applies to the generators of its second homology group. In particular we see that

$$\begin{aligned} \tau &= dim H^2_{\mathbf{C}}(\mathcal{M}) = dim H_2(\mathcal{M}) = \\ &= \text{rank of the corresponding Lie Algebra} = \\ &= \#\ \text{non trivial conj. classes in } \Gamma = |\mathcal{R}| \,. \end{aligned} \tag{41}$$

where τ is the Hirzebruch signature [6]. These results are summarized in Table 1.

Kronheimer construction

The HyperKähler quotient is performed on a suitable flat HyperKähler space $\mathcal{S}$ that now we define. Given any finite subgroup of SU(2), Γ, consider a space $\mathcal{P}$ whose elements are two-vectors of $|\Gamma| \times |\Gamma|$ complex matrices: $p \in \mathcal{P} = (A, B)$. The action of an element $\gamma \in \Gamma$ on the points of $\mathcal{P}$ is the following:

$$\begin{pmatrix} A \\ B \end{pmatrix} \stackrel{\gamma}{\longrightarrow} \begin{pmatrix} u_\gamma & i\bar{v}_\gamma \\ iv_\gamma & \bar{u}_\gamma \end{pmatrix} \begin{pmatrix} R(\gamma)AR(\gamma^{-1}) \\ R(\gamma)BR(\gamma^{-1}) \end{pmatrix} \tag{42}$$

where the two dimensional matrix in the r.h.s. is the realization of γ in the defining representation $\mathcal{Q}$ of Γ, while $R(\gamma)$ is the regular, $|\Gamma|$-dimensional representation [7]. This transformation property identifies $\mathcal{P}$, from the point of view of

[6] for ALE manifolds $\tau = \chi - 1$, χ being the Euler characteristic; it just counts the normalizable selfdual forms

[7] The basis vectors e_γ of the regular representation R are in one-to-one correspondence with the group elements γ and transform as $R(\gamma)e_\delta = e_{\gamma\cdot\delta}$, $\forall \gamma, \delta \in \Gamma$.

Table 1: *KLEINIAN GROUP versus ALE MANIFOLD properties*

Γ.	$W(x,y,z)$	$\mathcal{R}=\frac{\mathbf{C}[x,y,z]}{\partial W}$	$\lvert\mathcal{R}\rvert$	#c. c.	$\tau\equiv\chi-1$
A_k	$xy-z^{k+1}$	$\{1,z,\ldots,z^{k-1}\}$	k	$k+1$	k
D_{k+2}	$x^2+y^2z+z^{k+1}$	$\{1,y,z,y^2,z^2,\ldots,z^{k-1}\}$	$k+2$	$k+3$	$k+2$
$E_6=\mathcal{T}$	$x^2+y^3+z^4$	$\{1,y,z,yz,z^2,yz^2\}$	6	7	6
$E_7=\mathcal{O}$	$x^2+y^3+yz^3$	$\{1,y,z,y^2,z^2,yz,y^2z\}$	7	8	7
$E_8=\mathcal{I}$	$x^2+y^3+z^5$	$\{1,y,z,z^2,yz,z^3,yz^2,yz^3\}$	8	9	8

the representations of Γ, as $\mathcal{Q}\otimes\mathrm{End}(R)$. The space $\mathcal{P}$ can be given a quaternionic structure, representing its elements as "quaternions of matrices":

$$p\in\mathcal{P}=\begin{pmatrix}A & iB^\dagger\\ iB & A^\dagger\end{pmatrix} \qquad A,B\in\mathrm{End}(R)\ . \tag{43}$$

The space $\mathcal{S}$ is the subspace of Γ-invariant elements in $\mathcal{P}$:

$$\mathcal{S}\stackrel{\mathrm{def}}{=}\{p\in\mathcal{P}/\forall\gamma\in\Gamma,\gamma\cdot p=p\}\ . \tag{44}$$

Explicitly the invariance condition reads:

$$\begin{pmatrix}u_\gamma & i\bar{v}_\gamma\\ iv_\gamma & \bar{u}_\gamma\end{pmatrix}\begin{pmatrix}A\\ B\end{pmatrix}=\begin{pmatrix}R(\gamma^{-1})AR(\gamma)\\ R(\gamma^{-1})BR(\gamma)\end{pmatrix}\ . \tag{45}$$

The space $\mathcal{S}$ is elegantly described for all Γ's using the associated Dynkin diagram.

A two-vector of matrices can be thought of also as a matrix of two-vectors: that is, $\mathcal{P}=\mathcal{Q}\otimes\mathrm{Hom}(R,R)=\mathrm{Hom}(R,\mathcal{Q}\otimes R)$. Decomposing into irreducible representations the regular representation, $R=\bigoplus_{\nu=0}^{r}n_\mu D_\mu$, using eq.(34) and the Schur's lemma, one gets

$$\mathcal{S}=\bigoplus_{\mu,\nu}A_{\mu,\nu}\mathrm{Hom}(\mathbf{C}^{n_\mu},\mathbf{C}^{n_\nu})\ . \tag{46}$$

The dimensions of the irrepses, n_μ are expressed in Fig.s (4,4). From eq.(46) the real dimension of $\mathcal{S}$ follows immediately: $\dim\mathcal{S}=\sum_{\mu,\nu}2A_{\mu\nu}n_\mu n_\nu$ implies, recalling that $A=2\mathbf{1}-\bar{c}$ [see eq.(34)] and that for the extended Cartan matrix $\bar{c}n=0$,

$$\dim\mathcal{S}=4\sum_\mu n_\mu^2=4|\Gamma|\ . \tag{47}$$

The quaternionic structure of $\mathcal{S}$ can be seen by simply writing its elements as in eq.(43) with A, B satisfying the invariance condition eq.(45). Then the HyperKähler forms and the metric are described by $\Theta = \mathrm{Tr}(d\bar{m} \wedge m)$ and $ds^2 \mathbf{1} = \mathrm{Tr}(d\bar{m} \otimes dm)$. The trace is taken over the matrices belonging to $\mathrm{End}(R)$ in each entry of the quaternion.

•Example *The space $\mathcal{S}$ can be easily described when Γ is the cyclic group A_{k-1}. The order of A_{k-1} is k; the abstract multiplication table is that of $\mathbf{Z}_k$ and from it we can immediately read off the matrices of the regular representation. One has $R(e_1)_{lm} = \delta_{l,m+1}$ and of course $R(e_j) = (R(e_1))^j$. Actually, the invariance condition eq.(45) is best solved by changing basis so as to diagonalize the regular representation. Let $\nu = e^{\frac{2\pi i}{k}}$, so that $\nu^k = 1$. The change of basis is performed by the matrix $S_{ij} = \frac{\nu^{-ij}}{\sqrt{k}}$; in the new basis $R(e_j) = \mathrm{diag}(1, \nu^j, \nu^{2j}, \ldots, \nu^{(k-1)j})$. The entries are the representatives of e_j in the unidimensional irrepses.*

The explicit solution of eq.(45) is given in the above basis by

$$(A)_{lm} = \delta_{l,m+1} u^l \quad ; \quad (B)_{lm} = \delta_{l,m-1} v^l \tag{48}$$

We see that these matrices are parametrized in terms of $2k$ complex, i.e. $4k = 4|A_{k-1}|$ real parameters. □

Consider the action of $SU(|\Gamma|)$ on $\mathcal{P}$ given, using the quaternionic notation for the elements of $\mathcal{P}$, by

$$\forall g \in SU(|\Gamma|), g : \begin{pmatrix} A & iB^\dagger \\ iB & A^\dagger \end{pmatrix} \longmapsto \begin{pmatrix} gAg^{-1} & igB^\dagger g^{-1} \\ igBg^{-1} & gA^\dagger g^{-1} \end{pmatrix} . \tag{49}$$

It is easy to see that this action is a triholomorphic isometry of $\mathcal{P}$: ds^2 and Θ are invariant. Let F be the subgroup of $SU(|\Gamma|)$ which *commutes with the action of* Γ *on* $\mathcal{P}$, the Γ-action described in eq.(42). Then the action of F descends to $\mathcal{S} \subset \mathcal{P}$ to give a *triholomorphic isometry*: the metric and HyperKähler forms on $\mathcal{S}$ are just the restriction of those on $\mathcal{P}$. It is therefore possible to take the HyperKähler quotient of $\mathcal{S}$ with respect to F.

Let $\{f_A\}$ be a basis of generators for $\mathcal{F}$, the Lie algebra of F. Under the infinitesimal action of $f = \mathbf{1} + \lambda^A f_A \in F$, the variation of $m \in \mathcal{S}$ is $\delta m = \lambda^A \delta_A m$, with

$$\delta_A m = \begin{pmatrix} [f_A, A] & i[f_A, B^\dagger] \\ i[f_A, B] & [f_A, A^\dagger] \end{pmatrix} \tag{50}$$

The components of the momentum map (see (7)) are then given by

$$\mathcal{D}_A = \mathrm{Tr}\,(\bar{m}\,\delta_A m) \stackrel{\mathrm{def}}{=} \mathrm{Tr} \begin{pmatrix} f_A \mathcal{D}_3(m) & f_A \mathcal{D}_-(m) \\ f_A \mathcal{D}_+(m) & f_A \mathcal{D}_3(m) \end{pmatrix} \tag{51}$$

so that the real and holomorphic maps $\mathcal{D}_3 : \mathcal{S} \to \mathcal{F}^*$ and $\mathcal{D}_+ : \mathcal{S} \to \mathbf{C} \times \mathcal{F}^*$ can be represented as matrix-valued maps: [8]

$$\begin{aligned} \mathcal{D}_3(m) &= -i\,([A, A^\dagger] + [B, B^\dagger]) \\ \mathcal{D}_+(m) &= ([A, B]) \,. \end{aligned} \tag{52}$$

[8] It is easy to see that indeed the matrices $[A, A^\dagger] + [B, B^\dagger]$ and $[A, B]$ belong to the Lie algebra of traceless matrices $\mathcal{F}$; practically we identify $\mathcal{F}^*$ with $\mathcal{F}$ by means of the Killing metric

Let $\mathcal{Z} \equiv \mathcal{Z}^*$ be the dual of the centre of $\mathcal{F}$. In correspondence with a level $\zeta = \{\zeta^3, \zeta^+\} \in \mathbf{R}^3 \otimes \mathcal{Z}$ we can form the HyperKähler quotient $\mathcal{M}_\zeta \stackrel{\text{def}}{=} \mathcal{D}^{-1}(\zeta)/F$. *Varying ζ and Γ every ALE space can be obtained as $\mathcal{M}_\zeta$.*

First of all, it is not difficult to check that $\mathcal{M}_\zeta$ is four-dimensional. As for the space $\mathcal{S}$, there is a nice characterization of the group F in terms of the extended Dynkin diagram associated with Γ:

$$F = \bigotimes_\mu U(n_\mu) \ . \tag{53}$$

One must however set the determinant of the elements to one, since $F \subset SU(|\Gamma|)$. F has a $U(n_\mu)$ factor for each dot of the diagram, n_μ being associated to the dot as in Fig.s 1,2. F acts on the various "components" of $\mathcal{S}$ [which are in correspondence with the edges of the diagram, see eq.(46)] as dictated by the structure diagram. From eq.(53) it is immediate to derive that $\dim F = \sum_\mu n_\mu^2 - 1 = |\Gamma| - 1$. It follows that

$$\dim \mathcal{M}_\zeta = \dim \mathcal{S} - 4 \dim F = 4|\Gamma| - 4(|\Gamma| - 1) = 4 \ . \tag{54}$$

•Example *The structure of F and the momentum map for its action are very simply worked out in the A_{k-1} case. An element f of F must commute with the action of A_{k-1} on $\mathcal{P}$, eq.(42), where the two-dimensional representation in the l.h.s. is given in eq.(35). Then f must have the form*

$$f = \mathrm{diag}(e^{i\varphi_0}, e^{i\varphi_1}, \ldots, e^{i\varphi_{k-1}}) \ ; \ \sum \varphi_i = 0 \ . \tag{55}$$

Thus $\mathcal{F}$ is just the algebra of diagonal traceless k-dimensional matrices, which is $k-1$-dimensional. Choose a basis of generators for $\mathcal{F}$, for instance $f_1 = \mathrm{diag}(1, -1, \ldots)$,$f_2 = \mathrm{diag}(1, 0, -1, \ldots)$,..,$f_{k-1} = \mathrm{diag}(1, 0, \ldots, -1)$. *From eq.(52) one gets directly the components of the momentum map:*

$$\begin{aligned}
\mathcal{D}_{3,A} &= |u^0|^2 - |v_0|^2 - |u^{k-1}|^2 - |v_{k-1}|^2 - |u^A|^2 - |v_A|^2 + |u^{A-1}|^2 - |v_{A-1}|^2 \\
\mathcal{D}_{+,A} &= u^0 v_o - u^{k-1} v_{k-1} - u^A v_A + u^{A-1} v_{A-1} \ . \quad \square
\end{aligned} \tag{56}$$

In order for $\mathcal{M}_\zeta$ to be a manifold, it is necessary that F acts freely on $\mathcal{D}^{-1}(\zeta)$. This happens or not depending on the value of $\zeta \in \mathcal{Z}$. Again, a simple characterization of $\mathcal{Z}$ can be given in terms of the simple Lie algebra $\mathcal{G}$ associated with Γ [9]. There exists an isomorphism between $\mathcal{Z}$ and the Cartan subalgebra $\mathcal{H}$ of $\mathcal{G}$. Thus we have

$$\dim \mathcal{Z} = \dim \mathcal{H} = \mathrm{rank}\, \mathcal{G} = \#\text{of non trivial conj. classes in } \Gamma \ . \tag{57}$$

The space $\mathcal{M}_\zeta$ turns out to be singular when, under the above identification $\mathcal{Z} \sim \mathcal{H}$, any of the level components $\zeta^i \in \mathbf{R}^3 \otimes \mathcal{Z}$ lies on the walls of a Weyl chamber. In particular, as the point $\zeta^i = 0$ for all i is identified with the origin in the root space *the space $\mathcal{M}_0$ is singular.* We will see in a moment that

$\mathcal{M}_0$ corresponds to the *orbifold limit* $\mathbf{C}^2/\Gamma$ of a family of ALE manifolds with boundary at infinity S^3/Γ.

To see that this is general, choose the natural basis $\{e_\delta\}$ for the regular representation R. Define then the space $L \subset \mathcal{S}$ as follows:

$$L = \left\{ \begin{pmatrix} C \\ D \end{pmatrix} \in \mathcal{S} \,/\, C, D \text{ are diagonal in the basis } \{e_\delta\} \right\} . \tag{58}$$

For every element $\gamma \in \Gamma$ there is a pair of numbers (c_γ, d_γ) given by the corresponding entries of C, D: $C \cdot e_\gamma = c_\gamma e_\gamma$, $D \cdot e_\gamma = d_\gamma e_\gamma$. Applying the invariance condition eq.(45), which is valid since $L \subset \mathcal{S}$, it results that

$$\begin{pmatrix} c_{\gamma\cdot\delta} \\ d_{\gamma\cdot\delta} \end{pmatrix} = \begin{pmatrix} u_\gamma & i\bar{v}_\gamma \\ iv_\gamma & \bar{u}_\gamma \end{pmatrix} \begin{pmatrix} c_\delta \\ d_\delta \end{pmatrix} . \tag{59}$$

We can identify L with $\mathbf{C}^2$ associating for instance $(C, D) \in L \longmapsto (c_0, d_0) \in \mathbf{C}^2$. Indeed all the other pairs (c_γ, d_γ) are determined in terms of eq.(59) once (c_0, d_0) are given. By eq.(59) the action of Γ on L induces exactly the action of Γ on $\mathbf{C}^2$ that we considered in eq.s (32,33).

Note that we can directly realize $\mathbf{C}^2/\Gamma$ as an affine algebraic surface in $\mathbf{C}^3$ (see eq. (37)) by expressing the coordinates x, y and z of $\mathbf{C}^3$ in terms of the matrices $(C, D) \in L$.

•<u>Example</u> *The explicit parametrization of the matrices in $\mathcal{S}$ in the A_{k-1} case (which was given in eq.(48) in the basis in which the regular representation R is diagonal), can be conveniently rewritten in the "natural" basis $\{e_\gamma\}$ via the matrix S^{-1}. The subset L of diagonal matrices (C, D) is given by*

$$C = c_0 \,\mathrm{diag}(1, \nu, \nu^2, \ldots, \nu^{k-1}), \qquad D = d_0 \,\mathrm{diag}(1, \nu^{k-1}, \nu^{k-2}, \ldots, \nu), \tag{60}$$

where $\nu = e^{\frac{2\pi i}{k}}$. This is nothing but the fact that $\mathbf{C}^2 \sim L$. The set of pairs $(\nu^m c_0, \nu^{k-m} d_0)$, $m = 0, 1, \ldots, k-1$ is an orbit of Γ in $\mathbf{C}^2$ and determines the corresponding orbit of Γ in L. To describe $\mathbf{C}^2/A_{k-1}$ we identify $(x, y, z) \in \mathbf{C}^3$, such that $xy = z^k$, as

$$x = \det C \quad ; \quad y = \det D, \quad ; \quad z = \frac{1}{k} \mathrm{Tr}\, CD. \qquad \square \tag{61}$$

It is quite easy to show the following fundamental fact: *each orbit of F in $\mathcal{D}^{-1}(0)$ meets L in one orbit of* Γ. Because of the above identification between L and $\mathbf{C}^2$, this leads to the proof that $\mathcal{M}_0$ *is isometric to* $\mathbf{C}^2/\Gamma$.

•<u>Example</u> *Let us show explicitely in the case of the cyclic groups the one-to-one correspondence between the orbits of F in $\mathcal{D}^{-1}(0)$ and those of Γ in L. Choose the basis where R is diagonal. Then $(A, B) \in \mathcal{S}$ has the form of eq. (48). Now, the relation $xy = z^k$ (eq. (37)) holds true also when, in eq. (61), the pair $(C, D) \in L$ is replaced by an element $(A, B) \in \mathcal{D}^{-1}(0)$. Indeed the elements $(A, B) \in \mathcal{D}^{-1}(0)$ can be described solving eq. (52) at zero r.h.s.. It gives $u_j = |u_0| e^{i\phi_j}$ and $v_j = |v_0| e^{i\psi_j}$ and $\psi_j = \Phi - \phi_j$ $\forall j$ for a certain phase Φ. Then we immediately check that such a pair $(A, B) \in \mathcal{D}^{-1}(0)$ satisfies $xy = z^k$ if $x = \det A$, $y = \det B$ and $z = (1/k)\,\mathrm{Tr}\, AB$. We are left with $k+3$*

parameters (the k phases ϕ_j, $j = 0, 1, \dots k-1$, plus the absolute values $|u_0|$ and $|v_0|$ and the phase Φ). Indeed $\dim \mathcal{D}^{-1}(0) = \dim \mathcal{M} - 3 \dim F = 4|\Gamma| - 3(|\Gamma| - 1) = |\Gamma| + 3$, *where* $|\Gamma| = \dim \Gamma = k$.

Now we perform the quotient of $\mathcal{D}^{-1}(0)$ with respect to F. Given a set of phases f_i such that $\sum_{i=0}^{k-1} f_i = 0 \bmod 2\pi$ and given $f = \mathrm{diag}(e^{if_0}, e^{if_1}, \dots, e^{if_{k-1}}) \in F$, the orbit of F in $\mathcal{D}^{-1}(0)$ passing through (A, B) is given by (fAf^{-1}, fBf^{-1}). Choosing $f_j = f_0 + j\psi + \sum_{n=0}^{j-1} \phi_n$, $j = 1, \dots, k-1$, with $\psi = -\frac{1}{k} \sum_{n=0}^{k-1} \phi_n$, and f_0 determined by the condition $\sum_{i=0}^{k-1} f_i = 0 \bmod 2\pi$, one has

$$(fAf^{-1})_{lm} = a_0 \delta_{l,m+1} \quad ; \quad (fBf^{-1})_{lm} = b_0 \delta_{l,m-1} \tag{62}$$

where $a_0 = |u_0| e^{i\psi}$ and $b_0 = |v_0| e^{i(\Phi - \psi)}$. Since the phases ϕ_j are determined modulo 2π, it follows that ψ is determined modulo $\frac{2\pi}{k}$. Thus we can say $(a_0, b_0) \in \mathbf{C}^2/\Gamma$. This is the one-to-one correspondence between $\mathcal{D}^{-1}(0)/F$ and $\mathbf{C}^2/\Gamma$. □

5 Resolution of ALE singularities $W_\Gamma(t^\alpha)$ and Fayet-Iliopoulos parameters

So far we have reviewed the main points of the Kronheimer construction. In particular we have shown the constructive definition of the quaternionic flat space $\mathcal{S}$ and of the "gauge group" acting on it by triholomorphic isometries needed to retrieve an ALE space as a HyperKähler quotient. That is, we have described the necessary ingredients to specify, according to the procedure outlined in sec. 3, an N=4 renormalizable field theory (the *microscopic theory*) whose low-energy effective action (the *macroscopic theory*) is the sigma-model on the ALE space under consideration [9].

We do not insist on the mathematical proofs of the main statements of Kronheimer's work (in particular, the identification of *all* ALE spaces with $\mathcal{M}_\zeta$). We rather choose to illustrate, in the specific case of the cyclic subgroups, an *explicit* relation between the parameters $\zeta^i \in \mathcal{Z}, i = 1, 2, 3$ of the HyperKähler construction (the levels of the momentum map) and the deformation parameters t^α appearing in eq. (38). We divide the ζ parameters in r-parameters (the real levels of the $\mathcal{D}^3$ momentum map) and s-parameters (the complex levels of the $\mathcal{D}^+$ momentum map) since this was the notation utilized in sec. 3. This relation tells us explicitely which is the "deformed" potential describing an ALE space, obtained as a HyperKähler quotient with levels $\{r, s\}$, as an hypersurface in $\mathbf{C}^3$. We stress that the parameters r, s are coupling paramenters (the N=4 generalizations of Fayet-Iliopoulos parameters) in the "microscopic" N=4 lagrangian while the t^α are parameters in the σ-model (the "macroscopic" description), since they appear in the definition of the target space, and in particular of its complex structure. This gives a physical interest to the relation we describe.

---

[9] Of course, to carry out explicitely until the end computations analogous to those for the Calabi metrics is extremely complicated; indeed the form of the metric that would result from this quotient is in general not known, with the exception of the Eguchi-Hanson case.

To find the desired relation, we have in practice to find a "deformed" relation between the invariants x, y, z. To this purpose, we focus on the holomorphic part of the momentum map, i.e. on the equation $[A, B] = \Sigma_0$, where $\Sigma_0 = \mathrm{diag}(s_0, s_1, \ldots, s_{k-1})$ with $s_0 = -\sum_{i=1}^{k-1} s_i$. Recall the expression (48) for the matrices A and B. Calling $a_i = u_i v_i$, $[A, B] = \Sigma_0$ implies that $a_i = a_0 + s_i$ for $i = 1, \ldots, k-1$. Now, let $\Sigma = \mathrm{diag}(s_1, \ldots, s_{k-1})$. We have

$$xy = \det A \det B = a_0 \Pi_{i=1}^{k-1}(a_0 + s_i) = a_0^k \det\left(1 + \frac{1}{a}\Sigma\right) = \sum_{i=0}^{k-1} a_0^{k-i} S_i(\Sigma). \quad (63)$$

The $S_i(\Sigma)$ are the symmetric polynomials in the eigenvalues of Σ, defined by $\det(1 + \Sigma) = \sum_{i=0}^{k-1} S_i(\Sigma)$. In particular, $S_0 = 1$ and $S_1 = \sum_{i=1}^{k-1} s_i$. Define $S_k(\Sigma) = 0$, so that $xy = \sum_{i=0}^{k} a_0^{k-i} S_i(\Sigma)$, and note that $z = \frac{1}{k}\mathrm{Tr}AB = a_0 + \frac{1}{k}S_1(\Sigma)$. Then the desired deformed relation between x, y and z is obtained by substituting $a_0 = z - \frac{1}{k}S_1(\Sigma)$ in (63), obtaining finally

$$xy = \sum_{m=0}^{k} \sum_{n=0}^{k-m} \binom{k-m}{n} \left(-\frac{1}{k}S_1(\Sigma)\right)^{k-m-n} S_m(\Sigma) z^n = \sum_{n=0}^{k} t_n z^n. \quad (64)$$

$$\Longrightarrow \quad t_n = \sum_{m=0}^{k-n} \binom{k-n}{m} \left(-\frac{1}{k}S_1(\Sigma)\right)^{k-m-n} S(\Sigma)_n. \quad (65)$$

Notice in particular that $t_k = 1$ and $t_{k-1} = 0$, i.e. $xy = z^k + \sum_{n=0}^{k-2} t_n z^n$, which means that the deformation proportional to z^{k-1} is absent. This establishes a clear correspondence between the momentum map construction and the polynomial ring $\mathbf{C}[x, y, z]/\partial W$ where $W(x, y, z) = xy - z^k$ [compare with eq. (38)]. Moreover, note that we have only used one of the momentum map equations, namely $[A, B] = \Sigma_0$. The equation $[A, A^\dagger] + [B, B^\dagger] = R$ has been completely ignored. This means that the deformation of the complex structure is described by the parameters Σ, while the parameters R describe the deformation of the Kähler class.

The relation (65) can also be written in a simple factorized form, namely

$$xy = \Pi_{i=0}^{k-1}(z - \mu_i), \quad (66)$$

where

$$\begin{aligned} \mu_i &= \frac{1}{k}(s_1 + s_2 + \cdots + s_{i-1} - 2s_i + s_{i+1} + \cdots + s_k), \quad i = 1, \ldots, k-1 \\ \mu_0 &= -\sum_{i=1}^{k} \mu_i = \frac{1}{k}S_1(\Sigma). \end{aligned} \quad (67)$$

References

[1] D. Anselmi and P. Fre'. " Topological sigma models in four dimensions and tri–holo morphic maps ". *Nucl. Phys.*, B416:255, (1994).

[2] E. Witten and N. Seiberg. "Electric–magnetic duality, monopole condensation and confinement in N=2 supersymmetric Yang–Mills theory". *Nucl. Phys.*, B426:19, (1994).

[3] E. Witten and N. Seiberg. "Monopoles, duality and chiral supersymmetry breaking in N=2 QCD". *hep-th*, /9408013: , (1994).

[4] E. Witten. " Monopoles and four manifolds". *IASSNS-HEP-9496 hep-th*, /9411102: , (1994).

[5] A. Ceresole, R. D'Auria, and S. Ferrara. "On the geometry of moduli space of vacua in N=2 supersymmetric Yang–Mills theory". *CERN-TH 7384/94 POLFIS-TH*, 07/94: , (1994).

[6] M. Billo', P. Fre', L. Girardello, and A. Zaffaroni. " Gravitational instantons in heterotic string theory: the H–map and the moduli deformations of (4,4) superconformal theories ". *Int. Jour. Mod. Phys.*, A8:2351, (1993).

[7] D. Anselmi, M. Billo', P. Fre', L. Girardello, and A. Zaffaroni. " ALE manifolds and conformal field theoris ". *Int. Jour. Mod. Phys.*, A9:3007, (1994).

[8] M. Billo' and P. Fre'. " N=4 versus N=2 phases, HyperKähler quotients and the 2d topological twist ". *Class. and Quantum Grav.*, 11:785, (1994).

[9] P.B. Kronheimer. " The construction of ALE spaces as HyperKähler quotients ". *Jour Diff. Geo.*, 29:665, (1989).

[10] S. Hawking and C.N. Pope. " Symmetry breaking by instantons in supergravity ". *Nucl. Phys.*, B146:381, (1978).

[11] P. Fayet and J. Iliopulos. " Spontaneously broken supergauge symmetries and Goldstone spinors ". *Phys. Lett.*, 51B:46, (1974).

[12] D. Anselmi and P. Fre'. " Twisted N=2 supergravity as topological gravity in four dimensions ". *Nucl. Phys.*, B392:401, (1993).

[13] D. Anselmi and P. Fre'. " Topological twist in four dimensions, R duality and hyperinstantons ". *Nucl. Phys.*, B404:288, (1993).

[14] N.J. Hitchin, A. Karlhede, U. Lindström, and M. Rocek. "HyperKähler metrics and supersymmetry". *Comm. Math. Phys.*, 108:535, (1987).

[15] U. Lindstrom and M. Rocek. " Scalar-tensor dualities and N=1,2 non-linear sigma-models ". *Nucl. Phys.*, B222:285, (1983).

[16] P.B. Kronheimer. " A Torelli-type theorem for gravitational instantons ". *Jour Diff. Geo.*, 29:685, (1989).

[17] T. Eguchi and A.J. Hanson. "Self–dual solutions to Euclidean gravity". *Ann. Phys.*, 120:82, (1979).

[18] G.W. Gibbons and S. Hawking. " Classification of gravitational instanton symmetries ". *Comm. Math. Phys.*, 66:381, (1979).

[19] N.J.Hitchin. " Polygons and gravitons ". *Math. Proc. Cambridge Phylos. Soc.*, 85:465, (1979).

[20] E. Witten. " Phases of N=2 theories in two dimensions ". *Nucl. Phys.*, B403:159, (1993).

[21] K. Galicki. "A generalization of the momentum mapping construction for quaternionic Khaler manifolds". *Comm. Math. Phys.*, 108:117, (1987).

[22] E. Calabi. "Metriques Kähleriennes et fibres holomorphes". *Ann. Scie. Ec. Norm. Sup.*, 12:269, (1979).

[23] J. McKay. " Graphs, singularities and finite groups ". *Proc. Symp. Pure Math., Am. Math. Soc.*, 37:183, (1980).

[24] M. Artin. *Am. J. Math.*, 88:129, (1966).

[25] V.I. Arnold, S.M. Gusein-Zade, and A.N. Varchenko. *Singularities of differntiable maps.* Birkhäuser, 1975.

[26] E. Brieskorn. in *Actes Congrés Intern. Math. Ann. (t. 2)*, (1970).

[27] P. Slodowy. *Simple Singularities and Simple Algebraic Groups.* Lect. Notes in Math. 815, Springer Verlag, 1980.

Four Dimensional Integrable Theories [1]

Ch. Devchand and V. Ogievetsky

Joint Institute for Nuclear Research 141980 Dubna, Russia

Abstract. There exist many four dimensional integrable theories. They include self-dual gauge and gravity theories, all their extended supersymmetric generalisations, as well the full (non-self-dual) N=3 super Yang-Mills equations. We review the harmonic space formulation of the twistor transform for these theories which yields a method of producing explicit connections and metrics. This formulation uses the concept of harmonic space analyticity which is closely related to that of quaternionic analyticity.

1. Introduction

Many Lorentz invariant four dimensional exactly solvable nonlinear theories are known. The most remarkable of these are those admitting the Penrose-Ward twistor transform [1], which may be thought of as an analogue of the transformation to action-angle variables for hamiltonian dynamical systems, in the sense that it involves a transformation to variables in which the dynamics is trivial, reducing the problem to that of inverting the transformation. Further, the solution methods for many lower-dimensional completely integrable systems, like the inverse scattering transform for the KdV equation, may be thought of as reductions of the twistor transform [2], so the prospect has arisen, of a unification of the various existing methods of solving two dimensional systems as different manifestations of the twistor transform for self-dual Yang-Mills (SDYM).

The twistor transform, which takes its most dramatic form in its application to the solution of the self-dual Yang-Mills and Einstein equations has been found to have a remarkably clear realisation in the language of 'harmonic spaces' ([3]-[7]). In fact harmonic or twistor spaces admit supersymmetrisation, yielding a remarkably simple supersymmetrisation of the SDYM and Einstein equations, which is much more straightforward, and moreover independent of the N-extension (where

[1] Talk by V. Ogievetsky at the Gürsey Memorial Conference I, Istanbul, June 1994

N is the number of independent supersymmetries), than the supersymmetrisations of the corresponding full non-self-dual theories, for which the supersymmetrisation for each extension N has to be considered anew. All N-extended supersymmetric theories may therefore be treated on an equal footing [8, 9]. Moreover, for the self-dual super Yang-Mills theories, there exists a remarkable 'matreoshka'-like nested structure [8] in which the $N = 0$ solution data may be dressed-up to higher N solution data in a basically algebraic fashion using solutions of first-order equations.

The list of four-dimensional theories (which may equally well be considered to be in complexified space or in real spaces of Euclidean (4,0) or Kleinian (2,2) signature) amenable to the twistor transform is therefore quite large and includes

- Self-dual Yang-Mills (SDYM) equations, for any semisimple gauge group.
- All N-extended ($N = 1, .., 4$) supersymmetrisations of the latter.
- Self-dual Einstein equations, with or without cosmological constant.
- All N-extended Poincaré and conformal self-dual supergravities.
- The full (i.e. non-self-dual) N=3 super-Yang-Mills theory (even in the Minkowskian (3,1) signature).

In this talk, we shall describe the harmonic space versions of the twistor transform for all the above theories. The crucial feature allowing the applicability of the twistor transform to field theories is the possibility of presenting the equations of motion in the form of algebraic constraints amongst the components of some curvature tensor, the paradigmatic example being the Yang-Mills self-duality equations. In particular, the constraints take the form

$$[\mathcal{D}_{\alpha i}, \mathcal{D}_{\beta j}] = \epsilon_{\alpha\beta} F_{ij},$$

where α, β are spinor indices of some group having skew-symmetric invariant $\epsilon_{\alpha\beta}$, i, j are some other indices or labels, and F_{ij} are the non-zero curvatures representing the obstruction to Frobenius' integrability. Twistor or harmonic space is an auxiliary space in which the curvature is zero in some 'analytic' subspaces, allowing the use of 'Frobenius variables' to reduce the system. In the harmonic space setting a transformation to such variables converts the system to a set of Cauchy-Riemann-like (CR) equations, thereby reducing the problem to that of reconstructing the original variables from the 'analytic' data (satisfying these CR equations). The crucial idea of harmonic space analyticity is closely related to the concepts of quaternionic and Fueter analyticity, to which Feza Gürsey, whom we all loved so dearly, devoted so much attention. It is therefore especially appropriate to present these ideas at this meeting dedicated to his memory. In fact it was precisely in Feza's last paper (with V. Ogievetsky and M. Evans)[4] that the intimate relation between quaternionic and harmonic space analyticities was clarified. That paper was completed shortly after Feza's untimely death and we feel it appropriate to quote the dedication to Feza contained in its manuscript, which Physical Review refused to include in the published version.

"Feza Gürsey, a fine human being and outstanding physicist, passed away on April 13, 1992. He is a coauthor of the present paper, which is one of a series of his works devoted to quaternionic aspects of four-dimensional field theories, a field in which he was a pioneer. Feza enthusiastically participated in the writing of this paper, even as he fought the disease to which he finally succumbed. Sadly, he did not live long enough to approve the paper's final version, and so bears no responsibility for whatever shortcomings it may possess. It was a great joy and privilege to work with Feza, and to benefit from his fertile mind and keen intelligence. The experience of working with him and the wonderful personality of Feza Gürsey will abide forever in the memories of the two other authors. "

2. From 2D *complex* to 4D *quaternionic* analyticity

In two-dimensional Euclidean space the two *real* coordinates may be quite naturally combined into *a single complex number* $x^\mu = \{x^1, x^2\} \to z = x^1 + ix^2$ and the most general conformal coordinate transformation in two dimensions is the *analytic* transformation

$$z' = f(z), \quad \bar{z}' = \bar{f}(\bar{z}). \tag{1}$$

In virtue of the Cauchy-Riemann condition, $\frac{\partial}{\partial \bar{z}} f(z) = 0$, its d'Alembertian vanishes, $\frac{\partial}{\partial z}\frac{\partial}{\partial \bar{z}} f(z) = 0$.

Similarly naturally, four dimensional coordinates may be combined into a quaternion. In spinor notation we have

$$x^\mu \to q = x^{\alpha\dot{\alpha}} = \begin{pmatrix} x^0 - ix^3 & -ix^1 - x^2 \\ -ix^1 + x^2 & x^0 + ix^3 \end{pmatrix} = x^0 + e_a x^a \tag{2}$$

where the Pauli matrices represent the algebra of the quaternionic units, $e_a = -i\sigma_a$

$$e_a e_b = -\delta_{ab} + \epsilon_{abc} e_c. \tag{3}$$

Analytic transformations (1) are fundamental to 2D-conformal field theories. Feza Gürsey often wondered whether there exist 4D theories in which some form of quaternionic analyticity plays a correspondingly crucial rôle [10], [11]. However, the notion of quaternionic analyticity is rather delicate and there are several possible forms, some of which being too restrictive to be applicable to field theories. For instance, the straightforward generalisation of the Cauchy-Riemann condition

$$\frac{\partial}{\partial \bar{q}} f = \frac{1}{2}\left(\frac{\partial}{\partial x^0} + \frac{1}{3} e_a \frac{\partial}{\partial x^a}\right) f = 0 \tag{4}$$

where $\frac{\partial}{\partial \bar{q}}$ satisfies $\frac{\partial}{\partial \bar{q}} q = 0$ and $\frac{\partial}{\partial \bar{q}} \bar{q} = 1$ is well known (see e.g. [12]) to allow only a linear solution $f = a + qb$, with constant quaternions a and b, because of the noncommutativity of quaternions.

Fueter quaternion analyticity [13, 14], however, is less restrictive. This defines an analytic function of a quaternion q, as a Weierstrass-like series

$$f(q) = \sum a_n q^n, \tag{5}$$

where the coefficients a_n are real or complex numbers (or quaternions, but multiplying q^n on only one side, e.g. left as in (5)). Such a function obeys a Cauchy-Riemann-like condition, of the *third order* in derivatives and is therefore in general not a harmonic function ($\Box f(q) \neq 0$), although it is bi-harmonic ($\Box^2 f(q) = 0$). Moreover, it is not invariant under SO(4) rotations [14].

In self-dual and $N = 2$ supersymmetric theories, however, manifolds of quaternionic character namely quaternionic-Kähler and hyper-Kähler manifolds naturally arise [15]. In these theories hyper-Kähler and quaternionic structures are related to yet another notion of analyticity, namely harmonic-space analyticity, which we shall explain.

3. Harmonic space

Harmonic space [3] is essentially an enlargement of four dimensional space-time, which may be thought of in terms of the coset space $\frac{\text{Poincaré group}}{\text{Lorentz group}}$, to coset space $\frac{\text{Poincaré group}}{SU(2)\times U(1)} = \frac{\text{Poincaré group}}{\text{Lorentz group}} \times \frac{SU(2)}{U(1)}$ (for the case of signature (4,0)). This space has additional coordinates parametrising the two-sphere $S^2 = \frac{SU(2)}{U(1)}$. Of course, one could choose polar (θ, ϕ) or stereographic $(z, \bar{z})$ coordinates to describe this sphere. However, it is in practice very useful to use a more abstract parametrisation using two fundamental representations of the SU(2) algebra, $u^{\pm}_{\dot\alpha}$ (where $\dot\alpha$ is an SU(2) spinor index and $\pm$ denote U(1) charges), which are just spin $\frac{1}{2}$ spherical harmonics of S^2, defined up to the U(1) equivalence $u^{\pm}_{\dot\alpha} \sim e^{\pm\gamma} u^{\pm}_{\dot\alpha}$; $\gamma \in \mathbb{C}$ and satisfying the equations $\epsilon_{\dot\alpha\dot\beta} u^{+\dot\alpha} u^{-\dot\beta} = 1$. The further hermiticity condition $u^{-}_{\dot\alpha} = \overline{u^{+\dot\alpha}}$ yields two independent real variables. In the complexified setting, however, $u^{+}_{\dot\alpha}$ and $u^{-}_{\dot\alpha}$ are independent and an appropriate equivalence relation holds [4].

4. Self-dual Yang-Mills

The usual self-duality condition for the Yang-Mills field strength

$$F_{\mu\nu} = \frac{1}{2}\epsilon_{\mu\nu\rho\sigma}F_{\rho\sigma} \ , \tag{6}$$

basically says that the (0,1) part of the gauge field vanishes. This is better expressed in terms of 2-spinor notation in the form: $f_{\dot{\alpha}\dot{\beta}} = 0$ which is equivalent to the statement that the field strengths curvature only contains the (1,0) Lorentz representation, i.e.

$$[\mathcal{D}_{\alpha\dot{\alpha}} \ , \ \mathcal{D}_{\beta\dot{\beta}}] = \ \epsilon_{\dot{\alpha}\dot{\beta}}f_{\alpha\beta}. \tag{7}$$

Now multiplying (7) by the two commuting spinors $u^{+\dot{\alpha}}, u^{+\dot{\beta}}$, one can compactly represent it as the vanishing of a curvature

$$[\nabla^+_\alpha, \nabla^+_\beta] = 0 \ , \tag{8}$$

where $\nabla^+_\alpha \equiv u^{+\dot{\alpha}}\nabla_{\alpha\dot{\alpha}}$, with linear system

$$\nabla^+_\alpha \varphi = \ 0 \ . \tag{9}$$

This is precisely the Belavin-Zakharov-Ward linear system for SDYM. Now the $u^{+\dot{\alpha}}$ are actually harmonics [3] on S^2 and it is better to consider these equations in an auxiliary space with coordinates $\{x^{\pm\alpha} \equiv x^{\alpha\dot{\alpha}}u^\pm_{\dot{\alpha}} \ , \ u^\pm_{\dot{\alpha}} \ ; \ u^{+\dot{\alpha}}u^-_{\dot{\alpha}} = 1\}$, where the harmonics are defined up to a $U(1)$ phase, and gauge covariant derivatives in this harmonic space are

$$\nabla^+_\alpha = \partial^+_\alpha + A^+_\alpha = \frac{\partial}{\partial x^{-\alpha}} + A^+_\alpha. \tag{10}$$

In this space (8) is actually not equivalent to the self-duality conditions. We also need

$$[D^{++}, \nabla^+_\alpha] = 0 \ , \tag{11}$$

where D^{++} is a harmonic space derivative which acts on negatively-charged harmonics to yield their positively-charged counterparts, i.e. $D^{++}u^-_{\dot{\alpha}} = u^+_{\dot{\alpha}}$, whereas $D^{++}u^+_{\dot{\alpha}} = 0$. This operator, in a fixed parametrisation, has also been considered by Newman (e.g. [16]). In ordinary x-space, when the harmonics are treated as parameters, the condition (11) is actually incorporated in the definition of ∇^+_α as a *linear* combination of the covariant derivatives. The system (8,11) is now *equivalent* to SDYM and has been considered by many authors, e.g. [4, 5]; the equivalence holding in spaces of signature (4,0) or (2,2), or in complexified space. In this regard, we should note that for real spaces, our understanding is completely clear for the Euclidean signature. For the (2,2) signature, the situation is richer and more intricate due to the noncompact nature of the rotation group and our present considerations concern only those signature (2,2) configurations which may be obtained by Wick rotation of (4,0) configurations.

Now, in (11) the covariant derivative (10) has pure-gauge form

$$\nabla^+_\alpha = \partial^+_\alpha + \varphi \partial^+_\alpha \varphi^{-1}. \tag{12}$$

and D^{++} is 'short' i.e. has no connection. This choice of frame (the 'central' frame) is actually inherited from the four-dimensional x-space and is not the most natural one for harmonic space. We may however change coordinates to a basis in which ∇^+_α is 'short' and D^{++} is 'long' (i.e. acquires a Lie-algebra-valued connection) instead. Namely,

$$\begin{aligned} \nabla^+_\alpha &= \partial^+_\alpha \\ \mathcal{D}^{++} &= D^{++} + V^{++}, \end{aligned} \tag{13}$$

a change of frame tantamount to a gauge transformation by the 'bridge' φ in (9). In this basis (the 'analytic' frame) the SDYM system (8,11) remarkably takes the form of a Cauchy-Riemann (CR) condition

$$\frac{\partial}{\partial x^{-\alpha}} V^{++} = 0 \tag{14}$$

expressing independence of half the x-coordinates. In virtue of passing to this basis the nonlinear SDYM equations (6) are in a sense trivialised: Any 'analytic' (i.e. satisfying (14)) function $V^{++} = V^{++}(x^{+\alpha}, u^\pm)$ corresponds to some self-dual gauge potential. From any such V^{++}, by solving the *linear* equation

$$D^{++}\varphi = \varphi V^{++} \tag{15}$$

for the bridge φ, a self-dual vector potential may be recovered from the harmonic expansion:

$$\varphi \partial^+_\alpha \varphi^{-1} = u^{+\dot\alpha} A_{\alpha\dot\alpha}; \tag{16}$$

the linearity in the harmonics $u^{+\dot\alpha}$ being guaranteed by (11).

Solving (15) for an *arbitrary* analytic gauge algebra valued function V^{++} yields the *general* local self-dual solution. This correspondence between self-dual gauge potentials and holomorphic prepotentials V^{++} is a convenient tool for the explicit construction of local solutions of the self-duality equations.

Furthermore, in the analytic subspace of harmonic space (with coordinates $\{x^{+\alpha}, u^\pm_{\dot\alpha}\}$), there exists an especially simple presentation of the infinite-dimensional symmetry group acting on solutions of the self-duality equations. It is the (apparently trivial) transformation $V^{++} \to V^{++'} = g^{++}$, where g^{++} depends in an arbitrary way on V^{++} and its derivatives as well as on the analytic coordinates themselves, modulo gauge transformations $V^{++} \to e^{-\lambda}(V^{++} + D^{++})e^\lambda$, where λ is also an arbitrary analytic function.

5. Supersymmetric self-dual Yang-Mills theories

Yang-Mills theories can be supersymmetrised to couple successively lower spin fields to the vector field. Since extended super Yang-Mills theories are massless theories, the components are classified by helicity and we have the following representation content in theories up to N=3:

$$
\begin{array}{lcccccccc}
\textit{helicity}: & 1 & \frac{1}{2} & 0 & -\frac{1}{2} & \frac{1}{2} & 0 & -\frac{1}{2} & -1 \\
N=0 & f_{\alpha\beta} & & & & & & & f_{\dot\alpha\dot\beta} \\
N=1 & f_{\alpha\beta} & \lambda_\alpha & & & & & \lambda_{\dot\alpha} & f_{\dot\alpha\dot\beta} \\
N=2 & f_{\alpha\beta} & \lambda^i_\alpha & \overline{W} & & & W & \lambda_{\dot\alpha i} & f_{\dot\alpha\dot\beta} \\
N=3 & f_{\alpha\beta} & \lambda^i_\alpha & W_i & \chi_{\dot\alpha} & \chi_\alpha & W^i & \lambda_{\dot\alpha i} & f_{\dot\alpha\dot\beta}
\end{array}
\qquad (17)
$$

In real Minkowski space fields in the left and right triangles are related by CPT conjugation but in complexified space or in a space with signature (4,0) or (2,2), we may set fields in one of the triangles to zero without affecting fields in the other triangle. If we set all the fields in the right (left) triangle to zero, the equations of motion reduce to the super (anti-) self-duality equations. For instance, the self-duality equations for the N=3 theory take the form

$$
\begin{aligned}
\epsilon^{\beta\gamma}\mathcal{D}_{\gamma\dot\beta} f_{\alpha\beta} &= 0 \\
\epsilon^{\gamma\beta}\mathcal{D}_{\gamma\dot\beta} \lambda^i_\beta &= 0 \\
\epsilon^{\dot\gamma\dot\alpha}\mathcal{D}_{\alpha\dot\gamma} \chi_{\dot\alpha} &= -[\lambda^k_\alpha, W_k] \\
\mathcal{D}_{\alpha\dot\beta}\mathcal{D}^{\alpha\dot\beta} W_i &= \tfrac{1}{2}\epsilon_{ijk}\{\lambda^{\alpha j}, \lambda^k_\alpha\}.
\end{aligned}
\qquad (18)
$$

We see that the spin 1 source current actually factorises into parts from the two triangles, so it manifestly vanishes for super self-dual solutions. The first equation in (18) is just the Bianchi identity for self-dual field-strengths. So apart from the self-duality condition (6), we have one equation for zero-modes of the covariant Dirac operator in the background of a self-dual vector potential (having (6) as integrability condition) and two further non-linear equations. However, any given self-dual vector potential actually *determines* the general (local) solution of the rest of the equations. This is the most striking consequence of the matreoshka phenomenon: the N=0 core determining the properties of the higher-N theories. Another consequence is is that many conserved currents identically vanish in the super self-dual sector. For instance, since self-duality always implies the *source-free* second order Yang-Mills equations, the spin 1 source current vanishes for the entire matreoshka. Further, the usual Yang-Mills stress tensor clearly vanishes for self-dual fields:

$$
T_{\alpha\dot\alpha,\beta\dot\beta} \equiv f_{\dot\alpha\dot\beta} f_{\alpha\beta} = 0 \; ;
$$

and as a consequence of this, once one has put on further layers of the matreoshka, the supercurrents generating supersymmetry transformations, which contain the stress tensor as well as its superpartners also identically vanish for super self-dual fields.

In N-independent form, (18) can be conveniently written as the following super curvature constraints in chiral superspace:

$$\begin{aligned} \{\bar{\mathcal{D}}^i_{\dot\alpha}, \bar{\mathcal{D}}^j_{\dot\beta}\} &= \epsilon_{\dot\alpha\dot\beta} W^{ij} \\ \{\mathcal{D}_{\alpha i}, \mathcal{D}_{\beta j}\} &= 0 = [\mathcal{D}_{\alpha i}, \nabla_{\alpha\beta}] \\ \{\mathcal{D}_{\alpha j}, \bar{\mathcal{D}}^i_{\dot\beta}\} &= 2\delta^i_j \nabla_{\alpha\dot\beta} \, . \end{aligned} \tag{19}$$

Having expressed the super self-duality equations in this form, the supersymmetrisation of the harmonic-twistor construction is straightforward. In harmonic superspaces with coordinates

$$\{x^{\pm\alpha} \equiv u^{\pm}_{\dot\beta} x^{\alpha\dot\beta},\ \bar\vartheta^{\pm}_i \equiv u^{\pm}_{\dot\alpha} \bar\vartheta^{\dot\alpha}_i,\ \vartheta^{\alpha i},\ u^{\pm}_{\dot\alpha}\},$$

these take the form

$$\begin{aligned} \{\mathcal{D}_{\alpha i}, \mathcal{D}_{\beta j}\} = \ & 0 \ = \{\bar{\mathcal{D}}^{+i}, \bar{\mathcal{D}}^{+j}\} \\ [\nabla^+_\alpha, \nabla^+_\beta] = \ & 0 \ = [\bar{\mathcal{D}}^{+i}, \nabla^+_\alpha] \\ \{\mathcal{D}_{\alpha j}, \bar{\mathcal{D}}^{+i}\} \ &= \ 2\delta^i_j \nabla^+_\alpha \\ [\mathcal{D}_{\alpha i}, \nabla^+_\beta] \ &= \ 0, \end{aligned} \tag{20}$$

where the gauge covariant derivatives are given by

$$\mathcal{D}_{\alpha i} = D_{\alpha i} + A_{\alpha i}, \quad \bar{\mathcal{D}}^{+i} = u^{+\dot\alpha} \bar{\mathcal{D}}^i_{\dot\alpha} = \bar{D}^{+i} + \bar{A}^{+i}, \quad \nabla^+_\alpha = u^{+\dot\alpha} \nabla_{\alpha\dot\alpha} = \partial^+_\alpha + A^+_\alpha \, , \tag{21}$$

and satisfy the equations

$$[D^{++}, \mathcal{D}_{\alpha i}] = [D^{++}, \bar{\mathcal{D}}^{+i}] = [D^{++}, \nabla^+_\alpha] = 0 \, . \tag{22}$$

The equations (20,22) are *equivalent* to (19) and (20) are consistency conditions for the following system of linear equations

$$\begin{aligned} \mathcal{D}_{\alpha i}\varphi &= 0 \\ \bar{\mathcal{D}}^{+i}\varphi &= 0 \\ \nabla^+_\alpha \varphi &= 0, \end{aligned} \tag{23}$$

This system is extremely redundant, φ allowing the following transformation under the gauge group

$$\varphi \to e^{-\tau(x^{\alpha\dot\alpha}, \bar\vartheta^{\dot\alpha}_i \vartheta^{\alpha i})} \varphi e^{\lambda(x^{+\alpha}, \bar\vartheta^+_i, u^{\pm}_{\dot\alpha})} \, , \tag{24}$$

where τ and λ are arbitrary functions of the variables shown, without affecting the constraints (20). These constraints therefore allow an economic choice of chiral-analytic basis in which the bridge ϕ and the prepotential V^{++} depend only on *positively* $U(1)$*-charged, barred* Grassmann variables, viz. $\bar{\vartheta}_i^+$, being independent of $\vartheta^{i\alpha}$ and $\bar{\vartheta}_i^-$. In this basis, φ too is independent of $\vartheta^{i\alpha}$ and $\bar{\vartheta}_i^-$; its non-analyticity manifesting itself in its dependence on $x^{-\alpha}$. Moreover, consistently with the commutation relations (20), the covariant spinor derivatives take the form $\mathcal{D}_{\alpha i} = \frac{\partial}{\partial \vartheta^{\alpha i}}, \bar{\mathcal{D}}^i = 2\vartheta^{\alpha i}\nabla_\alpha^+$. The super self-duality conditions (20,22) are therefore equivalent to the *same* system of equations as the N=0 SDYM equations, viz. (9,11), except that now φ and A_α^+ are chiral superfields depending on $\{x^{\pm\alpha}, \bar{\vartheta}_i^+, u_{\dot{\alpha}}^\pm\}$ [8]. As for the N=0 case, we may express this system in the form of analyticity conditions for the harmonic space connection superfield V^{++}:

$$\frac{\partial}{\partial x^{-\alpha}} V^{++}(x^{+\alpha}, \bar{\vartheta}^{+i}, u_{\dot{\alpha}}^\pm) = 0. \tag{25}$$

and the bridge φ to the central basis may be found by solving (15). Fields solving (18) may then be obtained by inserting solutions φ of (15) into the expression

$$\varphi \partial_\alpha^+ \varphi^{-1} = u^{+\dot{\alpha}} A_{\alpha\dot{\alpha}}(x^{\alpha\dot{\alpha}}, \bar{\vartheta}_i^{\dot{\alpha}}), \tag{26}$$

(the left side being guaranteed to be linear in u^+), and expanding the superfield vector potential on the right to obtain the component multiplet satisfying (18) thus:

$$A_{\alpha\dot{\beta}}(x,\bar{\vartheta}) = A_{\alpha\dot{\beta}}(x) + \bar{\vartheta}_{\dot{\beta}i}\lambda_\alpha^i(x) + \epsilon^{ijk}\bar{\vartheta}_{\dot{\alpha}j}\bar{\vartheta}_i^{\dot{\alpha}}\nabla_{\alpha\dot{\beta}}W_k(x) + \epsilon^{ijk}\bar{\vartheta}_{\dot{\alpha}i}\bar{\vartheta}_j^{\dot{\alpha}}\bar{\vartheta}_k^{\dot{\gamma}}\nabla_{\alpha\dot{\gamma}}\chi_{\dot{\beta}}(x)\,. \tag{27}$$

It is remarkable that super self-duality implies the absence of higher-order terms in $\bar{\vartheta}$. In fact any N=0 solution completely and recursively determines its higher-N extensions [8].

The most general infinite-dimensional group of transformations of super-self-dual solutions acquires a transparent form in the analytic harmonic superspace with coordinates $\{x^{+\alpha}, \bar{\vartheta}^{+i}, u_{\dot{\alpha}}^\pm\}$. As for the $N = 0$ case, it is given by the transformation

$$V^{++} \to V^{++'} = g^{++}(V^{++}, x^{+\alpha}, \bar{\vartheta}^{+i}, u_{\dot{\alpha}}^\pm), \tag{28}$$

where g^{++} is an arbitrary doubly $U(1)$-charged analytic algebra-valued functional, modulo gauge transformations $V^{++} \to e^{-\lambda}(V^{++} + D^{++})e^\lambda$, where λ is also an arbitrary analytic function. This group has an interesting subgroup of transformations

$$V^{++} \to V^{++'} = V^{++}(x^{+'}, \bar{\vartheta}^{+'}, u'), \tag{29}$$

induced by diffeomorphisms of the analytic harmonic superspace

$$x^{+\alpha'} = x^{+\alpha'}(x^+, \bar{\vartheta}^+, u), \quad \bar{\vartheta}^{+i'} = \bar{\vartheta}^{+i'}(x^+, \bar{\vartheta}^+, u), \quad u' = u'(x^+, \bar{\vartheta}^+, u). \tag{30}$$

6. N=3 (non-self-dual) super Yang-Mills theory

As we have seen, the spin 1 source currents of all super self-dual theories vanish because they factorise into parts from the two triangles in (17). It turns out that we can restore these source currents and solve the full (i.e. non-self-dual) super Yang-Mills equations by intermingling self-dual and anti-self-dual holomorphic data [17]; and this works *exactly* for the N=3 case. Again the crucial feature is the presentability of the thrice-extended super Yang-Mills equations in the form of the super-curvature constraints [18]

$$\begin{aligned} \{\mathcal{D}_{i\alpha}\ ,\ \mathcal{D}_{j\beta}\} &= \epsilon_{\alpha\beta} W_{ij} \\ \{\bar{\mathcal{D}}^i_{\dot\alpha}\ ,\ \bar{\mathcal{D}}^j_{\dot\beta}\} &= \epsilon_{\dot\alpha\dot\beta} W^{ij} \\ \{\mathcal{D}_{i\alpha}\ ,\ \bar{\mathcal{D}}^j_{\dot\beta}\} &= 2\delta^j_i \nabla_{\alpha\dot\beta}, \end{aligned} \tag{31}$$

where $\mathcal{D}_A \equiv \partial_A + A_A = (\nabla_{\alpha\dot\beta}, \mathcal{D}_{i\alpha}, \bar{\mathcal{D}}^j_{\dot\beta}), i, j = 1, 2, 3,$ are gauge-covariant superderivatives. These constraints are purely kinematical for N=1,2 but are equivalent to the dynamical equations for the component fields for N=3 [19]. Now in order to present these as zero-curvature conditions in some harmonic space, the appearance of invariants of both simple parts of the Lorentz group $(\epsilon_{\alpha\beta}, \epsilon_{\dot\alpha\dot\beta})$ requires the 'harmonisation' of the entire Lorentz group. This allows the consideration of all possible signatures, with the corresponding harmonic spaces being given by:

$$\begin{aligned} &\textit{Euclidean} \quad (4,0): \quad \frac{\text{Poincaré group}}{\text{Lorentz group}} \times \frac{SU(2)}{U(1)} \times \frac{SU(2)}{U(1)} \\ &\textit{Lorentzian} \quad (3,1): \quad \frac{\text{Poincaré group}}{\text{Lorentz group}} \times \frac{SL(2,\mathbb{C})}{L(1,\mathbb{C})} \\ &\textit{Kleinian} \quad (2,2): \quad \frac{\text{Poincaré group}}{\text{Lorentz group}} \times \frac{SL(2,\mathbb{R})}{SO(2)} \times \frac{SL(2,\mathbb{R})}{SO(2)} \end{aligned}$$

To thus harmonise the entire Lorentz group, we need to introduce harmonics with both dotted and undotted indices: $u^+_{\dot\alpha}, u^-_{\dot\alpha}$ and $v^\oplus_\alpha, v^\ominus_\alpha$, satisfying the constraints

$$u^{+\dot\alpha} u^-_{\dot\alpha} = 1\ ,\ v^{\oplus\alpha} v^\ominus_\alpha = 1$$

and having the hermiticity condition $\overline{u^+_{\dot\alpha}} = v^{\ominus\alpha}$ for the Lorentzian signature. Now in harmonic space with coordinates $u^+_{\dot\alpha}, u^-_{\dot\alpha}$ and $v^\ominus_\alpha, v^\oplus_\alpha$ and

$$x^{\pm\oplus} = x^{\alpha\dot\alpha} u^\pm_{\dot\alpha} v^\oplus_\alpha\ ,\ x^{\pm\ominus} = x^{\alpha\dot\alpha} u^\pm_{\dot\alpha} v^\ominus_\alpha,$$

$$\vartheta^{i\oplus} = \vartheta^{i\alpha} v^\oplus_\alpha\ ,\ \vartheta^{i\ominus} = \vartheta^{i\alpha} v^\ominus_\alpha\ ,\ \bar\vartheta^\pm_i = \bar\vartheta^{\dot\alpha}_i u^\pm_{\dot\alpha};$$

The superspace constraints (31) are equivalent to the following system of equations in harmonic superspace

$$\begin{aligned}\{\bar{\mathcal{D}}^{+i}\,,\,\bar{\mathcal{D}}^{+j}\} &= 0 = \{\mathcal{D}^{\oplus}_i\,,\,\mathcal{D}^{\oplus}_j\}\\ \{\bar{\mathcal{D}}^{+j}\,,\,\mathcal{D}^{\oplus}_i\} &= 2\nabla^{+\oplus}\\ [D^{++},\bar{\mathcal{D}}^{+j}] &= 0 = [D^{++},\mathcal{D}^{\oplus}_i] = [D^{++},\nabla^{+\oplus}]\\ [D^{\oplus\oplus},\bar{\mathcal{D}}^{+j}] &= 0 = [D^{\oplus\oplus},\mathcal{D}^{\oplus}_i] = [D^{\oplus\oplus},\nabla^{+\oplus}]\\ [D^{++},D^{\oplus\oplus}] &= 0,\end{aligned} \tag{32}$$

where

$$D^{++} = u^{+\dot\alpha}\frac{\partial}{\partial u^{-\dot\alpha}}, \quad D^{\oplus\oplus} = v^{\oplus\alpha}\frac{\partial}{\partial v^{\ominus\alpha}}.$$

Now as before, we can go to an 'analytic frame' in which the covariant derivatives $(\bar{\mathcal{D}}^{+j},\mathcal{D}^{\oplus}_i,\nabla^{+\oplus})$ lose their connection parts and the derivatives $(D^{++},D^{\oplus\oplus})$ acquire connections $(V^{++},V^{\oplus\oplus})$ instead. For the latter, (32) are just the generalised Cauchy-Riemann 'analyticity' conditions

$$\begin{aligned}\frac{\partial}{\partial\bar\vartheta^-_i}V^{++} &= 0 = \frac{\partial}{\partial\bar\vartheta^-_i}V^{\oplus\oplus}\\ \frac{\partial}{\partial\vartheta^{\ominus i}}V^{++} &= 0 = \frac{\partial}{\partial\vartheta^{\ominus i}}V^{\oplus\oplus}\\ \frac{\partial}{\partial x^{-\ominus}}V^{++} &= 0 = \frac{\partial}{\partial x^{-\ominus}}V^{\oplus\oplus}\end{aligned}$$

together with the zero-curvature condition

$$D^{++}V^{\oplus\oplus} - D^{\oplus\oplus}V^{++} + [V^{++},V^{\oplus\oplus}] = 0,$$

which relates the two harmonic space connections $(V^{++},V^{\oplus\oplus})$. Analytic $(V^{++},V^{\oplus\oplus})$ satisfying this relationship therefore encode the solution of N=3 super Yang-Mills theory [20].

7. Self-dual gravity and supergravity

Analogously to (7) self-dual gravity may be described by the equation

$$[\mathcal{D}_{\alpha\dot\alpha}\,,\,\mathcal{D}_{\beta\dot\beta}] = \epsilon_{\dot\alpha\dot\beta}R_{\alpha\beta}, \tag{33}$$

where now the covariant derivative contains a vierbein as well as a connection,

$$\mathcal{D}_{\alpha\dot\alpha} = E^{\mu\dot\beta}_{\alpha\dot\alpha}\partial_{\mu\dot\beta} + \omega_{\dot\alpha\alpha}\,, \tag{34}$$

so (33) is not only a curvature constraint on the components of the connection, but also a zero-torsion condition on the vierbein. Moreover, since the Riemann tensor has irreducible components

$$\begin{aligned}R_{\alpha\beta} &\equiv C_{(\alpha\beta\gamma\delta)}\Gamma^{\gamma\delta} + R_{(\alpha\beta)(\dot\gamma\dot\delta)}\Gamma^{\dot\gamma\dot\delta} + \tfrac{1}{6}R\Gamma_{\alpha\beta},\\ R_{\dot\alpha\dot\beta} &\equiv C_{(\dot\alpha\dot\beta\dot\gamma\dot\delta)}\Gamma^{\dot\gamma\dot\delta} + R_{(\gamma\delta)(\dot\alpha\dot\beta)}\Gamma^{\gamma\delta} + \tfrac{1}{6}R\Gamma_{\dot\alpha\dot\beta},\end{aligned}$$

where $C_{(\dot{\alpha}\dot{\beta}\dot{\gamma}\dot{\delta})}(C_{(\alpha\beta\gamma\delta)})$ are the (anti-) self-dual components of the Weyl tensor, $R_{(\alpha\beta)(\dot{\gamma}\dot{\delta})}$ are the components of the tracefree Ricci tensor, R is the scalar curvature, $(\Gamma^{\gamma\delta}, \Gamma^{\dot{\gamma}\dot{\delta}})$ are generators of the tangent space gauge algebra, self-duality, i.e. the vanishing of $R_{\dot{\alpha}\dot{\beta}}$ clearly implies that the curvature takes values only in one SU(2) algebra with generators $\Gamma^{\gamma\delta}$, so we may work in a 'self-dual gauge' in which the connection also takes values only in this SU(2), i.e. only this half of the tangent space group is localised, while the second $SU(2)$ remains rigid. Restricting the holonomy group in this fashion, the curvature part of (33) is automatic, these equations reducing to the zero torsion conditions on the vierbein. Now, since we have to deal with the vanishing of torsions, the harmonic space system equivalent to (33) is rather different to that in the self-dual Yang-Mills case. It takes the form

$$\begin{aligned}
[\mathcal{D}_\alpha^+, \mathcal{D}_\beta^+] &= 0 \\
[\mathcal{D}^{++}, \mathcal{D}_\alpha^+] &= 0 \\
[\mathcal{D}_\alpha^+, \mathcal{D}_\beta^-] &= 0 \quad (\text{modulo } R_{\alpha\beta}) \\
[\mathcal{D}^{++}, \mathcal{D}_\alpha^-] &= \mathcal{D}_\alpha^+ .
\end{aligned} \tag{35}$$

Now, going to Frobenius coordinates

$$x^{\mu a} \rightarrow x_h^{\mu\pm} = x_h^{\mu\pm}(x^{\mu a}u_a^\pm, u_a^\pm) , \tag{36}$$

in which the covariant derivative $\mathcal{D}_\alpha^+ = \partial_\alpha^+$, the partial derivative, all the dynamics gets concentrated in the vielbeins and connection components of

$$\mathcal{D}^{++} = \partial^{++} + H^{++\mu+}\partial_{h\mu}^- + (x_h^{\mu+} + H^{++\mu-})\partial_{h\mu}^+ + \omega^{++}.$$

These may be solved for [6] in terms of an arbitrary analytic prepotential $\mathcal{L}^{+4}$ and the problem reduces to finding the explicit functions (36) for any specified choice of $\mathcal{L}^{+4}$. Inverting the transformation (36) the self-dual vierbein then allows itself to be decoded. An explicit illustration of the procedure may be found in [6], where the simplest monomial choice of prepotential, $\mathcal{L}^{+4} = g(x_h^{1+}x_h^{2+})^2$, where g is a dimensionful parameter, is shown to correspond to the self-dual Taub-NUT metric.

Remarkably, the N-extended supersymmetric self-duality equations allow themselves to be expressed in chiral superspace in the same form as (33),

$$[\mathcal{D}_{B\dot{\beta}}, \mathcal{D}_{A\dot{\alpha}}] = \epsilon_{\dot{\alpha}\dot{\beta}} R_{AB}, \tag{37}$$

except that now the indices A, B are 'superindices' of the superalgebra $OSp(N|1)$. The explicit construction of the self-dual super-vielbein therefore closely follows that for the non-supersymmetric case. This yields interesting non-trivial supersymmetrisations of hyper-Kähler manifolds. In [9] we construct some explicit examples of super deformations of flat space (with curvature only in the odd directions) and of Taub-NUT space.

8. Open problems

We have discussed a large class of four dimensional integrable systems allowing solution using the harmonic-twistor transform. Our considerations have raised a number of interesting questions. Integrability in two dimensions is known to imply remarkable constraints on the S-matrix yielding factorisation into two-particle amplitudes. Whether the integrability of the four dimensional theories discussed here has any analogous consequences, either for these thories themselves or for their dimensional reductions, remains an open question. This question is especially interesting for the full N=3 Yang-Mills theory, which is known to be an ultraviolet finite field theory. A further intriguing open question is what class of non-self-dual solutions to the usual N=0 Yang-Mills equations can be obtained by reduction of this supersymmetric construction; and whether the existence of two spectral parameters (corresponding to the two sets of harmonics) yields new classes of lower dimensional exactly solvable systems.

The remarkable conjunction of maximal supersymmetrisation, ultraviolet finiteness, and classical integrability in the sense described here, suggests the need to investigate the full (non-self-dual) Poincaré and conformal supergravity theories in this light as well. The corresponding super-twistor construction remains an open problem.

We should like to thank the Erwin Schrödinger Institute, Vienna, where this lecture was written up, for generous hospitality.

References

[1] R. Penrose, Gen. Rel. Grav. 7 (1976) 31; R.S. Ward and R.O. Wells, *Twistor geometry and field theory*, Camb. Univ. Press, Cambridge, 1990.

[2] R.S. Ward, Phil.Trans.Roy.Soc. A315 (1985) 451; L. Mason and G. Sparling, J. Geom. Phys. 8 (1992) 243.

[3] A. Galperin, E. Ivanov, S. Kalitzin, V. Ogievetsky, E. Sokatchev, Class. Quant. Grav. 1 (1984) 469, 2 (1985) 255.

[4] M. Evans, F. Gürsey, V. Ogievetsky, Phys.Rev. D47 (1993) 3496.

[5] S. Kalitzin and E. Sokatchev, Class. Quant. Grav. 4 (1987) L173; O. Ogievetsky, in *Group Theoretical Methods in Physics*, Ed. H.-D. Doebner et al, Springer Lect. Notes in Physics 313 (1988) 548; A. Galperin, E. Ivanov, V. Ogievetsky, E. Sokatchev, in *Quantum Field Theory and Quantum Statistics*, vol.2, 233 (Adam Hilger, Bristol, 1987); JINR preprint E2-85-363 (1985); A. Galperin, E. Ivanov, V. Ogievetsky, E. Sokatchev, Ann. Phys. (N.Y.) 185 (1988) 1; 11.

[6] C. Devchand and V. Ogievetsky, *Self-dual gravity revisited*, hep-th/9409160.

[7] A. Galperin, E. Ivanov, O. Ogievetsky, Ann. Phys. (N.Y.) 230, 201 (1994).

[8] C. Devchand and V. Ogievetsky, Phys. Lett. B297 (1992) 93; Nucl.Phys.B414 (1994), 763.

[9] C. Devchand and V. Ogievetsky, *Self-dual supergravities*, to appear.

[10] F.Gürsey, Nuovo Cim. 3 (1956) 988.

[11] F. Gürsey and H. C. Tze, Ann. Phys. (N.Y.) 128 (1980).

[12] A. Sudbery, Math. Proc. Camb. Phil. Soc. 85 (1979) 199.

[13] R. Fueter, Comment. Math. Helv. 7 (1935) 307, ibid. 8 (1936) 371; F. Gürsey, Conformal and quasi-conformal structures in space-time, Yale prep. YCTP - P34 - 91.

[14] F. Gürsey and W. X. Jiang, J. Math. Phys. 33 (1992) 682.

[15] G. W. Gibbons and S. W. Hawking, Phys. Lett. B 78 (1991) 430; L. Alvarez-Gaumé and D. Z. Freedman, Commun. Math. Phys. 80 (1981) 443; J. Bagger and E. Witten, Nucl. Phys. B 222 (1983) 1.

[16] E.T. Newman, J. Math. Phys. 27 (1986) 2797.

[17] E. Witten, Phys. Lett. 77B (1978) 394; P. Green, J. Isenberg and P. Yasskin, Phys. Lett. 78B (1978) 462; Yu. I. Manin, *Gauge Field Theory and Complex Geometry*, Nauka, Moscow, 1984 (English version, Springer-Verlag, Berlin, 1988).

[18] R. Grimm, M. Sohnius and J. Wess, Nucl. Phys. B133 (1978) 275; M. Sohnius, Nucl. Phys. B136 (1978) 461; L. Brink, J.H. Schwarz and J. Scherk, Nucl. Phys. B121 (1977) 77.

[19] J. Harnad, J. Hurtubise, M. Légaré and S. Shnider, Nucl. Phys. B256 (1985) 609;

[20] C. Devchand and V. Ogievetsky, Berezin Memorial Volume, Ed. R.L. Dobrushin, A.M. Vershik, M. A. Shubin (AMS , 1994); hep-th/9310071.

Regularisation, the BV Method, and the Antibracket Cohomology

Walter Troost[1] and Toine Van Proeyen[2]

Instituut voor Theoretische Fysica, Katholieke Universiteit Leuven, Belgium

Abstract We review the Lagrangian Batalin–Vilkovisky method for gauge theories. This includes gauge fixing, quantisation and regularisation. We emphasize the role of cohomology of the antibracket operation. Our main example is $d = 2$ gravity, for which we also discuss the solutions for the cohomology in the space of local integrals. This leads to the most general form for the action, for anomalies and for background charges.

1 Introduction.

The Batalin–Vilkovisky method for quantisation of general gauge theories has been introduced in [1]. It is applicable to all known gauge theories. Another advantage of the method is that it gives a comprehensive picture of the gauge fixing procedure. It uses a space of fields and antifields endowed with a symplectic structure defining an antibracket operation. The latter is a Poisson–like structure, and field transformations which are canonical with respect to this antibracket play an important part. Gauge fixing is essentially such a canonical transformation. We will summarize the essential ingredients and tools. In comparison with previous lectures [2, 3] we will put more emphasis here on the role of cohomology. We will use $d = 2$ gravity as our main example, and refer to [3] for chiral W_3.

In section 2 we will recall some definitions, introduce antifields, antibrackets and the 'extended classical action'. In the third section we introduce three different cohomologies and indicate the different roles they play. In section 4 we will review how gauge fixing is obtained by a canonical transformation. Then we will turn to the quantum theory. General principles will be given in section 5, where also the regularisation (based on Pauli–Villars) will be introduced [4, 5]. Anomalies are discussed in section 6, and illustrated with a calculation for $d = 2$ gravity. The latter will be treated also from a purely cohomological point of view in section 7. That section reports work in collaboration with F. Brandt [6]. Besides the most general form for the classical action and for anomalies, we also obtain the most general form for background charges.

This lecture is not self-contained. The reader is urged to have also [2, 3] at hand, as we will refer to specific formulas in these reviews. The relevant

[1]Onderzoeksleider, NFWO, Belgium. E–mail: Walter.Troost@fys.kuleuven.ac.be

[2]Onderzoeksleider, NFWO, Belgium. E–mail: Antoine.VanProeyen@fys.kuleuven.ac.be

equations and sections will be referred to as (I.*), resp. (II.*). A more complete text is still in preparation [7]. Many examples can also be found in [8].

2 The ingredients

Let us start by introducing the setup for $d = 2$ gravity with scalar matter. The classical fields are $\phi^i = \{X^\mu, g_{\alpha\beta}\}$, where X^μ are the D scalar fields, and $g_{\alpha\beta}$ is the two dimensional metric. There are two types of gauge symmetries, for which we introduce ghosts $c^a = \{\xi^\alpha, c\}$, namely ξ^α for the general coordinate transformations ($\alpha = +$ or $-$) and c for the local dilatations. Note that we do not introduce gauge fields for these transformations. For the diffeomorphisms they are nevertheless effectively included in the metric $g_{\alpha\beta}$, and in fact one could equally set up this same problem using the zweibeins instead of the metric. For the dilatations however they remain excluded.

We denote all these fields collectively as $\Phi^A = \{\phi^i, c^a\}$. We introduce antifields for all of them, Φ^*_A. They have opposite statistics. Then one defines an extended action $S(\Phi, \Phi^*)$, function of all the Fields[1]. The construction of this extended action was explained in detail in [2, 3], and leads to[2]

$$\begin{aligned} S &= -\tfrac{1}{2}\sqrt{g}\, g^{\alpha\beta}\partial_\alpha X^\mu \cdot \partial_\beta X_\mu \\ &\quad + X^*_\mu \xi^\alpha \partial_\alpha X^\mu + g^{*\alpha\beta}\left(\xi^\gamma \partial_\gamma g_{\alpha\beta} + 2 g_{\gamma(\alpha}\partial_{\beta)}\xi^\gamma + c g_{\alpha\beta}\right) \\ &\quad - \xi^*_\beta \xi^\alpha \partial_\alpha \xi^\beta - c^* \xi^\alpha \partial_\alpha c \,. \end{aligned} \tag{1}$$

The first line is the classical action, and the other lines are the BRST transformations of the fields multiplied by their corresponding antifields. In general, more terms may be present, for example expressing relations in an open gauge algebra. Introducing now an **antibracket**

$$(F, G) = F \frac{\overleftarrow{\partial}}{\partial \Phi^A} \cdot \frac{\overrightarrow{\partial}}{\partial \Phi^*_A} G - F \frac{\overleftarrow{\partial}}{\partial \Phi^*_A} \cdot \frac{\overrightarrow{\partial}}{\partial \Phi^A} G \,, \tag{2}$$

It is clear that the extended action satisfies

$$(S, S) = 0 \,. \tag{3}$$

When one checks the terms in this equation with a specific type of antifields, one will notice that it expresses the gauge invariance of the classical action, as well as the commutator algebra of the gauge transformations and their Jacobi identities. In general, for example with open gauge algebras, it will also include the other defining relations of the gauge theory.

---

[1]We need a common denomination for talking about fields and antifields together. For this we will use 'Fields' with a capital F. We will use the notation $z^\alpha = \{\Phi^A, \Phi^*_A\}$ (with $A = 1, ..., N$, and $\alpha = 1, ..., 2N$).

[2]We omit $\int d^2x$ for actions and anomaly expressions, here and in the sequel. (Anti-)Symmetrisation of indices is defined by $f_{(\mu\nu)} = \frac{1}{2}(f_{\mu\nu} + f_{\nu\mu})$, etc.

One assigns ghost numbers 0 for the classical fields ϕ^i, 1 for the ghosts c^a, and for the antifields a ghost number such that the sum of the ghost number of a field and its corresponding antifield always adds up to -1. As a consequence,by (2), this gives a ghost number assignment to (F, G) which is 1 lower than that of FG.

We invite the reader now to look to the example of W_3 in section II.1 and II.2. This example was treated in full using the BV method in [9].

Let us summarize the essential properties of the extended classical action. It is a function $S(z)$ of all the Fields z.

Classical limit : $S(\Phi, \Phi^* = 0)$ is the classical action.

Master equation : $(S, S) = 0$, which implies that[3] $R_{\alpha\beta} \equiv \frac{\partial}{\partial z^\alpha}\frac{\partial}{\partial z^\beta}S$ has rank $\leq N$.

Properness condition : the rank of R is equal to N. This property essentially means that we have to include all symmetries using ghosts, and also all zero modes. It will allow us to define the path integral.

It has been proven in several steps [10, 11, 9] that there is a local solution to these conditions for all classical actions (under certain regularity conditions satisfied by all reasonable gauge theories). The proof is very long, and will not be repeated here. The main tool is the 'Koszul–Tate' differential, which we will define below.

3 Cohomologies

We now introduce several different but related differentials and cohomologies. From the extended action one defines the nilpotent operator

$$\mathcal{S}F(z) \equiv (S, F) \ . \tag{4}$$

The nilpotency follows from the Jacobi identity of antibrackets and the master equation:

$$\mathcal{S}^2 F = (S, (S, F)) = \tfrac{1}{2}((S, S), F) = 0 \ . \tag{5}$$

The operator $\mathcal{S}$ raises the ghost number by 1, i.e.: $gh(\mathcal{S}F) = gh(F) + 1$. The cohomology of this operator in the space of local functions of ghost number 0 gives physically meaningful quantities. They are formed by arbitrary gauge invariant functionals, and are defined up to field equations.

We now introduce a second grading: the antifield number (*afn*). This is $-gh$ for the Fields of negative ghost number (antifields in the 'classical basis', the one used so far), and 0 for Fields of non-negative ghost number. Check that the extended action (1) has zero ghost number, but the 3 lines have antifield numbers 0,1 and 2 respectively. The action of $\mathcal{S}$ can be split according to its

[3] The argument was repeated in section I.3.1.

change in the antifield number:

$$\begin{array}{cccccccccccc} \mathcal{S} & = & \delta_{KT} & + & \Omega & + & D_1 & + & D_2 & + & \dots \\ afn & & -1 & & 0 & & 1 & & 2 & & \end{array} \tag{6}$$

Observe that the antifield number at most diminishes by one. This part defines the Koszul–Tate (KT) differential (introduced in BV quantisation in [11]). Explicitly, the main properties are:

$$\begin{array}{ll} \delta_{KT}\Phi = 0 & \text{vanishes on fields} \\ \delta_{KT}\phi^*_i = \frac{S^0 \overleftarrow{\partial}}{\partial\phi^i} & \text{field equations} \\ \delta_{KT}c^*_a = \phi^*_i R^i{}_a & \end{array} \tag{7}$$

(S^0 is the action without antifields, and $R^i{}_a$ determines its gauge transformations). Now the nilpotency property $\mathcal{S}^2 = 0$ implies

$$\begin{array}{ll} \delta^2_{KT} = 0 & \text{KT is nilpotent} \\ \Omega\delta_{KT} + \delta_{KT}\Omega = 0 & \\ \Omega^2 = -\delta_{KT}D_1 - D_1\delta_{KT} & \text{On-shell nilpotent BRST} \end{array} \tag{8}$$

The equation (6) also defines Ω, which will be called the BRST operator. Its action on functionals of the fields is given by

$$\Omega\Phi = \mathcal{S}\Phi|_{\Phi^*=0} \tag{9}$$

We will come back to the meaning of the last lines of (8) in a moment.

We have now *3 different cohomologies*, related to $\mathcal{S}$, δ_{KT} and Ω. We should also be careful to distinguish cohomology in the space of local functions, or in the space of integrals thereof (this is cohomology *mod d*, denoted $H^g(\mathcal{S} \mid d)$, where d is the space–time differential).

1. δ_{KT} acts in the space of functions of Φ^*_A and ϕ^i, i.e. functions of ghost number negative or zero: it vanishes on the ghosts, and also the image of δ_{KT} on the antifields or classical fields does not contain the ghosts. As a grading one uses the antifield number, and $H^k(\delta_{KT})$ then denotes the cohomology of δ_{KT} in local functions of antifield number k. Its main property is that

$$H^k(\delta_{KT}) = \delta^k_0 \times \text{ functions on the stationary surface,} \tag{10}$$

where the stationary surface refers to the surface in the space of fields of ghost number 0 ('classical fields' ϕ^i) where the classical field equations are satisfied[4]. Also for integrals over local functions of ghost number zero, or $H^0(\delta_{KT} \mid d)$, such a property holds. But there is not such a general statement for $H^k(\delta_{KT} \mid d)$

[4] To be more exact, in some cases one should define this concept using the concept of functions 'not proportional to field equations', see the discussion on 'evanescent functions' in section 3.1 of [9].

with $k > 0$ (in [12] some general results have been given, relating e.g. the $k = 1$ case to rigid symmetries).

2. The last line of (8) implies that Ω is nilpotent in the cohomology space of δ_{KT}. One can define a cohomology of Ω in the cohomology space $H^k(\delta_{KT})$. This is the BRST cohomology, which therefore is defined **on the stationary surface**. So a function F is in the cohomology of Ω, and nontrivial, if

$$\Omega F \approx 0 \qquad \text{and} \qquad F \not\approx \Omega G \ , \tag{11}$$

where $\approx$ means equality on the stationary surface, defined with the classical field equations, and F satisfies $\delta_{KT} F = 0$. This is denoted by $H^k\left(\Omega \mid H(\delta_{KT})\right)$. The grading of Ω is the ghost number. Therefore k refers here to the ghost number. For $k = 0$ the cohomology of local functions now identifies functions which differ by gauge transformations, and, combined with the result for δ_{KT}, this implies that we are left with the physical observables.

3. $\mathcal{S}$ acts on fields and antifields. Its grading is the ghost number. The cohomology of $\mathcal{S}$ combines the results of the previous two cohomologies:

$$\begin{array}{ll} \text{for } k < 0 : & H^k(\mathcal{S}) = 0 \\ \text{for } k \geq 0 : & H^k(\mathcal{S}) = H^k(\Omega) \equiv H^k\left(\Omega \mid H(\delta_{KT})\right) \ . \end{array} \tag{12}$$

Note that this statement for $k \geq 0$ applies also for cohomology *mod d* (local integrals). It is important that to define the cohomologies with the BRST method we have to use 'weak equalities' (even for a 'closed algebra'), while in the antibracket cohomology the inclusion of the antifields allows us to work with unqualified equalities.

4 Gauge fixing

Let us illustrate the gauge fixing with a simple example. Consider the theory with classical fields $\phi^i = \{X^\mu, h\}$ and classical action

$$S_0 = \partial X^\mu \bar{\partial} X_\mu \ . \tag{13}$$

Trivially, h does not occur in it, which we can interpret as a gauge invariance $\delta h = \epsilon$. The extended action is

$$S = \partial X^\mu \bar{\partial} X_\mu + h^* c \ . \tag{14}$$

For propagators, one wishes to invert the matrix of second derivatives of the action w.r.t. fields, but it is singular. Consider however the matrix of second derivatives w.r.t. Fields, $\partial_\alpha \partial_\beta S$, and make a slight rearrangement:

$$\begin{array}{c} \\ X^\mu \\ h \\ c \\ X^*_\mu \\ h^* \\ c^* \end{array} \begin{array}{c} \begin{array}{cccccc} X & h & c & X^* & h^* & c^* \end{array} \\ \begin{pmatrix} * & 0 & 0 & 0 & 0 & 0 \\ 0 & 0 & 0 & 0 & 0 & 0 \\ 0 & 0 & 0 & 0 & * & 0 \\ 0 & 0 & 0 & 0 & 0 & 0 \\ 0 & 0 & * & 0 & 0 & 0 \\ 0 & 0 & 0 & 0 & 0 & 0 \end{pmatrix} \end{array} \longrightarrow \begin{array}{c} \\ X^\mu \\ h^* \\ c \\ X^*_\mu \\ h \\ c^* \end{array} \begin{array}{c} \begin{array}{cccccc} X & h^* & c & X^* & h & c^* \end{array} \\ \begin{pmatrix} * & 0 & 0 & 0 & 0 & 0 \\ 0 & 0 & * & 0 & 0 & 0 \\ 0 & * & 0 & 0 & 0 & 0 \\ 0 & 0 & 0 & 0 & 0 & 0 \\ 0 & 0 & 0 & 0 & 0 & 0 \\ 0 & 0 & 0 & 0 & 0 & 0 \end{pmatrix} \end{array} \tag{15}$$

The stars in the matrix denote the non–zero entries. The rank of the matrix is indeed half its dimension (properness condition). The invertible part is just the left upper corner of the second matrix. (The other entries happen to be zero because of the simplicity of the model). If we reassign the antifield h^* to be a field, renaming it b, then h becomes an antifield b^*, and the action in terms of the new fields has no zero–modes. This is what we call gauge–fixed.

In general **gauge fixing is a canonical transformation such that the new 'fields' have no zero–modes.** Later we will choose these fields as integration variables of the path integral.

Also in $d = 2$ gravity gauge fixing is accomplished by just such a change of name. We define

$$g_{\alpha\beta} = \eta_{\alpha\beta} + b^*_{\alpha\beta} \; ; \qquad g^{*\alpha\beta} = -b^{\alpha\beta} \; . \tag{16}$$

We can now consider two bases:

classical basis	X^μ	$g_{\alpha\beta}$	ξ^α	c	X^*_μ	$g^{*\alpha\beta}$	ξ^*_α	c^*	(17)
statistics	$+$	$+$	$-$	$-$	$-$	$-$	$+$	$+$	
ghost number	0	0	1	1	-1	-1	-2	-2	
gauge fixed basis	X^μ	$b^*_{\alpha\beta}$	ξ^α	c	X^*_μ	$b^{\alpha\beta}$	ξ^*_α	c^*	

It is now easy to see that, substituting (16) in (1), the part of the extended action which is independent of antifields in the gauge–fixed basis has no gauge invariances any more, i.e. it is properly called gauge fixed. Note that $b^{\alpha\beta}$ plays the role of the antighost.

There are other types of canonical transformations that are often important. The general type of a canonical transformation $\{\Phi, \Phi^*\}$ to $\{\tilde{\Phi}, \tilde{\Phi}^*\}$ where the field–field part is invertible (this does not include the above transformation (16)), can be generated from a (fermionic) function $F(\Phi, \tilde{\Phi}^*)$:

$$\tilde{\Phi}^A = \frac{\partial F(\Phi, \tilde{\Phi}^*)}{\partial \tilde{\Phi}^*_A} \qquad \Phi^*_A = \frac{\partial F(\Phi, \tilde{\Phi}^*)}{\partial \Phi^A} \; . \tag{18}$$

We will here illustrate the use of such a transformation, not for gauge fixing (that will be done below), but to achieve a simplification in the extended action of $d = 2$ gravity. Consider the transformation from $\{\Phi\} = \{X^\mu, g_{\alpha\beta}, \xi^\alpha, c\}$

and their antifields to $\{\tilde{\Phi}\} = \{\tilde{X}^\mu, h_{++}, h_{--}, e, c^\alpha, \tilde{c}\}$ and the corresponding antifields generated by (an integral over d^2x is implied)

$$\begin{aligned} F &= \tilde{X}^*_\mu X^\mu + e^* \sqrt{g} + \tilde{c}^* (c\sqrt{g} + \partial_\alpha \xi^\alpha \sqrt{g}) \\ &\quad + h^{++*} \frac{g_{++}}{g_{+-} + \sqrt{g}} + h^{--*} \frac{g_{--}}{g_{+-} + \sqrt{g}} \\ &\quad + c^*_+ \left(\xi^+ + \frac{g_{--}}{g_{+-} + \sqrt{g}} \xi^- \right) + c^*_- \left(\xi^- + \frac{g_{++}}{g_{+-} + \sqrt{g}} \xi^+ \right) . \end{aligned} \tag{19}$$

This leads to

$$\begin{aligned} S &= \frac{1}{1 - h_{++} h_{--}} \left(-\nabla_+ X^\mu \cdot \nabla_- X^\nu \eta_{\mu\nu} + X^*_\mu c^\alpha \nabla_\alpha X^\mu \right) \\ &\quad + h^{++*} \nabla_+ c^- + h^{--*} \nabla_- c^+ + e^* \tilde{c} \\ &\quad - c^*_+ c^+ \partial_+ c^+ - c^*_- c^- \partial_- c^- , \end{aligned} \tag{20}$$

where $\nabla_+ \equiv \partial_+ - h_{++}\partial_- + \lambda(\partial_- h_{++} \cdot)$ (λ is the spin: lower + indices, or upper − indices ...). Here the term $e^*\tilde{c}$ exhibits **a trivial system**: $\mathcal{S}e = \tilde{c}$ and $\mathcal{S}\tilde{c}^* = e^*$. The cohomology in the quartet $\{e, \tilde{c}, e^*, \tilde{c}^*\}$ is trivial, it can therefore be omitted. The reader may also consult section I.3.2 for a review on the applications of canonical transformations, and section I.3.4 on the trivial systems. In the new basis gauge fixing is again performed just by the transformation $h^{++*} = b^{++}$ and $h_{++} = -b^*_{++}$.

The canonical transformation of the form (18) is often used for gauge fixing. Consider F of the particular form

$$F = \Phi^A \Phi^*_A + \Psi(\Phi) . \tag{21}$$

The first term by itself produces just the identical transformation. The second is the so-called gauge fermion Ψ, which was needed in [1], although it should be clear now that this method is a particular case. For Maxwell theory for example, the minimal extended action would be just $S = -\frac{1}{4}F_{\mu\nu}F^{\mu\nu} + A^*_\mu \partial^\mu c$. One can first add an auxiliary field term $-\frac{1}{2}b^*b^*$. In the cohomology language this is just again a trivial system $\mathcal{S}b = b^*$ and $\mathcal{S}b^* = 0$. Then the canonical transformation we need for gauge fixing is of the type (21) with $\Psi = b\partial_\mu A^\mu$. The reader can check that this gives the usual gauge-fixed action. Note that, although we had to introduce a trivial system before the canonical transformation, even in this simple example there is still a simplification w.r.t. the usual BRST procedure. In BRST one first introduces also an auxiliary field λ. This is necessary to have a nilpotent BRST-operator off shell. However the algebra of $\mathcal{S}$ is always closed.

We have related the physical observables to the cohomology of the operator $\mathcal{S}$. The essential principle is that we always use canonical transformations, and add or delete trivial systems. A canonical transformation does not change the cohomology of this operator, and neither do the trivial systems. In a gauge-fixed basis we may again define a BRST operator by (9), where the fields and antifields are now those of the gauge-fixed basis. Then one can prove that the cohomology of $\mathcal{S}$ is equivalent to the 'weak' cohomology of Ω (using the new field equations of all the fields). This proves that the BRST cohomology gives the physical observables again, but also shows the advantage of the $\mathcal{S}$–cohomology: it includes the information in the field equations, as $\mathcal{S}\Phi^* = \frac{S\overleftarrow{\partial}}{\partial\Phi}$.

5 Quantum theory and Pauli–Villars regularisation

The general principles of the quantum treatment in the BV framework have been described in section II.4, to which we refer for an exposition of our method. The advantage of the Pauli–Villars (PV) regularisation is that it has a Lagrangian formulation that interfaces nicely with the BV framework. We will now describe some additional aspects of the PV procedure, using $d = 2$ gravity as an illustration.[5] The quantum theory, in one loop approximation, is based on the action

$$S_T = S(z) + S_{PV}(z, \underline{z}) + S_M(z, \underline{\Phi}) \, . \tag{22}$$

In this formula, the massless PV extended action and the mass terms for the PV fields are given by

$$S_{PV} = \frac{1}{2}\underline{z}^\alpha \left(\frac{\overrightarrow{\partial}}{\partial z^\alpha} S \frac{\overleftarrow{\partial}}{\partial z^\beta} \right) \underline{z}^\beta \qquad \text{and} \qquad S_M = -\frac{1}{2} M^2 \underline{\Phi}^A T_{AB}(z) \underline{\Phi}^B \, , \tag{23}$$

where $T_{AB}(z)$ is largely arbitrary but must be invertible. Note that this last condition implies that, formally, in the limit $M \to \infty$, the PV Fields are cohomologically trivial since

$$(S, \underline{\Phi}^*_A) = M^2 T_{AB} \underline{\Phi}^B \, . \tag{24}$$

In the example of $d = 2$ gravity, we get from (1)

$$\begin{aligned}
S_{PV} \;=\; & -\tfrac{1}{2}\sqrt{g} g^{\alpha\beta} \partial_\alpha \underline{X}^\mu \cdot \partial_\beta \underline{X}_\mu \\
& + \underline{g}_{\gamma\delta} \sqrt{g} \left(g^{\alpha\gamma} g^{\beta\delta} - \tfrac{1}{2} g^{\alpha\beta} g^{\gamma\delta} \right) \partial_\alpha \underline{X}^\mu \cdot \partial_\beta X_\mu \qquad (25) \\
& - \tfrac{1}{2} \underline{g}_{\gamma\delta} \underline{g}_{\epsilon\phi} \sqrt{g} \left(g^{\alpha\gamma} g^{\delta\epsilon} g^{\phi\beta} - \text{traces} \right) \partial_\alpha X^\mu \cdot \partial_\beta X_\mu \\
& + \underline{X}^*_\mu \xi^\alpha \partial_\alpha \underline{X}^\mu + \underline{X}^*_\mu \underline{\xi}^\alpha \partial_\alpha X^\mu + X^*_\mu \underline{\xi}^\alpha \partial_\alpha \underline{X}^\mu \\
& + \underline{g}^{*\alpha\beta} \left(\xi^\gamma \partial_\gamma \underline{g}_{\alpha\beta} + 2 \underline{g}_{\gamma(\alpha} \partial_{\beta)} \xi^\gamma + c \underline{g}_{\alpha\beta} \right) \\
& + \underline{g}^{*\alpha\beta} \left(\underline{\xi}^\gamma \partial_\gamma g_{\alpha\beta} + 2 g_{\gamma(\alpha} \partial_{\beta)} \underline{\xi}^\gamma + \underline{c} g_{\alpha\beta} \right) + g^{*\alpha\beta} \left(\underline{\xi}^\gamma \partial_\gamma \underline{g}_{\alpha\beta} + 2 \underline{g}_{\gamma(\alpha} \partial_{\beta)} \underline{\xi}^\gamma + \underline{c} \underline{g}_{\alpha\beta} \right) \\
& - \underline{\xi}^*_\beta \underline{\xi}^\alpha \partial_\alpha \xi^\beta - \underline{\xi}^*_\beta \xi^\alpha \partial_\alpha \underline{\xi}^\beta - \xi^*_\beta \underline{\xi}^\alpha \partial_\alpha \underline{\xi}^\beta - \underline{c}^* \underline{\xi}^\alpha \partial_\alpha c - c^* \underline{\xi}^\alpha \partial_\alpha \underline{c} - \underline{c}^* \xi^\alpha \partial_\alpha \underline{c} \, . \qquad (26)
\end{aligned}$$

Gauge fixing is now performed by the canonical transformation (16) together with

$$\underline{g}^{*\alpha\beta} = -\underline{b}^{\alpha\beta} \, ; \qquad \underline{g}_{\alpha\beta} = \underline{b}^*_{\alpha\beta} \, . \tag{27}$$

[5] The notation used here and in the sequel differs slightly from the one used in (I) and (II), viz. we denote the PV–partner of a Field by underscoring its symbol.

Note that here we introduced the PV system already before gauge fixing. To keep the correspondence between the PV action and the action of the ordinary fields as in (23), their canonical transformations must be related. This question is treated in more detail in (I.42).

The PV mass term should be quadratic in the PV partners of the fields in the gauge–fixed basis. As an example we take

$$S_M = -\tfrac{1}{2}M^2 g^{t/2}\underline{X}^2 + \tfrac{1}{2}M g\epsilon_{\alpha\beta}\underline{\xi}^\alpha\underline{\xi}^\beta + \tfrac{1}{2}M\underline{b}^{\alpha\beta}\underline{b}^{\gamma\delta}g_{\alpha\gamma}\epsilon_{\beta\delta}\ . \tag{28}$$

To illustrate the arbitrariness in the choice of mass term, which will also play a role in section 6,we introduced an arbitrary real parameter t in the mass term for the scalars. If necessary one introduces several copies of PV fields with multiplicities C_i, and imposes the conditions $\sum_i C_i = 1$ and $\sum_i C_i(M_i)^{2n} = m^{2n}$ for $n = 1, 2, ..., n_{max}$. The divergences of the theory manifest themselves as terms of the form $\sum_i C_i(M_i)^{2n}\log M_i^2$, and have to be absorbed by renormalisation before the limit $M_i \to \infty$ is taken. An example will be seen in the next section.

6 Anomalies

How anomalies occur in the formal path integral has been explained in section (I.4) and the first part of section II.5.1. Concerning cohomology let us add that even in the presence of anomalies there is a nilpotent operator

$$\mathcal{S}_q F = e^{-\frac{i}{\hbar}W}\tfrac{\hbar}{i}\Delta e^{\frac{i}{\hbar}W}F = (W, F) + \tfrac{\hbar}{i}\Delta F + \mathcal{A}F\ , \tag{29}$$

where W is the quantum action $S + \hbar M_1 + \ldots$, and Δ is acting from the left: [6]

$$\Delta = (-)^A \frac{\vec{\partial}}{\partial\Phi^A}\frac{\vec{\partial}}{\partial\Phi^*_A}\ . \tag{30}$$

The nilpotency of Δ immediately implies the nilpotency of $\mathcal{S}_q$. The anomaly itself is $\mathcal{A} = \mathcal{S}_q 1$, and is trivially invariant under $\mathcal{S}_q$. If another quantity M can be found such that $\mathcal{A} = \mathcal{S}_q M$, then the anomaly can formally be removed by a redefinition $W' = W + \frac{\hbar}{i}\log(1-M)$. Note however that in general this additional contribution to the action can not be written as a *local* term. Note also that in $\hbar$ expansions of expressions like (29) it has often been assumed that ΔS gives only terms of order $\hbar^0$ [1, 5]. This may not always be true in a regulated theory.

In sections I.5.1, I.5.2 and II.5.1 it has been shown why these formal expressions need regularisation, and how this can be performed by using PV regulators [4, 5]. One ends up with a regulated determination of ΔS which depends on the choice of T in the PV mass term of (23). (i.e. on the regularisation), on the choice of basis (gauge fixing), ... but it always takes a value in the same cohomology class: for two such choices (a) and (b) we have

$$\Delta^{(b)}S = \Delta^{(a)}S + (F, S) \qquad \text{with } F \text{ local.} \tag{31}$$

[6] Important properties are $\Delta FG = \Delta F \cdot G + (-)^F F\,\Delta G + (-)^F (F, G)$ and $\Delta^2 = 0$.

Rather than repeating the general exposition, we will sketch here the application to the scalar loops in $d = 2$ gravity.

The anomaly at one loop arises from

$$\mathcal{A} = \tfrac{1}{2\hbar}(S_T, S_T) = \tfrac{1}{\hbar}(S + S_{PV}, S_M) \ . \tag{32}$$

We only treat here the integration over the "matter" PV fields $\underline{X}$, for the purpose of illustration. From (28) we then only take into account the first term, and will in particular consider two values for t:

$$\begin{aligned} \text{at } t = 0 \ : \quad & \mathcal{A}_0 = \frac{M^2}{\hbar}\left[\underline{X}^\mu \underline{X}_\mu \partial_\alpha \xi^\alpha - 2\underline{X}^\mu \underline{\xi}^\alpha \partial_\alpha X_\mu\right] \\ \text{at } t = 1 \ : \quad & \mathcal{A}_1 = \frac{M^2}{\hbar}\sqrt{g}\left[-c\underline{X}^\mu \underline{X}_\mu - 2\underline{X}^\mu \underline{\xi}^\alpha \partial_\alpha X_\mu\right] \ . \end{aligned} \tag{33}$$

After integration over the PV fields the off–diagonal terms will vanish and ($\Box = \partial_\alpha \sqrt{g} g^{\alpha\beta}\partial_\beta$)

$$\underline{X}^\mu \underline{X}_\mu \to \frac{\hbar}{M^2 g^{t/2} - \Box} \ . \tag{34}$$

We then use heat kernel techniques. The method was reviewed in section 5 of (I): one applies (I.47) and (I.50) with

$$\begin{array}{lll} t = 0 \ : & J = \partial_\alpha \xi^\alpha & \mathcal{R} = \Box \\ t = 1 \ : & J = -c & \mathcal{R} = \frac{1}{\sqrt{g}}\Box \ , \end{array} \tag{35}$$

leading to

$$\begin{aligned} \mathcal{A}_0 &= \frac{1}{8\pi}\left(M^2 \log M^2 - \tfrac{1}{6}\sqrt{g}\, R + \tfrac{1}{12}\Box \log g\right)\left(-\partial_\alpha \xi^\alpha\right) \\ \mathcal{A}_1 &= \frac{1}{8\pi}\left(M^2 \log M^2 \sqrt{g} - \tfrac{1}{6}\sqrt{g}\, R\right) c \ . \end{aligned} \tag{36}$$

At $t = 0$ our mass term is indeed invariant under Weyl transformations, and not under general $d = 2$ coordinate transformations. At $t = 1$ we have the reverse situation. This explains the occurrence of the Weyl ghosts for $t = 1$ and the diffeomorphism ghosts for $t = 0$. Note that for $M \to \infty$ there is a diverging part, which is however a total derivative for $t = 0$, and, as will be clear below, is even locally $\mathcal{S}$–exact for $t = 1$. The two expressions for the anomaly are related by $\mathcal{A}_1 - \mathcal{A}_0 = \mathcal{S} M_1$ with

$$M_1 = \frac{1}{8\pi}\left(M^2 \log M^2 \sqrt{g} - \tfrac{1}{12}\log g \sqrt{g}\, R + \tfrac{1}{48}\log g\, \Box \log g\right) \ , \tag{37}$$

showing explicitly that the change in regularisation preserves the cohomological class of the anomaly.

As already stressed in the previous lectures, the anomaly depends on $g_{\alpha\beta} = \eta_{\alpha\beta} + b^*_{\alpha\beta}$, i.e. on the antifields of the gauge–fixed basis.

Knowing that in general the anomaly is an element of the cohomology of $\mathcal{S}$, it is interesting to investigate what the possible classes are, a priori, without doing the actual anomaly calculation. To this we now turn.

7 Cohomological analysis of $d = 2$ gravity and background charges

In collaboration with F. Brandt [6], we have performed an analysis of the cohomology of $\mathcal{S}$ for local 2–forms modulo d, or $H^g(\mathcal{S} \mid d)$. The starting point of this analysis is the symmetry algebra, without assumptions about the classical action. Technically, we assumed knowledge of S^1 and S^2 only, i.e. the second and third line of (1).

There is a relation with the local cohomology of 0–forms [13]

$$H^g(\mathcal{S} \mid d) \sim H^{g+2}(\mathcal{S}) \ , \tag{38}$$

using descent equations (ω_f^g is an f–form of ghost number g)

$$\mathcal{S}\omega_2^g + d\omega_1^{g+1} = 0 \ ; \qquad \mathcal{S}\omega_1^{g+1} + d\omega_0^{g+2} = 0 \ ; \qquad \mathcal{S}\omega_0^{g+2} = 0 \ . \tag{39}$$

For this equivalence it is necessary that the reparametrisation ghosts are present. The rather general results of [13], implying e.g. that derivatives of ghosts are not present in the cohomology, can not be applied directly to the case of $d = 2$ gravity, because they rest on the assumption that all local symmetries have independent gauge fields, which is not the case for the Weyl symmetry in our setup. We will limit ourselves to a description of the results.

We find that a non–trivial cohomology exists for ghost numbers $g = -1, 0, \ldots 4$. Let us first mention the most general functional with ghost number 0 which does not depend on antifields. This functional can be used as a classical action, replacing the first line of (1):

$$S_{cl} = \tfrac{1}{2}\sqrt{g}\, g^{\alpha\beta} G_{\mu\nu}(X)\partial_\alpha X^\mu \cdot \partial_\beta X^\nu + B_{\mu\nu}(X)\partial_+ X^{[\mu} \cdot \partial_- X^{\nu]} \ . \tag{40}$$

Here $G_{\mu\nu}(X)$ and $B_{\mu\nu}(X)$ are arbitrary, but not all physically different: the following changes relate solutions that are cohomologically equivalent:

$$\begin{aligned} G \sim G'_{\mu\nu} &= G_{\mu\nu} + 2\partial_{(\mu} f_{\nu)} - \Gamma_{\mu\nu\rho} f^\rho \ , \\ B \sim B'_{\mu\nu} &= B_{\mu\nu} + 2\partial_{[\mu} B_{\nu]} + H_{\mu\nu\rho} f^\rho \ , \end{aligned} \tag{41}$$

where

$$f_\mu = G_{\mu\nu} f^\nu \ ; \qquad \Gamma_{\mu\nu,\rho} = \partial_{(\mu} G_{\nu)\rho} - \tfrac{1}{2}\partial_\rho G_{\mu\nu} \ ; \qquad H_{\mu\nu\rho} = 3\partial_{[\mu} B_{\nu\rho]} \ . \tag{42}$$

The functions B_μ and f^μ are arbitrary, and the latter can be interpreted as a target space reparametrisation $X^\mu \to X^\mu + f^\mu$.

In addition there are antifield–dependent solutions depending on arbitrary covariantly constant functions $f^{(\pm)\mu}(X)$,

$$D_\mu^\pm f_\nu^{(\pm)} \equiv \partial_\mu f_\nu^{(\pm)} - \Gamma^\pm_{\mu\nu,\rho} f^{(\pm)\rho} = 0 \qquad \text{with} \qquad \Gamma^\pm_{\mu\nu,\rho} = \Gamma_{\mu\nu,\rho} \pm \tfrac{1}{2} H_{\mu\nu\rho} \ . \tag{43}$$

These solutions are, using the notations of (19) and $y = h_{++}h_{--}$,

$$\begin{aligned} M(f^{(+)}, f^{(-)}) &= X^*_\mu \left(\partial_+\xi^+ + h_{--}\partial_+\xi^-\right) \cdot f^{(+)\mu} - 2\tfrac{1}{1-y}\partial_+ h_{--} \cdot \nabla_+ X^\mu \cdot f^{(+)}_\mu \\ &+ X^*_\mu \left(\partial_-\xi^- + h_{++}\partial_-\xi^+\right) \cdot f^{(-)\mu} - 2\tfrac{1}{1-y}\partial_- h_{++} \cdot \nabla_- X^\mu \cdot f^{(-)}_\mu \qquad (44) \end{aligned}$$

Before discussing the meaning of these terms, we give also the solutions for ghost number 1. There are 2 special solutions that contain no arbitrary parameters apart from an overall factor:

$$A_\pm = c^\pm \partial^3_\pm h_{\mp\mp} \,. \qquad (45)$$

Also, there are solutions depending on arbitrary functions $f^{(\pm)}_{\mu\nu}(X)$:

$$\omega(f^{(+)}_{\mu\nu}, f^{(-)}_{\mu\nu}) = \tfrac{1}{1-y}\nabla_+ X^\mu \cdot \nabla_- X^\nu \cdot \left[f^{(+)}_{\mu\nu}\left(\partial_+\xi^+ + h_{--}\partial_+\xi^-\right) + f^{(-)}_{\mu\nu}\left(\partial_-\xi^- + h_{++}\partial_-\xi^+\right)\right]. \qquad (46)$$

Again, some of these solutions are cohomologically equivalent: for arbitrary $H^\pm_\mu(X)$,

$$f^+_{\mu\nu} \sim f^+_{\mu\nu} + D^+_\nu H^+_\mu \; ; \qquad f^-_{\mu\nu} \sim f^-_{\mu\nu} + D^-_\mu H^-_\nu \,. \qquad (47)$$

All these solutions of ghost number 1 appear when discussing the anomalies of general σ–models with classical action (40). In our analysis for the action (1) we met the particular combination

$$A_+ + A_- \cong \tfrac{1}{2} c \sqrt{g} R \qquad (48)$$

where $\cong$ means an equality up to $\mathcal{S}$–exact terms. In chiral gravity one obtains separately A_+ and/or A_- as anomaly.

Now we come to the meaning of the other solutions of ghost number 0, eq.(44). First we give the antibracket[7]

$$\left(M(f^{(+)}, f^{(-)}), M(g^{(+)}, g^{(-)})\right) \cong -4f^{(+)\mu}g^{(+)}_\mu A_+ - 4f^{(-)\mu}g^{(-)}_\mu A_- \,. \qquad (49)$$

This implies that, although M is a local integral with ghost number zero, it cannot be used as an extra part of the extended action, as this breaks the master equation. However, the right hand side of (49) allows another interpretation. If we modify the action by adding an $M(f^{(+)}, f^{(-)})$–term that is formally of order $\hbar^{1/2}$, then (M, M) will contribute to the anomaly on the same level as the one–loop diagrams, modifying the coëfficient of the anomaly, and possibly canceling it. This term therefore becomes an important addition for the quantum theory. It is the proper generalisation of what is usually called a "background charge" term. This is apparent when, for the chiral case, one puts $h_{++} = 0$, drops the corresponding ξ^- ghost, and specializes to $G_{\mu\nu} = -\frac{1}{2}\delta_{\mu\nu}$. It is then another example of a term $M_{1/2}$, as treated in section II.5.3, presented there in the case of chiral W_3 to cancel the anomalies [14] (see [15, 9] for their inclusion in the BV formalism). For our non–chiral case, again one has to include both chiralities,

[7]Note that $f^{(+)\mu}g^{(+)}_\mu$ is a constant due to (43), which is valid also for g.

and add an appropriate $\mathcal{S}$–trivial term. This leads to the dilaton term in the σ–model.

8 Conclusions

In this review we described some of the ongoing work in the BV quantisation program, with emphasis on the cohomology of the operator $\mathcal{S} \equiv (S, \cdot)$. It appears in various places:

- the local cohomology at ghost number 0 gives the classical physical observables.
- the cohomology of integrals at ghost number 0 gives the possible actions, and is also important for counterterms in the renormalisation procedure [16].
- the cohomology of integrals at ghost number 1 gives the possible anomalies.

- Antifield dependent anomalies at ghost number 0 correspond to background charges, possibly allowing a cancellation of anomalies.

Acknowledgments

This work is carried out in the framework of the European Community Research Programme "Gauge theories, applied supersymmetry and quantum gravity", with a financial contribution under contract SC1-CT92-0789.

References

1. I.A. Batalin and G.A. Vilkovisky, Phys. Rev. **D28** (1983) 2567 (E:**D30** (1984) 508).
2. A. Van Proeyen, in Proc. of the Conference *Strings & Symmetries 1991*, Stony Brook, May 20–25, 1991, eds. N. Berkovits et al., (World Sc. Publ. Co., Singapore, 1992), pp. 388.
3. W. Troost and A. Van Proeyen, to be published in Proc. of the Conference *Strings 1993*, Berkeley, May 1993, preprint KUL-TF-93/32, hepth 9307126.
4. A. Diaz, W. Troost, P. van Nieuwenhuizen and A. Van Proeyen, Int. J. Mod. Phys. **A4** (1989) 3959.
5. W. Troost, P. van Nieuwenhuizen and A. Van Proeyen, Nucl. Phys. **B333** (1990) 727.
6. F. Brandt, W. Troost and A. Van Proeyen, NIKHEF–H 94–16, KUL–TF–94/17, hep-th/9407061, to appear in the proceedings of the *Geometry of Constrained Dynamical Systems* workshop, Isaac Newton Institute for Mathematical Sciences, Cambridge, June 15–18, 1994; and *A complete computation of the BRST-antibracket cohomology of 2d gravity conformally coupled to matter*, in preparation.

7. W. Troost and A. Van Proeyen, *An introduction to Batalin–Vilkovisky Lagrangian quantisation*, Leuven Notes in Math. Theor. Phys., in preparation.
8. J. Gomis, J. París and S. Samuel, preprint CCNY-HEP-94/03, KUL-TF-94/12, UB-ECM-PF 94/15, UTTG-11-94.
9. S. Vandoren, A. Van Proeyen, Nucl. Phys. **B411** (1994) 257.
10. G. Sterman, P. Townsend and P. van Nieuwenhuizen, Phys. Rev. **D17** (1978) 1501;
B. de Wit and J.W. van Holten, Phys. Lett. **79B** (1978) 389;
R. Kallosh, Nucl. Phys. **B141** (1978) 141.
11. J. Fisch, M. Henneaux, J. Stasheff and C. Teitelboim, Commun. Math. Phys. **120** (1989) 379;
J. Fisch and M. Henneaux, Commun. Math. Phys. **128** (1990) 627;
M. Henneaux, Nucl. Phys. **B** (Proc. Suppl.) **18A** (1990) 47;
M. Henneaux, Commun. Math. Phys. **140** (1991) 1.
12. G. Barnich, F. Brandt and M. Henneaux, preprint ULB-TH-94/06, NIKHEF-H 94-13, hep-th/9405109.
13. F. Brandt, N. Dragon and M. Kreuzer, Nucl. Phys. **B340** (1990) 187;
F. Brandt, preprint NIKHEF-H 93-21, hep-th/9310123.
14. L.J. Romans, Nucl. Phys. **B352** (1990) 829.
15. F. De Jonghe, R. Siebelink, W. Troost, Phys. Lett. B **306**, 295-300 (1993).
16. D. Anselmi, preprint SISSA/ISAS 147/93/EP, hep-th/9309085, to appear in Class. Quantum Grav.; preprint SISSA/ISAS 90/94/EP, hep-th/9407023.

E_{10} for beginners*

R.W. Gebert, H. Nicolai

IInd Institute for Theoretical Physics, University of Hamburg
Luruper Chaussee 149, D-22761 Hamburg, Germany

December 9, 1994

In this contribution, we summarize a recent attempt to understand hyperbolic Kac Moody algebras in terms of the string vertex operator construction [12] (which readers are also advised to consult for a comprehensive list of references). As is well known, Kac Moody algebras (see e.g. [21], [15]) fall into one of three classes corresponding to whether the associated Cartan matrices are positive definite, positive semi-definite and indefinite. Of these, the first two are well understood, leading to finite and affine Lie algebras (the latter being equivalent to current algebras in two space-time dimensions). Virtually nothing is known, however, about the last class of Kac Moody algebras, based on indefinite Cartan matrices. Nonetheless these have been repeatedly suggested as natural candidates for the still elusive fundamental symmetry of string theory (see e.g. [24], [28] for recent proposals in this direction). Being vastly larger than affine Kac Moody algebras, they are certainly "sufficiently big" for this purpose, but an even more compelling argument supporting such speculations is the intimate link that exists between Kac Moody algebras and the vertex operator construction of string theory which has been known for a long time. More specifically, it has been established that the elements making up a Kac Moody algebra of finite or affine type can be explicitly realized in terms of tachyon and photon emission vertex operators of a compactified open bosonic string [9], [14]. On the basis of these results, it has been conjectured that Kac Moody algebras of indefinite type might not only furnish new symmetries of string theory, but might themselves be understood in terms of string vertex operators associated with the higher excited (massive) states of a compactified bosonic string [14], [8].

Of what nature are these new symmetries then? In [17], it was argued that in the ultrahigh-energy limit of string theory, where the Planck mass goes to zero, an infinite number of linear relations exists between the scattering amplitudes of different string states that are valid order by order in perturbation theory. This suggests an enormous

---

* Invited talk presented by H. Nicolai at the Feza Gürsey Memorial Conference, Istanbul, June 1994.
Correspondence to: i02geb@dsyibm.desy.de or i02nic@dsyibm.desy.de

symmetry which is restored at high energies. It may seem unreasonable to study such a queer limit but, in fact, it is a very conservative approach. For example, if we had not known the spontaneously broken symmetry of the electroweak interactions, we could have in principle discovered it at high energies where all gauge particles become massless again. In agreement with this analogy it is indeed a reasonable hope to find hints of the unbroken string gauge algebra by studying the relations between high energy scattering amplitudes. But since the latter, according to the above result, are essentially unique for a given number of scattered physical states, it is tempting to regard the Lie algebra of physical states itself as part of some universal gauge algebra. Note that we obtain, by construction, different Lie algebras of physical states when we consider inequivalent string backgrounds. Moreover, due to the presence of infinitely many massive physical states, each Lie algebra would have to be spontaneously broken almost completely. If we take this picture for granted then our task will be to make a clever choice for the specific string background in order to find a Lie algebra of physical states as large as possible. 'Clever' here apparently means 'as symmetric as possible', and one is therefore naturally led to Minkowskian torus compactifications where *all* spacetime dimensions are chosen to be periodic (hence "finite in all directions" [24]). More specifically, for the 26-dimensional bosonic string there is a unique choice of maximal symmetry, namely the even selfdual Lorentzian lattice $II_{25,1}$ which indeed provides a "large" algebra — the infinite rank fake monster Lie algebra introduced by Borcherds [3]. To gain insight into the mathematical structure of these symmetry algebras it is instructive to study toy models such as the 10-dimensional bosonic string compactified with momentum lattice $II_{9,1}$.

The above, to some extent heuristic arguments were recently put on more solid ground in [28]. In this paper, it was established that the fake monster Lie algebra *is* a symmetry of string theory in the sense that every physical state leads to a symmetry of the string scattering amplitudes. In view of this result one could now pose the question to which extent the vertices are already fixed by stipulating the fake monster Lie algebra as symmetry algebra. The degree of uniqueness would then give us a clue of how small the algebra is in comparison with the universal string gauge algebra. Certainly, they cannot be the same. For on the one hand it is clear that the string vertices describe the string field theory, on the other hand we know (see [24]) that the fake monster Lie algebra does not contain all Lie algebras arising from other string backgrounds. It is worth mentioning that the calculations were carried out in the so-called group theoretical approach to string theory which seems to be a powerful formalism to analyze the issue of string symmetries.

Apart from possible relations to string theory, hyperbolic Kac Moody algebras might appear in the dimensional reduction of (extended) supergravity theories to one dimension [19]. Some evidence for this conjecture was presented in [25], where it was argued that the Matzner Misner group arising in the reduction of Einstein's theory to two dimensions can be generalized to a "Matzner Misner SL(3, $\mathbb{R}$)" group providing precisely the two nodes needed to extend the Dynkin diagram to a hyperbolic one. The null Killing reduction required for this investigation has been recently studied in [20]. We also mention that these hidden symmetries may be related to S,T and U duality symmetries arising in string theories (see [18] for recent progress in this direction).

Let us begin by reviewing how one constructs a Kac Moody algebra from a given Cartan matrix $A = (a_{ij})$, where the indices i, j are assumed to take d values (so d is the rank of this algebra), and where the matrix A may also be indefinite. The basic building blocks are the Chevalley generators e_i, f_i, h_i, which are subject to the following relations

$$[h_i, e_j] = a_{ij} e_j, \quad [h_i, f_j] = -a_{ij} f_j, \tag{1}$$
$$[e_i, f_j] = \delta_{ij} h_i. \tag{2}$$

The elements h_i then automatically obey $[h_i, h_j] = 0$ and constitute a basis of the Cartan subalgebra of $\mathfrak{g}(A)$. The free Lie algebra associated with A is obtained by forming multiple commutators of the elements $\{e_i, f_i, h_i \mid i\}$ in all possible ways taking into account the above relations. To obtain the Kac Moody algebra $\mathfrak{g}(A)$ itself we must still divide out by the Serre relations

$$(\mathrm{ad}\, e_i)^{1-a_{ij}} e_j = 0, \quad (\mathrm{ad}\, f_i)^{1-a_{ij}} f_j = 0. \tag{3}$$

It is a standard result [21] that this algebra can be written as a direct sum

$$\mathfrak{g}(A) = \mathfrak{n}_+ \oplus \mathfrak{h} \oplus \mathfrak{n}_- . \tag{4}$$

The subalgebras $\mathfrak{n}_-$ and $\mathfrak{n}_+$ are defined to consist of all linear combinations of multiple commutators of the form $[f_{i_1}, [f_{i_2}, \ldots [f_{i_{n-1}}, f_{i_n}] \ldots]]$ and $[e_{i_1}, [e_{i_2}, \ldots [e_{i_{n-1}}, e_{i_n}] \ldots]]$, respectively, modulo the multilinear Serre relations (3). Since the subalgebras $\mathfrak{n}_+$ and $\mathfrak{n}_-$ are conjugate to each other, it is in practice sufficient to analyze either of them. In this way one gets for positive definite or positive semi-definite A just the finite or affine Kac Moody algebras, respectively (for the affine algebras, one still has to add an outer derivation to $\mathfrak{h}$ due to the degeneracy of the Cartan matrix). When A is indefinite, on the other hand, the number of multiple commutators "explodes", and no manageable way to handle them is known that would be analogous to the realization of affine Kac Moody algebras in terms of current algebras. More specifically, the characteristic feature of indefinite Kac Moody algebras is the exponential growth in the number of Lie algebra elements associated with a root $\boldsymbol{\Lambda}$ as $\boldsymbol{\Lambda}^2 \to -\infty$. Thus for a given number of Chevalley generators $e_{i_1}, \ldots, e_{i_n}$, the number of inequivalent ways of arranging them into non-vanishing multiple commutators increases exponentially with $-\boldsymbol{\Lambda}^2$, where $\boldsymbol{\Lambda} = \mathbf{r}_{i_1} + \ldots + \mathbf{r}_{i_n}$ and $\mathbf{r}_i$ are the simple roots associated with e_i. This problem does not occur for either finite or affine Kac Moody algebras, for which $\boldsymbol{\Lambda}^2 = 2$ or 0 are the only possibilities. The problem here is not so much the enormous number of commutators, but rather the fact that all those multiple commutators which contain the Serre relation somewhere inside, or can be brought to such a form by use of the Jacobi identities, must be identified and discarded. Even the more modest question as to how many elements there are for a given root has defied all attempts at a general solution so far. For a limited number of cases, one knows explicit multiplicity formulas counting the dimensions of the root spaces, but the complete root multiplicities are not known for a single Kac Moody algebra of indefinite type (root multiplicities can be determined in principle from the Peterson recursion formula [23], but this formula quickly becomes too unwieldy for practical use).

We are here mainly interested in *hyperbolic* Kac Moody algebras which are based on indefinite A, but obey the additional requirement that the removal of any point from the Dynkin diagram leaves a Kac Moody algebra which is either of affine or finite type. One can then show that the maximal rank is ten and that the associated root lattice must be Minkowskian, i.e. with metric signature $(+\ldots+|-)$. There are altogether three hyperbolic algebras of maximal rank. Of these, E_{10} is not only the most interesting, containing E_8 and its affine extension E_9 as subalgebras, but also distinguished by the fact that it has only one affine subalgebra that can be obtained by removing a point from the E_{10} Dynkin diagram, while the other two rank 10 algebras contain at least two regular affine subalgebras. Furthermore, the root lattice $Q(E_{10})$ coincides with the (unique) 10-dimensional even unimodular Lorentzian lattice $II_{9,1}$ [5], whereas the root lattices of the other two maximal rank hyperbolic algebras are not self-dual. For the further discussion it is useful to introduce the notion of *level*. Denoting the "over-extended root", which turns an affine into a hyperbolic Kac Moody algebra, by $\mathbf{r}_{-1}$ (e.g. for E_{10}, this is the left-most point in the Dynkin diagram, see below), one defines the level $\ell \in \mathbb{Z}$ of a root such that positive ℓ counts the number of e_{-1} generators in $[e_{i_1}, [e_{i_2}, \ldots [e_{i_{n-1}}, e_{i_n}] \ldots]]$ (similarly, if ℓ is negative, $-\ell$ counts the number of f_{-1} generators in $[f_{i_1}, [f_{i_2}, \ldots [f_{i_{n-1}}, f_{i_n}] \ldots]]$). In terms of the corresponding root $\boldsymbol{\Lambda} = \mathbf{r}_{i_1} + \ldots + \mathbf{r}_{i_n}$, ℓ is defined by

$$\ell := -\boldsymbol{\Lambda} \cdot \boldsymbol{\delta}, \tag{5}$$

where $\boldsymbol{\delta}$ is the null vector of the affine subalgebra obtained by deleting the over-extended node from the Dynkin diagram (in principle one could also use the null vector of other regular affine subalgebras to define a level which would be different from the above; however, then not all of the results to be stated below remain valid, e.g. the level-one states would no longer form the basic representation of this affine subalgebra). The level derives its importance from the fact that it grades the hyperbolic Kac Moody algebra with respect to the affine subalgebra [7]. Consequently, the subspaces belonging to a fixed level can be decomposed into irreducible representations of the affine subalgebra, the level being equal to the eigenvalue of the central term on this representation (the full hyperbolic algebra contains representations of *all* integer levels). Multiplicities are known for levels $\ell = 0$ (corresponding to the affine subalgebra) and $\ell = 1$ (corresponding to the so-called basic representation). More precisely, we have [21]

$$\text{mult}(\boldsymbol{\Lambda}) = p_{d-2}(1 - \tfrac{1}{2}\boldsymbol{\Lambda}^2), \tag{6}$$

where d is the rank of the algebra and the function $p_{d-2}(n)$ counts the partitions of $n \in \mathbb{N}$ into "parts" of $d-2$ "colours". For $\ell = 2$, one knows general multiplicity formulas in some cases, and in particular for E_{10}. Beyond $\ell = 2$, no general formula seems to be known.

Now it is well known that, at least in principle, the string vertex operator construction can provide a more concrete realization of an indefinite Kac Moody algebra. To exploit it one interprets the root lattice as the momentum lattice of a fully compactified string, and tries to understand the multiple commutators in terms of string vertex operators associated with the excited string states. The real roots then correspond to spacelike (tachyon) momenta and the imaginary roots to either lightlike or timelike momenta. For the simple roots $\mathbf{r}_i$, all of which obey $\mathbf{r}_i{}^2 = 2$ (and hence are real), we have the following correspondence between Chevalley generators and tachyon and photon states:

$$e_i \quad \mapsto \quad |\mathbf{r}_i\rangle, \tag{7}$$

$$f_i \quad \mapsto \quad -|-\mathbf{r}_i\rangle, \tag{8}$$

$$h_i \quad \mapsto \quad \mathbf{r}_i(-1)|\mathbf{0}\rangle. \tag{9}$$

Here we use the shorthand notation $\mathbf{r}_i(-1) \equiv \mathbf{r}_i \cdot \boldsymbol{\alpha}_{-1}$ (where α^μ_{-1} is just the lowest string oscillator); furthermore, we define the states in such a way that cocycle factors have been absorbed and do not appear explicitly. The commutator between any two physical states φ and ψ is quite generally defined by the formula (cf. [1])

$$[\psi, \varphi] := \text{Res}_z \left(\mathcal{V}(\psi, z)\varphi\right), \tag{10}$$

where $\mathcal{V}(\psi, z)$ is the string vertex operator associated with the state ψ (the residue formula here is completely equivalent to the contour integrals employed down in [14]). An important consequence of this formula is that the physical string states always form a Lie algebra (not to be confused with the Kac Moody algebra, see remarks below).

Apart from yielding a concrete realization of an abstract algebraic structure, the string vertex operators construction has the advantage that the Serre relations (3) are built in from the outset: in this context they simply state that the string has no excited states "below" the tachyons. For finite and affine Kac Moody algebras, no other states beside tachyons and photons occur, whereas for indefinite A, excited string states of arbitrarily high levels must be taken into account. These will appear with certain polarizations, whose number increases rapidly with the mass level of the string state. Thus, in more physical parlance, the multiplicity of a root is equal to the number of linearly independent polarizations of the corresponding string state that can be reached by multiple commutators. This "easy" realization of the Kac Moody algebra might suggest that the problem of classifying the Lie algebra elements is essentially solved by the string vertex operator construction. Unfortunately, this is by no means the case because the problem of accounting for the Serre relations is now replaced by another one: not all physical states can be obtained in terms of multiple commutators. Denoting the Lie algebra of *all* physical states by $\mathfrak{g}_\Lambda$, we rather have the *proper* inclusion

$$\mathfrak{g}(A) \subset \mathfrak{g}_\Lambda . \tag{11}$$

In other words, the Lie algebra of physical states is well understood in physical terms, but the actual Kac Moody algebra $\mathfrak{g}(A)$ is only a subalgebra thereof, and all the complications now reside in the way in which $\mathfrak{g}(A)$ is embedded in the bigger, but simpler algebra $\mathfrak{g}_\Lambda$. In particular, there are "missing states", i.e. physical states that cannot be represented as multiple commutators of the Chevalley generators. A possible explanation for this phenomenon is the following. For continuous momenta it is well known that one can generate any physical state by multiple scattering of tachyons (multiple scattering is the equivalent of taking multiple commutators), so there can be no "missing states". This is no longer the case for the compactified string: when the tachyon momenta are on the lattice, certain "intermediate states" are no longer allowed, and therefore not all physical states are accessible in this way any more. The construction given in [12] is an attempt to make this intuitive argument more precise. As a further consequence of (11), the multiplicities are *not* given by the numbers of excited states of the string (which are of course well known), but only bounded above by them:

$$\text{mult}\boldsymbol{\Lambda} \leq p_{d-1}(1 - \tfrac{1}{2}\boldsymbol{\Lambda}^2) - p_{d-1}(-\tfrac{1}{2}\boldsymbol{\Lambda}^2). \tag{12}$$

Only for Euclidean lattices the two Lie algebras coincide, and we have equality in (11).

To summarize: the root system of the Kac Moody algebra $\mathfrak{g}(A)$ is well understood though its root multiplicities are not completely known for a single example; whereas the Lie algebra of physical string states $\mathfrak{g}_\Lambda$ has a much simpler structure (although it will not be easy to define a root system associated with it). Thus a complete understanding of (11) requires a "mechanism" which tells us how $\mathfrak{g}(A)$ has to be filled up with physical states to reach the complete Lie algebra of physical states.

We mention that recently Borcherds [2] has introduced a new type of generalized Kac Moody algebras by admitting "imaginary simple" roots (i.e. obeying $\mathbf{r}_i{}^2 \leq 0$). In the present interpretion this means that one adds "by hand" those physical states not obtainable as multiple commutators; the corresponding momenta will then be counted as new simple roots, whose multiplicity is simply given by the number of associated independent (missing) polarizations. However, this program has so far only been carried to completion for the 26-dimensional bosonic string, where special properties such as the no-ghost theorem play a crucial role. In this example, all missing states are under control (though not explicitly known): one has to adjoin a certain (infinite) set of photonic states as new Lie algebra generators to the ordinary Kac Moody generators in order to get a complete set of generators for the Lie algebra of physical states. The resulting algebra constitutes an example of a generalized Kac Moody algebra and has been dubbed fake monster Lie algebra (for historical reasons). The imaginary simple roots corresponding to the extra generators are just the positive integer multiples of the (lightlike) Weyl vector for the lattice $II_{25,1}$, and their multiplicities are equal to the number of photon states (i.e. = 24). On the other hand, for algebras such as E_{10}, not much is gained by this change of perspective, because supplying the missing generators "by hand" presupposes knowledge of what the missing Lie algebra elements are (not to mention the potential arbitrariness as to the number of ways in which this can be consistently done). So the problem identifying the elements belonging to $\mathfrak{g}_\Lambda$ and not to $\mathfrak{g}(A)$ in (11) remains.

In [12] it is proposed to understand Kac Moody algebras of hyperbolic type, and in particular the maximally extended hyperbolic algebra E_{10}, from a more "physical" (i.e. pedestrian) point of view by focussing on the "missing states" rather than on the Serre relations. For this purpose, we make use of a lattice version of the DDF construction, which provides the most direct and explicit solution of the physical state constraints in string theory [6]. The physical states, which by definition are annihilated by the Virasoro constraints, are simply obtained in this scheme by acting on a tachyonic groundstate with the DDF operators, which commute with all Virasoro generators and form a spectrum generating algebra. Our key observation is that for Kac Moody algebras of "subcritical" rank (i.e. $d < 26$), there appear *longitudinal* string states and vertex operators beyond level one, whose significance in this context has so far not been recognized to the best of our knowledge. The appearance of longitudinal states is already obvious from the known multiplicity formulas for level $\ell = 2$: for sufficiently large (negative) $\boldsymbol{\Lambda}^2$, one can check that there are roots $\boldsymbol{\Lambda}$ such that $\text{mult}(\boldsymbol{\Lambda}) > p_{d-2}(1 - \frac{1}{2}\boldsymbol{\Lambda}^2)$ (however, there may also exist higher level roots whose multiplicity is *less* than the number of transversal states). This clearly implies that whereas transversal states are sufficient to characterize the elements of an affine Kac Moody algebra (see below for further explanations of this point), they are no longer sufficient for indefinite Kac Moody algebras.

The necessary adaptation of the DDF construction involves a discretization of the string vertex operator formalism. As is well known [14], the allowed momenta of the string excitations must be elements of the weight lattice of the corresponding (affine or indefinite) Kac Moody algebra. For the definition of DDF operators one must choose a special Lorentz frame, in terms of which one can distinguish transversal and longitudinal DDF operators. On the lattice it is no longer possible to find a frame such that the relevant DDF vectors (see below for details) are still on the lattice, and we therefore are forced to make use of a rational extension of the (self-dual) root lattice as an auxiliary device. This is a curious feature of our construction, not encountered in previous studies. Despite the fact that our DDF vectors are not on the lattice, we employ them in our analysis because we can still use the associated (transversal and longitudinal) DDF operators to construct a complete basis for any root space of the Lie algebra of physical states $\mathfrak{g}_{II_{9,1}}$, of which the corresponding root space of the Kac Moody algebra $\mathfrak{g}(A)$ is then a proper subspace.

As it turns out, longitudinal states are absent only for levels zero and one; this accounts for the comparative simplicity of the corresponding multiplicity formulas.

A central role in the DDF construction is played by the tachyon momentum $\mathbf{a}$ (i.e. $\mathbf{a}^2 = 2$) of the ground state $|\mathbf{a}\rangle$, on which the physical states are built by means of DDF operators, and a null vector $\mathbf{k}$, subject to the condition $\mathbf{k}\cdot\mathbf{a} = 1$. For continuous momenta $\mathbf{a}$, we can always find suitable $\mathbf{k} = \mathbf{k}(\mathbf{a})$; moreover, we can rotate these vectors into a convenient frame by means of a Lorentz transformation [26]. On the lattice, however, the full Lorentz invariance is broken to a discrete subgroup (containing the Weyl group generated by the fundamental Weyl reflections), and for generic roots $\boldsymbol{\Lambda}$, the associated DDF vectors $\mathbf{a}$ and $\mathbf{k}$ will *not* be elements of the root lattice in general. We can, however, still rotate the vectors $\mathbf{a} - n\mathbf{k}$ into the *fundamental Weyl chamber* for n sufficiently large. The lightlike momentum $\mathbf{k}$ is then proportional to the null-root $\boldsymbol{\delta}$ characterizing the affine subalgebra; the latter is always in the fundamental Weyl chamber. Now we invoke the (obvious) fact that any root $\boldsymbol{\Lambda}$ in the fundamental Weyl chamber can be represented in the form

$$\boldsymbol{\Lambda} = \ell\mathbf{r}_{-1} + M\boldsymbol{\delta} + \mathbf{b}, \tag{13}$$

where ℓ is the level of $\boldsymbol{\Lambda}$ and $\mathbf{b}$ an element of the E_8-root lattice $Q(E_8)$ ($\mathbf{b}$ need not be positive by itself as only $M\boldsymbol{\delta} + \mathbf{b}$ must be positive). We then define the **DDF decomposition** of $\boldsymbol{\Lambda}$ by

$$\boldsymbol{\Lambda} = \mathbf{a} - n\mathbf{k}(\mathbf{a}), \tag{14}$$

where

$$\mathbf{k}(\mathbf{a}) := -\tfrac{1}{\ell}\boldsymbol{\delta} \tag{15}$$

and

$$n = 1 - \tfrac{1}{2}\boldsymbol{\Lambda}^2 = 1 + (M-\ell)\ell - \tfrac{1}{2}\mathbf{b}^2. \tag{16}$$

By construction, $\mathbf{a}$ obeys $\mathbf{a}^2 = 2$ and is therefore associated with a tachyon state, and n is the number of steps required to reach the root $\boldsymbol{\Lambda}$ by starting from $\mathbf{a}$ and decreasing the momentum by $\mathbf{k}$ at each step (n is non-negative because $\boldsymbol{\Lambda}^2 \leq 2$; note also that $\mathbf{k}$ is always a *negative* root, so $\boldsymbol{\Lambda}$ is positive for all n). Obviously, for $\ell > 1$, neither $\mathbf{k}$ nor $\mathbf{a}$ belong to the lattice in general. As a consequence, the intermediate DDF states associated with momenta $\mathbf{a} - m\mathbf{k}$ not on the lattice will not correspond to elements of the algebra, but they are nonetheless indispensable for the construction of a complete basis for any given root space. On the other hand, states associated with the root $\boldsymbol{\Lambda}$ do belong to the algebra of physical states, and the DDF decomposition enables us to write down all possible polarization states associated with the root $\boldsymbol{\Lambda}$ in terms of transversal and longitudinal DDF states; the totality of these states constitutes the complete set of elements in the root space $\mathfrak{g}^{(\boldsymbol{\Lambda})}_{II_{9,1}}$, whose dimension equals $p_{d-1}(1-\frac{1}{2}\boldsymbol{\Lambda}^2) - p_{d-1}(-\frac{1}{2}\boldsymbol{\Lambda}^2)$. Explicitly, given a tachyon momentum $\mathbf{a}$, the physical states are

$$A^{i_1}_{-m_1}(\mathbf{a})\ldots A^{i_M}_{-m_M}(\mathbf{a})\mathfrak{L}_{-n_1}(\mathbf{a})\ldots\mathfrak{L}_{-n_N}(\mathbf{a})|\mathbf{a}\rangle, \tag{17}$$

explicitly indicating the dependence of the DDF operators and their polarizations on the tachyon momentum $\mathbf{a}$ and the associated lightlike vector $\mathbf{k}(\mathbf{a}) = -\frac{1}{\ell}\boldsymbol{\delta}$, and assuming $n_i \geq 2$ to exclude null states. The transversal DDF operators associated with a tachyonic momentum $\mathbf{a}$ and its light-like momentum $\mathbf{k}$ are defined by the well known formula

$$A^i_{-m} = \mathrm{Res}_z\left[\mathcal{V}(\boldsymbol{\xi}_i(-1)|-m\mathbf{k}\rangle, z)\right], \tag{18}$$

i.e. is given by the contour integral of the vertex operator corresponding to the photon state $\boldsymbol{\xi}_i \cdot \boldsymbol{\alpha}_{-1}|\frac{m}{l}\boldsymbol{\delta}\rangle$. The transversal DDF operators always form a $(d-2)$-fold Heisenberg algebra. Note, however, that we have to deal with a whole plethora of transversal Heisenberg algebras, namely one for each admissible pair $(\mathbf{a}, \mathbf{k})$; this is in contrast to the single set of primitive oscillators, $\{\alpha^\mu_m | 1 \leq \mu \leq d, m \in \mathbb{Z}\}$, which makes up the full Fock space when acting on the groundstates. The longitudinal DDF operators are given by more complicated (but also standard) expressions; they involve a logarithmic contribution (cf. [4]). Again, for each admissible pair $(\mathbf{a}, \mathbf{k})$, we end up with a different set of operators. An important technical point is that the longitudinal vertex operators cannot be associated with definite states, as their action cannot be defined on all of Fock space, including the vacuum state $|\mathbf{0}\rangle$. Put differently, they do not correspond to summable operators in the sense of [10]; in this respect, vertex algebras encompassing longitudinal states transcend the definition given in [1], [10].

We quickly summarize some pertinent results about E_{10}. It is defined from its Cartan matrix in terms of multiple commutators of the Chevalley generators as described above. The root lattice $Q(E_{10})$ of E_{10} is the unique even self-dual $II_{9,1}$ in ten dimensions, which can be defined as the set of points $\mathbf{x} = (x_1, \ldots, x_9|x_0)$ in 10-dimensional Minkowski space for which the x_μ's are all in $\mathbb{Z}$ or all in $\mathbb{Z} + \frac{1}{2}$ and which have integer inner product with the vector $\mathbf{l} = (\frac{1}{2}, \ldots, \frac{1}{2} \mid \frac{1}{2})$, all norms and inner products being evaluated in the Minkowskian metric $\mathbf{x}^2 = x_1^2 + \ldots + x_9^2 - x_0^2$ (cf. [27]). According to [5], a set of positive norm simple roots for $II_{9,1}$ is given by the ten vectors $\mathbf{r}_{-1}, \mathbf{r}_0, \mathbf{r}_1, \ldots, \mathbf{r}_8$ in $II_{9,1}$ for which $\mathbf{r}_i^2 = 2$ and $\mathbf{r}_i \cdot \boldsymbol{\rho} = -1$ where the Weyl vector is $\boldsymbol{\rho} = (0, 1, 2, \ldots, 8|38)$ with $\boldsymbol{\rho}^2 = -1240$. The corresponding Coxeter-Dynkin diagram looks as follows

(19)

and is associated with the Cartan matrix:

$$A \equiv (a_{ij}) = (\mathbf{r}_i \cdot \mathbf{r}_j) = \begin{pmatrix} 2 & -1 & 0 & 0 & 0 & 0 & 0 & 0 & 0 & 0 \\ -1 & 2 & -1 & 0 & 0 & 0 & 0 & 0 & 0 & 0 \\ 0 & -1 & 2 & -1 & 0 & 0 & 0 & 0 & 0 & 0 \\ 0 & 0 & -1 & 2 & -1 & 0 & 0 & 0 & 0 & 0 \\ 0 & 0 & 0 & -1 & 2 & -1 & 0 & 0 & 0 & 0 \\ 0 & 0 & 0 & 0 & -1 & 2 & -1 & 0 & 0 & 0 \\ 0 & 0 & 0 & 0 & 0 & -1 & 2 & -1 & 0 & -1 \\ 0 & 0 & 0 & 0 & 0 & 0 & -1 & 2 & -1 & 0 \\ 0 & 0 & 0 & 0 & 0 & 0 & 0 & -1 & 2 & 0 \\ 0 & 0 & 0 & 0 & 0 & 0 & -1 & 0 & 0 & 2 \end{pmatrix}.$$

Let us first describe the E_9 subalgebra in terms of physical states. The Cartan subalgebra of E_{10} (and also of E_9) is spanned by the states

$$\boldsymbol{\delta}(-1)|\mathbf{0}\rangle \quad =: \quad K, \tag{20}$$

$$(\mathbf{r}_{-1} + \boldsymbol{\delta})(-1)|\mathbf{0}\rangle \quad =: \quad d, \tag{21}$$

$$\boldsymbol{\xi}_i(-1)|\mathbf{0}\rangle \qquad \text{for } i = 1, \ldots, 8\,, \tag{22}$$

where K represents the central element, d is the derivation of E_9, and $\{\boldsymbol{\xi}_i(-1)|\mathbf{0}\rangle \mid i = 1, \ldots, 8\}$ span the Cartan subalgebra of E_8. This is the standard "light-cone" basis of

$\mathfrak{h}(E_9)$ in the sense that K and d are lightlike. The allowed (positive and negative) roots are all $\mathbf{r} \in II_{9,1}$ obeying $\mathbf{r}^2 = 2$ and $\mathbf{r}\cdot\boldsymbol{\delta} = 0$ (hence having no $\mathbf{r}_{-1}$ component), and $m\boldsymbol{\delta}$ for $m \in \mathbb{Z}^\times$. These correspond to the tachyonic states $|\mathbf{r}\rangle$ and the photonic states $\boldsymbol{\xi}_i(-1)|m\boldsymbol{\delta}\rangle$ (where $\boldsymbol{\xi}_i\cdot\boldsymbol{\delta} = \boldsymbol{\xi}_i\cdot\mathbf{r}_{-1} = 0$) with multiplicities 1 and 8, respectively. The following commutation relations are already enough for a complete characterization of E_9

$$\left[\boldsymbol{\eta}(-1)|\mathbf{0}\rangle, \boldsymbol{\zeta}(-1)|\mathbf{0}\rangle\right] = 0, \tag{23}$$

$$\left[\boldsymbol{\eta}(-1)|\mathbf{0}\rangle, \boldsymbol{\xi}_i(-1)|m\boldsymbol{\delta}\rangle\right] = m(\boldsymbol{\eta}\cdot\boldsymbol{\delta})\boldsymbol{\xi}_i(-1)|m\boldsymbol{\delta}\rangle, \tag{24}$$

$$\left[\boldsymbol{\eta}(-1)|\mathbf{0}\rangle, |\mathbf{r}\rangle\right] = (\boldsymbol{\eta}\cdot\mathbf{r})|\mathbf{r}\rangle, \tag{25}$$

$$\left[\boldsymbol{\xi}_i(-1)|m\boldsymbol{\delta}\rangle, \boldsymbol{\xi}_j(-1)|n\boldsymbol{\delta}\rangle\right] = m\delta_{m+n,0}(\boldsymbol{\xi}_i\cdot\boldsymbol{\xi}_j)\boldsymbol{\delta}(-1)|\mathbf{0}\rangle, \tag{26}$$

$$\left[\boldsymbol{\xi}_i(-1)|m\boldsymbol{\delta}\rangle, |\mathbf{r}\rangle\right] = (\boldsymbol{\xi}_i\cdot\mathbf{r})|\mathbf{r} + m\boldsymbol{\delta}\rangle, \tag{27}$$

$$\left[|\mathbf{r}\rangle, |\mathbf{s}\rangle\right] = \begin{cases} 0 & \text{if } \mathbf{r}\cdot\mathbf{s} \geq 0, \\ \epsilon(\mathbf{r},\mathbf{s})|\mathbf{r}+\mathbf{s}\rangle & \text{if } \mathbf{r}\cdot\mathbf{s} = -1, \\ -\mathbf{r}(-1)|m\boldsymbol{\delta}\rangle & \text{if } \mathbf{r}+\mathbf{s} = m\boldsymbol{\delta}, \end{cases} \tag{28}$$

for $\boldsymbol{\eta}, \boldsymbol{\zeta} \in \mathfrak{h}(E_9)$ and E_9 roots $\mathbf{r}, \mathbf{s}$. In the last formula ϵ denotes the cocycle factor. The bra and ket notation used here may appear a bit unusual (and is not really necessary at this point), but will prove invaluable as one goes to higher level elements of the hyperbolic algebra. The multiplicities of the corresponding Lie algebra elements can be read off directly, and are given by 1 and 8, respectively, for the tachyonic and lightlike roots in accord with the formula (6) above.

The level-one elements exhibit already a slightly more involved structure. Inspection of the inverse Cartan matrix shows that the only such roots in the fundamental Weyl chamber $\mathcal{C}$ are of the form (for $k_{-1} \in \mathbb{N}$)

$$\boldsymbol{\Lambda} = \mathbf{r}_{-1} + (2 + k_{-1})\boldsymbol{\delta}, \tag{29}$$

corresponding to the DDF decomposition (14) with $\mathbf{a} = \mathbf{r}_{-1}$, $\mathbf{k} = -\boldsymbol{\delta}$ and $n = 2 + k_{-1}$. Since all these vectors are elements of the lattice, we can straightforwardly apply the DDF construction to obtain the physical states

$$A^{i_1}_{-m_1} \cdots A^{i_N}_{-m_N}|\mathbf{r}_{-1}\rangle, \tag{30}$$

where $m_1 + \ldots + m_N = 2 + k_{-1}$ and with the polarization vectors chosen such that $\boldsymbol{\xi}_i\cdot\boldsymbol{\xi}_j = \delta_{ij}$ and $\boldsymbol{\xi}_i\cdot\boldsymbol{\delta} = \boldsymbol{\xi}_i\cdot\mathbf{r}_{-1} = 0$ for $i, j = 1, \ldots, 8$. In terms of multiple commutators, these states correspond to

$$\left[\boldsymbol{\xi}_{i_1}(-1)|m_1\boldsymbol{\delta}\rangle, \left[\ldots, \left[\boldsymbol{\xi}_{i_N}(-1)|m_N\boldsymbol{\delta}\rangle, |\mathbf{r}_{-1}\rangle\right]\ldots\right]\right] \qquad \in E_{10}^{(\boldsymbol{\Lambda})}. \tag{31}$$

All relevant level-one states can now be obtained by acting with the E_9 Weyl group on these states and polarizations. Denoting the rotated DDF operators by $A^{\mathfrak{w}(i)}_{-m} \equiv A^{\mathfrak{w}(\boldsymbol{\xi}_i)}_{-m}$, we obtain the new states

$$A^{\mathfrak{w}(i_1)}_{-m_1} \cdots A^{\mathfrak{w}(i_N)}_{-m_N}|\mathfrak{w}(\mathbf{r}_{-1})\rangle \tag{32}$$

in this fashion. The so-called basic representation is spanned by all elements of the form (32); the highest weight vector of the representation is easily seen to be $|\mathbf{r}_{-1}\rangle$.

At level two, a general multiplicity formula was derived in [22]; it reads

$$\mathrm{mult}(\boldsymbol{\Lambda}) = \xi(3 - \tfrac{1}{2}\boldsymbol{\Lambda}^2), \tag{33}$$

where

$$\sum_{n\geq 0} \xi(n) q^n = \frac{1}{\phi(q)^8}\left[1 - \frac{\phi(q^2)}{\phi(q^4)}\right], \tag{34}$$

and $\phi(q)$ is the Euler function. The special example we have investigated in [12], is the level-two root $\boldsymbol{\Lambda} = \boldsymbol{\Lambda}_7$, dual to the simple root $\mathbf{r}_7$, which has $\boldsymbol{\Lambda}_7^2 = -4$ and is explicitly given by

$$\boldsymbol{\Lambda}_7 = \left[\begin{matrix} & & & & & & 7 & & \\ 2 & 4 & 6 & 8 & 10 & 12 & 14 & 9 & 4 \end{matrix}\right] = (0,0,0,0,0,0,0,0,0\,|\,2), \tag{35}$$

where the first notation with square brackets refers to the simple roots in the above Dynkin diagram. We can now invoke the results of [7] which tell us that level ℓ states can be obtained as $(\ell-1)$-fold commutators of level-one states, for which we can use the representation (32). Our analysis reveals that the following states form a complete basis of the root space $E_{10}^{(\boldsymbol{\Lambda}_7)}$ (no summation convention!):

$$\begin{aligned}
A^i_{-2}A^j_{-1}\,|\mathbf{a}\rangle & \qquad \text{for } i,j \text{ arbitrary,}\\
A^i_{-1}A^j_{-1}A^k_{-1}\,|\mathbf{a}\rangle & \qquad \text{for } i\neq j\neq k\neq i,\\
\{A^i_{-3} - A^i_{-1}A^j_{-1}A^j_{-1}\}\,|\mathbf{a}\rangle & \qquad \text{for } i\neq j,\\
\{5A^i_{-3} + A^i_{-1}A^i_{-1}A^i_{-1}\}\,|\mathbf{a}\rangle & \qquad \text{for } i \text{ arbitrary,}\\
\{A^i_{-3} - A^i_{-1}\mathcal{L}_{-2}\}\,|\mathbf{a}\rangle & \qquad \text{for } i \text{ arbitrary.}
\end{aligned} \tag{36}$$

Altogether, we get $64 + 2\cdot 56 + 2\cdot 8 = 192$ states in agreement with the formula (33) predicting $\xi(3) = 192$ [22]. Despite the fact that this number coincides with the number of transversal states, our result explicitly shows the appearance of longitudinal as well as the disappearance of some transversal states. The above states form irreducible representations of the little group; in particular, the longitudinal DDF operator is inert under the little Weyl group. To appreciate the simplicity of this result readers need only contemplate the problem of classifying the states in terms of 75-fold multiple commutators of the Chevalley generators for this example.

Having a complete description of the root space $E_{10}^{(\boldsymbol{\Lambda}_7)}$, we can now in principle explore root spaces associated with other level-two roots of the form $\boldsymbol{\Lambda} = \boldsymbol{\Lambda}_7 + n\boldsymbol{\delta}$ (i.e. the **root string** associated with $\boldsymbol{\Lambda}_7$) by commuting the states (36) with the E_9 elements $\boldsymbol{\xi}_i(-1)|m\boldsymbol{\delta}\rangle$. From (18) it is evident that all states obtained by acting with a product $A^{i_1}_{-2m_1}\ldots A^{i_M}_{-2m_M}$ on any of the states (36) belong to the root space of $\boldsymbol{\Lambda} = \boldsymbol{\Lambda}_7 + (m_1 + \ldots m_M)\boldsymbol{\delta}$ (note that each operator $A^i_{-2m}(\mathbf{a})$ shifts the momentum by $m\boldsymbol{\delta}$!). However, it is also clear that we cannot obtain all root space elements in this way. For this, it is necessary to calculate DDF commutators between appropriate elements of the form (32). An alternative, more elucidating way might be to consider the action of the Sugawara generators defined by

$$\mathcal{L}^{\mathrm{Sug.}}_m := \frac{1}{2(\ell + h^\vee)}\left\{\sum_{n\in\mathbb{Z}}\sum_{i=1}^{8} :A^i_n A^i_{m-n}: + \sum_{\mathbf{s}\in\Delta^{\mathrm{real}}(E_9)} {}^{\times}_{\times}\mathrm{ad}_{|\mathbf{s}\rangle}\,\mathrm{ad}_{|-\mathbf{s}-m\boldsymbol{\delta}\rangle}{}^{\times}_{\times}\right\} \tag{37}$$

on the states (36); here, $h^\vee = 30$ is the dual Coxeter number of E_8, the level is $\ell = 2$, and the normal-ordering of the operators $\mathrm{ad}_{|\mathbf{r}\rangle}$ is chosen as

$$\,^\times_\times \mathrm{ad}_{|\mathbf{s}+m\boldsymbol{\delta}\rangle}\,\mathrm{ad}_{|\mathbf{t}+n\boldsymbol{\delta}\rangle}\,^\times_\times := \begin{cases} \mathrm{ad}_{|\mathbf{s}+m\boldsymbol{\delta}\rangle}\,\mathrm{ad}_{|\mathbf{t}+n\boldsymbol{\delta}\rangle} & \text{if } m \geq n, \\ \mathrm{ad}_{|\mathbf{t}+n\boldsymbol{\delta}\rangle}\,\mathrm{ad}_{|\mathbf{s}+m\boldsymbol{\delta}\rangle} & \text{if } m < n, \end{cases} \tag{38}$$

for E_8 roots $\mathbf{s}, \mathbf{t}$ and $m, n \in \mathbb{Z}$. We get

$$\mathcal{L}_m^{\text{Sug.}}|\mathbf{a}\rangle = 0 \tag{39}$$

for $m \geq 1$. Furthermore, when evaluating $\mathcal{L}_0^{\text{Sug.}}$ on the ground state $|\mathbf{a}\rangle$, we find $A_0^8|\mathbf{a}\rangle = -2|\mathbf{a}\rangle$, and obtain

$$\mathcal{L}_0^{\text{Sug.}}|\mathbf{a}\rangle = \tfrac{1}{16}|\mathbf{a}\rangle, \tag{40}$$

showing that the state $|\mathbf{a}\rangle$ is a highest weight vector of weight $h = \frac{1}{16}$ for the level-two Sugawara generators. In view of the results of [22], we therefore expect these states to belong to the irreducible Virasoro module with $c = \frac{1}{2}$ and $h = \frac{1}{16}$. The problem that remains is to relate the Sugawara generators to the longitudinal DDF operators. If this can be done, a completely explicit description of *all* level-two root spaces is within reach.

Let us finally return to the issue of missing states in more detail. Comparing (6) with (12), it becomes obvious that tachyonic and photonic physical states are necessarily transversal, so that

$$\mathfrak{g}_{II_{9,1}}^{(\boldsymbol{\Lambda})} \equiv E_{10}^{(\boldsymbol{\Lambda})} \quad \text{for } \boldsymbol{\Lambda}^2 \geq 0 \tag{41}$$

(of course, for $\boldsymbol{\Lambda}^2 > 2$, both spaces are empty). This means that there are no missing states for $\boldsymbol{\Lambda}^2 \geq 0$. But already for the massive spin 2 states, we encounter one longitudinal physical state that surely does not belong to the Kac Moody algebra E_{10}. It is clear that there is only one weight in $\mathcal{C}$ of norm -2, namely the fundamental weight $\boldsymbol{\Lambda}_0 = \mathbf{r}_{-1} + 2\boldsymbol{\delta}$. Since the latter is a level-one element, which we know to occur in E_{10} just with transversal polarizations, we infer that the longitudinal state $\mathfrak{L}_{-2}|\mathbf{r}_{-1}\rangle$ is not contained in the root space $E_{10}^{(\boldsymbol{\Lambda}_0)}$ and thus represents a missing state, so

$$\mathfrak{g}_{II_{9,1}}^{(\boldsymbol{\Lambda}_0)} \equiv E_{10}^{(\boldsymbol{\Lambda}_0)} + \mathbb{R}\cdot\mathfrak{L}_{-2}|\mathbf{r}_{-1}\rangle. \tag{42}$$

Acting with the full Weyl group on the missing state, we obtain the associated orbit of missing states in E_{10}. Our detailed analysis of the root space for $\boldsymbol{\Lambda}_7$ now enables us to discuss the case of norm -4, for $\boldsymbol{\Lambda}_7$ is the only weight in the fundamental Weyl chamber with this property. From the multiplicity formula we learn that there have to be $201 - 192 = 9$ missing states, and in view of our DDF basis (36) we write

$$\mathfrak{g}_{II_{9,1}}^{(\boldsymbol{\Lambda}_7)} \equiv E_{10}^{(\boldsymbol{\Lambda}_7)} + \mathrm{span}_{\mathbb{R}}\{\mathfrak{L}_{-3}|\mathbf{a}\rangle;\ A_{-3}^i|\mathbf{a}\rangle, i = 1,\ldots,8\}, \tag{43}$$

which can be also acted on with $\mathfrak{W}(E_{10})$ to find its analogue in other chambers.

The above formulas naturally suggest two ways of how to proceed. If we are primarily interested in E_{10}, we can try to systematize the way of splitting of $\mathfrak{g}_{II_{9,1}}$ into E_{10} states and missing states. In other words, we are seeking a mechanism which satisfactorily answers the following question: How do the missing states decouple from the E_{10} states? That this idea is not far-fetched shows the example of the 26-dimensional bosonic string. There we separate the longitudinal physical states from the transversal ones by introducing a positive semidefintite contravariant bilinear form which renders the former states to be null physical states. If one prefers the modern cohomological treatment then the decoupling is furnished by the nilpotent BRST operator and its cohomology. Thus we may rephrase the above question as: *Is there a cohomology describing how E_{10} is embedded in the Lie algebra of physical states, $\mathfrak{g}_{II_{9,1}}$?*

The other point of view, as advocated by Borcherds, involves a generalization of the framework of Kac Moody algebras. We know how a large part of $\mathfrak{g}_{II_{9,1}}$, namely the E_{10} part, can be formulated in terms of generators and relations. The idea then is to extend this approach to the whole Lie algebra. We would have to find an additional set of Chevalley generators which, when adjoined to the generators for E_{10}, produce all physical states as multiple commutators. For example, we certainly have to add $\mathfrak{L}_{-2}|\mathbf{r}_{-1}\rangle$ as such a new generator. This amounts to saying that Λ_0 constitutes an imaginary simple root with multiplicity one. Kac Moody algebras allowing for imaginary ($\equiv$ nonpositive norm) simple roots were invented by Borcherds [2]. So far, the introduction of this new generator seems to be very natural and appealing, but the second step of the procedure is subtle and becomes cumbersome when repeatedly done. In order to decide which missing states for the case of Λ_7 have to be chosen as new generators, we need to take into account the previous additional generator $\mathfrak{L}_{-2}|\mathbf{r}_{-1}\rangle$. Thus we ought to calculate the commutator $[|\mathbf{s}\rangle, \mathfrak{L}_{-2}|\mathbf{r}\rangle]$ (where $\mathbf{s}+\mathbf{r}+2\delta = \Lambda_7$) and express it in terms of the DDF basis for $\mathfrak{g}_{II_{9,1}}^{(\Lambda_7)}$ to see which missing states now do appear. We have not completed this calculation yet, for we are mainly interested in E_{10} itself and hence focus on the first approach. Alternatively, it is also possible to determine recursively the imaginary simple roots by anticipating $\mathfrak{g}_{II_{9,1}}$ as a Borcherds algebra and then plug its well-known root multiplicities, $p_9(1-\frac{1}{2}\mathbf{r}^2) - p_9(-\frac{1}{2}\mathbf{r}^2)$, into the Weyl–Kac–Borcherds denominator formula [2].

So, what have we learnt from our analysis of the root space $E_{10}^{(\Lambda_7)}$ and how may it be relevant for other hyperbolic Kac Moody algebras? Our approach suggests that root spaces of E_{10} and other algebras of that type carry an additional structure related to polarization; this differs from the conventional point of view that a root space is essentially, up to its dimension, a black box. The DDF framework, as developed here, provides adequate tools for the analysis of the complicated structure of hyperbolic algebras.

In particular, we now have a deeper understanding why Frenkel's conjecture [8] is wrong. Inspired by the example of the 26-dimensional bosonic string and the results about the canonical hyperbolic extension of $\mathfrak{su}(2)$ [7], he conjectured that for every hyperbolic algebra $\mathfrak{g}$ of rank d one has, for any root $\mathbf{r}$, $\dim \mathfrak{g}^{(\mathbf{r})} \le p_{d-2}(1-\frac{1}{2}\mathbf{r}^2)$ as an upper bound. This conjecture was disproved in [22] by establishing the level-two multiplicity formula for E_{10} as a counterexample. We argue that the 26-dimensional bosonic string represents a rather untypical example, because there the longitudinal states span the radical of the contravariant bilinear form which is divided out. Hence only transversal states survive and we end up indeed with the exact multiplicity formula $p_{24}(1-\frac{1}{2}\mathbf{r}^2)$. In the generic case, on the other hand, the longitudinal states do appear as Lie algebra elements. In terms of the DDF realization the following picture emerges. At level-zero and level-one we naturally obtain all transversal states giving the affine subalgebra and its basic representation, respectively. By commuting transversal level-one states, which is necessary for generating higher level elements, we cannot escape from producing longitudinal states, too. Hence there is no reason to expect a connection between higher level root multiplicities and the formula $p_{d-2}(1-\frac{1}{2}\mathbf{r}^2)$, which just counts the number of transversal states. Of course, we start off from the transversal level-one states, but the more commutators we take between them the more subtle the entanglement of longitudinal and transversal states becomes.

For example, look at the canonical hyperbolic extension of $\mathfrak{su}(2)$ whose level-two root multiplicities coincide with the number of transversal states, $p_1(1-\frac{1}{2}\mathbf{r}^2)$, up to $\mathbf{r}^2 \leq -36$ (see [21, Table H_3]) and then drop below this bound. We conjecture that, when we perform the DDF construction for this example, we shall see at level-two from the very beginning longitudinal states to appear and transversal states to be missed, even though the multiplicity superficially suggests the existence of transversal states alone. For higher levels we predict an increasing mixing of longitudinal and transversal states which manifests itself in an increasing deviation of the multiplicities from the number of transversal states. Thus the DDF analysis of a single level-two root space of E_{10} allows us to make some reasonable predictions for some features occurring in other hyperbolic algebras of that type.

References

1. R. E. Borcherds. Vertex algebras, Kac-Moody algebras, and the monster. *Proceedings of the National Academic Society USA*, **83** :3068–3071, 1986.
2. R. E. Borcherds. Generalized Kac-Moody algebras. *Journal of Algebra*, **115** :501–512, 1988.
3. R. E. Borcherds. The monster Lie algebra. *Advances in Mathematics*, **83** :30–47, 1990.
4. R. C. Brower. Spectrum-generating algebra and no-ghost theorem for the dual model. *Physical Review*, **D6** :1655–1662, 1972.
5. J. H. Conway. The automorphism group of the 26-dimensional even unimodular Lorentzian lattice. *Journal of Algebra*, **80** :159–163, 1983.
6. E. Del Giudice, P. Di Vecchia, and S. Fubini. General properties of the dual resonance model. *Annals of Physics*, **70** :378–398, 1972.
7. A. J. Feingold and I. B. Frenkel. A hyperbolic Kac-Moody algebra and the theory of Siegel modular forms of genus 2. *Mathematische Annalen*, **263** :87–144, 1983.
8. I. B. Frenkel. Representations of Kac-Moody algebras and dual resonance models. In *Applications of Group Theory in Theoretical Physics*, pages 325–353, American Mathematical Society, Providence, 1985. Lect. Appl. Math., Vol. 21.
9. I. B. Frenkel and V. G. Kac. Basic representations of affine Lie algebras and dual models. *Inventiones Mathematicae*, **62** :23–66, 1980.
10. I. B. Frenkel, J. Lepowsky, and A. Meurman. *Vertex Operator Algebras and the Monster. Pure and Applied Mathematics Volume 134*, Academic Press, San Diego, 1988.
11. R. W. Gebert. Introduction to vertex algebras, Borcherds algebras, and the monster Lie algebra. *International Journal of Modern Physics*, **A8** :5441–5503, 1993.
12. R.W. Gebert and H. Nicolai. On E_{10} and the DDF construction, to appear in *Communications in Mathematical Physics*.
13. P. Goddard, A. Kent, and D. Olive. Virasoro algebras and coset space models. *Physics Letters*, **152 B**:88–92, 1985.
14. P. Goddard and D. Olive. Algebras, lattices and strings. In J. Lepowsky, S. Mandelstam, and I. M. Singer, editors, *Vertex Operators in Mathematics and Physics – Proceedings of a Conference November 10-17, 1983*, pages 51–96, Springer, New York, 1985. Publications of the Mathematical Sciences Research Institute #3.
15. P. Goddard and D. Olive. Kac-Moody and Virasoro algebras in relation to quantum physics. *International Journal of Modern Physics*, **A 1** :303–414, 1986.
16. M. B. Green, J. H. Schwarz, and E. Witten. *Superstring Theory Vol. 1&2*. Cambridge University Press, 1988.
17. D. J. Gross. High-energy symmetries of string theory. *Physical Review Letters*, **60** :1229–1232, 1988.
18. C. M. Hull, and P. K. Townsend. Unity of Superstring Dualities. preprint QMW-94-30, R/94/33, 1994.
19. B. Julia. in: *Superspace and Supergravity*, eds. S. W. Hawking and M. Rocek Cambridge University Press, 1981; *Lectures in Applied Mathematics*, **21**: 355–373, 1985.

20. B. Julia and H. Nicolai. Null-Killing Vector Dimensional Reduction and Galilean Geometrodynamics. preprint LPTENS 94/21, DESY 94-156, 1994.
21. V. G. Kac. *Infinite dimensional Lie algebras*. Cambridge University Press, Cambridge, third edition, 1990.
22. V. G. Kac, R. V. Moody, and M. Wakimoto. On E_{10}. In K. Bleuler and M. Werner, editors, *Differential geometrical methods in theoretical physics. Proceedings, NATO advanced research workshop, 16th international conference, Como*, pages 109–128, Kluwer, 1988.
23. S. Kass, R. V. Moody, J. Patera, and R. Slansky. *Affine Lie Algebras, Weight Multiplicities, and Branching Rules, Volume 1&2*. University of California Press, Berkeley, 1990.
24. G. Moore. *Finite in all directions*. preprint HEP-TH/9305139, YCTP-P12-93, Yale University, New Haven, 1993.
25. H. Nicolai. *Physics Letters*, **B276**:333–340, 1992
26. J. Scherk. An introduction to the theory of dual models and strings. *Reviews of Modern Physics*, **47** :123–164, 1975.
27. J. Serre. *A Course in Arithmetics*. Springer, New York, 1973.
28. P. C. West. *Physical States and String Symmetries*. preprint HEP-TH/9411029, KCL-TH-94-19, 1994.

Symmetry Principles for String Theory

Mark Evans[1] and Ioannis Giannakis[2,3]

[1] Department of Physics, The Rockefeller University, 1230 York Av., New York, NY 10021.
[2] Center for Theoretical Physics, Texas A&M University, College-Station, TX 77843-4242.
[3] Houston Advanced Research Center, The Mitchell Campus, Woodlands, TX 77381

In this talk we shall review work [1], [2], [3] on deformations of conformal field theories and symmetries of string theory. For more details the reader is referred to the original papers, or the review contained in [4].

What is the nature of the problem? String theory is presented as a set of rules for calculating scattering amplitudes rather than as the the dynamical consequence of a symmetry principle. Nevertheless, there are persuasive reasons to believe that string theory *does* possess such an underlying symmetry, and it is surely important that we understand what it is. This problem is not yet solved, but we shall describe what we believe to be considerable progress on this problem, culminating in the identification of a new type of infinite-dimensional supersymmetry algebra, a *weighted tensor algebra*, as, at least, a subalgebra of the gauge symmetry of string theory.

To study symmetries, we seek transformations of the space-time fields that take one solution of the classical equations of motion to another that is physically equivalent. Since, "Solutions of the classical equations of motion," are two-dimensional conformal field theories, we are thus interested in *isomorphic conformal field theories.*

Any quantum mechanical theory (including a CFT) is defined by three elements: i) an algebra of observables, $\mathcal{A}$ (determined by the degrees of freedom of the theory and their equal-time commutation relations), ii) a representation of that algebra and iii) a distinguished element of $\mathcal{A}$—the Hamiltonian. Note that for the same $\mathcal{A}$ we may have many choices of Hamiltonian, so that $\mathcal{A}$ should more properly be associated with a *deformation class* of theories than with one particular theory. For a CFT, we further want $\mathcal{A}$ to be generated by *local* fields, $\Phi(\sigma)$ (operator valued distributions on a circle parameterized by σ), and we require not just a single distinguished operator, but two distinguished *fields*, $T(\sigma)$ and $\overline{T}(\sigma)$, which are the components of the stress tensor. They generate the conformal transformations, and so must satisfy the Virasoro×Virasoro algebra:

$$[T(\sigma),T(\sigma')] = \frac{-ic}{24\pi}\delta'''(\sigma-\sigma') + 2iT(\sigma')\delta'(\sigma-\sigma') - iT'(\sigma')\delta(\sigma-\sigma') \quad (1a)$$

$$[\overline{T}(\sigma),\overline{T}(\sigma')] = \frac{ic}{24\pi}\delta'''(\sigma-\sigma') - 2i\overline{T}(\sigma')\delta'(\sigma-\sigma') + i\overline{T}'(\sigma')\delta(\sigma-\sigma') \quad (1b)$$

$$[T(\sigma),\overline{T}(\sigma')] = 0. \quad (1c)$$

Also of interest are the *primary fields* of dimension $(d, \bar{d})$, $\Phi(\sigma) \in \mathcal{A}$, defined by the conditions

$$\begin{aligned}
[T(\sigma), \Phi_{(d,\bar{d})}(\sigma')] &= id\Phi_{(d,\bar{d})}(\sigma')\delta'(\sigma-\sigma') - (i/\sqrt{2})\partial\Phi_{(d,\bar{d})}(\sigma')\delta(\sigma-\sigma') \\
[\overline{T}(\sigma), \Phi_{(d,\bar{d})}(\sigma')] &= -i\bar{d}\Phi_{(d,\bar{d})}(\sigma')\delta'(\sigma-\sigma') - (i/\sqrt{2})\bar{\partial}\Phi_{(d,\bar{d})}(\sigma')\delta(\sigma-\sigma')
\end{aligned} \tag{2}$$

Clearly, then, two CFTs will be physically equivalent if there is an isomorphism between the corresponding algebras of observables, $\mathcal{A}$, that maps stress tensor to stress tensor. (The mapping of primary to primary is then automatic). The simplest such example is a similarity transformation:

$$\Phi(\sigma) \mapsto e^{ih}\Phi(\sigma)e^{-ih} \tag{3}$$

for any fixed operator h. *Thus the physics will be unchanged if we change a CFT's stress tensor by just such a similarity transformation.*

Now, the stress tensor is parameterized by the space-time fields of the string. For example,

$$T_{G_{\mu\nu}}(\sigma) = \tfrac{1}{2}G_{\mu\nu}(X)\partial X^{\mu}\partial X^{\nu} \tag{4}$$

corresponds to the space-time metric $G_{\mu\nu}$, with other fields vanishing. If a similarity transformation (3) applied to T generates a shift in the space-time fields, that shift will be a *symmetry* transformation.

So how do T and $\bar{T}$ deform as we change the background fields? (we now consider deformations which, while they preserve conformal invariance, *need not be symmetries, e.g.* we may deform flat empty space so that a weak gravitational wave propagates through it). It is straightforward to show that, to first order, the Virasoro algebras (1) are preserved by so-called *canonical deformations* [1] (see also [5]),

$$\delta T(\sigma) = \delta\overline{T}(\sigma) = \Phi_{(1,1)}(\sigma) \tag{5}$$

where $\Phi_{(1,1)}(\sigma)$ is a primary field of dimension (1,1). We reiterate: (5) does *not* in general correspond to a symmetry transformation, but it preserves conformal invariance. Since (1,1) primary fields are vertex operators for physical states, they correspond naturally to the space-time fields, and equation (5) makes transparent the connection between changes of the stress tensor and changes of the space-time fields. Returning now to the problem of symmetries, if we take the generator h in equation (3) to be the zero mode of an infinitesimal (1,0) or (0,1) primary field (a current), then it is a simple consequence of the definitions that its action on the stress tensor is necessarily a canonical deformation, and so may be translated directly into a change in the space-time fields (for examples, see [1]). It is well known that *conserved* currents generate symmetries [6], but within the formalism described here, conservation is *not* necessary, a fact that does not seem to have been widely appreciated. Indeed, it may be shown that a non-conserved current generates a symmetry that is spontaneously broken by the particular background being considered [1].

At this point it is convenient to assess the strengths and weaknesses of the analysis so far. We begin with the strengths:

- The fact that a current zero-mode generates a canonical deformation guarantees that we can translate the inner automorphism into a transformation on the physical space-time fields.
- In this way, we may exhibit symmetries both familiar (general coordinate invariance, among many [7]) and unfamiliar (an infinite class of spontaneously broken, level-mixing gauge supersymmetries) [1].

However, there are also deficiencies:

- Explicit calculations are hard to perform for a general configuration of the space-time fields. It would appear that we need to know the precise form of the stress-tensor and its currents for that general configuration (recall that a current is only a current relative to a particular stress-tensor). This is a very hard problem, essentially requiring us to find the general solution to the equations of motion before we can discuss symmetries.
- It is hard to say anything about the symmetry algebra. This is just the algebra of the generators, but it is *not* in general true that the commutator of two current zero-modes is itself a current zero-mode: our symmetry "algebra" does not close!
- Finally, by considering a few examples, it is easy to see that the canonical deformation of equation (5) corresponds to turning on space-time fields *in a particular gauge* (something like Landau or harmonic gauge). We would like to understand the gauge principle behind string theory without imposing gauge conditions.

It turns out that these three drawbacks are intimately related, and may be largely overcome by moving beyond canonical deformations, equation (5), which are *not* the most general infinitesimal deformation that preserve the Virasoro algebras (1). In [2] we showed that, for the massless degrees of freedom of the bosonic string in flat space, we could find a distinct deformation of the stress tensor for each solution of the linearized Brans-Dicke equations. This correspondence was found by considering the general translation invariant *ansatz* of naive dimension two for δT;

$$\begin{aligned}\delta T =& H^{\nu\lambda}(X)\partial X_\nu \overline{\partial} X_\lambda + A^{\nu\lambda}(X)\partial X_\nu \partial X_\lambda + \\ & B^{\nu\lambda}(X)\overline{\partial} X_\nu \overline{\partial} X_\lambda + C^\nu(X)\partial^2 X_\nu + D^\lambda(X)\overline{\partial}^2 X_\lambda, \end{aligned} \tag{6}$$

with a similar, totally independent *ansatz* for $\delta\overline{T}$. The fields $H^{\mu\nu}$ *etc.* turn out to be characterized in terms of solutions to the linearized Brans-Dicke equation when we demand that the deformation preserves (to first order) the Virasoro algebras (1).

By considering this more general *ansatz*, we get more than just covariant equations of motion—we also understand a larger set of symmetry generators, h. Indeed, any generator that preserves the form of the *ansatz* (6) will do.

The condition that δT be of naive dimension two (with which we shall soon dispense) is preserved if h is of naive dimension zero. The condition of translation invariance is preserved whenever h is a zero-mode.

The lesson to be drawn from this massless example is clear: the way to introduce space-time fields unconstrained by gauge conditions is to consider an *arbitrary translation invariant ansatz* for the deformation of the stress tensor, and to ask only that it preserve the Virasoro algebras. To move beyond the massless level, we simply drop the requirement that the naive dimension be two. Thus symmetries are generated by arbitrary zero-modes in $\mathcal{A}$. We argued in [2] that, as with the superfield formulation of supersymmetric field theories, this was likely to introduce auxiliary fields beyond the massless level, but so be it. (Indeed, this whole formulation of string theory is rather akin to a superspace approach, with T and $\overline{T}$ as superfields and derivatives of the world-sheet scalars playing the rôle of the odd coordinates of superspace).

This extension of the set of symmetry generators ameliorates each of the drawbacks mentioned above. It should also be emphasized that the symmetry algebra is now manifestly background independent, since being a zero-mode is a property independent of the choice of stress-tensor.

Nevertheless, difficulties remain. In particular, the operator algebra, $\mathcal{A}$ and its subalgebras require a more careful construction than is usually given. In [3] we showed that, even in free scalar field theory, normal ordering is inadequate to define composite operators, even vertices and currents. We demonstrated this by showing that the commutator of two such fields may be singular. Typically, the algebra $\mathcal{A}$ is constructed by combining two (non-local!) chiral subalgebras into what Zuckerman has called the (local!) "bilateral" algebra [8]. However, this construction is problematic because commutators in one chiral subalgebra will often be very local (involving δ-functions). This puts fields from the other chiral component at the same point which, of course, introduces singularities [3].

This problem is not yet fully resolved, but it is possible [3] that the correct approach is to analytically continue in space-time momenta—the same procedure used to calculate scattering amplitudes. However, we may avoid this difficulty by finding interesting infinite-dimensional subalgebras of the full symmetry where it simply does not arise. The simplest such examples are just local subalgebras of the chiral algebras which go to make up $\mathcal{A}$. For the uncompactified bosonic string, these are generated by operators of the form:

$$h = \int d\sigma\, \phi_{\mu\nu\cdots\rho} \partial^{w_1} X^\mu \partial^{w_2} X^\nu \cdots \partial^{w_n} X^\rho. \tag{7}$$

It is clear that each such generator is specified by a constant tensor, ϕ, with a positive-definite integer weight, w associated to each index. For this reason, we coined the name *weighted tensor algebra* for such structures. It is straightforward, if a little tedious, to calculate the commutators of such operators. If we label each generator by its tensor and weights, $\{\phi_{\mu_S}, w_S\}$, where $\mathcal{S}$ is some index set, then the commutator is

$$[\{\psi_{\mu_S}, w_S\}, \{\chi_{\nu_T}, v_T\}] =$$
$$\sum_{\substack{\mathcal{U}\subseteq\mathcal{S}\\ \mathcal{P}:\mathcal{U}\hookrightarrow\mathcal{T}}} C_{\mathcal{U},\mathcal{P}}\Delta_{|\mathcal{S}-\mathcal{U}|}^{\left(-1+\sum_{i\in\mathcal{U}} w_i+v_{\mathcal{P}(i)}\right)}\left\{\psi_{\mu_S}\chi_{\nu_T}\prod_{i\in\mathcal{U}}\eta^{\mu_i\nu_{\mathcal{P}(i)}},\ w_{\mathcal{S}-\mathcal{U}}\oplus v_{\mathcal{T}-\mathcal{P}(\mathcal{U})}\right\}. \quad (8)$$

The sum, over subsets $\mathcal{U}$ of $\mathcal{S}$ and injections $\mathcal{P}$, is just a formal way of summing over all possible contractions of the tensors. The operator Δ is defined by

$$\Delta_r\left\{\phi^{(k)}, w_i\right\} = \sum_{l=1}^{r}\left\{\phi^{(k)}, (w_i+\delta_{il})\right\}, \quad (9)$$

and differentiates the first r factors in the integrand of the generator, (7). The coefficients are given by

$$C_{\mathcal{U},\mathcal{P}} = \frac{\prod_{k\in\mathcal{U}}(-1)^{w_k}(w_k+v_{\mathcal{P}(k)}-1)!}{\left(\left(\sum_{k\in\mathcal{U}} w_k+v_{\mathcal{P}(k)}\right)-1\right)!} \quad (10)$$

It may not be pretty, but it's a gauge subalgebra of string theory, and it seems likely that the full algebra will have a similar form. Somehow, we shall have to learn to love it.

The work of M. E. was supported in part by DOE Contract Number DOE-ACO2-87ER40325, TASK B, and that of I. G. by DOE Contract Number DE-FG05-91-ER-40633.

References

[1] M. Evans and B. Ovrut Phys. Rev. **D41** (1990), 3149; Phys. Lett. **231B** (1989), 80.
[2] M. Evans and I. Giannakis, Phys. Rev. **D44** (1991), 2467.
[3] M. Evans, I. Giannakis and D. Nanopoulos, Phys. Rev. **D50** (1994), 4022, hep-th/9401075.
[4] M. Evans and I. Giannakis in S. Catto and A. Rocha, eds., *Proceedings of the XXth International Conference on Differential Geometric Methods in Theoretical Physics*, World Scientific, Singapore (1992), p. 759, hep-th/9109055.
[5] G. Pelts, Rockefeller University preprints RU-93-5-B, hep-th/9309141 and RU-94-5-B, hep-th/9406022 and contribution, this volume.
[6] T. Banks and L. Dixon Nucl. Phys. **B307** (1988), 93.
[7] M. Evans and B. Ovrut, Phys. Lett. **214B** (1988), 177; Phys. Rev. **D39** (1989), 3016; M. Evans, J. Louis and B. Ovrut, Phys. Rev. **D35** (1987), 3045.
[8] G. Zuckerman, this volume.

Are On-shell 3-String Amplitudes the Structure Constants of the Fake Monster Lie Algebra ?

S.Nergiz and C.Saçlıoğlu

Physics Dept., Boğaziçi University, Istanbul, Turkey

Abstract : They seem to be (up to a sign).

Keywords Vertex operators, the Monster Lie algebra

Hyperbolic and Lorentzian algebras are particularly relevant to this conference in that they literally **are** the symmetries of the string. Leaving the specific string theory representation for a little later, one might define a Lorentzian algebra of rank d as a Lie algebra generated by root vectors belonging to a d dimensional Minkowskian lattice.[1] The number of roots is no longer finite, as in ordinary Lie algebras, or even "polynomially" infinite as in the Kac-Moody case ($T_a^n, \forall n \in Z$) but increases roughly exponentially with the length of the root vector.

The subclasses of Lorentzian algebras of greatest interest to physics are obtained by imposing a number of restrictions. If one demands unitary representations, then $d \leq 26$. If the root system is such that the removal of any simple root yields the simple roots for a Lie or a Kac-Moody algebra, the algebra is said to be hyperbolic and $d \leq 10$. There are 136 hyperbolic algebras for $3 \leq d \leq 10$. The stronger restriction of obtaining only a Lie algebra upon the deletion of any simple root defines strictly hyperbolic algebras, for which $d \leq 4$. There are 11 strictly hyperbolic algebras for $3 \leq d \leq 4$ and infinitely many for $d = 2$. [2,3,4] The pattern of critical dimensions ($d = 26, 10, 4, 2$) is tantalizingly reminiscent of (super)string theory and physics. In particular, in the best understood $d = 26$ case the proof of unitarity essentially parallels the no-ghost theorem.

The connection with string theory can be made more concrete through the integrated generalized vertex operator (igvo) representation of Lorentzian algebras: The algebra of igvos (modulo appropriate cocycle factors) with momenta chosen from the root lattice of a Lorentzian algebra M is precisely M itself. If the unique even self dual lattice $\Pi^{25,1}$ is selected, the resulting so-called "Fake Monster Lie Algebra", [5] which we will denote by M_∞, contains all other rank 26 Lorenztian algebras [6].

Witten has conjectured that string field theory must be related to a hyperbolic or Lorentzian algebra G associated with all the massive states of string theory. It is tempting to identify G with M_∞ especially since the root multiplicity of M_∞ matches that of the bosonic string in 26 dimensions [6].

In order to investigate the validity of this conjecture it would be useful: (A) to characterize M_∞ explicitly, for example, by obtaining the structure constants; (B) to provide an example of a "physics" application for M_∞. The latter demand may at first appear difficult to meet since one does not usually think of restricting string theory to latticised momenta.

Here we would like to indicate a specific connection between (A) and (B). We have presented the structure constants of M_∞ elsewhere [8]. We have also argued that one already has an effective latticization of momenta in an on-shell string theory process $1+2 \rightarrow 3$: If the mass shell values of the momenta p_i are scaled so that $p_i^2 = 2, 0, -2, -4, \ldots$, the on-shell conditions for $i = 1, 2, 3$ and $p_1 + p_2 = p_3$ force all the $p_i \cdot p_j$ to assume integer values. Since $\Pi^{25,1}$ is the only lattice that matches the multiplicity of the bosonic string, we are essentially forced to assume the p_i belong to it. We should perhaps emphasize that this does not imply that a permanent lattice in momentum space actually exists. It however allows us to evaluate the corresponding commutator $[V_1(p_1), V_2(p_2)] = f_{123}V_3(p_3) + \sum f_{12k}V_k(p_k)$ **as if** it does, since all that the calculation requires is for $p_i \cdot p_j$ to be an integer.

Having argued that M_∞ could be of direct use in string physics without committing ourselves to a fixed lattice, it now behooves us to address (B), i.e., specify at least one such use. A plausible expectation is that after a suitable normalization the on-shell amplitude $A(1+2 \rightarrow 3)$ and the structure constant f_{123} will be identical. If true, this would imply that M_∞ contains information about a fundamental quantity in string field theory, namely the on-shell three-string vertex, in conformity with Witten's conjecture.

The investigation of the equality of $A(1+2 \rightarrow 3)$ and f_{123} unfortunately has to be carried out in a case-by-case fashion. The reason for this is the following: The f_{123} can only be obtained in a general way by using igvos which are not irreps of $SO(25)$ while the states which enter $A(1+2 \rightarrow 3)$ are. Since there is no general algorithm for finding the precise linear combination of the igvos corresponding to an $SO(25)$ irrep, only an inductive "verification" can be attempted.

We have checked our conjecture up to the fourth mass-level ($p^2 = -6$) for a large number of commutators and found 15 cases of agreement **up to a sign** [8]. It has been suggested [9] that the sign changes may indicate a connection with the finite sporadic Monster group rather than the Lie algebra since the passage from the latter to the former involves identification of some of the points in $\Pi^{25,1}$.

REFERENCES

[1] P.Goddard and D.Olive, in Proc. of the 1983 M.S.R. I Conf.,eds. J.Lepowsky, S. Mandelstam and I.M.Singer (Springer Verlag, New York,1985) p.5.

[2] V.Kac, Infinite Dimensional Lie Algebras, Cambridge University Press, Cambridge 1990, third edition.

[3] C.Saçlıoğlu, J. Phys. A: Math. Gen. **22** (1989),3753.

[4] J.Lepowsky and R.Moody, Math. Ann. **245** (1979) 63.

[5] R.E. Borcherds, Advances in Mathematics **83** (1990) 30.

[6] I.B.Frenkel, Proc. ICM (Am.Math.Soc. Berkeley,CA) 1986, p.281.

[7] E.Witten, Int. J. Mod. Phys., **A 1** (1986) 39.

[8] S.Nergiz and C.Saçlıoğlu, Int. J. Mod. Phys. **A 5** (1990) 2647.

[9] C.Vafa, private communication.

Auxiliary Fields for Super Yang-Mills from Division Algebras

Jonathan M. Evans

Theoretical Physics Division, CERN, CH-1211, Genève 23, Switzerland

Abstract. Division algebras are used to explain the existence and symmetries of various sets of auxiliary fields for super Yang-Mills in dimensions $d = 3, 4, 6, 10$.

1 Introduction

The simplest supersymmetric Yang-Mills theories are those in which the physical degrees of freedom are described by a vector gauge field A_μ ($\mu = 0, \ldots, d-1$) and a spinor field ψ, both defined on d-dimensional Minkowski space and taking values in the Lie algebra of some gauge group. A necessary condition for supersymmetry is that the number of physical bosons and fermions should be equal, which leads to the possibilities $d = 3, 4, 6, 10$ with the spinor being Majorana, Majorana or Weyl, Weyl, Majorana and Weyl respectively. To show that this equality of bosons and fermions is sufficient for supersymmetry as well as necessary, one can take the detailed expressions for the supersymmetry transformations—which are fixed up to irrelevant constants by gauge-invariance and dimensional considerations—and check that they obey the standard algebra up to terms involving field equations (this is in fact equivalent to requiring invariance of the natural action in which the spinor is minimally coupled to the gauge field). The condition for the supersymmetry algebra to close *on-shell* in this way is a certain gamma-matrix identity which is indeed satisfied precisely for the values of d and the types of spinor listed above [1].

It is always desirable in a supersymmetric theory to try to promote the on-shell symmetry algebra to one which holds *off-shell*, that is, without the use of field equations. For $d = 3$ super Yang-Mills the superalgebra closes automatically; for $d = 4, 6$ one can introduce auxiliary fields to close the superalgebra in a Lorentz-covariant way [2,3]; but for $d = 10$ the best which can be done with a finite set of auxiliary fields is to partially close the superalgebra, breaking the manifest $d = 10$ Lorentz symmetry to some subgroup in the process. An interesting new perspective on these matters was provided recently in [4] with the introduction of more general fermionic transformations which include the conventional supersymmetry transformations of $d = 10$ super Yang-Mills as special cases. It was then shown in [5] that all previously known sets of auxiliary fields for the $d = 10$ theory could be recovered within this framework.

Our aim here is to show how the possible auxiliary fields for each of the allowed super Yang-Mills theories can be understood using the language of division algebras [6-8]. Connections between super Yang-Mills and the division algebras have been established in the past by interpreting the gamma-matrix identity necessary for on-shell supersymmetry from a number of closely related points of view using division algebra-valued spinors [9,10], Jordan algebras [10,11] and trialities [12]. The role of division algebras in understanding off-shell supersymmetry was emphasized first in [7] and then in [4] where octonions were used to find a new set of auxiliary fields in $d = 10$. We shall indicate here how the solutions given in [4,5] for $d = 10$, together with analogous solutions for the lower-dimensional theories, can all be understood using this language; we shall also explain from this point of view the residual symmetries of these various sets of auxiliary fields.

2 Division Algebra Notation

We denote by $\mathbb{K}_n$ with $n = 1, 2, 4, 8$ the division algebras $\mathbb{R}$, $\mathbb{C}$, $\mathbb{H}$, $\mathbb{O}$ of real numbers, complex numbers, quaternions and octonions respectively; for background see papers such as [6-12]. We denote by e_a $(a = 1, \ldots, n)$ an orthonormal basis for $\mathbb{K}_n$ with $e_n = 1$ and all other basis elements pure-imaginary. Bars will denote conjugation and daggers will denote hermitian conjugation.

The basic idea is to define the action of certain spin representations of $\mathrm{SO}(n+1, 1)$ on two-component objects with entries in $\mathbb{K}_n$ [7,8]. The cases $n = 1, 2, 4, 8$ obviously correspond to $d = 3, 4, 6, 10$ and the spinor appearing in each of the allowed super Yang-Mills theories in these dimensions can then be written as an object ψ^α $(\alpha = 1, 2)$ or as its conjugate $\bar{\psi}^{\dot{\alpha}}$ $(\dot{\alpha} = 1, 2)$. There are also dual spinor representations acting on objects which we denote by χ_α and $\bar{\chi}_{\dot{\alpha}}$. When spinor indices are suppressed we will regard ψ, $\bar{\psi}$ as row vectors and χ, $\bar{\chi}$ as column vectors. Spinor indices are never raised or lowered and the usual Lorentz-invariant inner-product is given by the real expression $\psi\chi + \chi^\dagger\psi^\dagger = \psi^\alpha\chi_\alpha + \bar{\chi}_{\dot{\alpha}}\bar{\psi}^{\dot{\alpha}}$.

The gamma-matrices needed to construct the spin representations are invariant tensors $(\Gamma_\mu)_{\alpha\dot{\alpha}}$ and $(\Gamma^\mu)^{\dot{\alpha}\alpha}$. We define these in a particular basis in which their components are equal and are given by the hermitian matrices:

$$\begin{pmatrix} 1 & 0 \\ 0 & 1 \end{pmatrix} \text{ for } \mu = 0; \quad \begin{pmatrix} 1 & 0 \\ 0 & -1 \end{pmatrix} \text{ for } \mu = n+1; \quad \begin{pmatrix} 0 & e_a \\ \bar{e}_a & 0 \end{pmatrix} \text{ for } \mu = a = 1, \ldots, n.$$

The order of indices on gamma-matrices indicates whether they should multiply spinors from the left or from the right with the convention that only adjacent upper and lower indices of the same type can be contracted. For example, the matrices $(\Gamma_\mu)_{\alpha\dot{\alpha}}$ act by right-multiplication on ψ^α and left-multiplication on $\bar{\psi}^{\dot{\alpha}}$.

It will be important for us to understand how certain subgroups of $\mathrm{SO}(n+1, 1)$ appear in this formalism. With the basis introduced above, the two $\mathbb{K}_n$-valued components of any spinor clearly carry irreducible representations of the light-cone subgroup SO(1,1)×SO(n). In fact these two copies of $\mathbb{K}_n$ carry the inequivalent spin representations of SO(n) for $n = 2, 4, 8$. If we restrict further to the subgroup SO(n−1) which fixes the direction $\mu = n$ then these spin representations become equivalent. The resulting representation is realized on $\mathbb{K}_n$ in two particularly simple ways: the relevant gamma-matrices or invariant tensors are just the imaginary basis elements e_i ($i = 1, \ldots, n-1$) and these multiply components of type ψ^α from the right and components of type χ_α from the left.

Notice that the elements of unit modulus in $\mathbb{C}$ and $\mathbb{H}$ form groups U(1) and SU(2) respectively. They act naturally by multiplication on complex and quaternionic spinors so as to commute with Lorentz transformations; we shall therefore refer to these operations on spinors as internal transformations. The generators of such transformations are unit imaginary elements e_i multiplying components of type ψ^α from the left and components of type χ_α from the right (compare with the last paragraph). The internal groups U(1) and SU(2) appear as symmetries of the super Yang-Mills theories in $d = 4$ and $d = 6$; no such symmetries arise in $d = 3$ or $d = 10$.

Finally, it can be shown that the algebras $\mathbb{H}$ and $\mathbb{O}$ have non-trivial continuous automorphism groups SO(3) and G_2 respectively [6,8]. The action of these automorphisms on spinor components can always be reproduced by combinations of Lorentz and internal transformations. The special feature of the octonionic case is that the automorphism group is contained entirely within the Lorentz group—in fact within the SO(7) subgroup mentioned above.

3 Generalized and Conventional Supersymmetry

The generalized supersymmetry transformations proposed in [4] involve real bosonic auxiliary fields G_i ($i = 1, \ldots, d-3$) which balance the off-shell bosonic and fermionic degrees of freedom. They can be written

$$\begin{gathered} \delta A_\mu = \epsilon(\Gamma_\mu \psi^\dagger) + (\psi \Gamma_\mu)\epsilon^\dagger, \quad \delta\psi = \tfrac{1}{2}(\epsilon\Gamma_\nu)\Gamma_\mu F^{\mu\nu} + G_i v_i, \\ \delta G_i = -v_i(\Gamma^\mu D_\mu \psi^\dagger) - (D_\mu \psi \Gamma^\mu) v_i^\dagger, \end{gathered} \tag{1}$$

where $D_\mu = \partial_\mu + A_\mu$ is the usual covariant derivative and $F_{\mu\nu} = [D_\mu, D_\nu]$ is the field strength. (We have defined δ so as to remove some factors of i compared to [4,5] and we have normalized spinors so as to remove some factors of $\frac{1}{2}$.) The parameters of the transformations are commuting spinors ϵ^α and v_i^α which must satisfy the additional relations

$$\begin{gathered} \epsilon(\Gamma_\mu v_i^\dagger) + (v_i \Gamma_\mu)\epsilon^\dagger = 0, \\ v_i(\Gamma_\mu v_j^\dagger) + (v_j \Gamma_\mu) v_i^\dagger = \delta_{ij}\left(\epsilon(\Gamma_\mu \epsilon^\dagger) + (\epsilon\Gamma_\mu)\epsilon^\dagger \right). \end{gathered} \tag{2}$$

These ensure that the standard supersymmetry algebra still holds up to field equations despite the introduction of extra parameters.

To recover conventional supersymmetry transformations from those written in (1) above one must solve the equations (2) with v_i depending linearly on ϵ and with the value of ϵ restricted to some subspace if necessary. The subset of conventional supersymmetry transformations obtained in this way will automatically obey a closed algebra. However, such a solution will, in general, break the full Lorentz invariance of equations (1) and (2) down to a subgroup determined both by the subspace to which ϵ is confined and by the precise definitions of the quantities v_i. These points are explained in more detail in [5].

4 Solutions and Their Symmetries

Solutions to (2) can be written very simply in division algebra notation. We consider first the possibility

$$v_i = e_i \epsilon \tag{3}$$

with the spinor ϵ confined to some subspace. The transformations (1) can now be re-expressed in a more compact way by combining the $n-1$ auxiliary fields into the pure-imaginary object $G = G_i e_i$ with the result

$$\begin{gathered} \delta A_\mu = \epsilon(\Gamma_\mu \psi^\dagger) + (\psi \Gamma_\mu)\epsilon^\dagger\,, \quad \delta\psi = \tfrac{1}{2}(\epsilon\Gamma_\nu)\Gamma_\mu F^{\mu\nu} + G\epsilon\,, \\ \delta G = \epsilon(\Gamma^\mu D_\mu \psi^\dagger) - (D_\mu \psi \Gamma^\mu)\epsilon^\dagger\,. \end{gathered} \tag{4}$$

Before discussing the individual solutions it will be best to make some general remarks concerning the possible symmetries of these equations.

In the associative cases the equations (4) are invariant under any Lorentz transformation if the auxiliary fields behave as scalars (and provided that the transformation respects whatever restriction is placed on ϵ of course). This can be checked by calculating explicitly the transformation of the expression given for δG and by noting also that if G is inert then the term $G\epsilon$ transforms in the same way that ϵ does; all other terms are covariant by construction. In the octonionic case, however, this statement must be qualified: we find that a Lorentz transformation is a symmetry of (4) with G inert only if it is constructed from gamma-matrices whose entries have vanishing associators with the components of ϵ. Aside from such Lorentz transformations, the equations (4) are also invariant under the internal transformations which arise in the complex and quaternionic cases in the manner discussed earlier. Lastly, the equations (4) are invariant under any automorphism of $\mathbb{K}_n$ (provided once again that it respects the restriction on ϵ) and then the auxiliary fields will transform in some non-trivial way. From the remarks made previously we know that automorphisms give genuinely new symmetries only in the octonionic case. We are now in a position to explain the symmetries which appear in each of our solutions.

It is easy to show that in the associative cases (3) provides a solution of (2) for any spinor ϵ and the formulas (4) then reproduce the standard, Lorentz-covariant, off-shell supersymmetry transformations for $d = 4$ [2] and $d = 6$ [3,7] super Yang-Mills with scalar auxiliary fields. In $d = 4$ the off-shell supersymmetries are also invariant under the U(1) internal group with the auxiliary field being inert. In $d = 6$ they are invariant under the SU(2) internal group with the auxiliary fields transforming as a triplet.

The lack of associativity of the octonions means that in $d = 10$ the formula (3) is no longer a solution of (2) for all ϵ. However, there are at least two interesting ways of restricting ϵ in (3) so as to obtain closed subalgebras of supersymmetry transformations of type (4). Both solutions require for their verification several lines of octonionic manipulations which we shall omit.

The first case in which (3) provides a solution of (2) is where one of the components of ϵ is restricted to be real, giving a closed algebra of nine supersymmetries [4]. To preserve the form of ϵ under a Lorentz transformation we must confine attention to a subgroup SO(1,1)×SO(7). The generator of SO(1,1) is real and so never gives rise to an associator. But the additional transformations are symmetries of (4) only if they lie in the subgroup of automorphisms G_2 within SO(7), with the auxiliary fields transforming in its seven-dimensional representation. In this way we find exactly the residual invariance SO(1,1)×G_2 and the pattern of representations given in section 3 of [5].

The second case in which (3) provides a solution of (2) is where both components of ϵ lie in some copy of the complex numbers $\mathbb{C}$ within $\mathbb{O}$, yielding a closed algebra of four supersymmetries. This solution is clearly invariant under a subgroup SO(3,1) generated by combining gamma-matrices with entries in this same copy of $\mathbb{C}$, since all the relevant associators then vanish. The other pairs of gamma-matrices which lead to vanishing associators give $d = 10$ Lorentz generators which coincide when acting on ϵ and which, when exponentiated, correspond simply to multiplication by an arbitrary phase within our chosen copy of $\mathbb{C}$. Lastly, the solution is invariant under the subgroup of G_2 which sends our particular copy of $\mathbb{C}$ to itself, and it is well-known that the subgroup of automorphisms which fix a given imaginary element is SU(3) [6]. The residual invariance is therefore SO(3,1)×U(1)×SU(3) and closer examination of the details of the representations gives complete agreement with those found in section 4 of [5].

So far we have discussed solutions of (2) of type (3); it is natural to ask whether there exist similar solutions of the form

$$v_i = \epsilon e_i \tag{5}$$

with ϵ restricted to some subspace. The transformations (1) can once again be re-written, after a little work, in terms of $G = G_i e_i$ and the result is

$$\begin{gathered} \delta A_\mu = \epsilon(\Gamma_\mu \psi^\dagger) + (\psi \Gamma_\mu)\epsilon^\dagger\,, \quad \delta\psi = \tfrac{1}{2}(\epsilon\Gamma_\nu)\Gamma_\mu F^{\mu\nu} + \epsilon G\,, \\ \delta G = (\overline{D_\mu \psi \Gamma^\mu})\bar{\epsilon}^\dagger - \bar{\epsilon}(\overline{\Gamma^\mu D_\mu \psi^\dagger})\,. \end{gathered} \tag{6}$$

In $d = 4$ the formulas (5) and (6) coincide with (3) and (4) because the complex numbers are commutative; but for $d = 6$ and $d = 10$ we obtain new possibilities.

In the non-commutative cases (5) fails to satisfy (2) for general ϵ, but it does provide a solution if we restrict ϵ to having only one non-zero component. This restriction clearly breaks the Lorentz group at least to the light-cone subgroup SO(1,1)×SO(n) and in fact the compact part of the surviving symmetry is SO($n-1$). This can be seen from the way in which the pure-imaginary elements e_i

appear in (6) through G, because these quantities act as invariant tensors for the subgroup SO($n-1$) when multiplying ϵ from the right, as we have already mentioned. Thus in $d = 6$ we obtain a closed algebra of four supersymmetries of type (6) with residual Lorentz invariance SO(1,1)×SO(3) and auxiliary fields transforming as a three-dimensional vector. There is also an SU(2) internal symmetry under which the auxiliary fields are inert. In $d = 10$ we find a closed algebra of eight supersymmetries of type (6) with residual invariance SO(1,1)×SO(7) and auxiliary fields transforming as a seven-dimensional vector. This is just the solution presented in section 2 of [5].

5 Comments

We have seen that division algebras provide an elegant language in which to understand the occurrence and symmetries of bosonic auxiliary fields for super Yang-Mills, with the lack of a covariant set in $d = 10$ being traced directly to the non-associativity of the octonions [4]. It is natural to wonder to what extent the solutions we have discussed here are exhaustive. It would also be interesting to investigate whether division algebras could be used to understand fermionic auxiliary fields (see the papers cited in [5]) in a similar fashion.

Acknowledgements

I thank Nathan Berkovits for interesting discussions and correspondence.

References

[1] L. Brink, J. Scherk, J.H. Schwarz: Nucl. Phys. B **121** 77 (1977);
M.B. Green, J.H. Schwarz, E. Witten: *Superstring Theory*, C.U.P. (1987).

[2] M.F. Sohnius: Phys. Rep. **128** 39 (1985).

[3] P. Howe, G. Sierra, P.K. Townsend: Nucl. Phys. B **221** 331 (1983).

[4] N. Berkovits: Phys. Lett. B **318** 104 (1993).

[5] J.M. Evans: Phys. Lett. B **334** 105 (1994).

[6] M. Günaydin, F. Gürsey: J. Math. Phys. **14** 1651 (1973).

[7] T. Kugo, P.K. Townsend: Nucl. Phys. B **221** 357 (1983).

[8] A. Sudbery: J. Phys. A **17** 939 (1984);
K.-W. Chung, A. Sudbery: Phys. Lett. B **198** 161 (1987).

[9] C.A. Manogue, A. Sudbery: Phys. Rev. D **40** 4073 (1989);
D.B. Fairlie, C.A. Manogue: Phys. Rev. D **36** 475 (1987).

[10] R. Foot, G.C. Joshi: Mod. Phys. Lett. A **3** 999 (1988).

[11] G. Sierra: Class. Quantum Grav. **4** 227 (1987).

[12] J.M. Evans: Nucl. Phys. B **298** 92 (1988).

Automorphism Groups of Discrete Octonions and Possible Applications in Physics

Mehmet Koca

Physics Department, Çukurova University, 01330 Adana, Turkey

Abstract

Octonions constituting the root system of E_8 form a closed algebra where the root system of E_7 can be represented by pure imaginary octonions. It has been shown that the automorphism group of the octonionic root system of E_7 is the adjoint Chevalley group $G_2(2)$ of order 12096. The maximal subgroups of $G_2(2)$ preserve the octonionic root systems of the maximal Lie algebras of E_7 with one exception .

It has been shown that the automorphism group of the octonionic set $\pm e_i$ $(i=1,...,7)$ (called Moufang loop) which represents the roots of $SU(2)^7$ is the non-split extension of the elementary abelian group 2^3 of order 8 by the Klein's group $L_2(7)$ of order 168. Matrix generators of the 7-dimensional irreducible representations of the split and non-split extensions of the groups of order 1344 have been constructed. The Weyl group of D_4 turns out to be maximal in the split 1344 and the Demazure-Tits' subgroups of A_2, G_2 and B_2 are naturally embedded in the groups of concern.

There are indications that the parity, P, CP, and CPT violations can be explained in a group theoretical way as C, P, T generate an elementary abelian group of order 8 while acting on the bilinear Dirac fields.

Construction of the E_8 root system with the icosians is also briefly mentioned.

1. Introduction

It is a great honor to have been invited at the conference commemorating the late professor Feza Gürsey, a great physicist who has deeply influenced my perception of physics. Octonions have been introduced to physics community through his pioneering work with Günaydın [1] in which they have tried to attribute prominent roles to octonions.

Following his path, in this talk, I will summarize some aspects of discrete octonions associated with the exceptional groups E_8 and E_7 and explain some features of automorphism groups of discrete octonions. It is all well known from the Coxeter's work [2] that the root system of E_8 can be represented by octonions and the 240 discrete octonionic roots form a closed algebra generated by three simple roots. We also know through the works of Du-Val [3] and Coxeter [4] that the discrete quaternions which represent the finite subgroups of SU(2) can be used for the description of certain Coxeter graphs. The peculiar one is D_4 where the 24 roots correspond to the elements of the binary tetrahedral group represented by quaternions and abbreviated by the set A_0. Together with the quaternionic weights of the three

If we analyze the root system of S0(12)XSU(2) we observe that it involves the roots $\pm e_1, \pm e_i (i=2,\dots,7)$ as well as other imaginary roots. The group $2^2.S_4{:}2$ preserve separately $\pm e_1$ and the remaining six imaginary units with positive and negative signs. One worries whether there exists a larger group which preserves the octonionic set $\pm e_i (i=1,2,\dots,7)$. It turns out that the larger group can be generated from the group $2^2.S_4{:}2$ by adjoining a generator which transforms the octonionic units in the following manner:
$e_1 \to e_2 \to e_4 \to e_3 \to e_6 \to e_5 \to e_7 \to e_1$ This transformation not only preserves the octonion algebra but also shuffles the set $\pm e_i$ which corresponds the roots of $SU(2)^7$ in E_7. We have shown [8] that the group is of order 1344. It is the non-split extension of the elementary abelian group 2^3 by the Klein's simple group $L_2(7)$ of order 168 so that it is abbreviated by the notation $2^3.L_2(7)$. Interestingly enough it has two maximal subgroups of order 192, one of which is the usual group $2^2.S_4{:}2$ and the other group has a structure $2^3.S_4$ which has 13 conjugacy classes preserving the distinct sets $(\pm e_1, \pm e_2, \pm e_3)$ and $(\pm e_4, \pm e_5, \pm e_6, \pm e_7)$. The third maximal subgroup is of order 168 with 8 conjugacy classes possessing the structure $2^3{:}7{:}3$ which is also one of the maximal subgroup in the sporadic Janko group J_1.

There exists a split extension $2^3{:}L_2(7)$, of the same order with $2^3{:}L_2(7)$ both sharing the same character table. But the structure of $2^3{:}L_2(7)$ is completely different from the non-split group and does not preserve the octonion algebra. It turns out that the group $2^3{:}L_2(7)$ possesses five maximal subgroups. One of them is the group $2^3{:}7{:}3$. Two groups of order 192 have analogous structures to those in $2^3.L_2(7)$ but they are not isomorphic. A surprising fact is that the group $2^3{:}L_2(7)$ has two non conjugate maximal subgroups of $L_2(7)$. The split and non split groups of order 1344 have been explicitly worked out in the reference[8]

As to how the groups of order 1344 we note the following possibility of applications in particle physics. The usual discrete symmetries C,P, T when acting on the bilinear Dirac fields $\overline{\psi}\Gamma_i\psi$, they generate the elementary abelian group 2^3. Since either $2^3.L_2(7)$or $2_3{:}L_2(7)$ are the extensions of 2^3 by its automophism group $L_2(7)$ it is suggestive that their subgroups can be used for the P, CP and CPT violations. We have noted in [8] that the invariance of a general Hamiltonian which is quartic in the Dirac fields under a subgroup of order 16 leads to the usual parity violating weak interaction. CP and CPT violating Hamiltonians can be constructed in similar manners.

We have also noted that [8] the Demazure-Tits subgroups DT(SU(3)), DT(G_2) and DT(SO(5)) can be naturally embedded in the finite groups of order 1344.

Another remark before we conclude is about the possibility of construction of the root system of E_8 by icosians, the quaternionic units which form the binary icosahedral subgroup of SU(2) of order 120. In an earlier publication [9] we had noted that a different pairing of the quaternionic roots of F_4 in the following manner leads to the root system of E_8.

8-dimensional representations A_1, A_2, A_3 one can constitute the root system of F_4:

A_0	A_1	A_2	A_3	
$\pm1, \pm e_1, \pm e_2, \pm e_3$	$1/2(\pm1 \pm e_1)$	$1/2(\pm1 \pm e_2)$	$1/2(\pm1 \pm e_3)$	(1)
$1/2(\pm1 \pm e_1 \pm e_2 \pm e_3)$	$1/2(\pm e_2 \pm e_3)$	$1/2(\pm e_3 \pm e_1)$	$1/2(\pm e_1 \pm e_2)$	

Here $e_i\,(i = 1,2,3)$ are the quaternionic imaginary units satisfying the relation $e_i e_j = -\delta_{ij} + \varepsilon_{ijk} e_k$, where ε_{ijk} is the Levi-Civita symbol. Using the well known Cayley-Dixon procedure, namely, pairing two sets of quaternionic roots of F_4 one can construct the octonionic roots of E_8 [5]:

$$[A_0, 0] + [0, A_0] + [A_1, A_1] + [A_2, A_3] + [A_3, A_2] \qquad (2)$$

Here we mean by $[p,q] = p + e_7 q$,where p and q are quaternions. If one represents the roots of SU(2) by ±1 then those roots which arc orthogonal to ±1 consist of pure imaginary octonions constituting the roots of E_7. Let p be the octonionic root belonging to the coset $E_8 / E_7 X SU(2)$. Then the transformations $e_i' = p e_i \bar{p}$ $(i = 1,2,\ldots,7)$ generate the derived Chevalley group $G_2(2)'$ of order 6048. With the additional transformation $e_i \to e_i (i = 1,2,3)$, $e_j \to -e_j$(j=4,5,6,7) which preserve the root system of E_7 and the octonion algebra, the automorphism group of the octonionic root system of E_7 is extended to the adjoint Chevalley Group $G_2(2)$ of order 12096 [6]. The $G_2(2)$ is one of the maximal subgroups of the Weyl group $W(E_7) \approx 2XO_7(2)$with index 240. Since $G_2(2)$ consists of matrices of det +1 it is indeed maximal in $O_7(2)$ with index 120.

2. Maximal Subgroups of $G_2(2)$ and The Groups of Order 1344

The $G_2(2)$ has four maximal subgroups which preserve the octonionic roots of maximal Lie algebras of E_7 with one exception. Details of this analysis can be found in the ref [7]. Below we give the list of the maximal subgroups of $G_2(2)$ and the corresponding Lie algebras:

Octonionic roots of Subalgebras of E_7	Maximal Subgroups of $G_2(2)$ and their orders
$E_6XU(1)$	3_+^{1+2}:8:2 , 432
SO(12)XSU(2)	$2^2.S_4$:2 , 192
$SO(8)XSU(2)^3$	$4^2.S_3$:2 , 192
SU(8)	$L_2(7)$:2 , 336

$$(A_o,0)\oplus(0,A_o)\oplus(A_1,A_2)\oplus(A_2,A_3)\oplus(A_3,A_1) \tag{3}$$

Here we define (p,q) = p +σq with $\sigma=(1-\sqrt{5})/2$. If one imposes the reduced scalar product $<p,q>=1/2(\bar{p}q+\bar{q}p)$ = a+σb → a then the set (3) is a copy of the E_8 roots represented by icosians. Half of the roots in (3) form one of the 2-dimensional representation of the finite subgroup of SU(2) of order 120. As we know that the finite subgroup of SU(2) of order 120 has another 2 dimensional representation we note that (3) can be replaced by $(A_o,0)\oplus(0,A_o)\oplus$ $(A_1,A_3)\oplus(A_3,A_2)\oplus(A_2,A_1)$ which still represents the roots of E_8 as well as half of them constitute the other 2-dimensional irreducible representation of the binary icosahedral group. As a side remark we observe that the quaternionic elements of the binary icosahedral group themselves constitute the roots of the Coxeter graph H_4 whose Coxeter group is of order 14400. By this method one can embed H_4 in the Weyl group of E_8.

References

[1] Günaydın M and Gürsey F 1973 J.Math.Phys. **14** 1651
Gürsey F 1987 Mod.Phys.Lett.**A2** 967

[2] Coxeter H S M 1946 Duke Math. J.**13** 561

[3] DuVal P. 1964 Homographies, Quaternions and Rotations, Clarendon Press, Oxford

[4] Coxeter H S M 1974 Regular Complex Polytopes, Camb.Univ.press

[5] Koca M 1986 Preprint IC/86/224 ICTP
Koca M and Ozdeş N 1989 J.Phys.A: Math.Gen.**22** 1469
Koca M 1992 J.Math.Phys **33** 497

[6] Karsch F and Koca M 1990 J.Phys.A:Math and Gen. **23** 4739

[7] Koca M and Koç R 1993 J.Phys.A:Math and Gen. **27** 2429

[8] Koca M and Koç R 1994 Octonions and the groups of order 1344, Çukurova Üniv. Preprint

[9] Koca M 1989 J.Phys. **A22** 1947
Koca M 1989 J.Phys. **A22** 4125

Spatial Geometry and the Wu-Yang Ambiguity

Daniel Z. Freedman and Ramzi R. Khuri

CERN, Theory Division, CH-1211 Geneva 23 Switzerland

Abstract: We display continuous families of SU(2) vector potentials $A_i^a(x)$ in 3 space dimensions which generate the same magnetic field $B^{ai}(x)$ (with det $B \neq 0$). These Wu-Yang families are obtained from the Einstein equation $R_{ij} = -2G_{ij}$ derived recently via a local map of the gauge field system into a spatial geometry with 2-tensor $G_{ij} = B^a{}_i B^a{}_j \det B$ and connection Γ^i_{jk} with torsion defined from gauge covariant derivatives of B.

The Wu-Yang ambiguity [1] is the phenomenon that two or more gauge inequivalent non-abelian potentials $A_i^a(x)$ generate the same field strength $F_{ij}^a(x)$. Although the original example is 3-dimensional, it was mainly the 4-dimensional case which was of past interest. Many examples of a discrete ambiguity have been exhibited, specifically two potentials A and $\bar{A}$ giving the same F. The few examples of a continuous ambiguity were degenerate in some way: for example, they were effectively 2-dimensional. In this talk we summarize previously published work [2,3] and display examples of continuous families of potentials which generate the same magnetic field

$$B^{ai} = \epsilon^{ijk}\left[\partial_j A_k^a + \frac{1}{2}\epsilon^{abc} A_j^b A_k^c\right] \tag{1}$$

In 3 dimensions there is no "algebraic obstruction" to an ambiguity. However, this fact is not sufficient to demonstrate that (1), viewed as a partial differential equation for A_j^b given B^{ai}, has multiple solutions, and it is this which we wish to explore here.

The 3-dimensional case is relevant for the Hamiltonian form of gauge field dynamics in 3+1 dimensions and especially for an attempt [2] to transform from A_i^a to B^{ai} as the basic field variables. An intermediate step is to replace the A, B system by a set of gauge-invariant spatial geometric variables, namely a metric G_{ij} and connection Γ_{ij}^k with torsion. It turns out that the information we find on the Wu-Yang ambiguity invalidates the proposed form of Hamiltonian dynamics [2]. But the geometry is valid, and it is through the geometrical equations that our Wu-Yang information is obtained.

We begin with our first example. Consider the smooth, algebraically non-singular (*i.e.* $\det B \neq 0$) magnetic field $B^{ai} = \delta^{ai}$, in Euclidean space with Cartesian coordinates x, y, z. It is easy to show explicitly that, for any real parameter β with $|\beta| > 1$, the 1-parameter family of potentials

$$A_i^a = \begin{pmatrix} \beta \pm \sqrt{\beta^2 - 1}\cos(z/\beta) & \pm\sqrt{\beta^2-1}\sin(z/\beta) & 0 \\ \pm\sqrt{\beta^2-1}\sin(z/\beta) & \beta \mp \sqrt{\beta^2-1}\cos(z/\beta) & 0 \\ 0 & 0 & 1/\beta \end{pmatrix} \tag{2}$$

all reproduce the same B^{ai}. Gauge inequivalence is demonstrated by the fact that the invariants $B^{aj} D_i B^{ak}$ depend on β and z. The particular magnetic field $B^{ai} = \delta^{ai}$ is invariant under rotations and translations of the configuration space $\mathbb{R}^3$ (the spatial rotations must be combined with a suitably chosen SO(3) gauge transformation, constant in this case, as is well known). Since (1) is also covariant, each such isometry which does not leave A_i^a invariant produces another Wu-Yang related potential. In this way one can extend the potentials displayed in (2) to a 4-parameter family, in which the wave has an arbitrary phase, $z \rightarrow z - z_0$ and direction $(0,0,1) \rightarrow \hat{k}$. (1) is covariant under diffeomorphisms. Thus examples of the Wu-Yang ambiguity automatically extend to entire orbits of the diffeomorphism group, and one can find a diffeomorphism under which the field $B^{ai} = \delta^{ai}$, which has infinite energy, transforms to a configuration $B'^{a\alpha}(y)$ which falls sufficiently fast as $y^\alpha \rightarrow \infty$ that it has finite energy.

Let us now review the spatial geometry which is the main tool used in this work. Let $B^a{}_i(x)$ denote the matrix inverse of the magnetic field of SU(2) gauge theory, and $\det B = \det B^{ai}$, which is gauge invariant. Then $G_{ij}(x) = B^a{}_i(x) B^a{}_j(x) \det B$ is a gauge invariant symmetric tensor (under diffeomorphisms). The following geometry, which obviously uses the fact that the gauge group SU(2) is also the tangent space group of a 3-manifold, emerged from the physical aim of studying the action of the electric field on gauge invariant state functionals $\psi[G]$.

The quantity $b_i^a = |\det B|^{\frac{1}{2}} B_i^a$ is essentially a frame for G_{ij}. One may apply a Yang-Mills covariant derivative and define a quantity Γ_{ij}^k as follows:

$$D_i b_j^a = \Gamma_{ij}^k b_k^a . \tag{3}$$

It then follows immediately that Γ_{ij}^k transforms as a connection under diffeomorphisms and that it is well defined in terms of A_i^a and B^{ai} for all non-singular magnetic fields, $\det B \neq 0$. If we multiply (3) by b_k^a and symmetrize in j, k, we easily derive

$$\partial_i G_{jk} - \Gamma_{ij}^l G_{lk} - \Gamma_{ik}^l G_{jl} = 0 . \tag{4}$$

This means that Γ is a metric compatible connection for G, and can be written as

$$\Gamma_{ij}^k = \dot{\Gamma}_{ij}^k(G) - K_{ij}{}^k \tag{5}$$

where $\dot{\Gamma}$ is the Christoffel connection and K is the contortion tensor, which is antisymmetric in the last pair of indices, $K_{ijk} = -K_{ikj}$.

Next one studies the anti-symmetrized Yang-Mills derivative of (3)

$$[D_i, D_j]b_k^a = b_l^a \partial_i \Gamma_{jk}^l + \Gamma_{jk}^l \partial_i b_l^a - (i \leftrightarrow j) \tag{6}$$

One uses (3) again to find that the curvature tensor of Γ appears. On the left one uses the gauge field Ricci identity and a standard formula for the inverse of a 3×3 matrix. The result is

$$R_{kij}^l = \delta_j^l G_{ik} - \delta_i^l G_{jk} \ , \tag{7}$$

the statement that the spatial geometry associated with the gauge fields A, B is maximally symmetric. However, in 3 dimensions, with or without torsion, the Rieman tensor R_{kij}^l is fully determined by its Ricci contraction $R_{kj} = \delta_l^i R_{kij}^l$, so that no information is lost by restricting consideration to the contracted form of (7), namely

$$R_{ij}(\Gamma) = -2G_{ij} \tag{8}$$

which defines an Einstein geometry with torsion. One may show using the second Bianchi identity of curvatures with torsion, that an integrability condition for (8) is that the contortion tensor is traceless, ${K_{kj}}^k = 0$, and that this is also a direct requirement of the gauge field Bianchi identity, $D_i B^{ai} = 0$, applied to the definition (3) of Γ.

The discussion above defines the forward map from Yang-Mills fields A and B, always related by (1), to geometric variables G and Γ defined by explicit local formulas above. The gauge field Ricci and Bianchi identities then imply that G and Γ are related by the Einstein condition (8) with traceless contortion. The fundamental reason for the Einstein geometry is that the magnetic field is simultaneously the curvature (1) of the gauge connection A and also essentially the frame of the spatial geometry.

One may also ask about the inverse map from tensor $G_{ij}(x)$ and connection $\Gamma_{ij}^k(x)$ on $\mathbb{R}^3$ to gauge fields. Suppose that a frame b_i^a, with $\det b > 0$, is constructed for G_{ij} by any standard method, then (3) can be written out as

$$\partial_i b_j^a - \Gamma_{ij}^k b_k^a + \epsilon^{abc} A_i^b b_j^c = 0 \ . \tag{9}$$

This is just the "dreibein postulate" with A essentially the spin connection, and one can solve for A, obtaining

$$A_i^a = -\frac{1}{2}\epsilon^{abc} b^{bj} \left(\partial_i b_j^c - \Gamma_{ij}^k b_k^c\right) \tag{10}$$

while the magnetic field is defined from the inverse frame by

$$B^{ai}(x) = |\det G_{jk}|^{\frac{1}{2}} b^{ai}. \tag{11}$$

Thus given a frame one obtains the magnetic field from (11), while both b and Γ are required to define the potential via (10). Since the frame is unique up to a local $SO(3)$ rotation, these maps define A and B uniquely up to an $SU(2)$ gauge transformation. Furthermore, A and B defined in this way satisfy the gauge theory relation (1) if Γ and G satisfy (8).

Thus the gauge theory Wu-Yang ambiguity will appear whenever the Einstein equation (8) viewed as a partial differential equation for K, given G, has multiple solutions. To investigate this it is useful to use the representation

$$K^i{}_{jk} = \epsilon_{jkn} S^{ni} \frac{1}{|\det G|^{1/2}} \tag{12}$$

which automatically satisfies the antisymmetry and tracelessness requirements if S^{ni} is a symmetric tensor. When (8) is expanded out using (5) and (12), one finds that the Einstein equation is equivalent to

$$\frac{\epsilon^{jkl}}{|\det G|^{1/2}} \dot{\nabla}_k S_{li} - \left(S^j_k S^k_i - S^k_k S^j_i\right) = \dot{R}^j_i + 2\delta^j_i . \tag{13}$$

In (13) $\dot{\nabla}_k$ indicates a spatial covariant derivative with Christoffel connection $\dot{\Gamma}$ and $\dot{R}_{ij}$ is the conventional symmetric Ricci tensor. The $\epsilon\dot{\nabla}S$ term is non-symmetric, so that (13) comprises 9 equations for the 6 components of S_{ij}. However it was shown explicitly [2] that there is a Bianchi identity which imposes 3 constraints on the 9 equations, so there is no reason to think that (13) is an overdetermined system. From (5), (10) and (12), A can be expressed in terms of S as

$$A^a_i = -\frac{1}{2}\epsilon^{abc} b^{bj} \dot{\nabla}_i b^c_j - b^{ak} S_{ki}. \tag{14}$$

Our approach to the Wu-Yang ambiguity is to take an input metric $G_{ij}(x)$ and study the solutions of (13) for the torsion $S_{ij}(x)$. It is not clear why this should be a simpler method than to study directly whether (1) has multiple solutions for A, given B. Perhaps it is because an equation for the 6 components of S_{ij} is simpler to handle than an equation for the 9 components of A^a_i, but it may just be an historical accident that has led us to approach the Wu-Yang ambiguity via the spatial geometry.

Before beginning to study applications of (13), it is perhaps useful to note that (3) indicates that Γ is completely determined by first covariant derivatives $D_i B^{aj}$ of the magnetic field. It then follows from properties of the inverse map discussed above that there is no Wu-Yang ambiguity for the potential A^a_i, if we require that both B and DB are preserved[1]. In 4 dimensional $SU(2)$ gauge theory, the field strength and its first two covariant derivatives determine the potential locally uniquely.

The Wu-Yang ambiguity indicates that the potential $A^a_i(x)$ contains gauge invariant information beyond that in the magnetic field $B^{ai}(x)$. Therefore the change of field variable $A^a_i \to B^{ai}$ which was the basis of the version of gauge invariant Hamiltonian dynamics presented in [2] is invalid. The discrete 2:1 ambiguity envisaged there could be handled, but it is probably impossible to deal with a continuous ambiguity without serious revision of the proposal.

It is the tensor $G_{ij}(x) = \delta_{ij}$ that corresponds to the magnetic field $B^{ai} = \delta^{ai}$, and it can be seen without difficulties that the torsion solutions of (13) are related

[1] Actually it is sufficient to require that $D_{[i} b^a_{j]}$ is preserved, because this determines the torsion tensor from (3).

in this simple case to the potentials of (2) by $A_i^a(z) = -S_{ai}(z)$. Note that at $\beta = 1$, the solution (2) reduces to $S_{ij} = \delta_{ij}$. We found the family of solutions (2) by first linearizing about $S_{ij} = \delta_{ij}$ and using Fourier analysis to find linearized modes of wave number $k^2 = 1$. This led us to investigate the single variable ansatz $S_{ij}(z)$ which reduces (13) to a non linear system of ordinary differential equations. Some fiddling then led to (2), which is unique within this ansatz (except for translation $z \to z - z_0$). One can show that the only spherically symmetric solutions of (13) for input $G_{ij} = \delta_{ij}$ are the solutions $S_{ij} = \pm\delta_{ij}$. There is a heuristic argument that the potentials displayed in (2) together with those obtained from them by translation and rotation are the only potentials for the field $B^{ai} = \delta^{ai}$ which continuously limit to potentials $\bar{A}_i^a = \delta_i^a$ with $\beta = 1$ in (2). The reason is that one can show using the Fourier transform that the set of linear perturbations about $\bar{A}_i^a$ obtained in the $\beta \to 1$ limit of our Wu-Yang family are complete.

Another set of Wu-Yang examples emerges from the 3-dimensional hyperbolic metrics

$$ds^2 = \frac{1}{c^2 z^2}\left(dx^2 + dy^2 + dz^2\right) \tag{15}$$

for which $R_{ij} = -2c^2 G_{ij}$. One may anticipate that the case $c^2 = 1$ is especially simple because the right side of (13) vanishes. It turns out that one can also make the $\epsilon\dot{\nabla}S$ and SS terms vanish separately. There is no integrability constraint when $\dot{\nabla}_j$ is applied to the former condition, while the second condition implies that S_{ij} is a rank 1 dyadic matrix. With this structure in view one can easily find that within the two variable ansatz $S_{ij}(z,x)$, there is the family of solutions

$$S_{ij}(z,x) = \delta_{i1}\delta_{j1}\frac{1}{z}h(x) \tag{16}$$

which involves an arbitrary function of the variable x. The solution can be rotated by an angle θ in the x, y plane to obtain

$$\begin{aligned} S_{ij} &= \frac{1}{z}h(x\cos\theta + y\sin\theta)V_i V_j, \\ V_i &= (\cos\theta, \sin\theta, 0). \end{aligned} \tag{17}$$

We have not studied the application of the full $SO(2,1)\times SO(2,1)$ isometry group of the metric (15), but more solutions seem likely. In this frame the magnetic field is given simply by $B^{ai} = \delta^{ai}/z^2$ while the gauge potential corresponding to (16) is obtained from (17):

$$A_1^1 = -h(x), \quad A_2^1 = -A_1^2 = 1/z, \tag{18}$$

with the rest of the components vanishing. In this frame, the magnetic field B^{ai} is singular on the plane $z = 0$. It is straightforward to transform our configuration to a frame in which both the magnetic field and gauge potential are manifestly regular over all of $\mathbb{R}^3$ (see [3]).

When $c^2 \neq 1$ the full nonlinear equations are very difficult to handle, so we restrict ourselves to a perturbative expansion about the symmetric solution

$\bar{S}_{ij} = \sqrt{1-c^2}G_{ij}$ of (13) by setting $S_{ij} = \bar{S}_{ij} + \hat{\Sigma}_{ij}$. The perturbation $\hat{\Sigma}_{ij}$ satisfies the linear equation

$$(cz)\epsilon^{jkl}\dot{\nabla}_k\hat{\Sigma}_{li} + \sqrt{1-c^2}(\hat{\Sigma}_{ji} + \hat{\Sigma}_{kk}\delta_{ji}) = 0. \tag{19}$$

(the placement of the j index reflects the removal of the conformal factor.) The 9 equations for the 6 components of $\hat{\Sigma}$ cannot all be independent, and the fact that we find a consistent solution below is a practical test of the exact Bianchi identity satisfied by (13). Note that the ij contraction of (19) immediately tells us that the trace $\hat{\Sigma}_{kk} = 0$.

Because of the x-translation symmetry of the metric (15) we look for a solution of the form $\hat{\Sigma}_{ij}(z,x) = \Sigma_{ij}(z,k)e^{ikx}$. The 9 equations of (19) can be manipulated to obtain a second order differential equation for the component Σ_{23}

$$\left(z^2\frac{d^2}{dz^2} + z\frac{d}{dz} - k^2z^2 + \frac{1-c^2}{c^2}\right)\Sigma_{23} = 0, \tag{20}$$

while the other components are related to Σ_{23} by

$$\begin{aligned}
\Sigma_{33} &= \frac{-ikc}{\sqrt{1-c^2}}z\Sigma_{23},\\
\Sigma_{13} &= \frac{c}{\sqrt{1-c^2}}z\frac{d}{dz}\Sigma_{23},\\
\Sigma_{12} &= \frac{i}{k}(\frac{d}{dz} - \frac{1}{z})\Sigma_{23},\\
\Sigma_{11} &= \frac{-ic}{k\sqrt{1-c^2}}z\frac{d^2}{dz^2}\Sigma_{23},\\
\Sigma_{22} &= -(\Sigma_{11} + \Sigma_{33}).
\end{aligned} \tag{21}$$

Note that (20) is the differential equation for Bessel functions of imaginary argument ikz and index $p = \sqrt{(c^2-1)/c^2}$ which is also imaginary when $c^2 < 1$ and the symmetric torsion $\bar{S}_{ij}$ is real.

Note that the wave number k of the linear perturbation is not restricted in contradistinction to the flat metric where, as can be seen from the small amplitude limit $\beta \to 1$ in (2), the wave number $k = 1$ is required. This means that the general real superposition

$$\int dk\varphi(k)\Sigma_{ij}(kz)e^{ikx} + \text{c.c.} \tag{22}$$

is also a solution, so we have the freedom of an arbitrary function at the linear level. We expect that (22) can be used as the "input" to the system of differential equations determining second and higher order perturbative solutions of (13), and that the functional freedom of $\varphi(k)$ remains. Thus the qualitative picture of the torsion solutions for the hyperbolic metrics for all values of c is that they contain an arbitrary function of a single variable and the additional parametric freedom obtained from isometries. The case $c = 1$ is special only because exact solutions can be easily obtained. A similar analysis to the above follows for the case of $2+1$ product metrics, in which functional freedom is again expected to persist in higher order perturbative solutions.

In summary, what we have discussed in this talk are several examples of a continuous Wu-Yang ambiguity for $SU(2)$ gauge fields in 3 dimensions and a new technique, namely the Einstein space condition (13) for obtaining such field configurations. It is intriguing to ask about the systematics of the ambiguity; namely what properties of the B-field determine the degree of ambiguity in the associated potentials A. Our examples provide at least a limited view of this systematics. Certainly an ambiguity is generated whenever there is a symmetry transformation of B which acts nontrivially on A, but this is not enough to explain the free parameter β in (2), nor the arbitrary functions such as $F(x)$ in the example (16-18) or in the linear solutions. Gauge field topology does not seem to be the issue here for two reasons. First of all the ambiguity can be exhibited in any compact subset of the configuration space $\mathbb{R}^3$. Second, if we are given in some gauge a Wu-Yang family with suitable behavior at spatial infinity one can apply a gauge transformation to change the topological class at will. Of course one does expect that, except for singularities of the map (1), the degree of ambiguity in A will not change as the parameters of B are smoothly varied. Our examples appear to be consistent with this requirement, although the discrete ambiguity found when $\dot{R}_{ij} = 0$ must be understood as a singular limit of the case of non-zero curvature. It is interesting that the Riemannian curvature $\dot{R}_{ij}$ of the metric G_{ij} obtained from B plays a role both in the ease of obtaining solutions for A and in the qualitative nature of the ambiguity.

This talk by R.K. is dedicated to the memory of Feza Gursey, teacher, mentor and dear friend. Rarely has an individual scientist so positively influenced the physics community in either personal or professional endeavours. Feza possesses the unique characteristic of having done so in both. Feza will always be greatly missed by family, friends, students and colleagues.

References

1. T.T. Wu and C.N. Yang: Phys. Rev. D **12** 3845 (1975).
2. D.Z. Freedman, P.E. Haagensen, K. Johnson and J.I. Latorre: hep-th 9309045 CERN-TH.7010/93, CTP #2238 (1993).
3. D.Z. Freedman and R.R. Khuri: Phys. Lett. B **329** 263 (1994).

The Classification of Third Order PDE with Respect to Symmetry

Atalay Karasu

Physics Department, Middle East Technical University, Ankara, Turkey

Abstract. We show that the the integrable subclassess of the equations $q_t = f(x,t)q_3 + H(x,t,q,q_1)$ are the same as the integrable subclassess of the equations $q_t = q_3 + F(q,q_1)$.

Recently ,there has been increasing interest in classifying integrable nonlinear partial differential equations [1, 2, 4, 5]. Up to now the complete classification has been done for some autonomous evolution equations[1, 3, 4, 7]. There are some partial attempts of the classification of nonautonomous evolution equations [1, 6, 8, 5, 9]. In this work we show that the most general nonautonomous evolution equations of the form $q_t = f(x,t)q_3 + H(x,t,q,q_1)$ are transformable to the autonomous equations $q_t = q_3 + F(q,q_1)$. It is well known that under the symmetry classification the autonomous evolution equations $q_t = q_3 + F(q,q_1)$, up to coordinate transformations, give the following integrable equations

$$q_t = q_3 + (aq^2 + bq)q_1, \tag{1}$$

$$q_t = q_3 + cq_1^3 + e(q)q_1, \qquad 3\frac{d^3e}{dq^3} + 4c\frac{de}{dq} = 0, \tag{2}$$

where a,b,c are constans and e depends only on q and $q_i = (\frac{\partial}{\partial x})^i q$

Now we shall classify third order nonautonomous evolution equations of the form

$$q_t = f(x,t)q_3 + H(x,t,q,q_1), \tag{3}$$

using the Gürses' approach [10] based on the compatibility of the linearized equation and an eigenvalue equation. Here in this work,instead of representing whole classification of (3),we shall give a few examples to prove our assertion. The complete classification of (3) is given in [11]. We divide the classification procedure into three cases:

i) f depends only on t. One of the integrable subclass is

$$q_t = v^3 q_3 + (\frac{hw^2}{2}q^2 + awhvq + bhv)q_1 + (\frac{\dot{h}}{2h} - \frac{\dot{v}}{v})xq_1 - (\frac{\dot{w}}{w})q \tag{4}$$

where $f(t) = v(t)^3$, h, w depend only on t and a, b are constants. The dot appearing over a quantity denotes the t derivatives. First we observe that equation

(4) is transformed into the equation (2) through the transformation

$$q = w^{-1}u(\xi, \tau),$$
$$\xi = x\frac{h^{1/2}}{v}, \qquad \tau = \int^t h^{3/2}dt' \tag{5}$$

ii) f depends only on x. In this case one integrable class turns out to be simple

$$q_t = q_3 + q_1^2 + ax + b, \tag{6}$$

where a and b are constants. Now,differentiate equation (5) with respect to x and substitute $q = u_1$ and use transformations

$$x = \xi - \frac{at^2}{2}, \qquad t = \tau \tag{7}$$

then equation (5) is transformed into equation (1).

iii) f depends on both x and t. Here we have a simple equation

$$q_t = v^3 q_3 + [\frac{a}{2v}q^2 + \frac{3}{2}v(v_1^2 - 2vv_2)](q_1 - \frac{v_1}{v}q), \tag{8}$$

where a is a constant. Here $f(x,t) = v(x,t)^3$ and $v(x,t)$ satisfies Harry Dym equation

$$v_t = v^3 v_3. \tag{9}$$

Equation (8) together with (9) is transformed into the equation (2). To show this, introduce a new dynamical variable z related to q by $q = vz$ then we have

$$z_t = v^3 z_3 + 3v^2 v_1 z_2 + (\frac{a}{2}vz^2 + \frac{3}{2}vv_1^2)z_1. \tag{10}$$

Now let us perform the following transformation

$$\xi = \int^x \frac{dx'}{v(x',t)} \qquad , \tau = t. \tag{11}$$

It is straightforward to show that under this transformation equation (10) belongs to equation (2). As a conclusion, this work shows that there are no generic integrable nonautonomous equations (3) and given an exceptional equation of the form (3) there exists a coordinate transformation which transforms it either into equation (1) or (2).

This work is partially supported by the Scientific and Technical Research Council of Turkey (TUBITAK).

References

[1] Mikhailov, A.V. , Shabat,A.B., Sokolov,V.V. The Symmetry Approach to Classification of Integrable Equations. (What is Integrability) Springer-Verlag,Berlin (1991).

[2] Ibragimov N.H. Transformation Groups Applied to Mathematical Physics Reidel, Boston (1985)

[3] Ibragimov,N.H.,Shabat,A.B. Evolutionary Equations with Nontrivial Lie-Backlund Groups. Functional Anal. Appl. **14** (1980),19-28

[4] Fokas,A.S., Symmetries and Integrability. SIAM **77** (1987), 253-297

[5] Olver, P.J. Application of Lie Groups to Differential Equations. Springer- Verlag, Berlin Heidelberg (1993)

[6] Svinolupov, S.I., Sokolov,V.V. Weak Nonlocalities in Evolution Equations. Math.Notes,**48** (1990),1234-1239

[7] Svinolupov, S.I., Sokolov, V.V. Evolution Equations with Nontrivial Conservative Laws. Functional Anal.Appl. **16** (1982), 317-319

[8] Svinolupov, S.I., Russ. Math. Surv. **40** (1985) 241-242

[9] Hernández Heredero,R., Sokolov,V.V., Svinolupov, S.I. Toward the Classification of Third Order Integrable Evolution Equations. J.Phys.A:Math.Gen.**27** (1994), 4557- -4569

[10] Gürses, M., Karasu, A., Satır, A. Linearization as a New Test for Integrability (Nonlinear Evolution Equations and Dynamical Systems) World Scientific, Singapura (1992)

[11] Gürses, M., Karasu, A., Variable Coefficient Third Order KdV type of Equations summited for publication,1994.

Symmetries of Large Systems of Differential Equations

Louis Marchildon

Département de physique, Université du Québec
Trois-Rivières, Qc, Canada G9A 5H7
e-mail: marchild@uqtr.uquebec.ca

Abstract. Hereman, Marchildon and Grundland have earlier reported on an investigation of Lie point symmetries of two systems of nonlinear partial differential equations, both representing classical field theories. Determining equations (more than 200 of them) associated with the first system (coupled electromagnetic and complex scalar fields) were solved completely. Here I develop the general solution of determining equations associated with the second system (coupled electromagnetic and Weyl spinor fields). The method used to solve these determining equations can likely be adapted to other related problems.

Keywords. Symmetry, Differential Equations, Classical Fields

1 Introduction

Consider a general system of nth order nonlinear partial differential equations, with x denoting the independent and U the dependent variables:

$$\Delta_\nu\left(x, U, \partial U, \dots, \partial^{(n)} U\right) = 0 . \tag{1}$$

By a symmetry of the system, we shall mean any transformation of the independent and dependent variables that maps an arbitrary solution of the system into a solution. The infinitesimal generator of a symmetry transformation can be written as

$$V = \sum_\mu H^\mu(x, U) \frac{\partial}{\partial x^\mu} + \sum_a \Phi_a(x, U) \frac{\partial}{\partial U_a} , \tag{2}$$

where μ and a label the independent and dependent variables, respectively. We point out that, in this paper, summations over repeated indices are *not* implicit.

The infinitesimal generators can be obtained by a well-defined procedure, which goes back to Lie and is explained in detail by Olver.[1] It involves computing the so-called n^{th} prolongation of V:

$$\mathrm{pr}^{(n)} V = V + \sum_{a,\{J\}} \Phi_a^{\{J\}}\left(x, U, \partial U, \dots, \partial^{(n)} U\right) \frac{\partial}{\partial U_a^{\{J\}}} . \tag{3}$$

Here $\{J\}$ is a multi-index denoting all partial derivatives (up to n^{th} order) of the dependent with respect to the independent variables. The $\Phi_a^{\{J\}}$ are functions which can be written in closed form.[1] The nth prolongation often involves a very large number of terms, but its significance is simple: Given the fact that V generates a transformation of x^μ and U_a, $\mathrm{pr}^{(n)}V$ generates the corresponding transformation of x^μ, U_a and $U_a^{(J)}$. With rather weak restrictions on Eqs. (1), one can show that necessary and sufficient conditions for V to generate a symmetry of (1) is that

$$\left[\mathrm{pr}^{(n)} V\right]\left[\Delta_\nu\right] = 0 \quad \text{whenever} \quad \Delta_\nu = 0 \,. \tag{4}$$

These so-called determining equations are a system of nth order *linear* partial differential equations for the H^μ and Φ_a.

2 Two-component spinor in an electromagnetic field

The method just outlined has been applied to two systems of classical field equations.[2] The first system, involving an electromagnetic field coupled with a complex scalar field, was investigated in detail. For the second system, involving an electromagnetic field coupled with a two-component complex spinor field, only results were given. That system is here described in more detail.

The Lagrangian of an electromagnetic field minimally coupled with a two-component complex spinor is given by

$$\mathfrak{L} = -\frac{1}{4}\sum_{\mu,\nu}\left\{\partial_\mu A_\nu - \partial_\nu A_\mu\right\}\left\{\partial^\mu A^\nu - \partial^\nu A^\mu\right\} + i\sum_\mu \psi^\dagger \sigma^\mu \left\{\partial_\mu + ieA_\mu\right\}\psi \,. \tag{5}$$

Here σ^1, σ^2 and σ^3 are Pauli matrices, and σ^4 is the identity in two dimensions. There are four independent variables x^μ ($\mu = 1 \ldots 4$), the Minkowski metric having signature (-1,-1,-1,+1). There are eight dependent variables A^μ, U_a ($a = 5 \ldots 8$), with

$$\psi = \begin{pmatrix} U_5 + iU_6 \\ U_7 + iU_8 \end{pmatrix} . \tag{6}$$

The field equations are given by

$$\begin{gathered} \sum_\mu \partial_\mu\left\{\partial^\mu A^\nu - \partial^\nu A^\mu\right\} - e\psi^\dagger \sigma^\nu \psi = 0 \,, \\ i\sum_\mu \sigma^\mu\left\{\partial_\mu + ieA_\mu\right\}\psi = 0 \,. \end{gathered} \tag{7}$$

3 Determining equations

The computation of the nth prolongation is entirely algorithmic. It was effected by means of the symbolic package SYMMGRP.MAX, which also calculates determining equations.[3] In the case of Eqs. (7), the package eventually yielded 200 determining equations for the functions

$$H^\mu\left(x^\lambda\right) , \quad \Phi^\mu\left(x^\lambda, A^\nu\right) \quad \text{and} \quad \Phi_a\left(x^\lambda, U_b\right) . \tag{8}$$

The determining equations are not manifestly Lorentz covariant. Nevertheless, it is very useful to write them, as far as possible, in relativistic notation. There are a number of equations that involve H^μ and Φ^ν only. After elimination of redundant equations and minor rearrangements, they are given by

$$\frac{\partial^2\Phi^\mu}{\partial A^\kappa \partial A^\lambda} = 0 \; ; \quad \frac{\partial H^\mu}{\partial x^\mu} - \frac{\partial H^\nu}{\partial x^\nu} = 0 = \frac{\partial \Phi^\mu}{\partial A^\mu} - \frac{\partial \Phi^\nu}{\partial A^\nu} \; ; \tag{9}$$

$$\frac{\partial \Phi^\mu}{\partial A_\nu} + \frac{\partial H^\nu}{\partial x_\mu} = 0 = \frac{\partial \Phi^\mu}{\partial A_\nu} - \frac{\partial H^\mu}{\partial x_\nu} \, , \quad \mu \neq \nu \; ; \tag{10}$$

$$2\,\frac{\partial^2\Phi^\mu}{\partial A_\lambda \partial x_\nu} - \frac{\partial^2\Phi^\nu}{\partial A_\lambda \partial x_\mu} + \frac{\partial^2 H^\nu}{\partial x_\lambda \partial x_\mu} = 0 \, , \quad \begin{matrix} \lambda \neq \mu \\ \nu \neq \mu \end{matrix} \; ; \tag{11}$$

$$\frac{\partial^2\Phi^\kappa}{\partial x^\kappa \partial A^\kappa} + \frac{\partial^2\Phi^\lambda}{\partial x^\lambda \partial A^\kappa} + \frac{\partial^2\Phi^\mu}{\partial x^\mu \partial A^\kappa} - \frac{\partial^2 H^\nu}{\partial x^\nu \partial x^\kappa} = 0 \, , \quad \kappa, \lambda, \mu, \nu \neq \; ; \tag{12}$$

$$2\,\frac{\partial^2\Phi^\nu}{\partial x_\mu \partial A^\nu} - \frac{\partial^2\Phi^\mu}{\partial x_\nu \partial A^\nu} - \frac{\partial^2 H^\mu}{\partial x_\kappa \partial x^\kappa} - \frac{\partial^2 H^\mu}{\partial x_\lambda \partial x^\lambda} - \frac{\partial^2 H^\mu}{\partial x_\mu \partial x^\mu} = 0 \, , \quad \kappa, \lambda, \mu, \nu \neq . \tag{13}$$

Other equations involve H^μ and Φ_a ($a = 5 \ldots 8$):

$$\frac{\partial \Phi_5}{\partial U_8} = \frac{1}{2}\frac{\partial}{\partial x^2}\left\{H^4 + H^3\right\} = -\frac{\partial \Phi_6}{\partial U_7} \; ; \quad \frac{\partial \Phi_5}{\partial U_7} = \frac{1}{2}\frac{\partial}{\partial x^1}\left\{H^4 + H^3\right\} = \frac{\partial \Phi_6}{\partial U_8} \; ; \tag{14}$$

$$\frac{\partial \Phi_7}{\partial U_6} = \frac{1}{2}\frac{\partial}{\partial x^2}\left\{H^3 - H^4\right\} = -\frac{\partial \Phi_8}{\partial U_5} \; ; \quad \frac{\partial \Phi_7}{\partial U_5} = \frac{1}{2}\frac{\partial}{\partial x^1}\left\{H^4 - H^3\right\} = \frac{\partial \Phi_8}{\partial U_6} \; ; \tag{15}$$

$$\frac{\partial \Phi_8}{\partial U_7} + \frac{\partial \Phi_5}{\partial U_6} = \frac{\partial H^2}{\partial x^1} \; ; \quad \frac{\partial \Phi_7}{\partial U_7} - \frac{\partial \Phi_5}{\partial U_5} = -\frac{\partial H^4}{\partial x^3} \; ; \tag{16}$$

$$\frac{\partial \Phi_8}{\partial U_7} + \frac{\partial \Phi_7}{\partial U_8} = 0 = \frac{\partial \Phi_5}{\partial U_6} + \frac{\partial \Phi_6}{\partial U_5} \; ; \tag{17}$$

$$\frac{\partial \Phi_8}{\partial U_8} - \frac{\partial \Phi_7}{\partial U_7} = 0 = \frac{\partial \Phi_6}{\partial U_6} - \frac{\partial \Phi_5}{\partial U_5} \, . \tag{18}$$

Finally, there are equations involving H^μ, Φ^ν and Φ_a:

$$\begin{aligned} &\psi^\dagger \sigma_\nu \Xi + \Xi^\dagger \sigma_\nu \psi + \sum_\mu \psi^\dagger \sigma_\mu \psi \frac{\partial H^\mu}{\partial x^\nu} + \psi^\dagger \sigma_\nu \psi \left(\frac{\partial H^\nu}{\partial x^\nu} - \frac{\partial \Phi^\nu}{\partial A^\nu} \right) \\ &\quad = \frac{1}{e}\left(\sum_\mu \frac{\partial^2 \Phi_\nu}{\partial x_\mu \partial x^\mu} - \frac{\partial}{\partial x^\nu} \sum_\mu \frac{\partial \Phi^\mu}{\partial x^\mu} \right), \end{aligned} \tag{19}$$

and

$$\begin{aligned} &\sum_\mu A^\mu \sigma_\mu \Xi + \sum_\mu \Phi^\mu \sigma_\mu \psi + \sum_\nu \left(\partial_\nu H^4\right) \sigma^\nu \left(\sum_\mu A^\mu \sigma_\mu \psi \right) \\ &\quad + i \left(\mathrm{Im}\left[\sum_\mu A^\mu \psi^\dagger \sigma_\mu \nabla \right] \right) \Xi = \frac{i}{e} \sum_\mu \sigma_\mu \partial^\mu \Xi \, , \end{aligned} \tag{20}$$

where we have introduced the following definitions:

$$\Xi = \begin{pmatrix} \Phi_5 + i\Phi_6 \\ \Phi_7 + i\Phi_8 \end{pmatrix}, \qquad \nabla = \begin{pmatrix} \dfrac{\partial}{\partial U_5} + i\dfrac{\partial}{\partial U_6} \\ \dfrac{\partial}{\partial U_7} + i\dfrac{\partial}{\partial U_8} \end{pmatrix}. \tag{21}$$

4 Solution of determining equations

As shown in Ref. (2), the most general solution of Eqs. (9) - (13) is given by

$$\begin{aligned} \Phi^\mu &= \sum_\nu f^\mu{}_\nu(x^\lambda) A^\nu + f(x^\lambda) A^\mu + f^\mu(x^\lambda) , \\ \frac{\partial H^\mu}{\partial x^\nu} &= f^\mu{}_\nu + (C - f)\, \delta^\mu{}_\nu , \end{aligned} \tag{22}$$

where $f_{\mu\nu} = - f_{\nu\mu}$, $\delta^\mu{}_\nu$ is the Kronecker delta, the f^μ are 4 arbitrary functions, C is an arbitrary constant and

$$f = -c_{11} - 2xc_{12} - 2yc_{13} - 2zc_{14} + 2tc_{15} , \tag{23}$$

$$\begin{aligned} f^1{}_2 &= c_5 - 2c_{12}y + 2c_{13}x , \quad & f^1{}_3 &= c_6 - 2c_{12}z + 2c_{14}x , \\ f^1{}_4 &= c_8 + 2c_{12}t - 2c_{15}x , \quad & f^2{}_3 &= c_7 - 2c_{13}z + 2c_{14}y , \\ f^2{}_4 &= c_9 + 2c_{13}t - 2c_{15}y , \quad & f^3{}_4 &= c_{10} + 2c_{14}t - 2c_{15}z . \end{aligned} \tag{24}$$

We have written x, y, z and t for x^1, x^2, x^3 and x^4; The c_i stand for arbitrary constants.

With some work, one can now show that Eqs. (14) - (19) imply that

$$\begin{aligned} \Phi_5 = \frac{1}{2}\left\{f^4{}_3 - 2C + 3f\right\} U_5 + \frac{1}{2}\left\{f^2{}_1 + 2G\right\} U_6 \\ + \frac{1}{2}\left\{f^4{}_1 + f^3{}_1\right\} U_7 + \frac{1}{2}\left\{f^4{}_2 + f^3{}_2\right\} U_8 , \end{aligned} \tag{25}$$

$$\begin{aligned} \Phi_6 = \frac{1}{2}\left\{f^4{}_3 - 2C + 3f\right\} U_6 - \frac{1}{2}\left\{f^2{}_1 + 2G\right\} U_5 \\ + \frac{1}{2}\left\{f^4{}_1 + f^3{}_1\right\} U_8 - \frac{1}{2}\left\{f^4{}_2 + f^3{}_2\right\} U_7 , \end{aligned} \tag{26}$$

$$\begin{aligned} \Phi_7 = \frac{1}{2}\left\{-f^4{}_3 - 2C + 3f\right\} U_7 + \frac{1}{2}\left\{-f^2{}_1 + 2G\right\} U_8 \\ + \frac{1}{2}\left\{f^4{}_1 - f^3{}_1\right\} U_5 - \frac{1}{2}\left\{f^4{}_2 - f^3{}_2\right\} U_6 , \end{aligned} \tag{27}$$

$$\begin{aligned} \Phi_8 = \frac{1}{2}\left\{-f^4{}_3 - 2C + 3f\right\} U_8 + \frac{1}{2}\left\{f^2{}_1 - 2G\right\} U_7 \\ + \frac{1}{2}\left\{f^4{}_1 - f^3{}_1\right\} U_6 + \frac{1}{2}\left\{f^4{}_2 - f^3{}_2\right\} U_5 , \end{aligned} \tag{28}$$

where G is an arbitrary function of x^μ. Substituting (22) - (28) into (20) yields

$$C = 0 , \quad f^\mu = \frac{1}{e}\frac{\partial G}{\partial x_\mu} . \tag{29}$$

Integration of derivatives of H^μ gives, finally, four more arbitrary constants c_1, c_2, c_3 and c_4.[2]

It is not difficult to see that c_1, c_2, c_3 and c_4 correspond to space-time translations; that c_5, c_6, c_7, c_8, c_9 and c_{10} correspond to Lorentz transformations; that c_{11} corresponds to dilatations; that c_{12}, c_{13}, c_{14} and c_{15} correspond to special conformal transformations; and that G corresponds to gauge transformations. Provided that, as is likely, Eqs. (7) are nondegenerate in the sense of Ref. (1), this constitutes a proof that the most general symmetry transformations of Eqs. (7) (excluding discrete transformations) are the product of general conformal transformations and local gauge transformations.

We should remark in closing that, although determining equations in general do not manifestly share the symmetries of the original equations, it is nevertheless very useful to write the former in a way that takes advantage of the symmetries of the latter.

It is a pleasure to thank W. Hereman and A.M. Grundland for sustained collaboration on the problems here discussed. This work was supported by the Natural Sciences and Engineering Research Council of Canada.

References

[1] P.J. Olver, *Applications of Lie Groups to Differential Equations* (Springer-Verlag, New York, 1986).

[2] W. Hereman, L. Marchildon and A.M. Grundland, *Proc. of the XIXth International Colloquium on Group Theoretical Methods in Physics*, M.A. del Olmo, M. Santander and J. Mateos Guilarte, Eds. (Anales de Física, Monografías 1, CIEMAT/Real Sociedad Española de Física, Madrid), Vol. 1, pp. 402-405.

[3] B. Champagne, W. Hereman and P. Winternitz, Computer Phys. Comm. **66**, 319 (1991).

q-Deformed Path Integral and Generalized Grassmann Variables*

M. Chaichian ** *and* **A.P.Demichev**†

Abstract

Using differential and integral calculi on the quantum plane which are invariant with respect to quantum inhomogeneous Euclidean group $E(2)_q$, we construct path integral representation for the quantum mechanical evolution operator kernel of q-oscillator.

Introduction. Deformations of different groups and algebras [1] have attracted great attention during the last few years. These mathematical objects were originated in quantum inverse scattering method [2] and have found many interesting and important physical applications. A partial list of possible applications in quantum field theory and particle physics includes: *i)* q-deformations of the space-time symmetry groups and Lie algebras (see e.g. [3] - [4] and refs. therein) with the hope to obtain naturally regularized field theory; *ii)* attempts in q-deformations of internal (gauge) symmetries [5]; *iii)* q-deformations of Heisenberg algebra of raising and lowering operators (see e.g. [6] and refs. therein).

In this paper we deal with the last topic and consider the problem of the path integral representation of the quantum mechanical evolution operator kernel for q-deformed oscillator. The corresponding raising and lowering operators $\mathbf{b}^+$ and $\mathbf{b}$ obey the following commutation relation (CR)

$$\mathbf{b}\mathbf{b}^+ - q^2\mathbf{b}^+\mathbf{b} = 1, \qquad q \in \mathbf{R} . \tag{1}$$

There are two possible ways to consider the relation (1). In the first case one considers $\mathbf{b}, \mathbf{b}^+$-operators as constructed from usual canonical variables. This leads to Macfarlane's representation [6] of (1),

$$\mathbf{b}^+ = \bar{\alpha}\left[e^{2isx} - e^{is\partial}e^{isx}\right] \ , \ \mathbf{b} = \alpha\left[e^{-2isx} - e^{-isx}e^{is\partial}\right] \ , \tag{2}$$

where $q = e^{-s^2}$, $\alpha\bar{\alpha} = (1-q^2)^{-1}$. As is seen from (2) one must identify the coordinates x and $x + 2\pi s$ so that the configuration space of q-oscillator is compact and topologically equivalent to a circle. The basic CR is the canonical

*Talk given by M. Chaichian

**High Energy Physics Laboratory, Department of Physics and Research Institute for High Energy Physics, P.O.Box 9 (Siltavuorenpenger 20 C), SF-00014, University of Helsinki, Finland; E-mail address: chaichian@finuhcb.helsinki.fi

†Nuclear Physics Institute, Moscow State University, 119899, Moscow, Russia; E-mail address: demichev@theory.npi.msu.su

one and CR (1) has an auxiliary meaning. Correspondingly, one can represent operators $\mathbf{b}, \mathbf{b}^+$ in terms of the ordinary differential operators and operators of multiplication by coordinate.

In the second case one considers CR (1) as the basic relation and represents raising and lowering q-operators in terms of differential operators and coordinates [7] on quantum plane .

It is well known that for the quantization of field theories the path integral in holomorphic representation is most suitable (see e.g. [8]). From the latter point of view and taking into account the analogy with fermionic (Berezin) path integral, it would be desirable to construct path integral over "classical" analog of operators $\mathbf{b}, \mathbf{b}^+$, that is variables $z, \bar{z}$, satisfying the CR $z\bar{z} - q^2\bar{z}z = 0$, or in other words, over variables on a quantum plane. Such an attempt was made in [9]. However, the authors of [9] aimed to construct the evolution operator for modified Schrödinger equation

$$iD_q\Psi(t) = \mathbf{H}(\mathbf{b}, \mathbf{b}^+)\Psi(t) \ , \tag{3}$$

where

$$D_q f(t) = \frac{f(q^2t) - f(t)}{t(q^2 - 1)} \tag{4}$$

is the q-derivative. First of all we note that such a choice of time evolution equation is completely independent of q-CR (1), that is one could assume q-deformed form of Scrödinger equation for usual oscillator and vice versa. Operator solution of eq. (3) has the following form [9]:

$$\Psi(t) = \mathbf{U}(t, t_0)\Psi(t_0) \ , \tag{5}$$

$$\mathbf{U}(t, t_0) = \exp_{q^{-1}}\{-it_0\mathbf{H}\}\exp_q\{it\mathbf{H}\} \ , \tag{6}$$

where

$$\exp_q A = \sum_{n=0}^{\infty} \frac{A^n}{[n]!} \tag{7}$$

is the q-deformed exponential [10], and

$$[X] = \frac{q^{2X} - 1}{q^2 - 1} \ . \tag{8}$$

Due to the property of the q-exponents, $\exp_q\{A\}\exp_{q^{-1}}\{-A\} = 1$, the operator $\mathbf{U}$ has the necessary properties of an evolution operator: $\mathbf{U}(t, t) = 1$, $\mathbf{U}(t_1, t_2)\mathbf{U}(t_2, t_3) = \mathbf{U}(t_1, t_3)$, but, unfortunately, $\mathbf{U}$ is not a unitary operator

$$\mathbf{U}^+(t, t_0)\mathbf{U}(t, t_0) = exp_q\{-it\mathbf{H}\}exp_{q^{-1}}\{it_0\mathbf{H}\}exp_{q^{-1}}\{-it_0\mathbf{H}\}exp_q\{it\mathbf{H}\} \neq 1 \ , \text{ if } q \neq 1 \ .$$

This contradicts the basic principle of quantum mechanics (probability interpretation, etc.). Note that from the point of view of Heisenberg equation of motion the absence of unitarity for q-evolution was shown in [11]. Thus we will not proceed in this way, but instead we shall consider the usual (non-deformed in time) Schrödinger equation. We must add also that in a part of their paper the authors of [9] consider q^2 as a pure phase [1]: $q^2 = e^{is}, s \in \mathbf{R}$. In this case operators $\mathbf{b}^+$ and $\mathbf{b}$ can not be conjugated to each other since one would obtain after conjugation of (1) meaningless relation: $\mathbf{b}^+\mathbf{b} = 0$. Throughout this paper we shall consider $q \in \mathbf{R}$. Most of the useful formulas from [9], however, remain valid also for this case. The central point in [9] is the construction of the deformed Bargmann-Fock representation (BFR) of (1) in the space of antianalytic functions on the q-plane. The differential calculus used for the construction of BFR is not invariant with respect to any deformed $GL(N)_q$ group (in contrast to the author's statement) and part of the CR for derivatives and coordinates was chosen arbitrarily. The final formula of [9] is the discrete approximation for path integral or, in other words, convolution of finite number of (nonunitary) evolution operators.

The aim of the present paper is to obtain the path integral in holomorphic representation for q-oscillator in full analogy with usual (bosonic and fermionic) case.

Path integral for q-oscillators. The BFR for CR (1) is constructed in the Hilbert space $\mathcal{H}$ of antianalytic functions $f(\bar{z})$ on the q-plane with help of $E(2)_q$-invariant differential and integral calculi (for details see [12]) with scalar product of the form

$$< g, f >= \int d\bar{z}dz \; e_q^{-\bar{z}z}\overline{g(\bar{z})}f(\bar{z}) \; . \tag{9}$$

The monomials $\psi^n(\bar{z}) = \frac{\bar{z}^n}{\sqrt{[n]!}}$ form the orthonormal complete set of vectors in $\mathcal{H}$. Lowering and raising operators are represented as derivative and coordinate $\mathbf{b} = \bar{\partial}$, $\quad \mathbf{b}^+ = \bar{z}$, and using the formulas

$$\bar{\partial} e_q^{a\bar{z}z} = az e_q^{a\bar{z}z} \; , \; \partial e_q^{a\bar{z}z} = aq^{-2}\bar{z}e_q^{aq^{-2}\bar{z}z} \; , \tag{10}$$

one can check that b^+ and b are hermitian conjugated to each other with respect to the scalar product (9).

As usual the action of any operator A in $\mathcal{H}$ can be represented with the help of its kernel

$$(\mathbf{A}f)(\bar{z}_1) = \int d\bar{z}_2 dz_2 \; e_q^{-\bar{z}_2 z_2} A(\bar{z}_1, z_2) f(\bar{z}_2) \; , \tag{11}$$

where

$$A(\bar{z}_1, z_2) = \sum_{m,n} A_{mn} \frac{\bar{z}_1^m}{\sqrt{[m]!}} \frac{z_2^n}{\sqrt{[n]!}} \; . \tag{12}$$

[1] This pure phase in the case of q^2 a root of unity of order k has been associated in [20] with the very interesting possibility of obtaining an extension of Berezin calculus as $z^k = 0, \bar{z}^k = 0$. We would like to notice that in the case of the deformation parameter being a pure phase, one ought to use other q-oscillator operators: $a = q^{-N/2}b, a^+ = b^+q^{N/2}$. Then the whole construction of Bargmann-Fock representation has to be revised in this case [13].

Here one more pair of q-commuting coordinates is introduced. So we have to define the CR for coordinates on different copies of q-planes. In [9] the simplest choice was made and was postulated that coordinates on different q-planes commute with each other. From our point of view the choice of the CR depends on the concrete meaning of different planes. In our case these q-planes will correspond to different time slices in the process of time evolution. In the continuous limit they become infinitesimally close to each other and it would be quite unnatural if coordinates on them would commute. We postulate that any copies of coordinates $\bar{z}_i$, z_i $(i = 1, 2, \ldots)$ on q-planes have the following CR:

$$z_i \bar{z}_j = q^2 \bar{z}_j z_i \ , \qquad \bar{z}_i \bar{z}_j = \bar{z}_j \bar{z}_i \ , \ z_i z_j = z_j z_i \ , \tag{13}$$

i.e. they do not depend on the indices which distinguish the copies. The same is true for CR with derivatives and differentials. Note that this is more similar to the fermionic case than the case of commuting copies of coordinates.

Now we can express A_{mn} through the scalar product: $A_{mn} = q^{2m(n+1)} < \psi_m || \mathbf{A} || \psi_n >$. The next formula which is necessary for constructing the path integral, is the convolution of operator kernels. Let us consider the action of two operators on some function $f(\bar{z})$ from $\mathcal{H}$,

$$\mathbf{A_2 A_1} = \int d\bar{z}_1 dz_1 e_q^{-\bar{z}_1 z_1} A_2(\bar{z}_2, z_1) \int d\bar{z}_0 dz_0 e_q^{-\bar{z}_0 z_0} A_1(\bar{z}_1, z_0) f(\bar{z}_0) \ . \tag{14}$$

General $A_2(\bar{z}_2, z_1)$ does not commute with $\exp_q\{-\bar{z}_0 z_0\}$ and one can not express the kernels of operator $\mathbf{A_2 A_1}$ through those of $\mathbf{A_2}$ and $\mathbf{A_1}$. For example, if $A_2(\bar{z}_2, z_1)\bar{z}_0 z_0 = q^k \bar{z}_0 z_0 A_2(\bar{z}_2, z_1)$ for some integer k, one has

$$\mathbf{A_2 A_1} f(\bar{z}_2) = \int d\bar{z}_0 dz_0 exp_q\{-q^k \bar{z}_0 z_0\} \left[\int d\bar{z}_1 dz_1 exp_q\{-\bar{z}_1 z_1\} A_2(\bar{z}_2, z_1) A_1(\bar{z}_1, z_0) \right] f(\bar{z}_0) \ . \tag{15}$$

To avoid this problem, we restrict our consideration to the operators with the kernels of the form $A(\bar{z}_i, z_i) = A(\bar{z}_i z_i)$. Such kernels commute with $\bar{z}_m z_n$, so that the integer k is zero and from (15) one has the usual convolution formula. This is analogous to even operators restriction in the fermionic case. If one has some operator in the normal form, e.g. the monomial $\mathbf{M}_k = (\mathbf{b}^+)^k \mathbf{b}^k$ for some integer k, one obtains from (12) the expression for integral kernel

$$M_k(\bar{z} z) = \sum_{n=0}^{\infty} q^{2(n+k)(n+k+1) - 2nk} \frac{\bar{z}^n z^n}{[n]!} \bar{z}^k z^k \ . \tag{16}$$

Using the relation $\bar{z}^n z^n = q^{-n(n-1)} (\bar{z} z)^n$, the expression (16) can be written in the form

$$M_k(\bar{z} z) = q^{2k(k+1)} \left(\sum_{n=0}^{\infty} q^{n(n-1)} \frac{\left(q^{2(k+2)} \bar{z} z\right)^n}{[n]!} \right) \bar{z}^k z^k \ . \tag{17}$$

The sum in (17) is nothing but the second basic exponential function [10], $\exp_{1/q}\{x\}$, so that the final expression for integral kernel of normal operator is

$$M_k(\bar{z} z) = q^{2k(k+1)} \exp_{1/q}\{q^{2(k+2)} \bar{z} z\} \bar{z}^k z^k \ . \tag{18}$$

According to our discussion in the Introduction we consider the usual Schrödinger equation

$$i\frac{d}{dt}\Psi(\bar{z},t) = \mathbf{H}(\mathbf{b}^+,\mathbf{b})\Psi(\bar{z},t)\ , \tag{19}$$

with the simplest nontrivial Hamiltonian [2]

$$\mathbf{H}(\mathbf{b}^+,\mathbf{b}) = \omega\mathbf{b}^+\mathbf{b}\ . \tag{20}$$

The integral kernel for infinitesimal operator $\mathbf{U} \approx \mathbf{1} - i\mathbf{H}\Delta t = \mathbf{1} - i\omega\mathbf{b}^+\mathbf{b}\Delta t$, takes the form

$$U(\bar{z}z) \approx e_{1/q}^{q^4\bar{z}z} exp\{-iH_{eff}\Delta t\}\ , \tag{21}$$

where

$$H_{eff} = q^4 e_q^{-q^4\bar{z}z} e_{1/q}^{q^6\bar{z}z}\omega\bar{z}z = q^4 e_q^{-q^4\bar{z}z+q^6\bar{z}z}\omega\bar{z}z = \frac{q^4\omega\bar{z}z}{1+q^4(1-q^2)\bar{z}z}\ .$$

Here we have used the summation theorem for q-exponentials of commuting arguments [10], together with Exton's notations $e_q^A e_{1/q}^B = e_q^{A+B}$, $AB = BA$. Now we can write the convolution of K infinitesimal evolution operator kernels

$$\begin{gathered} U(\bar{z}_K z_{K-1}) * U(\bar{z}_{K-1}z_{K-2}) * \ldots * U(\bar{z}_1 z_0)\ = \\ \int d\bar{z}_{K-1}dz_{K-1}\ldots d\bar{z}_1 dz_1 e_q^{-\bar{z}_{K-1}z_{K-1}}\ldots e_q^{-\bar{z}_1 z_1} \\ \times\, e_{1/q}^{-q^4\bar{z}_K z_{K-1}}\ldots e_{1/q}^{-q^4\bar{z}_1 z_0} e^{-iH_{eff}(\bar{z}_K z_{K-1})\Delta t}\ldots e^{-iH_{eff}(\bar{z}_1 z_0)\Delta t}\ . \end{gathered} \tag{22}$$

Using the product representation for the second q-exponent [10], $e_{1/q}^x = \prod_{r=0}^{\infty}\{1+ xq^{2r}(1-q^2)\}$, and introducing Δz_{K-l} as $z_{K-l-1} = z_{K-l} - \Delta z_{K-l}$, one can write

$$\begin{gathered} e_{1/q}^{q^4\bar{z}_K z_{K-1}}\ldots e_{1/q}^{q^4\bar{z}_1 z_0} = \exp\left\{\sum_{l=1}^{K-1}\ln\, e_{1/q}^{q^4\bar{z}_{K-l}z_{K-l-1}}\right\} = \exp\left\{\sum_{l=1}^{K-1}\sum_{r=0}^{\infty}\ln(1+q^{2r+4}(1-q^2)\bar{z}_{K-l}z_{K-l-1}\right\} \\ \approx e_{1/q}^{q^4\bar{z}_{K-1}z_{K-1}}\ldots e_{1/q}^{q^4\bar{z}_1 z_1}\exp\left\{\sum_{l=1}^{K-1}\left(q^4\sum_{r=0}^{\infty}\frac{q^{2r}(1-q^2)}{1+q^{2r+4}(1-q^2)\bar{z}_{K-l}z_{K-l}}\right)\bar{z}_{K-l}\Delta z_{K-l}\right\}\ . \end{gathered} \tag{23}$$

Substituting (23) into (22) and taking the continuous limit $\Delta t \to 0$ as usual, we finally obtain the path integral representation for the evolution operator kernel:

$$\begin{gathered} U(\bar{z},z;t''-t') = \int\left(\prod_t d\bar{z}(t)dz(t)(1+(1-q^2)\bar{z}(t)z(t))^{-1}(1+q^2(1-q^2)\bar{z}(t)z(t))^{-1}\right) \\ \times e_{1/q}^{q^4\bar{z}(t'')z(t'')}\exp\left\{-\int_{t'}^{t''}(\phi(\bar{z}(t)z(t))\bar{z}(t)\dot{z}(t) + iH_{eff}(\bar{z}(t)z(t)))\,dt\right\}\ , \end{gathered} \tag{24}$$

[2]It is obvious that for the q-deformed oscillator the trivial "free" Hamiltonian is proportional to the particle number operator i.e. $H_0 \sim N$ (remind that $b^+b = [N]$); the spectrum of H_0 is the equidistant one.

where the dot means time derivative and $\phi(\bar{z}(t)z(t)) = \sum_{r=0}^{\infty} \frac{q^{2r}}{(q^4(1-q^2))^{-1}+q^{2r}\bar{z}z}$. Note that in the $q^2 \to 1$ limit $\phi \to 1$, $H_{eff} \to H_{cl} \equiv \omega\bar{z}z$, so that in this limit one obtains the usual expression for the harmonic oscillator path integral (cf., e.g., [8]).

Finally, the case of q-deformed path integral with q being a k-th root of unity, as presented in this talk, leads to the generalization of Grassmann variables and will be given elsewhere [13].

References

[1] V.G.Drinfel'd, Proc.of the Berkley Inst. Congress of Math. 1(1987)798; Y. Manin, Quantum group and non-commutative geometry, Center des Recherches Mathématiques, Montreal (1988); L.D.Faddeev,N.Yu.Reshetikhin and L.A.Takhtadjan, Algebra i Analis 1 (1989) 178;S.L.Woronowicz, Comm.Math.Phys. 111 (1987) 613.

[2] L.Faddeev, in: Les Houches XXXIX, eds. J.-B.Zuber and R.Stora (North Holland, Amsterdam, 1984); P.Kulish and E.Sklyanin, Lecture Notes in Phys., vol.151 (1984).

[3] W.B.Schmidke, J.Wess and B.Zumino, Z.Phys. C52 (1991) 471; O.Ogievetsky, W.B.Schmidke, J.Wess and B.Zumino, q-Deformed Poincare Algebra, pr.MPI-Ph/91-98, 1991; J.Lukierski, H.Ruegg, A.Nowicki and V.N.Tolstoy, Phys.Lett. 264B (1991) 331.

[4] M.Chaichian and A.P.Demichev, Phys. Lett. B 304 (1993) 220;M.Chaichian and A.P.Demichev, Quantum Poincaré Group, Algebra and Quantum geometry of Minkowski Space, Helsinki Univ. prep. HU-TFT-93-24, 1993.

[5] D.Bernard, Prog.Theor.Phys. Suppl., 102 (1990) 49; L.Castellani, Phys.Lett., B279 (1992) 291; I.Ya.Aref'eva and I.V.Volovich, Mod.Phys.Lett. A6 (1991) 893; Y.Frishman, J.Lukierski and W.J.Zakrzewski, J.Phys.A26 (1993) 301; T.Brzezinski and S.Madjid, Quantum Group Gauge Theory on Classical Spases, prep.DAMTP/92-51.

[6] A.Macfarlane, J.Phys. A22 (1989) 4581; L.Biedenharn, J. Phys. A22 (1989) L873; M. Chaichian and P. Kulish, Phys. Lett. B 234 (1990) 72; M.Chaichian, P.Kulish and J.Lukierski, Phys.Lett. B262 (1991) 43; W.Pusz and S.L.Woronowicz, Rep. Math. Phys. 27 (1989) 231; A.Kempf, J.Math.Phys. 34 (1993) 969.

[7] J.Wess and B.Zumino, Nucl.Phys.Suppl. 18B (1990) 302.

[8] A.A.Slavnov and L.D.Faddeev, Introduction to quantum theory, Reading, Mass., Benjamin 1980.

[9] L.Baulieu and E.G.Floratos, Phys.Lett. B258 (1991) 171.

[10] H.Exton, q-Hypergeometric functions and applications, Ellis Horwood, Chichester, 1983.

[11] M.Chaichian and D.Ellinas, J.Phys. A23 (1990) L291.

[12] M.Chaichian and A.P.Demichev, Phys.Lett. B320 (1994) 273.

[13] M. Chaichian and A.P. Demichev (in preparation).

QUANTUM MECHANICAL QUANTUM GROUP IN TWO-DIMENSIONAL GRAVITY

Jean-Loup GERVAIS

Laboratoire de Physique Théorique de l'École Normale Supérieure[1],
24 rue Lhomond, 75231 Paris CEDEX 05, France.

Abstract. These notes summarize the operator approach to Liouville theory where the underlying quantum group symmetry is instrumental. Both weak and strong coupling regimes are discussed, displaying the most recent progress.

Keywords: Conformal theories, two dimensions, gravity, quantum group.

Foreword
It is appropriate to dedicate this talk to the memory of Feza Gürsey, who did so much to show the fundamental importance of group theory in Physics.

1 Introduction

The operator approach to Liouville theory, which originated more than ten years ago, has come a long way starting from the analysis of the simplest Liouville field — the inverse square root of the metric — which corresponds to the $J = 1/2$ representation, by Neveu and myself[1]. The unravelling of its quantum group structure[2, 3, 4, 5] has recently led to striking progress which we plan to summarize in these notes. On the one hand, the most general Liouville operators in the standard (weak coupling) regime has been constructed[6, 7, 8], corresponding to arbitrary highest/lowest weight representations of the quantum group. On the other hand, the operator approach to Liouville theory seems at present to be the only method applicable to the strong coupling regime[3, 9, 19]. Although the Liouville exponentials loose meaning in the strongly coupled regime, since their operator product algebra involves operators (and/or highest weight states) with complex Virasoro weights, the general operator-family of their chiral components may still be used, as a result of the truncation theorems[11, 3, 9]. Indeed at Liouville central charges $C = 7,\ 13,\ 19$, there exist subfamilies of these chiral operators with real Virasoro weights which form closed operator algebras. They

[1] Unité Propre du Centre National de la Recherche Scientifique, associée à l'École Normale Supérieure et à l'Université de Paris-Sud.

are used to construct new local fields which replace the Liouville exponentials. Since the old and the new local families are constructed out of the same free Bäcklund fields, they may be considered as related by a new type of quantum Bäcklund transformation, that connect the weak and strong coupling regimes of two-dimensional gravity.

These notes are organized as follows. After recalling the properties of the classical solution which are helpful to understand the quantum case (section 2), we recall the fusion and braiding properties of the chiral vertex operators, display their connection with quantum group symbols, and exhibit the quantum group symmetry of the operator algebra (section 3). This is next used to recall how the present approach deals with the weak coupling regime, and confirms the matrix-model results on the sphere (section 4). Section 5 recalls the recent advances on the strong coupling regime, concerning in particular new topological models that appear sovable explicitly contrary to the standard Liouville string models. Some further general remarks are made in section 6 as a conclusion.

2 The classical case

In order to set the stage, let us first recall some relevent features of the classical Liouville dynamics. In the conformal gauge, it is governed by the action:

$$S = \frac{1}{4\pi}\int d_2x\sqrt{\hat{g}}\Big\{\frac{1}{2}\hat{g}^{ab}\partial_a\Phi\partial_b\Phi + e^{2\sqrt{\gamma}\Phi} + \frac{1}{2\sqrt{\gamma}}R_0\Phi\Big\} \tag{2.1}$$

$\hat{g}_{ab}$ is the fixed background metric. We work for fixed genus, and do not integrate over the moduli. As is well known, one can choose a local coordinate system such that $\hat{g}_{ab} = \delta_{ab}$. Thus we are reduced to the action

$$S = \frac{1}{4\pi}\int d\sigma d\tau\,\Big(\frac{1}{2}(\frac{\partial\Phi}{\partial\sigma})^2 + \frac{1}{2}(\frac{\partial\Phi}{\partial\tau})^2 + e^{2\sqrt{\gamma}\Phi}\Big) \tag{2.2}$$

where σ and τ are the local coordinates. The complex structure is assumed to be such that the curves with constant σ and τ are everywhere tangent to the local imaginary and real axis respectively. In a typical situation, one may work on the cylinder $0 \leq \sigma \leq 2\pi$, $-\infty \leq \tau \leq \infty$ obtained by an appropriate mapping from one of the handles of a general Riemann surface, and we shall do so in the present article. The action 2.2 corresponds to a conformal theory such that $\exp(2\sqrt{\gamma}\Phi)d\sigma d\tau$ is invariant. The classical equivalent of the chiral vertex operators may be obtained very simply[1] by using the fact that the field $\Phi(\sigma, \tau)$ satisfies the equation

$$\frac{\partial^2\Phi}{\partial\sigma^2} + \frac{\partial^2\Phi}{\partial\tau^2} = 2\sqrt{\gamma}\, e^{2\sqrt{\gamma}\Phi} \tag{2.3}$$

if and only if

$$e^{-\sqrt{\gamma}\Phi} = i\sqrt{\frac{\gamma}{2}}\sum_{j=1,2} f_j(x_+)g_j(x_-); \quad x_\pm = \sigma \mp i\tau \tag{2.4}$$

where f_j (resp.(g_j), which are functions of a single variable, are solutions of the same Schrödinger equation

$$-f_j'' + T(x_+)f_j = 0, \quad (\text{ resp. } -g_j'' + \overline{T}(x_-)g_j\,). \tag{2.5}$$

The solutions are normalized so that their Wronskians $f_1'f_2 - f_1f_2'$ and $g_1'g_2 - g_1g_2'$ are equal to one. The proof of this basic fact is straightforward [1]. The potentials $T(x_+)$ and $\overline{T}(x_-)$ are the two components of the stress-energy tensor, and, after quantization, Eqs.2.5 become the Virasoro Ward-identities associated with the vanishing of the singular vector at the second level. As a result the Liouville theory also describes minimal models provided the coupling constant γ is taken to be negative. This is how, we shall treat the matter fields. For the dynamics associated with the action Eq.2.2, τ is the time variable, and the canonical Poisson brackets are

$$\{\Phi(\sigma_1,\tau), \frac{\partial}{\partial\tau}\Phi(\sigma_2,\tau)\}_{\text{P.B.}} = 4\pi\,\delta(\sigma_1-\sigma_2), \quad \{\Phi(\sigma_1,\tau),\Phi(\sigma_2,\tau)\}_{\text{P.B.}} = 0 \tag{2.6}$$

The cylinder $0 \le \sigma \le 2\pi$, $-\infty \le \tau \le \infty$ may be mapped on the complex plane of $z = e^{\tau + i\sigma}$, and the above Poisson brackets lead to the usual radial quantization. Any two pairs f_j and g_j of linearly independent solutions of Eq.2.5 are suitable as well, a priori. In this connection, it is convenient to rename the functions g_j by letting $\overline{f}_1 = -g_2$, $\overline{f}_2 = g_1$. Then one easily sees that Eq.2.4 is left unchanged if f_j and $\overline{f}_j$ are replaced by $\sum_k M_{jk}f_k$ and $\sum_k M_{kj}\overline{f}_k$, respectively, where M_{jk} is an arbitrary constant matrix with determinant equal to one. Eq.2.4 is $sl(2,C)$-invariant with f_j transforming as a representation of spin 1/2. At the classical level, it is trivial to take Eq.2.4 to any power. For positive integer powers $2J$, and letting $\beta = i\sqrt{\frac{\gamma}{2}}$, one get a formula of the form

$$e^{-2J\sqrt{\gamma}\Phi} = \beta^{2J} \sum_{M=-J}^{J} (-1)^{J+M} f_M^{(J)}(x_+)\overline{f}_{-M}^{(J)}(x_-). \tag{2.7}$$

where $J \pm M$ run over integer. The $sl(2)$-structure has been made transparent by letting

$$f_M^{(J)} \equiv \sqrt{\binom{2J}{J+M}}\,(f_1)^{J-M}(f_2)^{J+M}, \quad \overline{f}_M^{(J)} \equiv \sqrt{\binom{2J}{J+M}}\,(\overline{f}_1)^{J-M}(\overline{f}_2)^{J+M}. \tag{2.8}$$

The notation anticipates that $f_M^{(J)}$ and $\overline{f}_M^{(J)}$ form representations of spin J. This is indeed true since f_1, f_2 and $\overline{f}_1$, $\overline{f}_2$ span spin 1/2 representations, by construction, as we just explained. Explicitly one finds

$$\begin{aligned} I_\pm f_M^{(J)} &= \sqrt{(J\mp M)(J\pm M+1)}f_{M\pm1}^{(J)}, \quad I_3 f_M^{(J)} = M f_M^{(J)} \\ \overline{I}_\pm \overline{f}_M^{(J)} &= \sqrt{(J\mp M)(J\pm M+1)}\overline{f}_{M\pm1}^{(J)}, \quad \overline{I}_3 f_M^{(J)} = M\overline{f}_M^{(J)}, \end{aligned} \tag{2.9}$$

where I_ℓ and $\overline{I}_\ell$ are the infinitesimal generators of the x_+ and x_- components respectively. Moreover, one sees that

$$(I_\ell + \overline{I}_\ell)\,e^{-2J\sqrt{\gamma}\Phi} = 0 \tag{2.10}$$

so that the exponential of the Liouville field are group invariants.

Next there exists the so-called Bäcklund transformation which relates the Liouville field Φ, and its conjugate momentum $\partial_\tau \Phi$ to a free field. It may be achieved as follows. Consider, for instance the f's. Since Eq.2.5 only involves a single variable, we may consider it at $\tau =$ constant, say zero. The potential is periodic in σ, and hence there exist (Bloch wave) solution which are periodic up to multiplicative constants. We denote them by $V_m^{(J)}$. Then one may write (primes mean derivatives)

$$V_{-J}^{(J)} = (\frac{1}{\sqrt{A'}})^{2J} = e^{2J\sqrt{\gamma}\vartheta_L}, \tag{2.11}$$

where

$$A(\sigma) = \left\{ e^{-4\pi p_0\sqrt{\gamma}} \int_0^\sigma V_1^{(-1)}(\rho)d\rho + \int_\sigma^{2\pi} V_1^{(-1)}(\rho)d\rho \right\} \Big/ \left(e^{-4\pi p_0\sqrt{\gamma}} - 1 \right). \tag{2.12}$$

The last expression is chosen such A is periodic up to a constant in agreement with Eq.2.11. The dynamical variable p^0 is defined from the following Fourier decomposition of ϑ_L

$$\vartheta_L(\sigma) = q_0 + p_0\sigma + i\sum_{n\neq 0} e^{-in\sigma}\, p_n^{(j)}/n, \tag{2.13}$$

The first two terms are the zero-mode "position" and "momentum". Then we may rewrite Eq.2.8 in the following Coulomb-gas form

$$V_m^{(J)} = \sqrt{\binom{2J}{J+m}} V_{-J}^{(J)} A^{J+m}. \tag{2.14}$$

This Bäcklund transformation is such that[1] the canonical Poisson brackets Eqs.2.6 are equivalent to the following free-field P.B. for the ϑ_L fields:

$$\left\{ \vartheta_L'(\sigma_1), \vartheta_L'(\sigma_2) \right\}_{\mathrm{P.B.}} = 2\pi\, \delta'(\sigma_1 - \sigma_2). \tag{2.15}$$

Replacing these by (i times) commutators is[1] the starting point of the operator approach to Liouville theory.

In general, the $sl(2,C)$ action of the chiral components does not take a Bloch wave into a Bloch wave, since it mixes different values of m. The only exception is the transformation $V_{1/2}^{(1/2)} \to V_{-1/2}^{(1/2)}$ which gives $V_m^{(J)} \to V_{-m}^{(J)}$ for arbitrary J. This leads[1] to formulae similar to Eqs.2.11-2.15, albeit with another equivalent free field (see ref.[8] for a recent review).

3 The chiral operator algebra

At the quantum level, the $f_m^{(J)}$'s and $\overline{f}_m^{(J)}$'s become operators that do not commute, and the group $sl(2)$ is deformed to become the quantum group $U_q(sl(2))$.

This structure plays a crucial role at the quantum level, and we now summarize it without going through the derivations for brevity. The parameter h that characterizes the quantum group structure $U_q(sl(2))$ is related to the central charge C of Virasoro algebra by the equation

$$C = 1 + 6(\frac{h}{\pi} + \frac{\pi}{h} + 2). \tag{3.1}$$

The Hilbert space of states is made up with Verma modules $\mathcal{H}_J$, span by states[2] $|J, \{\nu\}>$ obtained by the action of the Virasoro generators on the highest weight state noted $|J, 0>$ (or equivalently $|J>$) with weight $\Delta_J = -hJ(J+1)/\pi - J$. The following form of the operator algebra for the quantum Bloch wave operators $V_m^{(J)}$ was derived in refs.[4, 5]. Using the world sheet variable $z = \exp(\tau + i\sigma)$ it takes the Moore Seiberg form. The fusing (operator-product) algebra reads

$$< J_{123}, \{\nu_{123}\}|V^{(J_1)}_{J_{23}-J_{123}}(z_1)V^{(J_2)}_{J_3-J_{23}}(z_2)|J_3, \{\nu_3\} >=$$

$$\sum_{J_{12}=|J_1-J_2|}^{J_1+J_2} \frac{g^{J_{12}}_{J_1J_2}\, g^{J_{123}}_{J_{12}J_3}}{g^{J_{23}}_{J_2J_3}\, g^{J_{123}}_{J_1J_{23}}} \left\{ \begin{matrix} J_1 & J_2 \\ J_3 & J_{123} \end{matrix} \middle| \begin{matrix} J_{12} \\ J_{23} \end{matrix} \right\}_q \times$$

$$\sum_{\{\nu_{12}\}} < J_{123}, \{\nu_{123}\}|V^{(J_{12},\{\nu_{12}\})}_{J_3-J_{123}}(z_2)|J_3, \{\nu_3\}>$$

$$< J_{12}, \{\nu_{12}\}|V^{(J_1)}_{J_2-J_{12}}(z_1 - z_2)|J_2> . \tag{3.2}$$

The g's which are called coupling constants contain the contributions that are not trigonometric functions of h. They are given by

$$g^{J_{12}}_{J_1J_2} = (g_0)^{J_1+J_2-J_{12}} \prod_{k=1}^{J_1+J_2-J_{12}} \sqrt{F(1 + (2J_1 - k + 1)h/\pi)} \times$$

$$\sqrt{\frac{F(1 + (2J_2 - k + 1)h/\pi)F(-1 - (2J_{12} + k + 1)h/\pi)}{F(1 + kh/\pi)}}, \tag{3.3}$$

where we define $F(z) \equiv \Gamma(z)/\Gamma(1-z)$, and g_0 is a constant. In Eq.3.2 there appear the 6j symbols of $U_q(sl(2))$, noted $\left\{ \begin{smallmatrix} J_1 & J_2 \\ J_3 & J_{123} \end{smallmatrix} \middle| \begin{smallmatrix} J_{12} \\ J_{23} \end{smallmatrix} \right\}_q$ to emphasize that they are the Racah-Wigner 6-j coefficients which do not have the full tetrahedral symmetry. Next the braiding is given by

$$< J_{123}, \{\nu_{123}\}|V^{(J_1)}_{J_{23}-J_{123}}(z_1)V^{(J_2)}_{J_3-J_{23}}(z_2)|J_3, \{\nu_3\} >=$$

$$\sum_{J_{13}} e^{\pm i\pi(\Delta_{J_{123}} + \Delta_{J_3} - \Delta_{J_{23}} - \Delta_{J_{13}})} \times$$

[2] the notation $\{\nu\}$ represents a multi-index.

$$\frac{g_{J_1J_3}^{J_{13}}g_{J_{13}J_2}^{J_{123}}}{g_{J_2J_3}^{J_{23}}g_{J_1J_{23}}^{J_{123}}}\left\{\begin{matrix}J_1 & J_3\\ J_2 & J_{123}\end{matrix}\middle|\begin{matrix}J_{13}\\ J_{23}\end{matrix}\right\}_q < J_{123},\{\nu_{123}\}|V^{(J_2)}_{J_{13}-J_{123}}(z_2)V^{(J_1)}_{J_3-J_{13}}(z_1)|J_3,\{\nu_3\}> . \tag{3.4}$$

The formulae just given are the basic tool of the present scheme. For the standard case (half-integer J's), their derivation is carried out in ref.[4] which completes the earlier discussions given in refs.[1] and [2]. The extension to semi-infinite representations with arbitrary J's is given in refs[6, 8, 9]. The g factors in Eqs.3.2 and 3.4 show that the fusing and braiding matrices are not purely group theoretical. These factors may be removed from these expressions by changing the normalization of the V operators (so far, it is such that the highest weight matrix elements are equal to one or zero). However, working out the actual value of the coefficients of the operator-product expansion from Eq.3.2, one sees that they are sensitive to the normalizations due to the last term there. Thus[3] the g factors cannot be forgotten.

Although the relevance of quantum group is clear by now, one does not yet see the quantum group action, since the 6j's only involve total spins while the quantum group generators act on the magnetic quantum numbers. In this connection there exists[2, 5] an equivalent set of fields noted $\xi_M^{(J)}$, where M is a magnetic quantum number. Their fusion algebra is given by

$$\xi^{(J_1)}_{M_1}(z_1)\xi^{(J_2)}_{M_2}(z_2) = \sum_{J_{12}=|J_1-J_2|}^{J_1+J_2} g_{J_1J_2}^{J_{12}}(J_1,M_1;J_2,M_2|J_{12})_q\times$$
$$\sum_{\{\nu\}}\xi^{(J_{12},\{\nu\})}_{M_1+M_2}(z_2) < J_{12},\{\nu\}|V^{(J_1)}_{J_2-J_{12}}(z_1-z_2)|J_2> . \tag{3.5}$$

where $(J_1, M_1; J_2, M_2|J_{12})_q$ denotes the q-Clebsch-Gordan (3j) symbols. The braiding properties were derived in ref.[2]. One has (at this point it is simpler to return to the σ, τ variables, and to stay on the unit circle $\tau = 0$; for definiteness, we consider the case where $0 \le \sigma_1 \le \sigma_2 \le \pi$)

$$\xi^{(J_1)}_{M_1}(\sigma_1)\xi^{(J_2)}_{M_2}(\sigma_2) = \sum_{-J_1\le N_1\le J_1;\ -J_2\le N_2\le J_2} (J_1,J_2)_{M_1\,M_2}^{N_2\,N_1}\xi^{(J_2)}_{N_2}(\sigma_2)\xi^{(J_1)}_{N_1}(\sigma_1). \tag{3.6}$$

The symbol $(J_1, J_2)_{M_1\,M_2}^{N_2\,N_1}$ denotes the following matrix element of the universal R-matrix of $U_q(sl_2)$):

$$(J_1,J_2)_{M_1\,M_2}^{M_2'\,M_1'} = \Big(<< J_1,M_1|\otimes << J_2,M_2|\Big)\ R\ \Big(|J_1,M_1' >> \otimes|J_2,M_2' >>\Big), \tag{3.7}$$

where $|J, M >>$ are group theoretic states which span the representation of spin J of $U_q(sl(2))$. The universal R-matrix R is given by

$$R = e^{-2ihJ_3\otimes J_3}\sum_{n=0}^{\infty}\frac{(1-e^{2ih})^n\, e^{ihn(n-1)/2}}{\lfloor n\rfloor!_q}e^{-ihnJ_3}(J_+)^n\otimes e^{ihnJ_3}(J_-)^n. \tag{3.8}$$

[3] although they are irrelevent from the viewpoint of the polynomial equations (associativity of operator algebra, link-invariance in three dimensions) since they are "pure gauges" from the three-dimensional viewpoint,

As usual, we let in general

$$\lfloor n \rfloor !_q = \prod_1^n \lfloor r \rfloor_q, \quad \lfloor x \rfloor_q = \frac{\sin(hx)}{\sin h} \tag{3.9}$$

$J_\pm$, and J_3 are the quantum-group generators. Note that we did not specify the states of the Hilbert space considered in the braiding and fusing of the ξ fields. This is not needed since **the braiding and fusing matrices of the ξ fields do not depend on them**. This is in sharp contrast with the V fields. Now the quantum group action become appearent, since the formulae just given show that, as a consequence of standard properties of the 3j symbols and universal R-matrix, the operator algebra is covariant under action of the quantum group. Indeed[3], define the quantum group action on the ξ fields by

$$J_3(\xi_M^{(J)}) = M\xi_M^{(J)}, \quad J_\pm(\xi_M^{(J)}) = \sqrt{\lfloor J \mp M \rfloor_q \lfloor J \pm M + 1 \rfloor_q}\, \xi_{M\pm1}^{(J)}. \tag{3.10}$$

Then the operator-product $\xi_{M_1}^{(J_1)}(\sigma_1)\,\xi_{M_2}^{(J_2)}(\sigma_2)$ also gives a representation of the quantum group algebra Eq.3.10 with generators

$$\Lambda(J_\pm) := J_\pm \otimes e^{ihJ_3} + e^{-ihJ_3} \otimes J_\pm, \quad \Lambda(J_3) := J_3 \otimes 1 + 1 \otimes J_3, \tag{3.11}$$

where the tensor product is defined so that

$$(A \otimes B)\left(\xi_{M_1}^{(J_1)}(\sigma_1)\,\xi_{M_2}^{(J_2)}(\sigma_2)\right) := \left(A\left(\xi_{M_1}^{(J_1)}(\sigma_1)\right)\right)\left(B\left(\xi_{M_2}^{(J_2)}(\sigma_2)\right)\right), \tag{3.12}$$

and where each term in the expansion over J transforms according to a representation of spin J. Of course Eq.3.11 coincides with one of the two standard coproducts of $U_q(sl_2)$). The other one appears with the other ordering on the half unit circle: $0 \le \sigma_2 \le \sigma_1 \le \pi$. The operator algebra of the ξ fields is covariant under the quntum group action just recalled. Indeed, first the covariance of the fusion equation Eq.3.5 directly follows from the fact that 3j symbols are the coefficients of the decomposition of (q) tensor products of irreducible representations into irreducible representations. Second, the basic property of the universal R matrix that it exchanges the two coproducts is responsible for the covariance of the braiding equation Eq.3.6. The relationship between ξ and V fields may be neatly written as

$$< J_1, \{\nu\} | \xi_{M_2}^{(J_2)}(z) = \sum_{J_{12}} g_{J_2 J_{12}}^{J_1} (J_1, \infty; J_2, M_2 | J_{12}) < J_1, \{\nu\} | V_{J_{12}-J_1}^{(J_2)}(z). \tag{3.13}$$

The symbol $(J_1, \infty; J_2, M_2 | J_{12})$ denotes a limit of 3j symbols suitably defined[4].

So far we only discussed a subset of the relevent chiral primaries. Indeed, first, the relationship between C and h being quadratic is one-to-two. There exits another quantum group parameter $\widehat{h}$ such that

$$h\widehat{h} = \pi^2 \tag{3.14}$$

which corresponds to the same C. This reflects the existence of two screening charges. Letting

$$Q_L = \sqrt{(C-1)/3}, \quad Q_M = \sqrt{(25-C)/3} \tag{3.15}$$

they are given by

$$\alpha_\pm = Q_L/2 \pm iQ_M/2. \tag{3.16}$$

All the quantum numbers double, and the general operators have the form $V^{(J\widehat{J})}_{m\widehat{m}}$ and so on. Moreover, there are the bar components which are similar, and commute with the above. The Bäcklund free field associated with the Liouville field Φ has of course two chiral components ϑ_L, and $\overline{\vartheta}_L$. Altogether, the building blocks of the quantum group approach to Liouville theory are chiral fields of the form[4]

$$\widetilde{V}^{(J\widehat{J})}_{m\widehat{m}} \propto V^{(J\widehat{J})}_{-J\,-\widehat{J}}\, S^{J+m}\, \widehat{S}^{\widehat{J}+\widehat{m}}, \quad \widetilde{\overline{V}}^{(\overline{J}\widehat{\overline{J}})}_{\overline{m}\widehat{\overline{m}}} \propto \overline{V}^{(\overline{J}\widehat{\overline{J}})}_{-\overline{J}\,-\widehat{\overline{J}}}\, \overline{S}^{\overline{J}+\overline{m}}\, \widehat{\overline{S}}^{\widehat{\overline{J}}+\widehat{\overline{m}}}, \tag{3.17}$$

The V fields are functions of z, and the $\overline{V}$ fields functions of $\bar{z}$. The fields $V^{(J\widehat{J})}_{-J-\widehat{J}}$ are simple exponentials which may be directly re-expressed in terms of the Bäcklund free fields.

$$\widetilde{V}^{(J\widehat{J})}_{-J-\widehat{J}} = :\exp\left[(J\alpha_- + \widehat{J}\alpha_+)\vartheta_L\right]:, \quad \widetilde{\overline{V}}^{(\overline{J}\widehat{\overline{J}})}_{-\overline{J}-\widehat{\overline{J}}} = :\exp\left[(\overline{J}\alpha_- + \widehat{\overline{J}}\alpha_+)\overline{\vartheta}_L\right]: . \tag{3.18}$$

S, $\overline{S}$, and $\widehat{S}$, $\widehat{\overline{S}}$ are the screening operators[6] associated with the two screening charges Eq.3.16. The first two are the quantum version of the function A given by Eq.2.12. The J's are quantum group spins, and the general structure is of the type $U_q(sl(2)) \odot U_{\widehat{q}}(sl(2))$ introduced in ref[3, 13]. We leave further details to the original articles. For half-integer spins, the symbol $\odot$ stands for a particular graded tensor product, but for general J's, the two factors become non trivially entangled[9]. It is interesting that, according to Eq.3.14, $U_q(sl(2))$ and $U_{\widehat{q}}(sl(2))$ are dual in a way similar to electricity and magnetism. In practice the J's also determine the weight of the V operators which are of the type $(2\widehat{J}+1, 2J+1)$ in the BPZ classification. We deal with irrational theories, since this will be the case in the strong coupling regime. Thus the range of J and $\widehat{J}$ is unbounded. Concerning the Hilbert space, the Verma modules is of course charaterized by four quantum group spins, and may be written as $|J, \widehat{J}>|\overline{J}, \widehat{\overline{J}}>$.

[4] From now on, for notational simplicity, we switch to another normalization such that the fusing and braiding of the $\widetilde{V}$ fields are exactly equal to 6j symbols.

4 The weak coupling regime

4.1 The quantum Liouville exponentials

Next we recall some basic points of the weak-coupling discussion[13, 6, 4, 5, 8]. The local Liouville exponentials are given by[3, 6]

$$e^{-(J\alpha_- + \widehat{J}\alpha_+)\Phi(z,\bar{z})} = \sum_{m,\widehat{m}} \widetilde{V}^{(J,\widehat{J})}_{m\widehat{m}}(z)\,\widetilde{\overline{V}}^{(J,\widehat{J})}_{m\widehat{m}}(\bar{z}) \tag{4.1}$$

This form is dictated by locality and closure under fusion. These two basic properties neatly follow from Eqs.3.2, and 3.4 together with the standard orthogonality properties of 6j symbols[4, 6, 8]. For the following, it is important to stress that this expression only involves V and $\overline{V}$ fields with equal quantum numbers ($J = \overline{J}$, $m = \overline{m}$, and so on), while m, amd $\widehat{m}$ are summed over independently. Thus the Liouville exponentials have zero conformal spins. On the other hand, it follows from the formulae just summarized that

$$< J_2,\, \widehat{J}_2|\widetilde{V}^{(J\widehat{J})}_{m\widehat{m}}|J_1,\, \widehat{J}_1 > \propto \delta_{J_1-J_2-m,0}\,\delta_{\widehat{J}_1-\widehat{J}_2-\widehat{m},0}, \tag{4.2}$$

so that the Liouville exponential applied to a highest-weight state with $J = \overline{J}$, $\widehat{J} = \widehat{\overline{J}}$ only gives states satisfying the same condition. Thus we may restrict ourselves to the subsector with zero winding number. We stress this well known fact, since this will not be true any more in the strong coupling regime. In ref.[8], Eq.4.1 was derived for arbitrary $J, \widehat{J}$, then by letting $\widehat{J} = 0$, and taking the derivative with respect to J at $J = 0$, one defines the quantum Liouville field itself, obtaining

$$\Phi(\sigma,\tau) = -(\vartheta_L(x_+) + \bar{\vartheta}_L(x_-)) + \frac{2h}{\alpha_- \sin h}\times$$

$$\sum_{n=1}^{\infty} \frac{1}{\lfloor n \rfloor_q} \prod_{k=1}^{n} \frac{1}{\lfloor \frac{2ip_0}{\alpha_-} + 2n - k\rfloor_q \lfloor \frac{2ip_0}{\alpha_-} + k\rfloor_q} S(x_+)^n \overline{S}(x_-)^n. \tag{4.3}$$

Then[8], one may show that it satisfies the quantum version of the classical field equation Eq.2.3, that is[5]

$$\frac{\partial^2\Phi}{\partial\sigma^2} + \frac{\partial^2\Phi}{\partial\tau^2} = \alpha_-\, e^{\alpha_-\Phi}. \tag{4.4}$$

Moreover, one may derive the equal time commutation relations that replace Eq.2.6. In terms of the ξ fields the Liouville exponentials also have a neat form[3]. For $\widehat{J} = 0$, and ignoring unessential complications one has

$$e^{-J\alpha_-\Phi(\sigma,\tau)} = \sum_{M=-J}^{J} (-1)^{J-M}\, e^{ih(J-M)}\, \xi^{(J)}_M(x_+)\,\overline{\xi}^{(J)}_{-M}(x_-) \tag{4.5}$$

[5] This assumes a special choice of the cosmological constant, which is not made in Eq.4.1 for simplicity.

The quantum group transformations of the $\overline{\xi}$ fields are similar to Eq.3.10, namely

$$\overline{J}_3\left(\overline{\xi}_M^{(J)}\right)=M\overline{\xi}_M^{(J)},\quad \overline{J}_\pm\left(\overline{\xi}_M^{(J)}\right)=\sqrt{\lfloor J\mp M\rfloor_q\lfloor J\pm M+1\rfloor_q}\left(\overline{\xi}_{M\pm 1}^{(J)}\right). \quad (4.6)$$

It is convenient to define the coproduct of bar and unbar generators as follows

$$\mathcal{J}_\pm = J_\pm e^{-ih\overline{J}_3}+e^{ihJ_3}\otimes\overline{J}_\pm,\quad \mathcal{J}_3=J_3+\overline{J}_3, \quad (4.7)$$

which does give a representation of $U_q(sl(2))$. Then one easily checks that

$$\mathcal{J}_\pm\left(\exp(-J\alpha_-\Phi)\right)==\mathcal{J}_3\left(\exp(-J\alpha_-\Phi)\right)=0, \quad (4.8)$$

so that the quantized Liouville field is a quantum-group invariant. Such is the quantum version of Eq.2.10.

4.2 The weak coupling regime

Coupling Liouville theory (two-dimensional gravity) with some world sheet matter is the next step. In the weak coupling regime, this gives back the result of matrix models, albeit for fixed genus. According to refs.[13, 9], this comes out as follows. For $C>25$, Q_L is real and Q_M pure imaginary, so that $\alpha_\pm$ are real. The above formulae are directly useful. One represents matter by another copy of the theory summarized above, now with central charge $c=26-C$, so that Q_M is its background charge. One constructs local fields in analogy with Eq.4.1:

$$e^{-(J\alpha'_-+\widehat{J}\alpha'_+)\Phi'(z,\bar{z})}=\sum_{m,\widehat{m}}\widetilde{V}'^{(J,\widehat{J})}_{m\,\widehat{m}}(z)\,\widetilde{\overline{V}}'^{(J,\widehat{J})}_{m\,\widehat{m}}(\bar{z}). \quad (4.9)$$

Symbols pertaining to matter are distinguished by a prime[6]. In particular $\Phi'(z,\bar{z})$ is the matter field (it commutes with $\Phi(z,\bar{z})$), and $\alpha'_\pm$ are the matter screening charges

$$\alpha'_\pm=\mp i\alpha_\mp. \quad (4.10)$$

The correct dressing of these operators by gravity is achieved by considering the vertex operators

$$\mathcal{W}^{J,\widehat{J}}\equiv e^{-((-\widehat{J}-1)\alpha_-+J\alpha_+)\Phi-(J\alpha'_-+\widehat{J}\alpha'_+)\Phi'} \quad (4.11)$$

which is an operator of weights $\Delta=\bar{\Delta}=1$. In particular for $J=\widehat{J}=0$, we get the cosmological term $\exp(\alpha_-\Phi)$. The three-point function was computed in refs[13, 9]. The corresponding product of coupling constants gives the correct leg factors after drastic simplifications. After that, one may follow the line of ref.[14] and derive the higher point function. We will come back to this in the coming section.

[6] or, if more convenient by the index M.

5 The strong coupling regime

5.1 The new local fields.

At this point we turn to the strong coupling regime. Now $1 < C < 25$, and Q_L Q_M are real. The screening charges $\alpha_\pm$ are complex and related by complex conjugation. Thus complex weights appear in general. There are two types of exceptional cases. The states $|J, J>$ (resp. $|-J-1, J>$) have highest weights which are real and negative (resp. positive). One could try to work with the corresponding Liouville exponentials $\exp[-J(\alpha_- - \alpha_+)\Phi]$ (resp $\exp[((J+1)\alpha_- - \alpha_+)\Phi]$), but this would be inconsistent, since these operators do not form a closed set under fusing and braiding. Moreover, as is clear from Eqs.4.1, 4.2, they do not preserve the reality condition for highest weights just recalled. The basic problem is that Eq.4.1 involves the V operators with arbitrary $m\ \widehat{m}$, while the reality condition forces us to only use V operators of the type[7]

$$V^{(J)}_{m,+} \equiv \widetilde{V}^{(-J-1,J)}_{-m\,m}, \quad V^{(J)}_{m,-} \equiv \widetilde{V}^{(J,J)}_{m\,m}. \tag{5.1}$$

Now is a good time to recall the truncation theorems which hold for

$$C = 1 + 6(s+2), \quad s = 0, \pm 1. \tag{5.2}$$

First define the physical Hilbert space

$$\mathcal{H}^\pm_{\text{phys}} \equiv \bigoplus_{r=0}^{1\mp s} \bigoplus_{n=-\infty}^{\infty} \mathcal{H}^\pm_{r/2(2\mp s)+n/2} \tag{5.3}$$

where $\mathcal{H}^\pm_J$ denotes the Verma modules with highest weights $|\mp(J+1/2)-1/2, J>$. The physical operators $\chi^{(J)}_\pm$ are defined for arbitrary[8] $2J \in \mathcal{Z}/(2\mp s)$, and $2J_1 \in \mathcal{Z}/(2\mp s)$. to be such that[9]

$$\chi^{(J)}_\pm \mathcal{P}_{\mathcal{H}^\pm_{J_1}} = \sum_{\nu \equiv J+m \in \mathcal{Z}_+} (-1)^{(2\mp s)(2J_1+\nu(\nu+1)/2)} V^{(J)}_{m,\pm} \mathcal{P}_{\mathcal{H}^\pm_{J_1}}, \tag{5.4}$$

where $\mathcal{P}_{\mathcal{H}^\pm_{J_1}}$ is the projector on $\mathcal{H}^\pm_{J_1}$. Denote by $\mathcal{A}^\pm_{\text{phys}}$ the set of fields $\chi^{(J)}_\pm$, with $2J \in \mathcal{Z}/(2\mp s)$. The basic properties of the special values Eq.5.2 is the TRUNCATION THEOREM:

For $C = 1+6(s+2)$, $s = 0, \pm1$, and when it acts on $\mathcal{H}^+_{\text{phys}}$ (resp. $\mathcal{H}^-_{\text{phys}}$), the above set $\mathcal{A}^+_{\text{phys}}$ (resp. $\mathcal{A}^-_{\text{phys}}$) is closed by braiding and fusion and only gives states that belong to $\mathcal{H}^+_{\text{phys}}$ (resp. $\mathcal{H}^-_{\text{phys}}$).

---

[7] We could also introduce mixed operators of the type $V^{(J,J)}_{-m\,m}$ as done in ref.[3], or $V^{(-J-1,J)}_{m\,m}$, but they are not needed at present.

[8] By the symbol $\mathcal{Z}/(2\mp s)$, we mean the set of numbers $r/(2\pm s)+n$, with $r = 0, \cdots, 1\pm s$, n integer; $\mathcal{Z}$ denotes the set of all positive or negative integers, including zero.

[9] $\mathcal{Z}_+$ denotes the set of non negative integers.

Note that the operators $V_{m,\pm}^{(J)}$ themselves are not closed by fusing and braiding, contrary to the very specific combinations Eq.5.4. The proof is given in refs.[3, 9]. It follows from a neat mathematical property of the quantum group structure $U_q(sl(2)) \odot U_{\widehat{q}}(sl(2))$ recalled above. At the special values, one has, in addition $h + \widehat{h} = s\pi$. Then the q-6j symbols of the two dual quantum groups become equal up to a sign[9]. Using the orthogonality relation of the q-6j's this leads to the truncation theorems.

Next we construct local fields out of the chi fields. The braiding of the chi fields is a simple phase. On the unit circle, one has

$$\chi_{\pm}^{(J_1)} \chi_{\pm}^{(J_2)} = e^{2i\pi\epsilon(2\mp s)J_1 J_2} \chi_{\pm}^{(J_2)} \chi_{\pm}^{(J_1)}, \tag{5.5}$$

where $\epsilon = \pm 1$ is fixed by the ordering of the operator on the left-hand side in the usual way. From the spectrum of the J's, it follows that the phase factor is of the form $\exp(i\pi N/2(2 \mp s))$, where $N \in \mathcal{Z}$. Thus, we have parafermions. As shown in ref.[9], simple products of the form $\chi_{\pm}^{(J)} \overline{\chi}_{\pm}^{(\overline{J})}$, with $J - \overline{J} \in \mathcal{Z}$ are local. In such a product, the summations over m, and $\overline{m}$ are independent, while the summations over m, $\widehat{m}$, and $\overline{m}$, $\widehat{\overline{m}}$ are correlated. Now we have a complete reversal of the weak coupling situation summarized by Eq.4.1: the new fields preserve the reality condition, but **do not preserve the equality between J and $\overline{J}$ quantum numbers**. Thus we observe a sort of deconfinement of chirality in the strong coupling regime.

5.2 The Liouville string

One may consider two different problems. First, one may build a full-fledged string theory, by coupling, for instance, the above with $26 - C$ free fields $\vec{X}$. A typical string vertex is of the form $\exp(i\vec{k}.\vec{X}) \chi_{+}^{(J)} \overline{\chi}_{+}^{(\overline{J})}$, where $\vec{k}$, J, and $\overline{J}$ are related so that this is a $1,1$ operator. Here obviously, the restriction to real weight is instrumental. Moreover, since one wants the representation of Virasoro algebra to be unitary, one only uses the chi+ fields. This line was already persued with noticable success in refs.[12]. However, the N-point functions seem to be beyond reach at present. Second a simpler problem seems to be tractable, namely, we may proceed as in the construction of topological models just recalled. We consider another copy of the present strongly coupled theory, with central charge $c = 26 - C$. Since this gives $c = 1 + 6(-s + 2)$, we are also at the special values, and the truncation theorems applies to matter as well. This "string theory" has no transverse degree of freedom, and is thus topological. The complete dressed vertex operator is now

$$\mathcal{V}^{J,\overline{J}} = \chi_{+}^{(J)} \overline{\chi}_{+}^{(\overline{J})} \chi'^{(J)}_{-} \overline{\chi}'^{(\overline{J})}_{-} \tag{5.6}$$

As in the weak coupling formula, operators relative to matter are distinguished by a prime. The definition of the $\overline{\chi}$ is similar to the above, with an important difference. Clearly, the definition Eq.5.1 of $V_{m,+}^{(J)}$ is not symmetric between α_+, and α_-. The truncation theorems also holds if we interchange the two screening charges. We re-establish some symmetry between them by taking the other possible definition for $\overline{\chi}$, namely, we let

$$\overline{V}^{(\overline{J})}_{\overline{m},+} \equiv \widetilde{\overline{V}}^{(\overline{J},-\overline{J}-1)}_{\overline{m}\,-\overline{m}}, \quad V^{(\overline{J})}_{\overline{m},-} \equiv V^{(\overline{J},\overline{J})}_{\overline{m}\,\overline{m}}. \tag{5.7}$$

Our results will then be invariant by complex conjugation provided we exchange J's and $\overline{J}$'s. Thus left and right movers are interchanged, which seems to be a sensible requirement. For $J = \overline{J} = 0$, we get the new cosmological term

$$\mathcal{V}^{0,0} = \chi_+^{(0)}(z)\overline{\chi}_+^{(0)}(\bar{z}). \tag{5.8}$$

Thus the area element of the strong coupling regime is $\chi_+^{(0)}(z)\overline{\chi}_+^{(0)}(\bar{z})dzd\bar{z}$. It is factorized into a simple product of a single z component by a $\bar{z}$ component. From this expression one may compute the string susceptibility using the operator version of the DDK argument developed in ref.[13] for the weak coupling regime. For this we introduce the cosmological constant — so far it was set equal to one. In ref.[13], the weak coupling string susceptibility was rederived from the following ansatz[10]

$$\begin{aligned}\widetilde{V}^{(J\,\widehat{J})}_{m\,\widehat{m}}\Big|_{(\mu)} &= \mu^{J+\widehat{J}\alpha_+/\alpha_-}\mu^{-ip_0/\alpha_-}V^{(J\,\widehat{J})}_{m\,\widehat{m}}\,\mu^{ip_0/\alpha_-} \\ \widetilde{\overline{V}}^{(\overline{J}\,\widehat{\overline{J}})}_{\overline{m}\,\widehat{\overline{m}}}\Big|_{(\bar{\mu})} &= \bar{\mu}^{\overline{J}+\widehat{\overline{J}}\alpha_+/\alpha_-}\bar{\mu}^{-i\overline{p}_0/\alpha_-}\widetilde{\overline{V}}^{(\overline{J}\,\widehat{\overline{J}})}_{\overline{m}\,\widehat{\overline{m}}}\,\bar{\mu}^{i\overline{p}_0/\alpha_-}.\end{aligned} \tag{5.9}$$

Recall that p_0 is the zero-mode momentum of the ϑ_L free field (see Eq.2.13). We take a priori two different parameters μ and $\bar{\mu}$. In ref.[19] this ansatz was used to compute the string suceptibility, obtaining:

$$\gamma_{\mathrm{str}} = (2-s)/2 \equiv (c-1)/12. \tag{5.10}$$

The last expression is the real part of the weak coupling formula of KPZ[18]. This seems to agree with recent approximate calculations[15], which, however are performed for lower values of c. For the special values one gets

$$\left\{\begin{array}{cccc} s & c & C & \gamma_{\mathrm{str}} \\ 1 & 7 & 19 & 1/2 \\ 0 & 13 & 13 & 1 \\ -1 & 19 & 7 & 3/2 \end{array}\right., \qquad \left\{\begin{array}{cccc} s & c & C & \gamma_{\mathrm{str}} \\ 2 & 1 & 25 & 0 \\ -2 & 25 & 1 & 2 \end{array}\right. \tag{5.11}$$

The last two are the extreme points of the strong coupling regime. The values at $c = 1$, and $c = 25$ agree with the weak-coupling formula. The result is always positive, contrary to the weak-coupling regime. At $c = 7$, we find the value $\gamma_{\mathrm{str}} = 1/2$ of branched polymers.

[10] The previous discussion was actually somewhat different, since V and $\overline{V}$ operators were treated differently. This does not make any difference for the weak coupling regime, but matters at present.

5.3 The N-point functions

The method of ref.[19], is to concentrate on the N-point function with one incoming and N-1 outgoing legs. The relation between incoming and outgoing momenta is as follows. First in general[4, 5] the two-point function of two $\widetilde{V}$ fields with spins J_1, $\overline{J}_1$, and J_2, $\overline{J}_2$ vanishes unless $J_1+J_2+1=0$, and $\overline{J}_1+\overline{J}_2+1=0$. Thus conjugation involves the transformation $J \to -J-1$. Taking account of the exchange between J and $\widehat{J}$, due to complex conjugation, yields the following vertex operator for the conjugate representation:

$$\mathcal{V}^{J,\overline{J}}_{conj} = \chi_+^{(J)}\,\overline{\chi}_+^{(\overline{J})}\,\chi'^{(-J-1)}_-\,\overline{\chi}'^{(-\overline{J}-1)}_-. \tag{5.12}$$

Next, one determines the dependence of the N-point functions on the cosmological constant μ_c:

$$\left\langle \mathcal{V}^{J_1,\overline{J}_1}_{conj}\mathcal{V}^{J_2,\overline{J}_2}\ldots\mathcal{V}^{J_N,\overline{J}_N}\right\rangle_{\mu_c}$$

$$= \mu_c^{iQ_M\left(\sum_{i=1}^N[(J_i+1)\alpha_+-(\overline{J}_i+1)\alpha_-)]\right)/4}\,\mu_c^{-(N-2)}\left\langle \mathcal{V}^{J_1,\overline{J}_1}_{conj}\mathcal{V}^{J_2,\overline{J}_2}\ldots\mathcal{V}^{J_N,\overline{J}_N}\right\rangle. \tag{5.13}$$

The three-point function was computed in ref.[9] for three outgoing legs. It is a product of leg factors. In ref[19], the same was shown to hold with two outgoing and one incoming lines. With the new cosmological term, one uses the effective action

$$S_{\mu_c} = S_0 + \mu_c\int\mathcal{V}^{0,0} \tag{5.14}$$

to generate the N-point functions. Hence

$$-\frac{\partial}{\partial\mu_c}\left\langle\mathcal{V}^{J,\overline{J}}_{conj}\mathcal{V}^{J,\overline{J}}\right\rangle_{\mu_c} = \left\langle\mathcal{V}^{J,\overline{J}}_{conj}\mathcal{V}^{0,0}\mathcal{V}^{J,\overline{J}}\right\rangle_{\mu_c}. \tag{5.15}$$

and, from the expression of the three-point function,

$$\left\langle\mathcal{V}^{J,\overline{J}}_{conj}\mathcal{V}^{J,\overline{J}}\right\rangle = \frac{2}{Q_M}\left(\alpha'_-(J+\frac{1}{2})+\alpha'_+(\overline{J}+\frac{1}{2})\right)^{-1} \tag{5.16}$$

For correlation functions with removed external legs, the two-point function is the inverse of the propagator. The correct normalization turns out to involve an extra 1/2 factor so that the propagator is

$$P(J,\overline{J}) = \frac{Q_M}{4}\left(\alpha'_-(J+\frac{1}{2})+\alpha'_+(\overline{J}+\frac{1}{2})\right) \tag{5.17}$$

In ref.[19], it is shown that the above is enough to determine the N-point functions, assuming they are given by graphical rules of the Feynman type, with higher one-particle irreducible vertices determined reccursively from the generalization of Eq.5.15 to more than two point. A strong consistency check is that the irreducible vertices so obtained are symmetric between all legs, although the actual derivation is not.

5.4 connection with $c = 1$ string

In the present topological models, both matter and gravity have a background charge. By construction, the stress-energy tensor takes the usual free-field form after Bäcklund transformation. It is thus clearly possible to recombine the Liouville Bäcklund field ϑ_L, with its matter counterpart ϑ_M so that the background charge appears in one of the free fields only. For this we let

$$\vartheta_L = -\vec{X}.\vec{\mu}_L, \quad \vartheta_M = \vec{X}.\vec{\mu}_M. \tag{5.18}$$

where $\vec{X} \equiv (\varphi, X)$, and we introduce the two orthonormalized vectors

$$\vec{\mu}_L = \left(\frac{Q_L}{2\sqrt{2}}, \frac{Q_M}{2\sqrt{2}}\right), \quad \vec{\mu}_M = \left(\frac{Q_M}{2\sqrt{2}}, -\frac{Q_L}{2\sqrt{2}}\right). \tag{5.19}$$

Our conventions for $\vec{X}$ coincide with the one of ref.[16], and the present fields are identical **apart from the zero-mode spectrum.** Let $\vec{X}_0$ be the center-of-mass position. It is easy to see that $V^{(J)}_{m,\pm} \propto \exp(im\vec{k}_\pm.\vec{X}_0)$, and $V'^{(J)}_{m,\pm} \propto \exp(im\vec{k}'_\pm.\vec{X}_0)$, where $\vec{k}_- = -iQ_L\vec{\mu}_L$, $k_+ = Q_M\vec{\mu}_L$, $\vec{k}'_- = -iQ_M\vec{\mu}_M$, and $\vec{k}'_+ = Q_L\vec{\mu}_M$. Remembering the condition $J + m \in \mathcal{Z}_+$ of Eq.5.4, we see that these momenta lie on the lattice generated by the vectors

$$\frac{-i}{2\sqrt{2}}\left(1, \sqrt{\frac{2-s}{2+s}}\right), \quad \frac{1}{2\sqrt{2}}\left(\sqrt{\frac{2+s}{2-s}}, 1\right),$$
$$\frac{-i}{2\sqrt{2}}\left(1, -\sqrt{\frac{2+s}{2-s}}\right), \quad \frac{1}{2\sqrt{2}}\left(\sqrt{\frac{2-s}{2+s}}, -1\right), \tag{5.20}$$

which is itself embedded in a four dimensional space with signature 2, 2. Thus there may exist a connection between our topological theories and the $N = 2$ superstring[11]. Since $Q_M\vec{\mu}_L/2 - Q_L\vec{\mu}_M/2 = (0, \sqrt{2})$, it follows that the momenta of the $SU(2)$ generators $\exp \pm i\sqrt{2}X$ belong to the above lattice. Moreover, $k_{+X} = (2-s)/\sqrt{2}$, $k'_{+X} = -(2+s)/\sqrt{2}$. Thus, for integer $s = 0, \pm 1$ there are points of the lattice which differ by the momenta $(0, \pm\sqrt{2})$ of the $SU(2)$ generators. Of course, only the $\vec{X}_0$ dependence of the χ fields is simple. In general Eqs.5.4, and 5.6 show that $\mathcal{V}^{J,\bar{J}}$ is a rather involved function of $\vec{X}$ involving momenta of the form $m(\vec{k}_+ + \vec{k}'_-)$. Since that $(\vec{k}_+ + \vec{k}'_-)^2 = 0$, all our on-shell string states are massless, and orthogonal to each other.

5.4.1 A speculative remark

The redefinition of the cosmological term led us to modify the KPZ formula[18]. On the other hand, in standard studies of the matrix models or KP flows, one first derives γ_{str} and deduces the value of the central charge by assuming that the KPZ formula holds. In this way of thinking, one would start from our

[11] See, e.g. ref.[17].

formula Eq.5.10 and apply KPZ, which would lead to a different value of the central charge, say d. It is easy to see that for $c = 1 + 6(-s + 2)$ one gets $d = 1 - 6(2 - s)^2/2s$. This is the value of a $2, s$ minimal model! What happens is that in terms of d, we have $\gamma_{str} = (d - 1 + \sqrt{(d-1)(d-25)})/12$, in contrast with the KPZ formula $(d - 1 - \sqrt{(d-1)(d-25)})/12$. Thus the new topological theories may be another branch of $d < 1$ theories.

6 Concluding remarks

There are many additional remarks to make about the summary presented here. In particular, the vertex $\mathcal{V}_{0,0}$ of Eq.5.8 seems to define a cosmological term which differs drastically from the one (the Liouville exponential) which is relevent for the weak-coupling regime. Its study should throw light on the nature of the "c=1" barrier. From the viewpoint of conformal theories, the cosmological term is the marginal operator that takes us away from free-field theory. Thus at $c = 1$ a new marginal operator replaces the standard one, and this is why the theory for $c > 1$ looks so different. We have seen that the barrier seems to be related to a deconfinement of chirality. This is also clear on the expression of $\mathcal{V}_{0,0}$: it corresponds to a metric tensor which is a simple product of one analytic function by its anti-analytic counterpart. In a way, the surface becomes degenerate. Another remark is that the present new topological models may be simple enough so that their n-point functions are derivable in closed form.

The quantum group technology we have reviewed is clearly interesting in itself. It should be helpful, to make progress for the Liouville string theories in full-fledged space-times. In the weak coupling regime, going to continuous J also seems a key step. For instance, it allows to define the Liouville field itself[8].

The present discussion should be extendable to the $N = 1$ super-Liouville theory, making use of the quantum group structure exhibited in ref.[19]. This will be useful to study the Liouville superstrings[12], which are very interesting physically.

References

[1] J.-L. Gervais, A. Neveu, *Nucl. Phys.* **B199** (1982) 59; *Nucl. Phys.* **B209** (1982) 125; *Nucl. Phys.* **B224** (1983) 329; *Nucl. Phys.* **B238** (1984) 125; *Nucl. Phys.* **B238** (1984) 396; *Nucl. Phys.* **B257[FS14]** (1985) 59; *Comm. Math. Phys.* **100** (1985) 15; *Phys. Lett.* **B151** (1985) 271; *Nucl. Phys.* **B264** (1986) 557.

[2] J.-L. Gervais, *Comm. Math. Phys.* **130** (1990) 257.

[3] J.-L. Gervais, *Comm. Math. Phys.* **138** (1991) 301.

[4] E. Cremmer, J.-L. Gervais, J.-F. Roussel *Nucl. Phys.* **B413** (1994) 244.

[5] E. Cremmer, J.-L. Gervais, J.-F. Roussel, *Comm. Math. Phys.* **161** (1994) 597.

[6] J.-L. Gervais, J. Schnittger, *Phys. Lett.* **B315** (1993) 258;

[7] J.-L. Gervais, J. Schnittger *Nucl. Phys.* **B413** (1994) 433.

[8] J.-L. Gervais, J. Schnittger, "Continous spins in 2D gravity: chiral vertex operators and construction of the local Liouville field", hep-th/9405136.

[9] J.-L. Gervais, J.-F. Roussel, *Nucl. Phys.* **B426** (1994) 140.

[10] J.-L. Gervais, J.-F. Roussel, *Phys. Lett.* **B338** (1994) 437.

[11] J.-L. Gervais and A. Neveu, *Phys. Lett.* **B151** (1985) 271.

[12] A. Bilal, J.-L. Gervais, *Nucl. Phys.* **B284** (1987) 397, *Phys. Lett.* **B187** (1987) 39, *Nucl. Phys.* **B293** (1987) 1, *Nucl. Phys.* **B295 [FS21]** (1988) 277.

[13] J.-L. Gervais, *Nucl. Phys.* **B391** (1993) 287.

[14] P. Di Francesco, D. Kutasov, *Nucl. Phys.* **B375** (1992) 119.

[15] E. Brézin, S. Hikami, *Phys. Lett.* **B295** (1992) 209; S. Hikami, *Phys. Lett.* **B305** (1993) 327; *Prog. Theor. Phys.* **92-3** (1994) .

[16] E. Witten, *Nucl. Phys.* **B373** (1992) 187.

[17] H. Ooguri, C. Vafa, *Nucl. Phys.* **B361** (1991) 469.

[18] V. Knizhnik, A. Polyakov, A.A. Zamolodchikov, *Mod. Phys. Lett.* **A3** (1988) 819.

[19] J.-L. Gervais, B. Rostand, *Comm. Math. Phys.* **143** (1992) 175.

Electromagnetism and Gravity in Noncommutative Geometry

Kameshwar C. Wali

Physics Department, Syracuse University
Syracuse, New York, 13244-1130

It is indeed a privilege and a great honor to participate in this first symposium to commemorate the memory of my dear friend and mentor, Feza Gürsey. Feza was the foremost among our contemporaries, who continually brought new mathematical ideas into theoretical physics. Through Feza and his students, octonians, exceptional groups and such became familiar and nonexceptional. He saw far far reaching connections between deep mathematical results and fundamental physical ideas. I recall one of his many wonderful colloquia titled "Episodes in the History of Marital Relations between Mathematics and Physics; Periods of passionate involvement followed by estrangement." With a more formal title, the contents of his thoughts on this topic were presented at a conference on Symmetries in Physics (1600-1980) in Barcelona in 1983 and published in its Proceedings[1].

I was fortunate to be around him in May of 1990 when both of us were visitors at the Institut des Hautes Etudes Scientifiques in Bure-sur-Yvette, France. We attended a seminar on the application of non-commutative geometry (NCG, hereafter) to the Standard Model in Particle Physics by John Lott[2]. It was his joint work with Alain Connes, the chief proponent of NCG. In striving to understand this new development, Feza realized a more familiar way of expressing Connes' ideas at least as far as applications in particle physics were concerned. This led to our joint work and a fulfillment of my long cherished dream of working with him[3]. This paper is dedicated to his memory.

The work I am going to describe is done jointly with Giovanni Landi and Nguyen Ai Viet[4]. Within the general framework of NCG, several apparently different approaches have been utilized in constructing models in particle physics[5]. The main advantages of these varied approaches over the conventional approach in constructing the Standard Model of electroweak interactions as well as Grand Unified models are: 1) Higgs field finds a geometric origin along with the gauge fields 2) A spontaneous symmetry breaking potential arises naturally 3) One can predict the Weinberg angle, the mass of the surviving Higgs and other parameters that are normally arbitrary in the conventional approaches . In what follows, I shall adopt Connes approach in which the basic objects are an algebra $\mathcal{A}$ possibly noncommutative), a Hilbert space on which the algebra acts, and a "Dirac Operator" on the Hilbert space. The chosen algebra generalizes the idea of an underlying manifold and the operator prescribes the rule for differentiation and provides the metric structure.

In applying this general framework to the standard and grand unified models, the starting point for Balakrishna, Gürsey and myself[3] was a non-commutative algebra with the Dirac operator D,

$$D = d + Q, \tag{1}$$

where d is the usual differential operator in Minkowski space and Q is a BRST-like operator in the internal symmetry space. In Connes approach on the other hand, one has two or more copies of space-time and different abelian or nonabelian algebras in these spaces. While the desired gauge field connections are introduced in these separate spaces, the connection across the two discrete space-times brings in the Higgs fields. Thus a unified description of gauge and Higgs fields emerges and what is remarkable is that the action constructed on such a generalized space-time structure contains a spontaneous symmetry-breaking potential.

An alternate way of picturing this generalized space-time is along the lines of Kaluza -Klein theory[6], in which the extra continuous fifth dimension is replaced by two or more discrete points. The resulting theory contains in general tensor, vector and scalar fields and has the advantage of containing only a finite number of massive modes along with a single zero-mass mode. This is to be contrasted with the infinite number of massive modes in the conventional Kaluza -Klein theory associated with the extra continuous dimension and the necessity of truncating the spectrum. We begin with a brief description of the underlying mathematical scheme in the form of a Dictionary that provides a correspondence between the Riemannian geometrical and the noncommutative geometrical concepts á la Connes:

A DICTIONARY

RIEMANNIAN COMMUTATIVE GEOMETRY	*CONNES NONCOMMUTATIVE GEOMETRY*
ALGEBRA:	
$\mathcal{A} = C^\infty(\mathcal{M})$	$\mathcal{A} = C^\infty(\mathcal{M}) \otimes Z_2$
SMOOTH FUNCTIONS ON A DIFFERENTIAL MANIFOLD $f(x) \in \mathcal{A}$	$e, r \in Z_2$ $e^2 = e, er = re = r, r^2 = e$ *A* 2×2 *MATRIX REPRESENTATION* $\Pi(e) = \begin{pmatrix} 1 & 0 \\ 0 & 1 \end{pmatrix}, \Pi(r) = \begin{pmatrix} 1 & 0 \\ 0 & -1 \end{pmatrix}$ $F(x) = \tilde{f}_1 e + \tilde{f}_2 r$ $\Pi(F) = \begin{pmatrix} f_1(x) & 0 \\ 0 & f_2(x) \end{pmatrix} = \tilde{f}_1 \begin{pmatrix} 1 & 0 \\ 0 & 1 \end{pmatrix} + \tilde{f}_2 \begin{pmatrix} 1 & 0 \\ 0 & -1 \end{pmatrix}$
TANGENT BASIS: DERIVATIONS	
∂_μ	$D_N : (D_\mu,\ D_5)$
$\partial_\mu = \dfrac{\partial}{\partial x^\mu}$	$D_\mu = \begin{pmatrix} \partial_\mu & 0 \\ 0 & \partial_\mu \end{pmatrix}$
	$D_5 = \begin{pmatrix} 0 & m \\ -m & 0 \end{pmatrix}$ m is a C – number param. with dim. of mass
$\partial_\mu f = [\partial_\mu, f]$	$D_N(F) = [D_M, F]$
NEWTON – LEIBNITZ RULE	
$\partial_\mu(fg) = (\partial_\mu f)g + f(\partial_\mu g)$	$D_N(FG) = D_N(F)G + FD_N(G)$

DIFFERENTIAL ELEMENTS

$$DX^N : DX^\mu,\ DX^5$$

$$DX^\mu = \begin{pmatrix} dx^\mu & 0 \\ 0 & dx^\mu \end{pmatrix}$$

dx^μ

$$DX^5 = \begin{pmatrix} \theta & 0 \\ 0 & -\theta \end{pmatrix}$$

θ : A Clifford Element

$$\theta^2 = 1 \ ,\ \theta dx^\mu = -dx^\mu \theta$$

EXTERIOR DERIVATIVE

$$d = dx^\mu \partial_\mu \qquad D = DX^N D_N = \begin{pmatrix} d & \theta m \\ \theta m & d \end{pmatrix}$$

$$df = dx^\mu \partial_\mu f \qquad DF = DX^N D_N F = \begin{pmatrix} df_1 & \theta m(f_2 - f_1) \\ \theta m(f_1 - f_2) & d \end{pmatrix}$$

VECTOR FIELD

$$V^\mu \partial_\mu \qquad V = V^N D_N = \begin{pmatrix} v_1^\mu \partial_\mu & m v_1 \\ -m v_2 & v_2^\mu \partial_\mu \end{pmatrix}$$

1 – FORM (COVECTOR FIELD)

$$u = dx^\mu u_\mu \qquad U = DX^N U_N = \begin{pmatrix} dx^\mu u_{1\mu} & \theta u_1 \\ \theta u_2 & dx^\mu u_{2\mu} \end{pmatrix}$$

The above "Dictionary" provides the basic elements of the algebra and its differentials in the case of two discrete points in the additional "fifth" dimension.

To construct higher differential forms and a differential algebra, we define the wedge product

$$\begin{aligned} DX^\mu \wedge DX^\nu &\doteq \begin{pmatrix} dx^\mu \wedge dx^\nu & 0 \\ 0 & dx^\mu \wedge dx^\nu \end{pmatrix} \equiv -DX^\nu \wedge DX^\mu, \\ DX^5 \wedge DX^\mu &\doteq \begin{pmatrix} \theta dx^\mu & 0 \\ 0 & \theta dx^\mu \end{pmatrix} \equiv -DX^\mu \wedge DX^5, \\ DX^5 \wedge DX^5 &\doteq 0. \end{aligned} \tag{2}$$

Our choice in the last equation implies that we treat the fifth coordinate on an equal footing with the space-time coordinates. A general p-form $Wp \in \Omega^p$ is defined as follows:

$$W_p \doteq DX^{N_1} \wedge ... DX^{N_p} W_{N_1 ... N_p}. \tag{3}$$

The exterior derivative $DWp \in \Omega^{p+1}$ of a p-form $Wp \in \Omega^p$ and the wedge product of two forms W_{1p} and W_{1q} are defined to be

$$DW_p = DX^M \wedge DX^{N_1} \wedge \ldots \wedge DX^{N_p} D_M W_{N_1 \ldots N_p},$$
$$W_{1p} \wedge W_{2q} = DX^{N_1} \wedge \ldots \wedge DX^{N_{p+1}} \wedge \ldots \wedge DX^{N_{p+q}} W_{1\, N_1 \ldots N_p} W_{2\, N_{p+1} \ldots N_{p+q}}. \quad (4)$$

The exterior derivative satisfies the required properties, namely,

$$D^2 W_p = 0\,, \quad \forall p\,,$$
$$D(W_p \wedge W_q) = DW_p \wedge DW_q + (-1)^p W_p \wedge DW_q\,. \quad (5)$$

We should note that the noncommutative character of the above definitions is reflected in the fact that $W_P \wedge W_q$ is not related to $W_q \wedge W_p$ by a simple numerical factor as in the case of ordinary commutative geometry. And although the geometrical objects we construct resemble those in the usual Riemannian geometry, the noncommutative nature of our underlying geometry requires that the order of the factors be strictly respected.

Next we introduce an orthonormal basis of vielbein $E^A(A = a, 5)$ in strict analogy with the usual Riemannian geometry. The E^A are 1-forms in the Z_2 noncommutative geometry and their general form is as follows:

$$E^a \doteq \begin{pmatrix} e_1^a & \theta f_1^a \\ \theta f_2^a & e_2^a \end{pmatrix}, \quad a = 1, \cdots, 4\,,$$

$$E^5 \doteq \begin{pmatrix} a_1 & \theta\phi_1 \\ \theta\phi_2 & a_2 \end{pmatrix}, \quad (6)$$

where e_1^a and e_2^a are vielbeins on $\mathcal{M}$, a_1 a_2 are 1-forms on $\mathcal{M}$ and $f_1^a, f_2^a, \phi_1, \phi_2$ are real functions on $\mathcal{M}$.

In the present report, we consider the particular case of a self-adjoint vielbein

$$E^a \doteq \begin{pmatrix} e^a & 0 \\ 0 & e^a \end{pmatrix} = DX^\mu e_\mu^a$$

$$E^5 \doteq \begin{pmatrix} a & \theta\phi \\ \theta\phi & a \end{pmatrix} = DX^\mu a_\mu + DX^5 \begin{pmatrix} 0 & 1 \\ -1 & 0 \end{pmatrix} \phi(x). \quad (7)$$

Having a vielbein, we can construct the metric tensor G which we can consider as a functional $G: \Omega^1 x \Omega^1 \to \mathcal{A}$, such that in the E^A-basis,

$$G(E^A, E^b) = \eta^{AB}\,,$$
$$\eta^{AB} = diag((-1, 1, 1, 1, 1). \quad (8)$$

In the DX^M-basis we will have the metric

$$G^{MN} = G(DX^M, DX^N) = E^M{}_A \eta^{AB} E^N{}_B, \quad (9)$$

where $E^M{}_A$ are the inverses of $E^A{}_M$.

In a more familiar notation,

$$G = E^A \otimes E^B G_{AB} = DX^M \otimes DX^N E^A{}_M E^B{}_N G_{AB} \tag{10}$$

and

$$G_{MN} = E^A{}_M E^B{}_N G_{AB} \tag{11}$$

The definition of a covariant derivative through a connection 1-form and Cartan's structure equations follow along the lines of Riemannian geometry, except for due attention to the ordering of the factors, which to some extent is arbitrary. Thus, the covariant derivative ∇ and the connection 1-form are defined by the relation

$$\nabla E^A = E^B \otimes \Omega^A{}_B. \tag{12}$$

The Cartan structure equations define torsion and curvature for a given connection and the torsion free condition results in the equation

$$DE^A = E^B \wedge \Omega^A{}_B\ . \tag{13}$$

The above structure equation and metric compatibility condition (known as Levi-Civita connection) enable us to determine uniquely the connection 1-forms and consequently to derive the curvature 2-forms $R^A{}_B$ and their components $R^A{}_{BCD}$, $R^A{}_B = E^C \wedge E^D R^A{}_{BCD}$. After some algebra, the scalar curvature R is found to be

$$R = R_4 - 2\frac{\Box\phi}{\phi} - \frac{1}{4}\Psi ab\Psi^{ab}\ , \tag{14}$$

where

R_4 is the scalar curvature in the physical four dimensions,

$\Box = \nabla_u \partial_u,\ \nabla_u$ is the 4-dimensional covariant derivative,

$$\psi_{ab} = e^\mu_a e^\nu_b(\partial_\mu a_\nu - \partial_\nu a_\mu) + a_\mu \frac{\partial_\nu \phi}{\phi} - a_\nu \frac{\partial_\mu \phi}{\phi}.$$

Redifining the vector field $a_\mu \to \phi a_\mu$; we obtain

$$R = R_4 - \frac{2\Box\phi}{\phi} - \frac{1}{4}\phi^2 f_{\mu\nu} f^{\mu\nu}$$

where

$$f_{\mu\nu} = \partial_\mu a_\nu - \partial_\nu a_\mu. \tag{15}$$

The above expression for the scalar curvature R is identical to the one in Kaluza theory when one assumes that the physical fields are independent of the extra fifth dimension and when one retains only the zero-modes in the expansion of the compactified fifth dimension. By taking the trace of the 2 x 2 matrix. we effectively integrate over the discrete variables leading to the action

$$S \sim \int d^4x \sqrt{-det|g|}\phi R, \tag{16}$$

which reproduces the action in Kaluza's theory up to a proportionality constant. The dimensional parameter m replaces the radius of the circle in the compactified fifth dimension. Thus, with the simplified vielbein that we haveassumed, we are able to reproduce the dimensionally reduced Kaluza theory with massless tensor, vector and scalar fields. A more general, allowed vielbein provides a much richer theory with zero mass fields accompanied by massive ones.

Acknowledgements

Work supported in part by the U.S. Department of Energy under contract number DE-FG02-85ER40231. The author is also greatly indebted to Nguyen Ai Viet for many helpful discussions and for his help in preparing this manuscript.

References

[1] F. Gürsey, Proceedings of the 1st International Meeting on the History of Scientific Ideas (Published by Universitat Autonoma de Barcelona 1987), p.557-587.

[2] A.Connes and J.Lott, Nucl.Phys. **B18** (Proc.Suppl.) (1990) 29.

[3] B.S.Balakrishna, F.Gürsey and K.C. Wali, Phys. Letters **B254**, 430 (1991); Phys. Rev. **D44,** 3313 (1991), also the above authors and Nguyen Ai Viet, Phys. Rev. **D46**, 4698 (1992).

[4] G.Landi, Nguyen Ai Viet and K.C. Wali, Phys. Letters **B326** (1994) 45. See also A.E. Chamseddine,G.Felder and J.Fröhlich Comm. Math. Phys.**155** (1993) 201.

[5] The bibliography is extensive. See for instance, M.Dubois-Violette, R. Kerner and J. Madore, Class. Quant. Grav.**6**, (1989) 1709; J.Math.Phys.**31** (1990) 316. R. Coquereaux, G.Esposito-Farese and G. Vaillant, Nucl.Phys. **B353**, (1991) 689; R. Coquereaux, G. Esposito-Farese and F. Scheck, Int. J. Mod. Phys. **A7**,(1992) 6555; A. Sitarz Phys.Lett **308B** (1993) 311. A.E.Chamseddine, G. Felder and J. Fröhlich, Nucl. Phys. **B395** (1993) 672; for a detailed account of the standard model within Connes' framework of non-commutative geometry see D. Kastler Rev. Math. Phys. **5** (1993) 477 and articles to follow.

[6] Th. Kaluza, Sitzungsber. Preuss. Acad. Wiss. Phys. Math. Klasse 966 (1921); 0. Kleine, Z.F.Physik **37** (1926) 895; Y. Thirry, Comtes Rendus Paris) **226** 895.

The Geometry of W Symmetry Algebras

Paul McCloud

Department of Mathematics, Kings College London, Strand, London WC2R 2LS, UK

1 Introduction

The purpose of this presentation is to analyse the geometric conditions for the occurence of W symmetries in a general classical lagrangian field theory. The symmetries of such a theory are generated from conserved currents. It will be demonstrated that under certain specified conditions nonlinear combinations, in a sense to be made precise, of the currents may be taken. Some of these nonlinear combinations are also conserved, generating a greatly expanded set of symmetries of the theory. The possibility of nonlinear combinations of currents also allows the possibility of nonlinear current algebras. Since the principal defining feature of W algebras distinguishing them from the usual algebras occuring in classical field theories is nonlinearity, the conditions allowing for nonlinear combinations of currents are required if there are to be W symmetries in a theory.

Algebras of symmetries in a field theory are usually generated from a set of basic currents by taking linear combinations of these currents and finding the conditions on the coefficients for the resulting currents to be conserved. Thus for rigid symmetries the coefficients must be constant, whilst for gauge symmetries the coefficients may be any local functions. The algebra of symmetries may then be written as a poisson bracket algebra on the space of coefficients. When nonlinear combinations of the basic currents are permitted there are similarly conditions that need to be imposed on the combinations for the resulting currents to be conserved. In fact the W symmetries are in correspondence with the local functionals of the basic currents satisfying a chirality condition. The W algebra is thus a poisson bracket algebra on a space of chiral local functionals.

The natural framework for this work is the theory of jets and functional multivectors, where the precise nature of the underlying geometry of W symmetries is located. The W symmetries are dependent on an integrable split of the underlying manifold, such as the light cone decomposition of minkowski 2-space or the holomorphic-antiholomorphic decomposition of a riemann surface. The chiral local functionals may be defined on such a space, and these generate precisely the nonlinear combinations of the basic currents that are conserved. The W algebra is given by a chiral functional bivector, equivalently a skew adjoint matrix of chiral differential operators on the space of chiral local functionals.

Only the classical theory is considered in this presentation. When considering the quantum theory in such a general setup, the natural procedure to adopt is the BRST-BV quantisation procedure. In fact the method of jets and functional multivectors is readily exploited to generate solutions of the batalin-vilkovisky master equation, introducing chiral ghosts in a method paralleling the BRST quantisation of gauge theories.

Throughout this presentation the classical dynamical field $\phi = (\phi^\alpha)$ exists on an underlying space with coordinates $x = (x^i)$, $i = 1, \ldots, m$. Denote the partial derivatives by $\partial_i = \partial/\partial x^i$, and employ the multi-index notation for repeated derivatives: for the multi-index $I = (n_1, \ldots, n_m)$, a sequence of nonnegative integers, define $\partial_I = (\partial_1)^{n_1} \ldots (\partial_m)^{n_m}$. The summation convention extends to the multi-indices. The jet coordinates, i.e. the formal derivatives of the dynamical field, are denoted by $[\phi] = (\phi^\alpha_I)$, and the total derivatives d_I are defined to act as the partial derivatives ∂_I on functions on the underlying space and as $d_I \phi^\alpha_J = \phi^\alpha_{I+J}$ on the jet coordinates. In this way the total derivatives of functions of the dynamical field and its derivatives may be taken.

2 Symmetries and conserved currents

Consider a theory defined on the underlying space $\mathbb{R}^m$ by the local functional action $S[\phi]$. It will be assumed throughout this presentation that sufficient conditions are imposed on the integrands for the integral of a divergence to vanish, thus allowing for integration by parts. Boundary terms may be included in the analysis, but this makes the presentation unnecessarily complicated.

Symmetries of the action are given by expressions of the form:

$$\int v^\alpha \left(\frac{\delta S}{\delta \phi^\alpha} \right) d^m x = 0$$

for some differential operator $v^\alpha = v^{\alpha I} d_I$. Using integration by parts, the symmetry this corresponds to is $\delta\phi^\alpha = (-)^{|I|} d_I v^{\alpha I}$. The integrand above must be a divergence:

$$v^\alpha \left(\frac{\delta S}{\delta \phi^\alpha} \right) d^m x = dW$$

for some current $W = W^i[\phi] d_i \lrcorner\, d^m x$, equivalently:

$$v^\alpha \frac{\delta S}{\delta \phi^\alpha} = d_i W^i$$

This is the substance of the noether theorem: the current W is conserved as its divergence vanishes when the field equations are satisfied, $\delta S/\delta\phi^\alpha = 0$, and symmetries are always associated with conserved currents.

Consider now a set of basic conserved currents $W^S = W^{Si}[\phi] d_i \lrcorner\, d^m x$, satisfying:

$$v^{S\alpha} \frac{\delta S}{\delta \phi^\alpha} = d_i W^{Si}$$

for the differential operator $v^{S\alpha} = v^{S\alpha I} d_I$. The linear combination $h_S W^S$ for the functions h_S on the underlying space has divergence:

$$d(h_S W^S) = \{ W^{Si} \partial_i h_S + h_S v^{S\alpha} \frac{\delta S}{\delta \phi^\alpha} \} \, d^m x$$

and so is not necessarily conserved. By integrating the above expression the variation of the action under the field variation $\delta_h \phi^\alpha = (-)^{|I|} d_I(h_S v^{S\alpha I})$ is observed to be:

$$\delta_h S = -\int W^{Si} \partial_i(h_S)\, d^m x$$

The linear combination $h_S W^S$ generates a symmetry of the action only when this integral vanishes.

One way to achieve the vanishing of the above integral (in fact to achieve the vanishing of the integrand completely) is to have the basic currents W^S point in a perpendicular direction to the gradient of the coefficients dh_S. Split the underlying space $\mathbb{R}^m$ into chiral and antichiral components $\mathbb{R}^{m+} \times \mathbb{R}^{m-}$, not necessarily equal but with $m_+ + m_- = m$, with coordinates (x^+, x^-). Suppose this can be done in such a way that the basic currents only have antichiral components, $W^S = W^{S-}[\phi] d_- \lrcorner\, d^m x$. Then:

$$d(h_S W^S) = \{W^{S-} \partial_- h_S + h_S v^{S\alpha} \frac{\delta S}{\delta \phi^\alpha}\}\, d^m x$$

so the current $h_S W^S$ is conserved, and generates the symmetry of the action $\delta_h \phi^\alpha = (-)^{|I|} d_I(h_S v^{S\alpha I})$, when the coefficients are chiral, $\partial_- h_S = 0$.

Note that this analysis includes the cases of rigid and gauge symmetries, using the trivial decomposition $\mathbb{R}^m = \mathbb{R}^m \times 0$. For rigid symmetries the basic currents may point in any direction, $W^S = W^{Si}[\phi] d_i \lrcorner\, d^m x$, but the gradient of the coefficients must vanish, i.e. the coefficients are constant. For gauge symmetries the reverse applies, there are no conditions on the coefficients but the divergence of the basic currents must vanish, giving the gauge conditions:

$$v^{S\alpha} \frac{\delta S}{\delta \phi^\alpha} = 0$$

expressing a degeneracy in the action.

3 W symmetries

The crucial property resulting in the occurence of W symmetries is the existence of a split on the underlying space of the form $\mathbb{R}^{m+} \times \mathbb{R}^{m-}$ with the antichiral component 1-dimensional, $m_- = 1$. As in the previous section, suppose the split is achieved in such a way that the basic currents are antichiral. In this case the basic currents have only one component, and take the form $W^S d^{m+} x^+$ for the coefficient functions $W^S[\phi]$, with:

$$v^{S\alpha} \frac{\delta S}{\delta \phi^\alpha} = d_- W^S$$

Introduce an auxiliary field $b = (b^S)$. Note that now the basic currents determine a map from the dynamical field ϕ to the auxiliary field b, given by $\phi \mapsto b = W[\phi]$. In particular any function $P[b]$ of the auxiliary field and its derivatives may be evaluated at $b = W$, resulting in a function $P[W]$ of the dynamical field and its derivatives. This allows for nonlinear combinations of the basic currents to be constructed.

Evaluate the m_+-form $P[b]d^{m_+}x^+$ at $b = W$, and take the divergence:

$$d(P[W]d^{m_+}x^+) = \{\partial_- P[W] + d_- d_I(W^S)\frac{\partial P}{\partial b_I^S}[W]\}\, d^m x$$

The total derivatives commute, $d_- d_I = d_I d_-$, and $d_- W^S$ vanishes up to the field equations:

$$d(P[W]d^{m_+}x^+) = \{\partial_- P[W] + d_I v^{S\alpha}\left(\frac{\delta S}{\delta \phi^\alpha}\right)\frac{\partial P}{\partial b_I^S}[W]\}\, d^m x$$

so the current $P[W]d^{m_+}x^+$ is conserved when $\partial_- P = 0$. The symmetry generated by this current is obtained by integrating the above expression:

$$\int \frac{\delta \mathcal{P}}{\delta b^S}[W] v^{S\alpha}\left(\frac{\delta S}{\delta \phi^\alpha}\right)\, d^m x = 0$$

where $\mathcal{P}[b] = \int P[b]\, d^m x$ is a local functional of the auxiliary field, and integration by parts has been employed. The symmetry is:

$$\delta_{\mathcal{P}} \phi^\alpha = (-)^{|I|} d_I (\frac{\delta \mathcal{P}}{\delta b^S}[W] v^{S\alpha I})$$

The symmetries generated from the basic currents are in correspondence with the local functionals of the auxiliary field, satisfying the chirality condition $\partial_- P = 0$. This is a greatly expanded set of symmetries — the symmetries generated from linear combinations of the basic currents in this case arise from linear local functionals of the form $h[b] = \int h_S b^S\, d^m x$ with $\partial_- h^S = 0$.

4 The precise conditions for the existence of W symmetries

The previous analysis is a simplified version of the fully general result for arbitrary bundles on curved spaces. In the general situation the analysis is conducted locally, and care must be taken to include consideration of coordinate transformations and also the volume densities in integrals. The spirit of the analysis however is unchanged, and rather than reproduce the previous results in a more complex environment, the precise conditions for the occurence of W symmetries and the precise form these symmetries take will simply be stated here.

Consider a classical lagrangian field theory defined on an orientable, compact and boundaryless m-dimensional manifold M with local coordinates $x = (x^i)$. The conditions on the base space are required to ensure that global integrals may be defined, and that these integrals vanish if and only if the integrand is a divergence. These conditions may be relaxed provided that sufficient boundary conditions can be imposed on any integrands occuring. The dynamical field is a section $\phi = (\phi^\alpha)$ of a (not necessarily linear) bundle over M, and the field equations determine the staionary point of a local functional action $S[\phi]$. For W symmetries to occur, the required additional structures are:

1. There must be an integrable decomposition of the tangent bundle to the base space, $TM = T_+M \oplus T_-M$, into chiral and antichiral directions. Furthermore the antichiral space T_-M must be 1-dimensional. The integrability requirement implies that locally the base space may be decomposed, $M = M_+ \times M_-$, with $T_\pm M = TM_\pm$. In particular, the local coordinates may always be taken in the form $x = (x^+, x^-)$ with $\partial_\pm \in T_\pm M$. Note that the decomposition is not necessarily global. For example:

   - Minkowski 2-space — the split is determined by the light ray directions.
   - Riemann surfaces — the decomposition occurs in the complexified tangent bundle, into holomorphic and antiholomorphic vectors.

   Both these examples are important in string theory.

   - Hamiltonian manifolds — the split is into one timelike direction with the remaining spacelike directions.

   This example is important in the theory of evolution equations.

2. There is a set of locally defined antichiral conserved basic currents. Since there is only one antichiral direction, these currents must take the form $W^S[\phi]\Omega_+$ for a locally defined chiral volume form $\Omega_+ = \rho_+(x^+)d^{m_+}x^+$. The basic currents are conserved:

$$d(W^S\Omega_+) = v^{S\alpha}\left(\frac{\delta S}{\delta\phi^\alpha}\right)\Omega$$

   for a globally defined volume form $\Omega = \rho(x)d^m x$. The basic currents are only locally defined, the transformations for the S-index determine an auxiliary current bundle over the base space. Denote by the auxiliary field $b = (b^S)$ sections of this auxiliary bundle.

3. The auxiliary current bundle is a chiral vector bundle, i.e. the transformation functions of the bundle are chiral, with the transformations of the auxiliary field given by matrices of the form $\Lambda^R_T(x^+)$ via $b^R \mapsto \Lambda^R_T b^T$. The bundle twists only over the chiral component in any local decomposition of the base space. Thus it is possible to consider chiral sections of the auxiliary bundle, since the notion of chirality is preserved by transformations. For example, the chiral cotangent bundle T^*_+M (the current bundle for the antichiral component of the stress tensor) and its symmetric powers $\bigvee^s T^*_+M$ (the current bundle for higher spin generalisations) are chiral vector bundles.

Given these conditions are satisfied, the W symmetries are generated from conserved currents of the form $P[W]\Omega_+$, where $P[b]$ depends only on the chiral coordinates, the auxiliary field, and the chiral derivatives of the auxiliary field. The symmetries generated are in correspondence with the chiral local functionals of the auxiliary field, $\mathcal{P}[b] = \int P[b]\,\Omega_+$, via:

$$\int \frac{\delta \mathcal{P}}{\delta b^S}[W] v^{S\alpha} \left(\frac{\delta S}{\delta \phi^\alpha}\right) \Omega = 0$$

where the variational derivatives in the above expression for $\mathcal{P}$ and S are taken with respect to the corresponding volume forms Ω_+ and Ω. The symmetry generated by the chiral local functional $\mathcal{P}[b]$ is thus:

$$\delta_{\mathcal{P}} \phi^\alpha = (-)^{|I|} d_I \left(\frac{\delta \mathcal{P}}{\delta b^S}[W] v^{S\alpha I}\right)$$

This is the precise statement of the result. The full set of symmetries generated by the basic currents is greatly expanded by the existence of nonlinear combinations corresponding to the chiral local functionals of the auxiliary field.

5 W algebras

The existence of an expanded set of symmetries also gives greater scope for the closure of the symmetry algebra, resulting in the nonlinear W algebras. When the set of W symmetries generated by the basic currents closes, the correspondence between the chiral local functionals of the auxiliary field and the symmetries induces a poisson bracket structure on the space of chiral local functionals:

$$[v_{\mathcal{P}}, v_{\mathcal{Q}}] = v_{\{\mathcal{P},\mathcal{Q}\}}$$

The poisson bracket of chiral local functionals is given by a skew adjoint matrix of chiral differential operators D^{RT}, depending on the chiral coordinates, the auxiliary field, and the chiral derivatives of the auxiliary field, via:

$$\{\mathcal{P}, \mathcal{Q}\} = \int \frac{\delta \mathcal{P}}{\delta b^R} D^{RT} \left(\frac{\delta \mathcal{Q}}{\delta b^T}\right) \Omega_+$$

or equivalently in terms of the auxiliary field:

$$\{b^R(x_1^+), b^T(x_2^+)\} = D^{RT} \delta_{x_1^+}(x_2^+)$$

Note that if the algebra of symmetries were to close in the usual sense, i.e. on the space of linear chiral local functionals, then the operator D^{RT} would have to be linearly dependent on the auxiliary field. The possibility of nonlinear dependence of this operator on the auxiliary field results in nonlinear W algebras occuring in the classical field theory.

6 Conclusion

Using the correspondence between conserved currents and symmetries, symmetry algebras generated by a set of basic currents may be expressed in terms of the basic currents. The scope for such algebras is limited by the linearity condition — ordinarily only linear combinations of the basic currents may be taken. In order to obtain nonlinear algebras a method of combining the basic currents nonlinearly must be found. One way to achieve this is to consider basic currents that have only one component, the antichiral component of currents on a base space with a $(m-1,1)$ split in the tangent bundle. This single component function may be manipulated algebraically, in particular total derivatives and nonlinear combinations may be taken. In this way chiral local functionals of the basic currents generate symmetries of the theory. The W algebras result from closure of the symmetry algebras thus generated as expressed in the space of chiral local functionals of the basic currents.

It should be noted that W symmetries may exist on base spaces with arbitrary dimension, although the conditions for their occurence seem to be satisfied naturally only in 2-dimensional theories. The W algebras studied in 2-dimensional theories are the virasoro algebra, generated by the antichiral component of the stress tensor, and nonlinear extensions of this algebra generated by higher spin currents. Higher spin nonlinear combinations are essential for closure in these W algebras when there is only a finite set of higher spin basic currents. The possibility of higher spin, finitely generated algebras is the principal reason for the consideration of W algebras in two dimensional theories.

Quantization of a Gauge Theory of Quadratic Lie Algebras in 2d

Ömer F. Dayi

TÜBİTAK-Marmara Research Centre, Research Institute for Basic Sciences, Department of Physics, P.O.Box 21, 41470 Gebze, Turkey[1]

Abstract. A gauge theory of quadratic Lie algebras in two dimensions which possesses an open gauge algebra, is quantized by solving the master equation of Batalin and Vilkovisky in terms of the generalized field method. Moreover, W_3 algebra, formulated in terms of a continuous variable is exploit in the mentioned gauge theory to construct a W_3 topological gravity.

Keywords. Quantization, field theory.

Lie groups can be deformed by adding some quadratic terms to the commutator of their two generators. Obviously this is not a Lie algebra, nevertheless, it is known as "quadratic Lie algebra". A gauge theory Lagrangian of quadratic Lie algebras is proposed in [1]. Its gauge algebra is open i.e. it is an algebra on mass shell.

Quantization of the gauge systems in the Lagrangian formulation, possessing an open gauge algebra and/or irreducible gauge generators can be obtained in terms of the Batalin-Vilkovisky (BV)[2] scheme.

In the BV-method the main point is to find the proper solution of the master equation. For a vast class of first order systems a general solution is known[3]. It is the generalized field method. Here I present the application of this method to the gauge theory of quadratic Lie algebras[4].

The theory which we deal with is a topological quantum field theory[5]. Because it does not depend on the space-time metric and does not lead to local excitations.

W_3 algebra[6] is an example to quadratic Lie algebras. To write W_3 algebra I use continuous variables, and give a formulation of W_3 topological gravity.

A Lie algebra generated by T_a can be deformed to obtain the quadratic Lie algebra

$$[T_a, T_b] = f_{ab}{}^{c} T_c + V_{ab}^{cd} T_c T_d. \tag{1}$$

Constant coefficents f, and V, possess the symmetry properties

$$f_{ab}{}^{c} = -f_{ba}{}^{c},\ V_{ab}^{cd} = -V_{ba}^{cd},\ V_{ab}^{cd} = V_{ab}^{dc}. \tag{2}$$

[1]E-mail addresses: dayi@trmbeam.bitnet and dayi@yunus.mam.tubitak.gov.tr

The Jacobi identities should be satisfied, so that f, and V, should be chosen such that

$$f_{[ab}{}^{d}f_{c]d}{}^{e} = 0, V_{[ab}^{de}V_{c]d}^{fg} = 0,$$
$$f_{[ab}{}^{d}V_{c]d}^{ef} + V_{[ab}^{df}f_{c]d}{}^{e} + V_{[ab}^{ed}f_{c]d}{}^{f} = 0, \qquad (3)$$

are obeyed. Here, [ ] denote antisymmetrization in the indices which are within them.

In 2d space-time the Lagrange density[1]

$$\mathcal{L} = -\frac{1}{2}\epsilon^{\mu\nu}\Phi_a(\partial_\mu h_\nu^a - \partial_\nu h_\mu^a + f_{bc}{}^{a}h_\mu^b h_\nu^c + V_{bc}^{ad}\Phi_d h_\mu^b h_\nu^c) \qquad (4)$$

is invariant under the gauge transformations

$$\delta h_\mu^a = \partial_\mu \lambda^a + f_{bc}{}^{a}h_\mu^b\lambda^c + 2V_{bc}^{ad}\Phi_d h_\mu^b\lambda^c, \qquad (5)$$
$$\delta\Phi_a = f_{ba}{}^{c}\Phi_c\lambda^b + V_{ba}^{cd}\Phi_c\Phi_d\lambda^b, \qquad (6)$$

which yield an open gauge algebra

$$[\delta_\lambda, \delta_\kappa]h_\mu^a = \delta_\sigma h_\mu^a - 2\lambda^c\kappa^d V_{cd}^{ab}(\partial_\mu\phi_b + \phi_{c'}f_{ba'}^{c'}h_\mu^{a'} + \phi_{c'}\phi_{d'}V_{b'b}^{c'd'}h_\mu^{b'}),$$

where

$$\sigma^a \equiv (f_{bc}^a + 2V_{bc}^{ad}\phi_d)\lambda^b\kappa^c.$$

To find the proper solution of the (BV-) master equation whose classical limit is given by (4), introduce the generalized fields

$$\tilde{h} = h_{(1,0)} + \eta_{(0,1)} - \Phi^\star_{(2,-1)}, \qquad (7)$$
$$\tilde{\Phi} = -h^\star_{(1,-1)} - \eta^\star_{(2,-2)} + \Phi_{(0,0)}, \qquad (8)$$

where the first number in the parenthesis is the order of differential forms and the second is the ghost number. η^a are the ghost fields, and the star denotes the antifields as well as the Hodge map. Now by replacing the fields h, Φ with the generalized ones $\tilde{h}$, $\tilde{\Phi}$ one can obtain the proper solution of the master equation:

$$S = -\int d^2x\frac{1}{2}\tilde{\Phi}_a(d\tilde{h}^a + f_{bc}{}^{a}\tilde{h}^b\tilde{h}^c + V_{bc}^{ad}\tilde{\Phi}_d\tilde{h}^b\tilde{h}^c). \qquad (9)$$

It is the solution of the master equation, because S is invariant under the gauge transformation obtained as the generalization of the original ones (5)-(6).

W_3 algebra is defined in terms of the generators $G_A(z) = (T(z), W(z))$ as

$$[G_A(z), G_B(w)] = \int dv f_{AB}{}^{C}(z,w,v)G_C(v) + \int dvdr V_{AB}^{CD}(z,w,v,r)G_C(v)G_D(r), \qquad (10)$$

where ${f_{AB}}^C$, and V_{AB}^{CD} are given as

$$
\begin{aligned}
{f_{11}}^1(z,w,v) &= \partial_z\delta(z-w)\delta(z-v) - \partial_w\delta(w-z)\delta(w-v) && (11)\\
{f_{12}}^2(z,w,v) &= 3\partial_z\delta(z-w)\delta(z-v) + 2\delta(z-w)\partial_z\delta(z-v) && (12)\\
V_{22}^{11}(z,w,v,r) &= \delta(z-v)\delta(z-r)\partial_z\delta(z-w) \\
&\quad -\delta(w-v)\delta(w-r)\partial_w\delta(z-w). && (13)
\end{aligned}
$$

One can show that the Jacobi identities are satisfied. Moreover, f, and V possess the desired symmetry properties (2).

Now, to formulate a topological W_3 gravity let us introduce

$$h_\mu^A = (e_\mu, B_\mu),\ \Phi_A = (t, w).$$

Using these fields, and (11)-(13) in (4) yields

$$S_0 = \int d^2x \int dz \epsilon^{\mu\nu}(t\partial_\mu e_\nu + w\partial_\mu B_\nu + te_\mu\partial e_\nu + 3we_\mu\partial B_\nu + 2\partial we_\mu B_\nu + ttB_\mu\partial B_\nu), \tag{14}$$

where ∂ is the derivative with respect to z. Thus introduce the generalized fields

$$
\begin{aligned}
\tilde{h} &= (e_\mu, B_\mu \oplus \eta_1, \eta_2 \ominus t^\star, w^\star), \\
\tilde{\Phi} &= (\ominus e^{\star\mu}, B^{\star\mu} \oplus \eta_1^\star, \eta_2^\star \oplus t, w), && (15)
\end{aligned}
$$

to obtain the proper solution of the master equation as

$$
\begin{aligned}
S_{W_3} = S_0 &+ \int d^2x \int dz \{B^{\star\mu}[\partial_\mu\eta_2 + e_\mu\partial\eta_2 - 2\partial e_\mu\eta_2 - \partial B_\mu\eta_1 + 2B_\mu\partial\eta_1] \\
&+ e^{*\mu}[\partial_\mu\eta_1 + e_\mu\partial\eta_1 - \partial e_\mu\eta_1 + 2t(B_\mu\partial\eta_2 - \partial B_\mu\eta_2) - \epsilon_{\mu\nu}e^{\star\nu}\eta_2\partial\eta_2] \\
&+ \eta_1^\star[\eta_1\partial\eta_1 - 2t\eta_2\partial\eta_2] + \eta_2^\star[\eta_1\partial\eta_2 - 2\partial\eta_1\eta_2] \\
&- t^\star[2t\partial\eta_1 + \partial t\eta_1 + 3w\partial\eta_2 + 2\partial w\eta_2] \\
&- w^\star[\partial w\eta_1 + 3w\partial\eta_1 + 2t\partial t\eta_2 + 2tt\partial\eta_2]\}. && (16)
\end{aligned}
$$

For gauge fixing one enlarges the configuration space by introducing some new fields. Here, instead of giving the details which can be found in [4], I present the result in the gauge where the first components of the fields vanish. In this gauge the partition function is

$$Z = \sum_{[G]} \text{sign}_G(\Delta), \tag{17}$$

where

$$\Delta = \lim_{N\to\infty} \frac{\det^N \partial_1 \det^N \partial_1}{|\det^N \partial_1 \det^N \partial_1|}.$$

As a matter of fact, the main point is to clarify what is meant by the set $[G]$. This should be decided due to the properties of the target manifold[5].

References

[1] N. Ikeda and K.I. Izawa, Prog. Theor. Phys. 89 (1993) 1077.

[2] I.A. Batalin and G.A. Vilkovisky, Phys. Rew. 28 (1983) 2567.

[3] Ö.F. Dayi, Mod. Phys. Lett. A 8 (1993) 811; 2087.

[4] Ö.F. Dayi, Mod. Phys. Lett A 9 (1994) 2157.

[5] D. Birmingham, M. Blau, M. Rakowski and G. Thompson, Phys. Rep. 209 (1991) 129.

[6] A.B. Zamolodchikov, Teor. Mat. Fiz. 65 (1985) 1205.

THE BV-ALGEBRA STRUCTURE OF $\mathcal{W}_3$ COHOMOLOGY

Peter Bouwknegt and Krzysztof Pilch

Department of Physics and Astronomy, U.S.C., Los Angeles, CA 90089-0484

Abstract: We summarize some recent results obtained in collaboration with J. McCarthy on the spectrum of physical states in $\mathcal{W}_3$ gravity coupled to $c = 2$ matter. We show that the space of physical states, defined as a semi-infinite (or BRST) cohomology of the $\mathcal{W}_3$ algebra, carries the structure of a BV-algebra. This BV-algebra has a quotient which is isomorphic to the BV-algebra of polyvector fields on the base affine space of $SL(3,\mathbb{C})$. Details will appear elsewhere.

1. Introduction

Understanding the spectrum of physical states in theories of two-dimensional $\mathcal{W}$-gravity coupled to matter poses an interesting challenge. Unlike in the case of ordinary gravity, the computation of the relevant semi-infinite (or BRST) cohomology of the underlying $\mathcal{W}$-algebra appears to be very difficult, and only a small number of results have been rigorously established. One expects that by studying the structure of this cohomology space it might be possible to achieve a better understanding of (quantum) $\mathcal{W}$-geometry and string field theory. The problem is also mathematically quite interesting as it involves generalizing some of the standard techniques for computing semi-infinite cohomologies to non-linear algebras.

In this paper we summarize some recent work done in collaboration with J. McCarthy on the computation of physical states in $\mathcal{W}_3$-gravity coupled to two scalar fields, as the semi-infinite cohomology of a tensor product of two Fock space modules of the $\mathcal{W}_3$ algebra. A complete result for the cohomology is given in Conjecture 3.1, Theorem 3.2 and Corollary 3.3. We then discuss in some detail the structure of the space of physical states as a Batalin-Vilkovisky (BV) algebra and, in particular, show that it is modelled on the well-known BV-algebra of regular polyvector fields on the base affine space of $SL(3,\mathbb{C})$. The main result here is given in Theorem 4.6. For more details we refer to [1–3] and the forthcoming paper [4].

Throughout this paper we will use the notation $\mathfrak{h}$ for the Cartan subalgebra, $\mathfrak{h}^*_{\mathbb{Z}}$ for the set of integral weights, P_+ for the set of dominant integral weights, P_{++} for the set of strictly dominant integral weights, Δ_+ for the positive roots and W for the Weyl group of some Lie algebra $\mathfrak{g}$. $\mathcal{L}(\Lambda)$ will denote the finite dimensional irreducible representation of $\mathfrak{g}$ with highest weight $\Lambda \in P_+$ and $\ell(w)$ the length of $w \in W$. In the following $\mathfrak{g}$ will always refer to $\mathfrak{sl}_3$.

2. The $\mathcal{W}_3$ algebra and its modules

The $\mathcal{W}_3$ algebra with central charge $c \in \mathbb{C}$ (denoted simply by $\mathcal{W}$ in the sequel) is defined as the quotient of the free Lie algebra generated by L_m, W_m, $m \in \mathbb{Z}$, by the ideal generated by the following commutation relations (see *e.g.* the review on $\mathcal{W}$-algebras [5], and references therein).

$$\begin{aligned}
[L_m, L_n] &= (m-n)L_{m+n} + \tfrac{c}{12}m(m^2-1)\delta_{m+n,0}\,, \\
[L_m, W_n] &= (2m-n)W_{m+n}\,, \\
[W_m, W_n] &= (m-n)\left(\tfrac{1}{15}(m+n+3)(m+n+2) - \tfrac{1}{6}(m+2)(n+2)\right)L_{m+n} \\
&\quad + \beta(m-n)\Lambda_{m+n} + \tfrac{c}{360}m(m^2-1)(m^2-4)\delta_{m+n,0}\,,
\end{aligned} \tag{2.1}$$

where $\beta = 16/(22+5c)$ and

$$\Lambda_m = \sum_{n\leq -2} L_n L_{m-n} + \sum_{n>-2} L_n L_{m-n} - \tfrac{3}{10}(m+3)(m+2)L_m\,. \tag{2.2}$$

Notice that, due to the non-linearity of Λ_m in (2.1), $\mathcal{W}$ is *not* a Lie algebra. The Cartan subalgebra $\mathcal{W}_0$ of $\mathcal{W}$ is spanned by L_0 and W_0, but, because $(\operatorname{ad} W_0)$ is not diagonalizable, $\mathcal{W}$ does not admit a root space decomposition (a generalized root space decomposition, *i.e.* a Jordan normal form, does however exist). Nevertheless, it is still convenient to decompose the generators of $\mathcal{W}$ according to the $(-\operatorname{ad} L_0)$ eigenvalue, and define $\mathcal{W}_\pm = \{L_n, W_n \mid \pm n > 0\}$. However, this is not a triangular decomposition in the usual sense.

For physical applications the most interesting representations of $\mathcal{W}$ are the so-called positive energy modules, which are defined by the condition that (the energy operator) L_0 is diagonalizable with finite dimensional eigenspaces, and with the spectrum bounded from below. If the lowest energy eigenspace is one dimensional, we denote the eigenvalues of L_0 and W_0 on the highest weight state by h and w, respectively.

In particular, the Verma module $M(h,w,c)$ is defined as the (positive energy) module induced by $\mathcal{W}_-$ from an 1-dimensional representation of $\mathcal{W}_0$. By the standard argument, $M(h,w,c)$ contains a maximal submodule. We denote the corresponding irreducible quotient module by $L(h,w,c)$. The module contragradient to $M(h,w,c)$ will be denoted by $\overline{M}(h,w,c)$.

Another class of positive energy modules of $\mathcal{W}$ are the Fock space modules $F(\Lambda,\alpha_0)$, which arise in the free field realization of $\mathcal{W}$ in terms of two scalar fields (see *e.g.* [5], and references therein). The modules $F(\Lambda,\alpha_0)$ are labelled by the background charge $\alpha_0 \in \mathbb{C}$ and an $\mathfrak{sl}_3$ weight Λ.

The central charge c and the highest weights h and w of $F(\Lambda,\alpha_0)$ are given by

$$\begin{aligned}
c(\alpha_0) &= 2 - 24{\alpha_0}^2\,, \\
h(\Lambda) &= -(\theta_1\theta_2 + \theta_1\theta_3 + \theta_2\theta_3) - {\alpha_0}^2 = \tfrac{1}{2}(\Lambda, \Lambda + 2\alpha_0\rho)\,, \\
w(\Lambda) &= \sqrt{3\beta}\,\theta_1\theta_2\theta_3\,,
\end{aligned} \tag{2.3}$$

where

$$\theta_1 = (\Lambda + \alpha_0\rho, \Lambda_1)\,, \quad \theta_2 = (\Lambda + \alpha_0\rho, \Lambda_2 - \Lambda_1)\,, \quad \theta_3 = (\Lambda + \alpha_0\rho, -\Lambda_2)\,. \quad (2.4)$$

Here, Λ_1 and Λ_2 are the fundamental weights of $\mathfrak{sl}_3$, and $\rho = \frac{1}{2}\sum_{\alpha\in\Delta_+}\alpha$ is the Weyl vector. Note that $h(\Lambda)$ and $w(\Lambda)$ as in (2.3) determine Λ only up to a Weyl rotation $\Lambda \to w(\Lambda + \alpha_0\rho) - \alpha_0\rho$, $w \in W$.

The following theorem summarizes some of the known results on the structure of Fock space modules $F(\Lambda, \alpha_0)$:

Theorem 2.1 [1,2].

(i) Let $\imath'$ and $\imath''$ be the canonical ($\mathcal{W}$-) homomorphisms

$$M(h(\Lambda), w(\Lambda), c(\alpha_0)) \overset{\imath'}{\longrightarrow} F(\Lambda, \alpha_0) \overset{\imath''}{\longrightarrow} \overline{M}(h(\Lambda), w(\Lambda), c(\alpha_0))\,. \quad (2.5)$$

Then $\imath'$ (resp. $\imath''$) is an isomorphism if $i(\Lambda+\alpha_0\rho) \in \eta D_+$ (resp. $-i(\Lambda+\alpha_0\rho) \in \eta D_+$) and ${\alpha_0}^2 \leq -4$. Here $D_+ = \{\lambda \in \mathfrak{h}^* | (\lambda, \alpha) \geq 0 \ \forall \alpha \in \Delta_+\}$ denotes the fundamental Weyl chamber and $\eta \equiv \mathrm{sign}(-i\alpha_0)$.

(ii) For $c = 2$, the Fock space $F(\lambda, 0)$ is completely reducible. Explicitly, for all $\lambda \in \mathfrak{h}^*_{\mathbb{Z}}$, we have

$$F(\lambda, 0) \cong \bigoplus_{\Lambda\in P_+} m_\lambda^\Lambda \, L(h(\Lambda), w(\Lambda), 2)\,, \quad (2.6)$$

where m_λ^Λ is equal to the multiplicity of the weight λ in the irreducible finite dimensional representation $\mathcal{L}(\Lambda)$ of $\mathfrak{sl}_3$ with highest weight Λ.

3. Fock space cohomology of the $\mathcal{W}_3$ algebra

Despite the fact that $\mathcal{W}$ is not a Lie algebra, the analog of semi-infinite (or BRST-) cohomology can still be defined [6,7]. As usual, one introduces two sets of ghost operators $(b_m^{[j]}, c_m^{[j]})\,, j = 2, 3$ of conformal dimension $(j, -j+1)$, corresponding to the generators L_m and W_m, $m \in \mathbb{Z}$, respectively. These ghost operators satisfy anti-commutation relations $\{b_m^{[j]}, c_n^{[j']}\} = \delta_{m+n,0}\delta^{j,j'}$. Let F^{gh} denote the standard positive energy module. The ghost Fock space $F^{\mathrm{gh}} = \bigoplus_{n\in\mathbb{Z}} F^{\mathrm{gh},n}$ is graded by ghost number, where $\mathrm{gh}(c_m^{[j]}) = -\mathrm{gh}(b_m^{[j]}) = 1$ and the highest weight state (physical vacuum) is chosen to have ghost number 3 (*i.e.* such that states and their corresponding operators have identical ghost numbers). For any two positive energy modules V^M and V^L, such that $c^M + c^L = 100$, there exists a complex $(V^M \otimes V^L \otimes F^{\mathrm{gh},n}, d)$, graded by ghost number, and with a differential (BRST operator) d of degree 1. For an explicit formula for d, which is rather involved, we refer to [7,1,2]. We will denote the cohomology of this complex by $H(\mathcal{W}, V^M \otimes V^L)$. The cohomology relative to the Cartan subalgebra $\mathcal{W}_0$ will be denoted by $H(\mathcal{W}, \mathcal{W}_0; V^M \otimes V^L)$.

For $V^L \cong F(\Lambda^L, \alpha_0^L)$ this cohomology is interpretated as the set of physical states in $\mathcal{W}$-gravity coupled to some matter theory represented by V^M. One is interested mainly in two cases: where V^M is either a so-called minimal model $L(h^M, w^M, c^M)$ or a free field Fock space $F(\Lambda^M, \alpha_0^M)$. The minimal model case was discussed in [1,3]. The analysis of $H(\mathcal{W}, F(\Lambda^M, \alpha_0^M) \otimes F(\Lambda^L, \alpha_0^L))$ for generic α_+ (*i.e.* $\alpha_+{}^2 \notin \mathbb{Q}$ where we have parametrized $\alpha_0^M = \alpha_+ + \alpha_-$, $-i\alpha_0^L = \alpha_+ - \alpha_-$, $\alpha_+\alpha_- = -1$) was started in [7] and completed in [3]. Here we will complete the analysis, begun in [2], of a non-generic case, namely $\alpha_\pm = \pm 1$ (*i.e.* $\alpha_0^M = 0, -i\alpha_0^L = 2$ or $c^M = 2, c^L = 98$).

Because of Theorem 2.1 (ii) it suffices to compute the cohomology for the $c = 2$ irreducible $\mathcal{W}$-modules $L(\Lambda) \equiv L(h(\Lambda), w(\Lambda), 2)$

Conjecture 3.1 [4]. *Let $\Lambda \in P_+$.*

(i) The cohomology $H^n(\mathcal{W}, \mathcal{W}_0; L(\Lambda) \otimes F(\Lambda^L, 2i))$ is nontrivial only if there exist $w \in W$, $\sigma \in W \cup \{0\}$ such that

$$-i\Lambda^L + 2\rho = w^{-1}(\Lambda + \rho - \sigma\rho)\,. \tag{3.1}$$

(ii) For w, σ, Λ and Λ^L as in (3.1), the cohomology $H^n(\mathcal{W}, \mathcal{W}_0; L(\Lambda) \otimes F(\Lambda^L, 2i))$ is 1-dimensional in the following cases

$$\begin{array}{lll}
\sigma \in W\,, & \Lambda \in P_+\,, & w \in W\,,\; n = \ell(w^{-1}) - \ell(w^{-1}\sigma) + 3\,,\\
\sigma = 0\,, & \Lambda \in P_{++}\,, & w \in W\,,\; n = \ell(w^{-1}) + 1 \text{ or } \ell(w^{-1}) + 2\,,\\
\sigma = 0\,, & (\Lambda, \alpha_i) = 0\,, \Lambda \neq 0\,, & w \in \langle r_i \rangle\backslash W\,,\; n = \ell(w^{-1}) + 2\,,\\
\sigma = 0\,, & (\Lambda, \alpha_i) = 0\,, \Lambda \neq 0\,, & w \in r_i(\langle r_i \rangle\backslash W)\,,\; n = \ell(w^{-1}) + 1\,.
\end{array}$$

and vanishes otherwise.

In the case that certain weights $(\Lambda, -i\Lambda^L)$ and certain ghost number n satisfy (i) and (ii) for more than one choice of (w, σ), the above should be understood in the sense that the corresponding cohomology is nevertheless 1-dimensional.

Let us comment on the status of this conjecture. For $-i\Lambda^L + 2\rho \in P_+$ we have an isomorphism $F(\Lambda^L, 2i) \cong \overline{M}(h(\Lambda^L), w(\Lambda^L), 2)$ (see Theorem 2.1 *(i)*). By taking the (conjectured) resolutions of $L(\Lambda)$ in terms of generalized Verma modules $M(h, w, c = 2)_{N}$ [2] and using the known result for $H^n(\mathcal{W}, \mathcal{W}_0; M(h, w, c) \otimes \overline{M}(h', w', 100 - c))$, the conjecture follows (see [2] for details). [The resolution of $L(\Lambda)$ for $\Lambda \in P_{++}$ in [2] contains a minor misprint, see [4].]

For the other Weyl chambers, *i.e.* $w(-i\Lambda^L + 2\rho) \in P_+$, the conjecture is based on an analysis of the cohomology for generic α_+ in the limit $\alpha_+ \to 1$ (*i.e.* $c^M \to 2$) and passes various nontrivial consistency checks. Among others, it is consistent with duality

$$H^{6-n}(\mathcal{W}, \mathcal{W}_0; L(\Lambda) \otimes F(\Lambda^L, 2i)) \cong H^n(\mathcal{W}, \mathcal{W}_0; L(\Lambda) \otimes \bar{F}(\Lambda^L, 2i))\,, \tag{3.2}$$

where $\bar{F}(\Lambda, \alpha_0) \cong F(w_0(\Lambda + \alpha_0\rho) - \alpha_0\rho)$ denotes the module contragradient to $F(\Lambda, \alpha_0)$.

Both the conjectured resolutions of $L(\Lambda)$ as well as the result for the cohomology (Conjecture 3.1) have also been verified by extensive computer calculations using Mathematica™.

Let L be the lattice

$$L \equiv \{(\lambda,\mu) \in \mathfrak{h}^*_{\mathbb{Z}} \otimes \mathfrak{h}^*_{\mathbb{Z}} \,|\, \lambda - \mu \in \mathbb{Z}\cdot\Delta_+\}. \tag{3.3}$$

Note that, in particular,

$$(\lambda,\lambda') - (\mu,\mu') = (\lambda-\mu,\lambda') + (\mu,\lambda'-\mu') \in \mathbb{Z}, \tag{3.4}$$

for all pairs (λ,μ) and (λ',μ') in L. We will restrict the momenta $(\Lambda^M,-i\Lambda^L)$ to the lattice L. As a consequence, all the vertex operators $V_{(\Lambda^M,\Lambda^L)}(z) = \exp(i\Lambda^M\cdot\phi^M + i\Lambda^L\cdot\phi^L)(z)$ will become mutually local because of (3.4) and, moreover, one can find a set of cocycles turning the underlying BRST-complex into a Vertex Operator Algebra (VOA). This will be essential for the construction of the BV-algebra in Section 3. In addition, the most interesting cohomology happens to be situated at $(\Lambda^M,-i\Lambda^L)\in L$.

Now consider the cohomologies

$$\begin{aligned}\mathcal{H} &= \bigoplus_{(\Lambda^M,-i\Lambda^L)\in L} H(\mathcal{W}, F(\Lambda^M,0)\otimes F(\Lambda^L,2i)),\\ \mathcal{H}_{\rm rel} &= \bigoplus_{(\Lambda^M,-i\Lambda^L)\in L} H(\mathcal{W},\mathcal{W}_0; F(\Lambda^M,0)\otimes F(\Lambda^L,2i)).\end{aligned} \tag{3.5}$$

We recall

Theorem 3.2 [1,2].
(i) $\mathcal{H}$ *(and* $\mathcal{H}_{\rm rel}$*) carries the structure of a* $\mathfrak{g}\oplus\mathfrak{h}$ *module (*$\mathfrak{g}\cong\mathfrak{sl}_3$*). The action of* $\mathfrak{g}$ *is through the zero modes of the Frenkel-Kac-Segal vertex operator construction (in matter fields only), while* $\mathfrak{h}$ *acts as* $-ip^L$ *(with eigenvalues* $-i\Lambda^L$*). This module is completely reducible under* $\mathfrak{g}\oplus\mathfrak{h}$*.*
(ii) There exists a (non-canonical) isomorphism (as $\mathfrak{g}\oplus\mathfrak{h}$ *modules)*

$$\mathcal{H}^i \cong \mathcal{H}^i_{\rm rel}\oplus\mathcal{H}^{i-1}_{\rm rel}\oplus\mathcal{H}^{i-1}_{\rm rel}\oplus\mathcal{H}^{i-2}_{\rm rel}.$$

By combining the results of Theorems 2.1, 3.2 and Conjecture 3.1, we find

Corollary 3.3. *The cohomology* $\mathcal{H}_{\rm rel}$ *is isomorphic (as a* $\mathfrak{g}\oplus\mathfrak{h}$ *module) to the direct sum of irreducible modules* $\mathcal{L}(\Lambda)\otimes\mathbb{C}_{\Lambda'}$ *with momenta* $(\Lambda,\Lambda')\in\mathfrak{h}^*_{\mathbb{Z}}\otimes\mathfrak{h}^*_{\mathbb{Z}}$ *lying in a set of disjoint cones* $\{\mathcal{S}^n_w + (\lambda, w^{-1}\lambda)\,|\,\lambda\in P_+\}$*, i.e.*

$$\mathcal{H}^n_{\rm rel} \cong \bigoplus_{w\in W}\ \bigoplus_{(\Lambda,\Lambda')\in\mathcal{S}^n_w}\ \bigoplus_{\lambda\in P_+} \left(\mathcal{L}(\Lambda+\lambda)\otimes\mathbb{C}_{\Lambda'+w^{-1}\lambda}\right),$$

where the sets $\mathcal{S}^n_w$ *(tips of the cones) are given in Table 1.*

space is given by $\mathbb{C}[x^i, y_i]/\langle x^i y_i \rangle$ $(i = 1,2,3)$, *i.e.* polynomials in 6 variables x^i, y_i transforming in the **3** and $\bar{\mathbf{3}}$ of $\mathfrak{sl}_3$ respectively, with a single relation $x^i y_i = 0$ [9]. In fact, one can show that $\mathcal{H}^0 \cong \mathcal{P}^0(A)$ as algebras [4]. One might think that, just as in the Virasoro case (corresponding to $\mathfrak{g} \cong \mathfrak{sl}_2$) [10–12], part of the rest of $\mathcal{H}$ allow an interpretation in terms of polyvector fields on this base affine space. This turns out to be true and will be elaborated on in the next section.

4. The BV-structure of $\mathcal{H}$

To explain the algebraic structure of the cohomology $\mathcal{H}$ of Section 2 we will first need to recall the definition of a Gerstenhaber algebra (or G-algebra, for short) [13] and a BV-algebra (or coboundary G-algebra) [14–16,12] as well as some basic facts.

Definition 4.1. *A G-algebra* $(\mathcal{A}, \cdot, [\ ,\ ])$ *is a* $\mathbb{Z}$*-graded, supercommutative, associative algebra* $\mathcal{A} = \bigoplus_{i\in\mathbb{Z}} \mathcal{A}^i$ *(under* $\cdot$*) as well as a* $\mathbb{Z}$*-graded Lie superalgebra (under* $[\ ,\ ]$*), such that the (odd) bracket acts as a superderivation of the algebra, i.e.*

$$[x, y\cdot z] = [x,y]\cdot z + (-1)^{(|x|-1)|y|} y\cdot[x,z]\,, \qquad x,y,z\in\mathcal{A}\,. \tag{4.1}$$

For any commutative algebra $\mathcal{A}$ and $\mathcal{A}$-module $\mathcal{M}$, one defines the the set $\mathcal{D}(\mathcal{A},\mathcal{M})$ of derivations of $\mathcal{A}$ with coefficients in $\mathcal{M}$ as the set of elements $D \in \mathrm{Hom}(\mathcal{A},\mathcal{M})$ that satisfy the Leibniz rule

$$D(x\cdot y) = y(Dx) + x(Dy)\,. \tag{4.2}$$

The set $\mathcal{D}^n(\mathcal{A})$ of polyderivations of order n is defined by induction as those $D \in \mathrm{Hom}(\mathcal{A}, \mathcal{D}^{n-1}(\mathcal{A}))$ satisfying the Leibniz rule (4.2) as well as being completely antisymmetric when considered as elements of $\mathrm{Hom}(\mathcal{A}^{\otimes n},\mathcal{A})$. We recall

Theorem 4.2 [17]. *Let* $\mathcal{A}$ *be a commutative algebra. The set of polyderivations* $\mathcal{D}(\mathcal{A})$ *carries the structure of a G-algebra, with the bracket given by the Schouten bracket.*

Another example of a G-algebra is the Hochschild cohomology $H(\mathcal{A},\mathcal{A})$ of an associative algebra $\mathcal{A}$ [13].

Definition 4.3. *A BV-algebra* $(\mathcal{A}, \cdot, \Delta)$ *is a* $\mathbb{Z}$*-graded, supercommutative, associative algebra* $\mathcal{A}$ *with a second order derivation* Δ *(BV-operator) of degree* -1 *satisfying* $\Delta^2 = 0$.

Lemma 4.4 [18,12,16]. *For any BV-algebra* $(\mathcal{A}, \cdot, \Delta)$ *we may define an odd bracket by*

$$[x,y] = (-1)^{|x|}\left(\Delta(x\cdot y) - (\Delta x)\cdot y - (-1)^{|x|} x\cdot(\Delta y)\right)\,, \qquad x,y\in\mathcal{A}\,. \tag{4.3}$$

This will equip $\mathcal{A}$ *with the structure of a G-algebra. Moreover, the BV-operator acts as a superderivation of the bracket*

$$\Delta[x,y] = [\Delta x, y] + (-1)^{|x|-1}[x,\Delta y]\,. \tag{4.4}$$

n	w	$\mathcal{S}_w^n$
0	1	$(0,0)$
1	1	$(\Lambda_1, -\Lambda_1+\Lambda_2)$, $(\Lambda_1+\Lambda_2, 0)$, $(\Lambda_2, \Lambda_1-\Lambda_2)$
	r_1	$(0, -2\Lambda_1+\Lambda_2)$
	r_2	$(0, \Lambda_1-2\Lambda_2)$
2	1	$(2\Lambda_1, -\Lambda_1)$, $(0, -\Lambda_1-\Lambda_2)$, $(2\Lambda_2, -\Lambda_2)$
	r_1	$(\Lambda_1, -2\Lambda_1)$, $(\Lambda_2, -3\Lambda_1+\Lambda_2)$, $(0, -4\Lambda_1+2\Lambda_2)$
	r_2	$(\Lambda_2, -2\Lambda_2)$, $(\Lambda_1, \Lambda_1-3\Lambda_2)$, $(0, 2\Lambda_1-4\Lambda_2)$
	r_1r_2	$(0, -3\Lambda_1)$
	r_2r_1	$(0, -3\Lambda_2)$
3	1	$(\Lambda_1+\Lambda_2, -\Lambda_1-\Lambda_2)$
	r_1	$(\Lambda_2, -2\Lambda_1-\Lambda_2)$, $(\Lambda_1, -4\Lambda_1+\Lambda_2)$, $(\Lambda_2, -5\Lambda_1+2\Lambda_2)$
	r_2	$(\Lambda_1, -\Lambda_1-2\Lambda_2)$, $(\Lambda_2, \Lambda_1-4\Lambda_2)$, $(\Lambda_1, 2\Lambda_1-5\Lambda_2)$
	r_1r_2	$(\Lambda_1, -3\Lambda_1-\Lambda_2)$, $(0, -5\Lambda_1+\Lambda_2)$, $(\Lambda_1, -5\Lambda_1)$
	r_2r_1	$(\Lambda_2, -\Lambda_1-3\Lambda_2)$, $(0, \Lambda_1-5\Lambda_2)$, $(\Lambda_2, -5\Lambda_2)$
	$r_1r_2r_1$	$(0, -2\Lambda_1-2\Lambda_2)$
4	r_1	$(0, -4\Lambda_1-\Lambda_2)$
	r_2	$(0, -\Lambda_1-4\Lambda_2)$
	r_1r_2	$(\Lambda_1, -4\Lambda_1-2\Lambda_2)$, $(\Lambda_2, -5\Lambda_1-\Lambda_2)$, $(0, -6\Lambda_1)$
	r_2r_1	$(\Lambda_2, -2\Lambda_1-4\Lambda_2)$, $(\Lambda_1, -\Lambda_1-5\Lambda_2)$, $(0, -6\Lambda_2)$
	$r_1r_2r_1$	$(0, -3\Lambda_1-3\Lambda_2)$, $(2\Lambda_1, -4\Lambda_1-3\Lambda_2)$, $(2\Lambda_2, -3\Lambda_1-4\Lambda_2)$
5	r_1r_2	$(0, -5\Lambda_1-2\Lambda_2)$
	r_2r_1	$(0, -2\Lambda_1-5\Lambda_2)$
	$r_1r_2r_1$	$(\Lambda_1, -5\Lambda_1-3\Lambda_2)$, $(\Lambda_1+\Lambda_2, -4\Lambda_1-4\Lambda_2)$, $(\Lambda_2, -3\Lambda_1-5\Lambda_2)$
6	$r_1r_2r_1$	$(0, -4\Lambda_1-4\Lambda_2)$

Table 1. The sets $\mathcal{S}_w^n$

In particular we see that, as an $\mathfrak{sl}_3$ module, the 'ground ring' $\mathcal{H}^0$ decomposes as $\mathcal{H}^0 \cong \bigoplus_{\Lambda\in P_+} \mathcal{L}(\Lambda)$ and is therefore a so-called 'model space' for $\mathfrak{sl}_3$. It is well-known that this model space can be realized as the space $\mathcal{P}^0(A)$ of polynomial functions on the so-called 'base-affine space' $A \equiv N_+\backslash G$ [8]. For $\mathfrak{sl}_3$ this model

In general, given a commutative algebra $\mathcal{A}$, the G-algebra $\mathcal{D}(\mathcal{A})$ of polyderivations of $\mathcal{A}$ will not carry the structure of a BV-algebra. However, if $\mathcal{A}$ is the algebra of (smooth or polynomial) functions on some smooth manifold M, then $\mathcal{D}(\mathcal{A})$ is isomorphic to the set of polyvector fields $\mathcal{P}(M)$ on M [17]. If, moreover, M possesses a volume form, then we can in fact equip $\mathcal{D}(\mathcal{A})$ $(= \mathcal{P}(M))$ with the structure of a BV-algebra [18,12]. Another example of a BV-algebra is the Grassmann algebra $\bigwedge^* \mathfrak{g}$ of a Lie algebra $\mathfrak{g}$ [12].

Given a BV-algebra $(\mathcal{A}, \cdot, \Delta)$, let $\mathcal{A}^0$ be its 'ground ring.' It follows from equations (4.1), (4.3) and (4.4) that there exists a natural way to embed $\mathcal{A}$ into the G-algebra of polyderivations of $\mathcal{A}^0$, *i.e.* $\mathcal{D}(\mathcal{A}^0)$, namely

Theorem 4.5. *Let $(\mathcal{A}, \cdot, \Delta)$ be a BV-algebra. Suppose $\mathcal{A}^n = 0$ for all $n < 0$.*

(i) There exists a homomorphism of G-algebras $\pi : \mathcal{A} \to \mathcal{D}(\mathcal{A}^0)$ defined by

$$\pi(y)(x_1, x_2, \ldots, x_n) = [[\ldots[[y, x_1], x_2], \ldots], x_n], \tag{4.5}$$

for $y \in \mathcal{A}^n$, $x_1, x_2, \ldots, x_n \in \mathcal{A}^0$.

(ii) Suppose that the G-algebra $\mathcal{D}(\mathcal{A}^0)$ admits a BV-structure $(\mathcal{D}(\mathcal{A}^0), \cdot, \Delta')$ and that $\pi\Delta(x) = \Delta'\pi(x)$ for all $x \in \mathcal{A}^1$, then π is a BV-homomorphism and $\mathcal{I} \equiv \operatorname{Ker} \pi$ is a BV-ideal of $\mathcal{A}$.

We are now ready to state the main result of this paper

THEOREM 4.6. *Let $\mathcal{H}$ be the cohomology defined in (3.5). Then*

(i) $\mathcal{H}$ can be equipped with the structure of a BV-algebra.

(ii) There exists an ideal $\mathcal{I} \subset \mathcal{H}$ such that we have an exact sequence of BV-algebras

$$0 \longrightarrow \mathcal{I} \longrightarrow \mathcal{H} \stackrel{\pi}{\longrightarrow} \mathcal{D}(\mathcal{H}^0) \longrightarrow 0, \tag{4.6}$$

where $\mathcal{D}(\mathcal{H}^0)$ is isomorphic to the BV-algebra $\mathcal{P}(A)$ of polyvector fields on the base affine space $A = N_+ \backslash G$.

Let us make some comments on the proof. Quite generally, as has been shown in [10–12,16], BRST cohomologies of VOA's carry the structure of a BV-algebra. The product in this BV-algebra is given by the normal ordered product of the VOA while $\Delta = b_0^{[2]}$. The crucial part of the proof of *(i)* is therefore to show that the complex carries the structure of a VOA. This amounts to showing that one can find an appropriate set of cocycles for the lattice L. This is a straightforward exercise. [One might wonder whether there exists additional structure in $\mathcal{H}$ beyond that of a BV-algebra, in particular whether $b_0^{[3]}$ gives rise to a second BV-operator. It turns out however that, due to the non-diagonalizability of W_0, $b_0^{[3]}$ does *not* act on $\mathcal{H}$.] As we have seen in Section 2, there exists a canonical isomorphism of algebras $\mathcal{H}^0 \cong \mathcal{P}^0(A)$, where $\mathcal{P}^0(A)$ denotes the (commutative) algebra of polynomials on A. This implies $\mathcal{D}(\mathcal{H}^0) \cong \mathcal{P}(A)$ as algebras. That π is in fact a BV-epimorphism follows from Theorem 4.5 by explicitly checking that π intertwines the BV-operators on $\mathcal{H}^1$ and $\mathcal{P}^1(A)$ and that it acts onto.

We would like to remark here that, contrary to the Virasoro case [12], both the dot product and the bracket in $\mathcal{I}$ are not identically zero. Also, the exact sequence (4.6) splits both as an exact sequence of $\mathcal{D}(\mathcal{H}^0)$ and $\mathfrak{g} \oplus \mathfrak{h}$ modules, but *not* as an exact sequence of BV-algebras.

Details of this paper as well as a more detailed analysis of the BV-algebra structure of the entire $\mathcal{H}$ will appear elsewhere [4].

Acknowledgement: We would like to thank the organizers of the "1st Gürsey Memorial Conference" for the opportunity to present this talk, and C. Thielemans for making available to us his Mathematica package `OPEdefs` [19].

References

[1] P. Bouwknegt, J. McCarthy and K. Pilch, Lett. Math. Phys. **29** (1993) 91.
[2] P. Bouwknegt, J. McCarthy and K. Pilch, in "Perspectives in Mathematical Physics," Vol. III, *eds.* R. Penner and S.T. Yau, pp. 77–89, (International Press, 1994).
[3] P. Bouwknegt, J. McCarthy and K. Pilch, in the proceedings of the workshop "Strings, Conformal Models and Topological Field Theory," Cargèse 1993, *eds.* L. Baulieu et. al., (Plenum Press, New York, 1994).
[4] P. Bouwknegt, J. McCarthy and K. Pilch, in preparation.
[5] P. Bouwknegt and K. Schoutens, Phys. Rep. **223** (1993) 183.
[6] J. Thierry-Mieg, Phys. Lett. **197B** (1987) 368.
[7] M. Bershadsky, W. Lerche, D. Nemeschansky and N.P. Warner, Phys. Lett. **292B** (1992) 35; Nucl. Phys. **B401** (1993) 304.
[8] I.N. Bernstein, I.M. Gel'fand and S.I. Gel'fand, in "Lie groups and their representations," Proc. Summer School in Group Representations, Bolyai Janos Math. Soc., Budapest 1971, pp. 21-64, (New York, Halsted, 1975).
[9] D.P. Zhelobenko, "Compact Lie groups and their representations," (Providence, Amer. Math. Soc., 1973).
[10] E. Witten, Nucl. Phys. **B373** (1992) 187; E. Witten and B. Zwiebach, Nucl. Phys. **B377** (1992) 55.
[11] Y.-S. Wu and C.-J. Zhu, Nucl. Phys. **B404** (1993) 245.
[12] B.H. Lian and G.J. Zuckerman, Comm. Math. Phys. **154** (1993) 613.
[13] M. Gerstenhaber, Ann. Math. **78** (1962) 267.
[14] J.-L. Koszul, Astérisque (hors série) (1985) 257.
[15] E. Getzler, *Batalin-Vilkovisky algebras and two-dimensional topological field theories* (`hep-th/9212043`).
[16] M. Penkava and A. Schwarz, *On some algebraic structures arising in string theory* (`hep-th/9212072`).
[17] I.S. Krasil'shchik, Lect. Notes in Math. **1334** (1988) 79.
[18] E. Witten, Mod. Phys. Lett. **A5** (1990) 487.
[19] C. Thielemans, Int. J. Mod. Phys. **C2** (1991) 787.

Black Hole Solution of Topologically Massive Gravity

Y. Nutku

TÜBİTAK Marmara Research Center, Research Institute for Basic Sciences, Department of Physics, 41470 Gebze, Turkey

The dynamical theory of gravity in $2+1$ dimensions is Deser, Jackiw and Templeton's theory of topologically massive gravity [1]. In the DJT field equations the Einstein tensor is coupled to the Cotton tensor which is the conformal tensor for 3-dimensional manifolds

$$R^i{}_k - \frac{1}{2}\delta^i{}_k R + \frac{1}{\mu}\frac{1}{\sqrt{-g}}\epsilon^{imn}\left(R_{km} - \frac{1}{4}g_{km}R\right)_{;n} = 0 \tag{1}$$

where R_{ik} is the Ricci tensor and μ is the DJT coupling constant.

We have presented [2] an exact solution of eqs.(1) which can be interpreted as the field of an isolated spinning mass. This is given by the metric

$$\begin{aligned} ds^2 \;&=\; -\tfrac{1}{6}\,(2J-M)\,dt^2 \;+\; 2\left(\tfrac{1}{3}\mu r^2 - \tfrac{1}{2\mu}J\right)dt d\theta \\ &+\frac{dr^2}{\frac{\mu^2 r^2}{36} - \frac{M}{6} + \frac{J^2}{4\mu^2 r^2}} + r^2\left(1 - \frac{\mu^2 r^2}{2(2J-M)}\right)d\theta^2 \end{aligned} \tag{2}$$

where we require the parameters M, J to be positive and $1/2 < J/M < 1$ for a black hole solution. In this solution there is an inner and outer horizon which are located at

$$r_{\pm}{}^2 = \frac{3}{\mu^2}\left(M \pm \sqrt{M^2 - J^2}\right) \tag{3}$$

but there is no singularity anywhere. In fact this solution is a homogeneous space, the Vuorio solution [3], which has been re-identified as in the Bañados-Teitelboim-Zanelli black hole solution [4] of the 3-dimensional Einstein equations with a cosmological constant. The physical components of the Riemann tensor and therefore all the curvature scalars are constants which depend only on μ. In the coordinate system we have used above the metric is regular at the origin of polar coordinates but it is not asymptotically flat. In fact at the radius

$$r_\theta{}^2 = \frac{2}{\mu^2}\,(2J-M) \tag{4}$$

the circumference of circles shrinks back to zero.

References

[1] S. Deser, R. Jackiw and S. Templeton, Phys. Rev. Lett. **48** (1982) 975; Ann. Phys. (NY) **140** (1982) 372

[2] Y. Nutku, Class. and Quantum Grav. **10** (1993) 2657

[3] I. Vuorio, Phys. Lett. B **163** (1985) 91

[4] M. Bañados, C. Teitelboim and J. Zanelli, Phys. Rev. Lett. **69** (1992) 1849

Linearized Gravity in "Closed" Universes with Continuous Symmetries

Atsushi Higuchi

Institut für theoretische Physik, Universität Bern, Sidlerstrasse 5, CH-3012 Bern, Switzerland

Keywords. general relativity, quantum field theory of gravitation

It has been pointed out that all the states in linearized quantum gravity in spacetime with compact Cauchy surfaces and with continuous symmetries are required to be invariant under those symmetries [1]. For de Sitter spacetime this requirement is rather paradoxical because the vacuum is the only state that is invariant under the de Sitter group $SO(4,1)$ in the Fock space of linearized gravity. It has been shown that there are infinite-norm invariant states and that they can be used to construct a new Hilbert space of invariant states by factoring out infinity from the norm of these states [2]. Here, I will review how this requirement arises and argue that it is a natural consequence of the underlying diffeomorphism invariance of the full quantum gravity and explain briefly the results of Ref. [2].

It is well known that some solutions to linearized gravity in spacetime with continuous symmetries and with compact Cauchy surfaces are spurious in the sense that they do not correspond to any exact solutions [3]. This fact can be seen as follows. In the Hamiltonian formulation [4] one parametrizes the line element as $ds^2 = -(N^2 - N^a N_a)dt^2 + 2N_a dx^a dt + g_{ab}dx^a dx^b$, where indices are raised and lowered by g_{ab}. The lapse function N and the shift vector N_a have no conjugate momenta. Hence, there are corresponding constraints

$$\mathcal{H} = \frac{1}{\sqrt{g}}\left(\pi^{ab}\pi_{ab} - \frac{1}{2}\pi^2\right) - \sqrt{g}\,{}^{(3)}R \approx 0\,, \quad \mathcal{H}_a = -2D_b\pi^b{}_a \approx 0\,, \tag{1}$$

where π^{ab} are the conjugate momenta of g_{ab}. These constraints generate diffeomorphisms in the following sense. Let $g_{\mu\nu}$ be a solution to the Einstein equations, $R_{\mu\nu} - (1/2)g_{\mu\nu} + \Lambda g_{\mu\nu} = 0$. Then consider the corresponding initial data (g_{ab}, π^{ab}) on a Cauchy surface Σ. Let $\delta_V g_{ab}$ and $\delta_V \pi^{ab}$ be the variations of g_{ab} and π^{ab} corresponding to $\delta g_{\mu\nu} = \mathcal{L}_V g_{\mu\nu}$, where $\mathcal{L}_V$ is the Lie derivative with respect to the vector field V^μ. Then, decompose V^μ on Σ as $V^\mu = n^\mu V_\perp + V_{||}^\mu$, where n^μ is the unit normal to Σ and $V_{||}^\mu$ is tangent to Σ. Then define

$$\Psi_V \equiv \int d\Sigma\left(V_\perp \mathcal{H} + V_{||}^a \mathcal{H}_a\right)\,. \tag{2}$$

One finds

$$\{g_{ab}, \Psi_V\}_{\rm PB} \approx \delta_V g_{ab}\,, \quad \left\{\pi^{ab}, \Psi_V\right\}_{\rm PB} \approx \delta_V \pi^{ab}\,, \tag{3}$$

where $\{\ ,\ \}_{\rm PB}$ is the Poisson bracket. We have used '$\approx$' instead of '$=$' to emphasize that the equalities hold only if (g_{ab}, π^{ab}) satisfies the constraints (1).

Now, consider solving the constraint $\Psi_V = 0$ perturbatively about a solution $(g_{ab}^{(0)}, \pi^{(0)ab})$. We let $g_{ab} = g_{ab}^{(0)} + h_{ab}$, $\pi^{ab} = \pi^{(0)ab} + p^{ab}$ and expand the constraint as $\Psi_V = \Psi_V^{(1)}(h,p) + \Psi_V^{(2)}(h,p) + \cdots$, where $\Psi_V^{(n)}$ is of order n in (h_{ab}, p^{ab}). A first order solution $(h_{ab}^{(1)}, p^{(1)ab})$ satisfies $\Psi_V^{(1)}(h^{(1)}, p^{(1)}) = 0$. Then, the second order solution $(h_{ab}^{(2)}, p^{(2)ab})$ is obtained by requiring

$$\Psi_V^{(1)}(h^{(2)}, p^{(2)}) = -\Psi_V^{(2)}(h^{(1)}, p^{(1)}). \tag{4}$$

Now, let $V^\mu = X^\mu$ where X^μ is a Killing vector. Then, $\mathcal{L}_X g_{\mu\nu}^{(0)} = 0$ by definition. This implies that $\delta_X g_{ab}^{(0)} = 0$ and $\delta_X \pi^{(0)ab} = 0$. Thus, from (3) we find that the first order variation of Ψ_X with respect to g_{ab} and p^{ab} about $(g_{ab}^{(0)}, \pi^{(0)ab})$ vanishes. Then, from (4) we conclude that $(h_{ab}^{(1)}, p^{(1)ab})$ must satisfy

$$\Psi_X^{(2)}(h^{(1)}, p^{(1)}) = 0. \tag{5}$$

Now, regard the fields $(h_{ab}^{(1)}, p^{(1)ab})$ as fields in the background spacetime with the metric $g_{\mu\nu}^{(0)}$. One can show that the constraint $\Psi_X^{(2)}(h^{(1)}, p^{(1)})$ generates the symmetry corresponding to the Killing vector X^μ on these fields, i.e., $\left\{h_{ab}^{(1)}, \Psi_X^{(2)}\right\}_{\mathrm{PB}} \approx \delta_X h_{ab}^{(1)}$, $\left\{p^{(1)ab}, \Psi_X^{(2)}\right\}_{\mathrm{PB}} \approx \delta_X p^{(1)ab}$, where δ_X indicates the variation under the isometry X^μ.

The condition (5) is implemented naturally in linearized *quantum* gravity by requiring that the physically allowed states be annihilated by $\Psi_X^{(2)}$. That is,

$$\Psi_X^{(2)}|\mathrm{phys}\rangle = 0. \tag{6}$$

Thus, the physical states are required to have the symmetry generated by $\Psi_X^{(2)}$. This condition can be understood intuitively as follows. Since Einstein's theory of gravity is diffeomorphism invariant, exact wavefunctions of its quantum theory must be diffeomorphism invariant. When one expands it about a classical background spacetime without symmetry, the linearized wavefunction does not possess any invariance. However, if the background has symmetries, then there remains part of diffeomorphism group which is not "destroyed" by the choice of background spacetime. Then, these residual symmetries are reflected in linearized gravity as the requirement (6).

When the symmetry group of the background is of finite volume, one can directly impose the requirement (6) on the Fock space. But if it has an infinite volume, one needs a new Hilbert space. This situation can be illustrated by a simple model described by $H = (1/2)p_x^2 + V(x) + Np_y$. This theory has a constraint $p_y \approx 0$. At the quantum level, this implies that the wavefunction should not depend on y. The theory is nothing but the one-dimensional quantum mechanics with inner product

$$(\Phi_1|\Phi_2) = \int dx\, \overline{\Phi_1(x)}\Phi_2(x). \tag{7}$$

Now, pretend for a moment that we did not know that this was actually a one-dimensional theory. Then we would introduce an inner product

$$\langle\phi_1|\phi_2\rangle = \int dxdy\, \overline{\phi_1(x,y)}\phi_2(x,y). \tag{8}$$

(This can be finite only if $\phi_1(x,y)$ or $\phi_2(x,y)$ is noninvariant under y-translation generated by p_y.) One can obtain an invariant state as

$$\Phi(x) \equiv \int d\alpha\, e^{i\alpha p_y}\phi(x,y) = \int d\alpha\, \phi(x, y+\alpha)\,. \tag{9}$$

Let $\Phi_1(x)$ and $\Phi_2(x)$ be the invariant states obtained in this manner from finite-norm states $\phi_1(x,y)$ and $\phi_2(x,y)$. Then, define the new inner product by

$$(\Phi_1|\Phi_2) = \int d\alpha\, \langle\phi_1|e^{i\alpha p_y}|\phi_2\rangle\,. \tag{10}$$

It is clear that this inner product agrees with the correct one given by eq. (7).

This construction of invariant states and inner product can be generalized to the case where the constraints are more complicated, in particular, when they generate a nonabelian group. One takes a (noninvariant) state, $|\psi\rangle$, in the Fock space and "smear" it over the symmetry group as

$$|\Psi\rangle = \int dc\, U(c)|\psi\rangle, \tag{11}$$

where $U(c)$ is the unitary representation, acting on the Fock space, of the element of the symmetry group, c, and where dc is the invariant measure. Then, the inner product of two states $|\Psi_1\rangle$ and $|\Psi_2\rangle$ obtained from $|\psi_1\rangle$ and $|\psi_2\rangle$ in the original Fock space through (11) can be defined by

$$(\Psi_1|\Psi_2) = \int dc\, \langle\psi_1|U(c)|\psi_2\rangle. \tag{12}$$

(The Hilbert space obtained in this manner is closely related to the one recently proposed by Marolf [5] for quantum cosmology.) This construction of invariant Hilbert space has been shown to work for the cases of flat space with the T^3 topology and de Sitter spacetime [2].

The constraint (6) reflects the underlying diffeomorphism invariance of the full theory as we argued. Hence, it may be useful to investigate implications of this constraint in more detail in order to understand quantum gravity.

Acknowledgments

I thank the organizers for giving me the opportunity to present this talk in memory of Professor Feza Gürsey, who was my teacher. I was reminded once again that I still have much to learn from him. This work was supported in part by Schweizerischer Nationalfonds.

References

[1] V. Moncrief, Phys. Rev. D **18**, 983 (1978); Gen. Relativ. Gravit. **10**, 93 (1979).

[2] A. Higuchi, Class. Quantum Grav. **8**, 1961; 1983; 2005; 2023 (1991).

[3] D. Brill and S. Deser, Commun. Math. Phys. **32**, 291 (1973);
A. Fischer and J. Marsden, Bull. Am. Math. Soc. **79**, 997 (1973);
V. Moncrief, J. Math. Phys. **16**, 493 (1975); *ibid.* **17**, 1893 (1976).

[4] R. Arnowitt, S. Deser, and C.W. Misner, in *Gravitation: An Introduction to Current Research*, edited by L. Witten (Wiley, New York, 1962).

[5] D. Marolf, "Quantum Observables and Recollapsing Dynamics," gr-qc/9404053. See also R.P. Woodard, Class. Quantum Grav. **10**, 483 (1993).

Quantum Fluctuations in the Presence of Gravitational Waves

M.Hortaçsu [1,2], R.Kaya [2], N.Özdemir [1]

[1] Physics Dept.,Faculty of Science and Letters, I.T.U., 80826 ,Istanbul,Turkey

[2] Physics Dept., Institute of Basic Sciences, TÜBITAK,MRC, Gebze, Turkey

Keywords. Cosmic strings, quantum fluctuations.

Cosmic strings arise when a continuous symmetry is broken into a discrete subgroup and may exist as relics of the large symmetry group that existed during the big bang. They give a scenario to galaxy formation alternative to the inflationary scheme. There is still some possibility that the cosmics strings may be present in the universe. In fact it is expected to know whether or not they exist in the universe definitely with the data on mili pulsars. This data will be available in few years .

We know that plane waves do not polarize the vacuum/[1]. A similar theorem does not exist for spherical waves. Here we do a mathematical exercise and investigate whether vacuum fluctuations exist in the presence of two spherical gravitational that are due to snapped strings. The metrics used are exact solutions of Einstein's field equations.

The first metric was given by Nutku and Penrose /[2]. It was first cited in /[3]. Additional information about this metric is given in /[4]. The metric is given by

$$ds^2 = 2dudv + 2|ud\zeta + v\Theta(v)\overline{f}(v,\zeta)d\overline{\zeta}|^2 \qquad 1$$

Here $f = f(\zeta + \Theta(v))$ where f is the schwartzian derivative of an arbitrary holomorphic function h. v is a null coordinate which can be regarded as retarded time. u is a Bondi-type luminosity distance. ζ is the stereographic coordinate on the sphere. The only nonidentically zero component of the curvature tensor is $\Psi_4 = u^{-1}f(\zeta)\delta(v)$ with the characteristic behaviour for impulsive gravitational waves.

We take the simplest nontrivial form for h

$$h = (\zeta)^{\pm(1+\delta+i\epsilon)} \qquad 2$$

which gives f proportional to $\frac{\delta+i\epsilon}{\zeta^2}$ in the first order. By taking h in this form we take both the normal and the rotating string. Our aim is to calculate the stress-energy tensor of a massless scalar filed in the background of this metric. We compute the Greens' function of the Klein-Gordon equation in this background

$$LG_F = \frac{\partial}{\partial x^\mu}\left(g^{\mu\nu}\sqrt{-g}\frac{\partial}{\partial x^\nu}\right)G_F = -\delta(x,x') \qquad 3$$

and deduce the stress-energy tensor from G_F by differentiation. This method is standard and is given in /[5]. The work reported here also appeared partially in references 6. We use first order perturbation theory. Here δ and ϵ are of the order of 10^{-6}. This is an observational constraint.

We first calculate the eigenfunctions corresponding to

$$L\phi_\lambda = \lambda\phi_\lambda \qquad 4$$

and obtain the Greens' function G_F using

$$G_F(x, x') = \sum_\lambda \frac{\phi_\lambda(x)\phi_\lambda^*(x')}{\lambda} \qquad 5$$

The calculation is long and tedious. One has to expand ϕ_λ and λ in terms of the perturbation parameters δ and ϵ . We find the Greens' function in terms of Hankel functions, using an infrared parameter m, which disappears at the end of the calculation.

As for the details, we first expand L, ϕ_λ and λ in powers of the parameter δ and ϵ; $L = L_0 + \delta L_1^\delta + \epsilon L_1^\epsilon$, $\phi_\lambda = \phi_0 + \delta\phi_1^\delta + \epsilon\phi_1^\epsilon + ...$ etc. Then

$$L_0\phi_0 = \lambda_0\phi_0 \qquad 6$$

$$L_1\phi_0 + L_0\phi_1 = \lambda_1\phi_0 + \lambda_0\phi_1. \qquad 7$$

Here

$$L_0 = 2u\frac{\partial}{\partial v} + 2u^2\frac{\partial^2}{\partial u\partial v} - \frac{\partial^2}{\partial z^2} - \frac{\partial^2}{\partial y^2} \qquad 8$$

We seperate terms that are proportional to δ and ϵ. The zeroth order solution is easy to find.

$$\phi_0^\lambda = \frac{1}{u\sqrt{2R}(2\pi)^2}\exp(\frac{-iK}{2uR})\exp(-iRv)\exp(ik_1z + ik_2y) \qquad 9$$

where $\lambda = k_1^2 + k_2^2 - K$. Here k_1, k_2, K, R are constants which will be integrated over to find G_F. It is easy to show that $\lambda_1 = (\phi_0, L_1\phi_0) = 0$. We have

$$L_1^\delta = \frac{2v}{u}\left(\frac{z^2 - y^2}{(z^2 + y^2)^2}\left(\frac{\partial^2}{\partial z^2} - \frac{\partial^2}{\partial y^2}\right) + \frac{4zy}{(z^2 + y^2)^2}\frac{\partial^2}{\partial z\partial y}\right) \qquad 10$$

$$L_1^\epsilon = \frac{8v}{u}\left(\frac{zy}{(z^2 + y^2)^2}\left(\frac{\partial^2}{\partial z^2} - \frac{\partial^2}{\partial y^2}\right) - \frac{(z^2 - y^2)}{(z^2 + y^2)^2}\frac{\partial^2}{\partial z\partial y}\right). \qquad 11$$

These operators suggest the ansatz $\phi_1^\lambda = g(z, y, u, v)\phi_0^\lambda$, where $g = vg_1(u, z, y) + g_2(u, z, y)$. We check that this set of solutions is complete. We then use this solution to find ϕ_1.

The problem seperates into two parts: Terms proportional to δ and terms proportional to ϵ. We find

$$g_1^\delta = \frac{1}{iu}\left(\frac{k_1 z + k_2 y}{z^2 + y^2}\right) \tag{12}$$

$$g_1^\epsilon = \frac{2}{iu}\left(\frac{k_1 y - k_2 z}{z^2 + y^2}\right) \tag{13}$$

We then turn the crank and calculate g_2^δ, g_2^ϵ. We find

$$g_2^\delta = \int_0^\infty \frac{dw}{uR}\left[\frac{2i}{(w+\frac{1}{u})} + \frac{K}{R}\right]\frac{B}{A} \tag{14}$$

where

$$B = \frac{w}{R} + \frac{k_1 z + k_2 y}{k_1^2 + k_2^2} \tag{15}$$

$$A = \frac{w^2}{R^2} + \frac{2w}{R}\frac{k_1 z + k_2 y}{k_1^2 + k_2^2} + \frac{z^2 + y^2}{k_1^2 + k_2^2} \tag{16}$$

and

$$g_2^\epsilon = \int_0^\infty \frac{dw}{uR}\left[\frac{2i}{w+\frac{1}{u}} + \frac{K}{K}\right]\frac{C}{A} \tag{17}$$

where $C = \frac{k_1 y - k_2 z}{k_1^2 + k_2^2}$.

$G_F^{(1)}$ is found by calculating

$$G_F^(1) = \int_{-\infty}^{\infty} dk_1 \int_{-\infty}^{\infty} dk_2 \int_{-\infty}^{\infty} dK \int_{-\infty}^{\infty} dR \frac{\phi_0^\lambda(x)\phi_1^{\lambda *}(x') + \phi_0^{\lambda *}(x')\phi_1^\lambda(x)}{K - k_1^2 - k_2^2} \tag{18}$$

The interesting point is that we get the same result from both g_1^δ, g_2^δ and $g_1^\epsilon, g_2^\epsilon$. Essentially we get the sum of different Hankel functions. For $z = z', y = y'$, the typical terms are given as

$$\frac{im^2 T^2 sgn(T)(z^2 + y^2)}{32S} H_2^{(2)}(2m\sqrt{s}),$$

$$\frac{im^2 T^2 sgn(T)(z^2 + y^2)}{32Y} H_2^{(2)}(2m\sqrt{Y}),$$

$$\frac{m}{4}\frac{1}{|T|}\left(\frac{1}{u} + \frac{1}{u'}\right)\frac{1}{\sqrt{S}} H_1^{(2)}(2m\sqrt{S})$$

where $S = (u - u')(v - v')$, $T = \frac{1}{u} - \frac{1}{u'}$, $Y = S + \frac{(z^2+y^2)u^3 u' T^2}{2}$ and m is an infrared cut-off which we put to zero at the end of the calculation.

We see that our Greens' functions behave as $\frac{1}{(x-x')^\alpha}$. We could not find the finite part of these expressions at the coincidence limit . Then we concluded that we could not get a finite part also for $< T_{\mu\nu} >$.

We then take a second solution by Nutku/[7]. Here

$$ds^2 = 2Pdudv + 2uP_\zeta d\zeta dv + 2uP_{\bar{\zeta}} d\bar{\zeta} dv - 2u^2 d\zeta d\bar{\zeta} \qquad 19$$

where $P = \frac{1}{|h_\zeta|}$, $h = h(\zeta + gv\Theta(v))$ where h is an arbitrary holomorphic function of its argument. The only non-zero component of curvature is

$$\Psi_4 = \frac{-g}{uP}(PP_{\zeta\zeta\zeta} - P_{\zeta\zeta}P_\zeta)\Theta(v) \qquad 20$$

which makes the shock wave character of this equation manifest.

We take

$$h = (\zeta + gv\Theta(v))^{(1+\delta+i\epsilon)} \qquad 21$$

Both δ and ϵ are of order 10^{-6}. The vacuum expectation value of the stress-energy tensor is calculated in the same manner. We first calculate G_F in the background of this new metric and obtain $< T_{\mu\nu} >$ from G_F by differentiation. We need a regularized G_F at the coincidence limit to get $< T_{\mu\nu} >$ from this expression.

Note that there are two types of expansion parameters. One is the set δ and ϵ. The other is the coupling constant g which has dimensions of mass. We note that to get a nontrivial solution we need to take the expression to first order both in δ or ϵ and in g.

We repeat exactly the same steps as in the first example. Now we have to perform the expansions both in δ, ϵ and g. Our L_0,ϕ_0 and λ_0 are exactly the same as before, since they correspond to the flat case in both metrics. We take $\phi_1 = \delta\phi_1^\delta + \epsilon\phi_1^\epsilon$. Then we get

$$(L_0 + \lambda_0)\phi_1^\delta =$$

$$\left[2i\left(\frac{iK}{2uR} - 1\right)\left(\frac{k_1 z + k_2 y}{z^2 + y^2}\right) + \frac{(k_1^2 + k_2^2)}{2}\left(2 + ln\left(\frac{z^2 + y^2}{2}\right) + \lambda_1\right)\right]\phi_0 \qquad 22$$

and

$$(L_0 + \lambda_0)\phi_1^\epsilon = \left[-2\frac{(k_1 y - k_2 z)}{z^2 + y^2}\left(\frac{iK}{2uR} - 1\right) + (k_1^2 + k_2^2)(tan^{-1}\frac{y}{z} + \lambda_1)\right]\phi_0 \qquad 23$$

We make the ansatz $\phi_1 = \phi_2(z, y, u)\phi_0(u, v, z, y)$ and then expand in g. We have to be first order both in g and (δ, ϵ) to get the non-flat case. This necessitates us finding and solving new inhomogenous differential equations. We find $\phi^{(1)}$ in both cases. The interesting thing is that our end result for both δ and ϵ ends up in the form

$$G_F^{(1)} = \frac{2igH_0^{(2)}(\chi)}{(z^2 + y^2)|u - u'|}\left[\left(\frac{u^2 + u'^2}{(u - u')^2}\right)(C_1\delta + E_1\epsilon) + C_2\delta + E_2\epsilon\right] \qquad 24$$

when $z = z', y = y'$. Here C_1, C_2, E_1, E_2 are constants of the same order, and $\chi^2 = 4M^2(u-u')(v-v')$. M is an infrared parameter which disappears when this expression is differentiated to get the stress-energy tensor. We could not get a finite part of also this expression in the coincidence limit.

Note that in both of the examples not only we get the same, namely the null result, but also we get the same functional form both for functions that are proportional to δ and to ϵ. This may seem reasonable since essentially we take the real and imaginary parts of the same function as the inhomogenous term of a differential equation. Since the functional form of the equations looked very different, this was still a surprise.

Another worthwhile question is whether we lose the nontrivial behaviour in these calculations because of our perturbative methods. Any nonperturbative phenomena will not be observed when our methods are used. This point should be investigated further.

REFERENCES

1. S. Deser, J. Phys. A: Math. Gen. 81 (1975) 972
   G.W. Gibbons, Comm. Math. Phys., 45 (1975) 191.
2. Y. Nutku and R. Penrose, Twistor Newsletter 34 (1992) 9.
3. M. Hortaçsu, Class. Quant. Grav. 7 (1990) L165.
4. P.A. Hogan, Phys. Rev. Lett. 70 (1993) 117.
5. N.D. Birrell and P.C.W. Davies, " Quantum Fields in Curved Space" Cambridge University Press (1982).
6. M. Hortaçsu, Class. Quant. Grav. 9 (1992) 1619,
   M. Hortaçsu, J. Math. Phys. 34 (1993) 690,
   M. Hortaçsu,R. Kaya, J. Math. Phys. 35 (1994) 3043,
   M. Hortaçsu, N. Özdemir, I.T.U. preprint.
7. Y.Nutku, Phys. Rev. 44 (1991) 3164.

On the Radiative Processes near Cosmic Strings

A. N. Aliev

TÜBİTAK Marmara Research Center, Research Institute for Basic Sciences, Department of Physics, 41470 Gebze, Turkey

A study of different mechanisims of radiation from a system of cosmic strings, or emitted by test particles in the gravitational field of cosmic strings is of importance both from a general theoretical viewpoint and for a possible implication of the cosmic string hypothesis. The main energy loss mechanism for the closed oscillating loops of cosmic strings is the gravitational radiation which can, in principle, be observable [1]. There is also a very interesting new exact solution of the Einstein equations [2] which describes impulsive gravitational waves from snapping cosmic strings. On the other hand, test particles freely moving in the gravitational field of a straight cosmic string become sources of the scalar, electromagnetic and gravitational radiation [3], [4]. Though the space-time around an infinite straight cosmic string is locally flat, the radiation arises due to the globally conical structure of the string's space-time(conical bremsstrahlung). Here in the framework of the perturbation theory the scalar, electromagnetic and gravitational radiation from test particles freely moving near a straight cosmic string is considered. Remarkably, the perturbation theory, the validity of which for the dynamically gravitating point-like systems was argued in a number of papers (see for ex.[5]), allows us to rederive the results of [3], [4] obtained within the formalisim of the spin-weighted Green's functions on the conical background.

We start with a representation

$$g_{\mu\nu} = \eta_{\mu\nu} + h_{\mu\nu} \tag{1}$$

where $\eta_{\mu\nu} = diag(1, -1, -1, -1)$ is the flat Minkowskian metric and $h_{\mu\nu}$ are the gravitational perturbations. Let us impose the flat space-time gauge conditions on the electromagnetic and gravitational potentials

$$\eta^{\mu\nu} A_{\mu,\nu} = 0 \qquad \eta^{\nu\lambda} \psi_{\mu\nu,\lambda} = 0 \tag{2}$$

where

$$\psi_{\mu\nu} = h_{\mu\nu} - \frac{1}{2}\eta_{\mu\nu} h, \qquad h = \eta^{\mu\nu} h_{\mu\nu} = -\eta^{\mu\nu}\psi_{\mu\nu} = -\psi$$

Substituting (1) and (2) into the equations of scalar, electromagnetic and gravitational fields we find

$$\Box\phi \;=\; 4\pi(J + S) = 4\pi I \tag{3}$$

$$\Box A_\mu = 4\pi\,(J_\mu + S_\mu) = 4\pi I_\mu \tag{4}$$
$$\Box \psi_{\mu\nu} = -16\pi G\,(T_{\mu\nu} + S_{\mu\nu}) = -16\pi G t_{\mu\nu} \tag{5}$$

where $\Box = \eta^{\mu\nu}\partial_\mu\partial_\nu$ is the usual d'Alambertian in the flat space-time, and the right-hand sides of these equations consist of two parts; the first one is coming from the matter sources and the second part presents all terms arising due to the gravitational stresses.

Let us consider a test point particle of mass m moving with constant velocity $U^\mu = (\gamma, \gamma v, 0, 0)$ on the plane which is orthogonal to string, where γ is the Lorentz-factor. In order to solve the field equations we apply an iterative scheme, i.e. we need a formal expansion of all quantities in (3) - (5) in terms of the gravitational constant G. The total gravitational field of the test particle and cosmic string in the first order of the iterative scheme is described by

$$\psi^{\mu\nu}(k) = \frac{64\pi^2 G}{k^2}\left[mU^\mu U^\nu e^{ikd_p}\delta(kU) + 2\pi\mu\,(V^\mu V^\nu - \Sigma^\mu\Sigma^\nu)\,e^{ikd_s}\delta(kV)\delta(k\Sigma)\right] \tag{6}$$

Here $\psi^{\mu\nu}(k)$ is the Fourier transform of $\psi^{\mu\nu}(x)$, k^μ is four momentum vector of the wave, $V^\mu = (1,0,0,0)$ and $\Sigma^\mu = (0,0,0,1)$ are the time-like and space-like vectors on the string world-sheet, $V^\mu V_\mu = 1$, $\Sigma^\mu\Sigma_\mu = -1$, d_p, d_s are impact parameters, μ is the linear mass density of the string which is assumed to be small ($G\mu \ll 1$) for the convergence of series in the iterative scheme. Clearly, unperturbed world-sheet of the string

$$x^\mu(\tau,\sigma) = V^\mu\,\tau + \Sigma^\mu\,\sigma + d_s^\mu$$

and the particle trajectory

$$x^\mu(s) = U^\mu\,s + d_p^\mu$$

satisfy zero order equations of motion of string and particle respectively. The dynamically perturbed trajectories of the particle and string in the first order of the approximation scheme are described by the Christofell symbols constructed by means of the potentials (6). As for the infinite self-interaction terms, they are referred to the normalization scheme and omitted.

For the scalar charge q the radiation power is given by

$$P^{(sc)} = \frac{1}{4\pi^2}\int \mid I^{(1)}(k)\mid^2\, d^3\vec{k} \tag{7}$$

Using the procedure described above, one can calculate the first order quantity $I^{(1)}(k)$ in the right-hand side of (7). After some lengthy calculations we obtain

$$P^{(sc)} = \int_0^{2\pi} d\varphi \int_0^{\pi} d\theta\,\sin\theta\, f(\theta,\varphi) \tag{8}$$

where the angular distribution of the intensity of radiation is given by

$$f(\theta,\varphi)=\frac{2\,v\,(G\mu qv)^2}{\gamma^2\,d}\frac{\sin^2\theta}{(1-v^2\sin^2\theta)^{\frac{3}{2}}\,(1-v\sin\theta\cos\varphi)^2} \tag{9}$$

In the non-relativistic limit of motion ($v \ll 1$) this formula gives

$$f(\theta,\varphi)=\frac{2\,v\,(G\mu qv)^2}{d}\sin^2\theta \tag{10}$$

while the total radiation power has the form

$$P^{(sc)}=\frac{16\pi}{3\,d}\,(G\mu)^2\,q^2v^3 \tag{11}$$

In case of the freely moving electric charge e the appropriate calculations are analogous to those for the scalar charge. When they are carried out, for two independent polarization states (π,σ) of the electromagnetic radiation we find

$$P_\pi^{(em)} = \frac{\pi}{d}\,(G\mu\gamma e)^2\left[(2v^2-1)\;v+\gamma\arctan(\gamma v)\right] \tag{12}$$

$$P_\sigma^{(em)} = \frac{\pi}{d}\,(2G\mu\gamma e)^2\,v^3\left[1+\frac{\gamma}{v}\arctan(\gamma v)\right] \tag{13}$$

For the non-relativistic motion

$$P_\pi^{(em)}=\frac{8\pi}{3d}\,(G\mu)^2\,e^2v^3,\qquad P_\sigma^{(em)}=\frac{8\pi}{d}\,(G\mu)^2\,e^2v^3$$

It is seen that the degree of polarization $\alpha=\frac{P_\pi}{P_\sigma}=\frac{1}{3}$, while the total radiated energy is two times more than the power of scalar radiation.

The gravitational radiation is described by the second order potentials $t_{\mu\nu}^{(2)}$ in Eq.(5). Carrying out the appropriate calculations we obtain the expression

$$P^{(grav)}=\frac{27}{2d}G\pi^2\,(G\mu)^2\,m^2\gamma^3 \tag{14}$$

which gives the total power of the radiated gravitational energy in the ultrarelativistic regime of motion ($\gamma \gg 1$).

References

[1] A. Vilenkin, Phys.Rept. **121** (1985) 263

[2] Y. Nutku, R. Penrose, Twistor Newsletter **34** (1992) 9

[3] V.P. Frolov, E.M. Serebriany, V.D.Skarzhinski, in "proceedings IV Moscow Seminar on Quantum Gravity", World Sci. Singapore (1988)

[4] A.N.Aliev, D.V. Gal'tsov, Ann. Phys. (NY) **193** (1989) 142

[5] L. Bel, T. Damour, N. Deruelle, J. Ibanez, J. Martin, Gen. Rel. Grav.**13** (1981) 963

Integrable Models in Multidimensional Cosmology with Multicomponent Perfect Fluid and Toda Lattices

V.R. Gavrilov, V. D. Ivashchuk and V. N. Melnikov

Center for Surface and Vacuum Research
8 Kravchenko Str. Moskow, 117331, Russia
e-mail: mel@cvsi.uucp.free.net

Abstract. The multidimensional cosmological model describing the evolution of n Einstein spaces in the presence of multicomponent perfect fluid is considered. When the vectors corresponding to the equations of state of the components are orthogonal with respect to the minisuperspace metric, the Einstein equations are integrated and a Kasner-like form of the solutions is presented. For special sets of parameters the cosmological model is reduced to the Euclidean Toda-like system connected with some Lie algebra. Some special integrable by quadrature vacuum model for two Einstein spaces with non-zero Ricci tensors is singled out.

Keywords. Multidimensional cosmology, integrable systems, Toda lattices.

1 Introduction

Here we study integrable pseudo-Euclidean Toda-like systems appearing in multidimensional cosmology (see [1,2,4-9,14] and refs. therein). As in [9], we consider the Einstein eqs. for the cosmological model, describing the evolution of n Einstein spaces in the presence of multicomponent perfect fluid.

We consider the Einstein eqs. $R^M_N - \frac{1}{2}\delta^M_N R = \kappa^2 T^M_N$ for the metric

$$g = -\exp[2\gamma(t)]dt \otimes dt + \sum_{i=1}^{n} \exp[2x^i(t)]g^{(i)}, \quad n \geq 2, \tag{1}$$

defined on the D-dimensional space-time manifold $M = R \times M_1 \times \ldots \times M_n$, where the manifold M_i is an Einstein space of the dimension N_i with the metric $g^{(i)}$, i.e. $R_{m_i n_i}[g^{(i)}] = \lambda^i g^{(i)}_{m_i n_i}, m_i, n_i = 1, \ldots, N_i$.

The energy-momentum tensor is taken in the following form

$$T^M_N = \sum_{\alpha=1}^{m} T^{M(\alpha)}_N, \tag{2}$$

$$(T^{M(\alpha)}_N) = \mathrm{diag}(-\rho^{(\alpha)}(t), p^{(\alpha)}_1(t)\delta^{m_1}_{k_1}, \ldots, p^{(\alpha)}_n(t)\delta^{m_n}_{k_n}), \tag{3}$$

We suppose that for any α-th component of the perfect fluid the pressures in all spaces are proportional to the density

$$p_i^{(\alpha)}(t) = (1 - h_i^{(\alpha)})\rho^{(\alpha)}(t), \; h_i^{(\alpha)} = \text{const}. \tag{4}$$

We impose the conservation law constraints: $\nabla_M T_N^{M(\beta)} = 0$, leading to the relations

$$\rho^{(\alpha)}(t) = A^{(\alpha)} \exp[-2\gamma_0 + u_i^{(\alpha)} x^i], \tag{5}$$

where $A^{(\alpha)}$ =const, $\gamma_0 = \sum_{i=1}^{n} N_i x^i$ and $u_i^{(\alpha)} = N_i h_i^{(\alpha)}$.

The Einstein eqs. are equivalent to the Lagrange-Euler eqs. for the Lagrangian

$$L = \frac{1}{2} \exp[-\gamma + \gamma_0] G_{ij} \dot{x}^i \dot{x}^j - \exp[\gamma - \gamma_0] V. \tag{6}$$

Here $G_{ij} = N_i \delta_{ij} - N_i N_j$ are the components of the minisuperspace metric. This metric has pseudo-Euclidean signature $(-, +, ..., +)$ [5,6]. The potential V contains the terms induced by the curvatures of M_i and perfect fluid

$$V = -\frac{1}{2} \sum_{i=1}^{n} \lambda^i N_i \exp[-2x^i + 2\gamma_0] + \kappa^2 \sum_{\alpha=1}^{m} A^{(\alpha)} \exp[u_i^{(\alpha)} x^i]. \tag{7}$$

2 Integrable models

After fixing the harmonic time gauge: $\gamma \equiv \gamma_0$ in (6) we are lead to the problem of integrability of the pseudo-Euclidean Toda-like systems described by the Lagrangian of the form

$$L = \frac{1}{2} < \dot{x}, \dot{x} > - \sum_{s=1}^{\bar{m}} a^{(s)} \exp(b_s, x), \tag{8}$$

where $b_s \in R^n$, $< x, y >= G_{ij} x^i y^j$ and $(b_s, x) = b_{si} x^i$. Such systems are algebraic generalizations of the well-known Toda lattices [13] to the case of indefinite bilinear form of the kinetic energy.

It is quite obvious that integrability depends on the set of scalar products $< b_s, b_{s'} >_*$, where $< u, v >_* = G^{ij} u_i v_j$ and $G^{ij} = \delta^{ij}/N_i + 1/(2 - D)$ are the components of the matrix inverse to (G_{ij}). For $\bar{m} = 1$ the system with the Lagrangian (8) is always integrable [5-9]. In the present paper we consider , as in [4], the multicomponent case: $\bar{m} \geq 2$.

2.1 Orthogonal set of vectors

We obtain the class of the exact solutions for the orthogonal case $< b_s, b_{s'} >_* = \eta_{ss'}$, ($b_1$ is time-like vector and other vectors are space-like). Such situation takes place for dilatonic charged black holes [14] in string induced gravity. The exact solution may be presented in the Kasner-like form [4]

$$\exp[x^i] = \prod_{s=1}^{m}[F_s^2(t-t_{0s})]^{-b_s^i/<b_s,b_s>_*}\exp[\alpha^i t+\beta^i],\ i=1,\ldots,n, \tag{9}$$

where we denoted

$$F_s(t-t_{0s}) = \sqrt{|a^{(s)}/E_s|}\Phi[\sqrt{|E_s<b_s,b_s>_*|/2}(t-t_{0s})], \tag{10}$$

where $\Phi(x) = \cosh x, \sin x, \sinh x, x$ for the cases a)$\eta_{ss}a^{(s)} > 0$, $\eta_{ss}E_s > 0$, b)$\eta_{ss}a^{(s)} < 0$, $\eta_{ss}E_s < 0$, c)$\eta_{ss}a^{(s)} < 0$, $\eta_{ss}E_s > 0$, d)$\eta_{ss}a^{(s)} < 0$, $\eta_{ss}E_s = 0$, respectively.

By t_{0s}, E_{0s} $(s = 1,\ldots,\bar{m})$ we denoted arbitrary integration constants. The Kasner-like parameters satisfy the relations

$$\sum_{i,j=1}^{n} G_{ij}\alpha^i\beta^j = 2(E_0-E_1-\ldots-E_m)\geq 0, \qquad \sum_{j=1}^{n}\alpha^j b_{sj} = \sum_{j=1}^{n}\beta^j b_{sj} = 0, \tag{11}$$

$s = 1,\ldots,\bar{m}$, where E_0 is the total energy of the system (9) ($E_0 \neq 0$ when the minimally coupled scalar field is considered).

2.2 Reducible to Toda lattices models

Let us suppose, that the vectors $b_1,\ldots,b_{\bar{m}}$ themselves and their arbitrary linear combinations are space-like vectors. Then the first components of these vectors must be zero in a suitably chosen orthonormal basis. This implies that in this basis we obtain the Lagrangian (8) in the following form

$$L = \frac{1}{2}\sum_{i,j=1}^{n}\eta_{ij}\dot{X}^i\dot{X}^j - \sum_{s=1}^{m}a^{(s)}\exp[\sum_{l=2}^{n}B_l^{(s)}X^l]. \tag{12}$$

Coordinate X^1 satisfies the eq.: $\ddot{X}^1 = 0$. Eqs. of motion for $X^2,\ldots,X^n$ are followed from the Euclidean Toda-like Lagrangian

$$L_E = \frac{1}{2}\sum_{k,l=2}^{n}\delta_{kl}\dot{X}^k\dot{X}^l - \sum_{s=1}^{m}a^{(s)}\exp[\sum_{l=2}^{n}B_l^{(s)}X^l]. \tag{13}$$

Thus, we obtained the reduction of a pseudo-Euclidean Toda-like system to the Euclidean one.

Nearly nothing is known about Euclidean Toda-like systems with arbitrary sets of vectors $(B_2^{(s)},\ldots,B_n^{(s)})$ $(s = 1,\ldots,\bar{m})$. But, if they form the set of admissible roots of a simple complex Lie algebra, then the system is completely integrable and possesses a Lax representation [3]. The explicit integration procedure of the eqs. of motion was developed in [10,11]. In [4] we presented the explicit solution for the 2-component cosmological model in the case , when it is reducible to the open Toda lattice connected with Lie algebra A_2.

2.3 Two curvatures case

Up till now only cases with at most one curvature were integrated. In this subsection, developing the procedure proposed in [1], we study the integrability of the vacuum model for $n = 2$, i.e. for $M = R \times M_1 \times M_2$, provided Einstein spaces M_1 and M_2 have non-zero Ricci tensors.

Let us consider the system with the Lagrangian (8) in the case of non-collinear and non-orthogonal time-like vectors

$$b_1 = \alpha e_1' + \beta e_2', \quad b_1 = \alpha e_1' + \gamma e_2', \quad \gamma \neq \beta, \quad \gamma\beta \neq \alpha^2. \tag{14}$$

Recall that vectors e_1' and e_2' form an orthonormal basis. For the coordinates X^1 and X^2 of the vector x in this basis we obtain the following eqs. of motion

$$\ddot{X}^1 = -\alpha \exp[-\alpha X^1](a^{(1)} \exp[\beta X^2] + a^{(2)} \exp[\gamma X^2]), \tag{15}$$

$$\ddot{X}^2 = -\exp[-\alpha X^1](\beta a^{(1)} \exp[\beta X^2] + \gamma a^{(2)} \exp[\gamma X^2]), \tag{16}$$

and the zero-energy constraint

$$-(\dot{X}^1)^2 + (\dot{X}^2)^2 = -2\exp[-\alpha X^1](a^{(1)} \exp[\beta X^2] + a^{(2)} \exp[\gamma X^2]). \tag{17}$$

Further we consider only solutions with the properties: $\dot{X}^2 \neq 0$ and $\dot{X}^1 \neq \pm\dot{X}^2$. The former condition is valid for all solutions when $a^{(1)} < 0$, $a^{(2)} < 0$ and the latter always holds, if $a^{(1)}a^{(2)} > 0$. For such solutions using (15)-(17) we get the Appel eq.

$$\frac{dg}{dX^2} = \frac{1}{2}(g^2 - 1)\{\frac{\beta a^{(1)} \exp[\beta X^2] + \gamma a^{(2)} \exp[\gamma X^2]}{a^{(1)} \exp[\beta X^2] + a^{(2)} \exp[\gamma X^2]} g - \alpha\}, \tag{18}$$

where we denoted $g \equiv dX^1/dX^2$. Note that one has no general method to integrate the Appel eq. However, for $\beta = -\gamma = \pm\alpha/3$ the eq. (18) is reducible to the Bernoulli equation. Therefore it is not difficult to obtain general solution of the eq. (18) and then by (17) to present the exact solution for the eqs. of motion (15),(16) by quadrature.

This special integrable by quadrature pseudo-Euclidean Toda-like system corresponds to the 2-component cosmological model when both components induce time-like vectors b_1 and b_2. In particular, we may consider the mentioned above $n = 2$ vacuum model with two curvatures. It can be easily checked that in this model the sufficient conditions of integrability of the Appel eq. (18): $\beta = -\gamma = \pm\alpha/3$ lead to the following dimension of the spaces M_1 and M_2: $\dim M_1 = \dim M_2 = 5$. Thus, we found the special 11-dimensional integrable model with two curvatures. General solution of the eq. (18) for $\lambda^1 = \lambda^2$ may be presented in the form

$$R_1^2 R_2^2 \{A^2(R_1^2 + R_2^2) \pm |R_1^2 - R_2^2| + A\sqrt{A^2(R_1^2 + R_2^2)^2 \pm 2|R_1^2 - R_2^2|(R_1^2 + R_2^2) + A^2(R_1^2 - R_2^2)^2}\} = B \tag{19}$$

where $R_1 \equiv \exp[x^1]$ and $R_2 \equiv \exp[x^2]$ are the scale factors of the spaces M_1 and M_2 correspondingly. By A and B we denoted the integration constants. To our knowledge, this is the first integrable by quadrature example with two curvatures.

References

[1] U.Bleyer, D.-E.Liebscher and A.G.Polnarev, Class. Quant. Grav. **8**, 477 (1991)

[2] U.Bleyer, V.D.Ivashchuk, V.N.Melnikov and A.I.Zhuk, Nucl. Phys. **B429**, 177 (1988) , (1994).

[3] O.I.Bogoyavlensky, Comm. Math. Phys. **51**, 201 (1976).

[4] V.R.Gavrilov, V.D.Ivashchuk and V.N.Melnikov, Multidimensional cosmology with multicomponent perfect fluid and Toda lattices, preprint RGA 009/94, gr-qc/9407019; submitted to J. Math. Phys.

[5] V.D.Ivashchuk and V.N.Melnikov, Phys. Lett. **A136**, 465 (1989).

[6] V.D.Ivashchuk, V.N.Melnikov and A.I.Zhuk. Nuovo Cim. **B104**, 575 (1989).

[7] V.D.Ivashchuk and V.N.Melnikov, Chines Phys, Lett. **7**, 97 (1990).

[8] V.D.Ivashchuk, Phys. Lett. **A170**, 16 (1992).

[9] V.D.Ivashchuk and V.N.Melnikov, Int. J. Mod. Phys. D, **3**, No 4, (1994).

[10] B.Kostant, Adv. in Math. **34**, 195 (1979).

[11] M.A.Olshanetsky and A.M.Perelomov, Invent. Math., **54**, 261 (1979).

[12] K.P.Stanyukovich and V.N.Melnikov, Hydrodynamics, Fields and Constants in the Theory of Gravitation (Moscow, Energoatomizdat, 1983) (in Russian).

[13] M.Toda, Progr. Theor. Phys. **45**, 174 (1970); Theory of Nonlinear Lattices (Springer-Verlag, Berlin, 1981).

[14] U. Bleyer and V. D. Ivashchuk, Phys. Lett. **B 332**, (1994) 292.
V.D.Ivashchuk and V.N.Melnikov, Class. and Quant. Grav. **11**, (1994) 1793.

Effective Field Theory

A.Zee

Institute for Theoretical Physics University of California Santa Barbara, CA 93106-4030 USA

Abstract. I give a brief review of effective field theory, discussing the contribution of Feza Gürsey in particular and focussing on the literature I am most familiar with.

I would like to begin by saying a few words about Feza Gürsey (1921-1992), whom I regard as one of the last "gentleman-physicists". Unlike many of the other speakers, I never had the pleasure of any direct physics interaction with him. But I have met him on a number of occasions and Feza and Suha have always been exceptionally nice to me, ever since the beginning of my career. The last time I saw them was a few years ago in a small town in eastern Hungary. Every time Feza saw me, he told me to come to Istanbul. Well, here I am finally, but unfortunately in his absence. The community has lost a true gentleman scholar.

My subject today is the low energy or long distance effective field theory, a concept that has pervaded throughout much of modern physics. In a sense, all of physics involves the concept of effective field theory. Hydrodynamics, for example, studies the behavior of a collection of particles on distance scales large compared to the separation between the particles. One can even say that all of known physics may be described by the effective low effective Lagrangian of string theory. In recent years, the concept of effective field theory has played an increasingly important role in condensed matter physics as well as in particle physics.

Perhaps the two most studied effective Lagrangians are the Landau-Ginzburg theory of superconductivity and the sigma model of the interaction between pions and nucleons. The Landau-Ginzburg theory went on to great glory as the prototype of a spontaneously broken gauge theory which underlies electroweak unification and grand unification. For its part, the non-linear sigma model has been studied intensively in recent years in connection with quantum spin systems, both ferromagnetic and anti-ferromagnetic. The discovery of high temperature superconductivity has thrust these studies into prominence since the relevant materials are known to be anti-ferromagnetic at low doping concentrations. In these applications, it is the non-linear sigma model, rather than the linear sigma model, that enters.

The ubiquitous non-linear sigma model first appeared in the work of Feza Gürsey.[1,2,3,4] Indeed, even the notation and the philosophical underpinning in Gürsey's first paper[1] were already remarkably close to what is used in modern times. Starting with the chiral transformation of the nucleon field

$$\psi(x) \rightarrow e^{2i\gamma_5 \vec{\tau}.\vec{\theta}}\psi(x)$$

Gürsey jumped to the non-linear transformation

$$\psi(x) \rightarrow e^{2i\gamma_5 \vec{\tau}.\vec{\varphi}(x)}\,\psi(x)$$

where φ denoted the pion field. Incidentally, Gürsey cited Nishijima[5] for this crucial step of replacing the parameter of a symmetry transformation, $\vec{\theta}$, by a field, $\varphi(x)$. he then identified the unitary matrix field $\Phi = e^{2i\,f\vec{\tau}.\vec{\varphi}}$, its kinetic energy term $\mathrm{tr}\partial\Phi\partial\Phi^{\dagger}$ and its interaction with nucleons. As another historical note, we mention here that Gürsey cited Glauber[6] as having written down, in 1951, a non-linear interaction of pions with nucleons in order to account for multiple pion production in nucleon nucleon collisions. Of course, back in 1951, chiral invariance was not yet appreciated.

Shortly afterwards, indeed, in the same volume of *Il Nuovo Cimento*, Gell-Mann and Lévy[7] wrote down the linear sigma model. They were able to make the theory renormalizable, at the cost of introducing another field, the sigma field. The meson described by the sigma field, while it is existence was, and remains controversial, has managed to give its name to this class of field theories. To put all this into perspective, we must remember that in the early sixties, renormalizability was considered "sacred", and non-renormalizable field theories, such as the non-linear sigma model, were regarded with distaste. More on this later.

Indeed at that time even the relevance of field theory for strong interaction physics was much in doubt, and the emphasis was decidely on the S-matrix and the dispersion theoretic approach. The notion of broken chiral invariance was established only with the successes of the current algebra approach[8] championed by Gell-Mann and others in the mid 1960's. Using current algebra Adler and Weisberger were able to calculate what amounts to the low-energy interaction between pions and nucleons, an approach developed further by a number of authors. In particular, Weinberg[9] showed how to calculate pion pion scattering at low energies, and in effect re-discovered the non-linear sigma model. In a series of influential papers, Weinberg not only brought respectability back to the non-linear sigma model, but to the entire Lagrangian field theory approach, thus sweeping away the S-matrix worship of the late fifties and early sixties and paving the way for electroweak unification.

In the early seventies, these considerations were extended to the interaction between pions and photons. Indeed, Schwinger's calculation of the pion decaying into two photons amounts to finding the effective Lagrangian at energies low compared to the nucleon mass, and in this sense represents an intellectual descendant of Euler and Heisenberg's calculation of the effective four-photon interaction at energies low compared to the electron mass. Adler, B. Lee, Treiman, and I, without ever mentioning the word Lagrangian or the word field, used various consistency requirements to determine[10,11,12] the amplitudes for the processes $\gamma \to 3\pi$ and for $2\gamma \to 3\pi$. At the same time, and completely independently, Wess and Zumino[13] found the effective Lagrangian describing these processes. The Lagrangian appeared quite strange: its action can only be written as a five dimensional integral.

Again, to put things into perspective, we should recall that the emphasis at that period in the history of particle physics was on momentum-space amplitudes, on the differential cross sections[14] that actually could be measured at various newly built electron accelerators. An effective Lagrangian was regarded as only a mnemonic device. Independently of Wess and Zumino, but during and shortly after their work had already appeared, Aviv and I used Schwinger's proper time method[15] to determine the effective Lagrangian describing the interaction of an arbitrary number of SU(3) octet mesons and photons[16]. Lacking the insight of Wess and Zumino, we wrote out the Lagangian as an infinite series. These days this exercise would be referred to as evaluating the fermion determinant.

Since as an intermediate step we had to work out the quark propagator in the presence of external meson and electromagnetic fields, we could close the fermion line to obtain not only the effective Lagrangian, but also other fermion bilinears as well. In particular, Aviv and I also wrote down the effective current.[17] Some years later, Goldstone and Wilczek[18] rediscovered that using the matrix field Φ introduced by Gürsey (with the Pauli matrices $\vec{\tau}$ promoted to the SU(3) Gell-Mann matrices $\vec{\lambda}$) the effective baryon current could be written in a compact form

$$J^{\mu} = \text{constant } \varepsilon^{\mu\nu\lambda\sigma} \text{tr} \, (\Phi^{\dagger}\partial_{\nu}\Phi\Phi^{\dagger}\partial_{\lambda}\Phi\Phi^{\dagger}\partial_{\sigma}\Phi)$$

Aviv and I used another representation of the non-linear sigma model, discussed by Gürsey and by Gell-Mann and Lévy, in which σ and φ were constrained by $\sigma^2 + \varphi^2 = f^2$, and thus failed to see the invariant group structure made apparent in the

Goldstone-Wilczek form. Various strands in the development of effective Lagrangian and effective currents were all interconnected in an interesting way.[19]

Years later, Witten[20] developed the subject further, starting with the five-meson scattering amplitude and realizing that this could not be described by a Lagrangian in four dimensional spacetime. Nowadays, this can be seen quite easily by noting that in analogy with the effective current written above, the effective Lagrangian would have the form $\mathcal{L} \sim \varepsilon^{\mu\nu\lambda\sigma\tau}\ \mathrm{tr}(\Phi^\dagger\partial_\mu\Phi\Phi^\dagger\partial_\nu\Phi\Phi^\dagger\partial_\lambda\Phi\Phi^\dagger\partial_\sigma\Phi\Phi^\dagger\partial_\tau\Phi)$ and an antisymmetric symbol with five indices is available only in five-dimensional spacetime. The five-pion amplitude is forbidden by G-parity, but in SU(3) X SU(3) five-meson scattering is allowed. Witten brought the theory into the modern form.

As mentioned earlier, the non-linear sigma model has been extensively studied in connection with ferromagnetism and anti-ferromagnetism in the condensed matter physics literature. Naively, it seems reasonable enough that in a ferrogmagnetic or anti-ferromagnetic system the magnetization (or the staggered magnetization) can be represented in the continuum limit by a 3-vector field $\vec{n}(x,t)$. We are typically not interested in the fluctuation of the magnitude of $\vec{n}(x,t)$ and thus $\vec{n}(x,t)$ may be taken to be a vector of unit length. The Lagrangian is naturally taken to be a non-linear sigma model

$$\mathcal{L} = \frac{1}{2f^2}\ \partial_\mu\vec{n}.\partial^\mu\vec{n}$$

For a long time, a number of field theorists puzzled over how quantum spin, even a single quantum spin with its non-commuting components, could be incorporated into the path integral formalism. I understand that Feynman himself was troubled over this point. In hindsight, as is often the case, the solution as now presented in textbooks[21] seems straighforward enough, and it appears as if one only needs to have "one's head screwed on straight" to be led by the formalism step-by-step towards the correct answer, namely that the Wess-Zumino-Witten term has to be included.

In fact, there is another approach which avoids having to write the action as a higher dimensional integral. To explain this, I must regrettably mention that particle physicists were traditionally confused by the difference between the ferromagnet and the anti-ferromagnet, as indicated by the discussion in the paragraph preceding the one above. A clue is provided by the fact, as shown in elementary solid state physics texts, that the spin wave disperses linearly (that is, $\omega \propto k$) in an antiferromagnet and quadratically (that is, $\omega \propto k^2$) in a ferromagnet. Thus, on the face of it, the ferromagnet canot be represented by the (relativistic) non-linear sigma model written above. This suggests that the Lagrangian has to involve only one derivative in time, rather than two, but there is no such term involving $\dot{\vec{n}}$! The term $\vec{n}.\partial_t \vec{n}$ is a total derivative.

The solution,[22] as it turns out, was to write $\vec{n} = z^\dagger \vec{\sigma} z$ with z a spinor such that $z^\dagger z = 1$. The desired Lagrangian is then

$$\mathcal{L} = -iz^\dagger \dot{z} - V(\vec{n})$$

with $V(\vec{n}) = (\nabla \vec{n})^2$ + possible other terms such as the coupling of $\vec{n}$ to an external magnetic field. Indeed, using the identity $\delta(z^\dagger \dot{z}) \propto \delta \vec{n}(\vec{n} \times \dot{\vec{n}})$ we obtain

$$\vec{n} \times \dot{\vec{n}} = \frac{\delta V}{\delta \vec{n}}$$

Taking the scalar product of this equation with n we obtain

$$\dot{\vec{n}} = \vec{n} \times \frac{\delta V}{\delta \vec{n}}$$

the familiar equation of motion for a spin. A straightforward exercise[22] shows that the dispersion laws around a ferromagnetic and an anti-ferromagnetic background are indeed different and as stated above.

My generation of physicists was taught the notion that quantum field theory, quantum electrodynamics, quantum chromodynamics, and so forth, were "fundamental," that these theories hold at arbitrarily short distance scales. We used renormalization group to study the ultraviolet flow towards short distances. We would write down a Lagrangian and use "renormalizability" and symmetry to limit the number of possible terms. A "more modern" view, which has emerged from condensed matter physics, in particular the theory of critical phenomena, and which may be called "Wilsonian", acknowledges and emphasizes that the short distance physics may be extremely complicated or unknown, or perhaps even "unknowable," depending on your philosophical persuasion. In condensed matter physics, the short distance physics may be described by lattice dynamics. In particle physics, the short distance physics is allegedly that of a string. Instead of the ultraviolet flow, we should study the infrared flow towards long distances and times and hope that most terms become *irrelevant* in that limit. We would then arrive at an effective field theory described by a small number of terms. The better among the more modern textbooks, such as the one by Polyakov, specifically emphasize this view. The relevant equations are essentially the same, but the mindset is different.

We are thus instructed to study the renormalization group flow of various operators.

In many cases, simple dimensional analysis, which may be regarded as "zeroth order renormalization group," suffices. In particle physics, this point of view was developed over a number of years. An early example is a "model independent" analysis of proton decay.[23,24]

Let us say that we believe only in SU(3) x SU(2) X U(1) and not in grand unification. We simply say that proton decay is due to some unknown short distance physics, but whatever this mechanism might be, we can still write down a long distance effective Lagrangian to describe proton decay. The most relevant operators are those with the lowest scaling dimensions. These dimension six operators have the form

$$\frac{1}{M_*^2}\, qql$$

where q and l represent quark and lepton fields respectively. By engineering dimensional analysis, an unknown mass M_* has to be introduced. Thus, the rate for proton decay is of course undetermined. However, by imposing SU(3) x SU(2) x U(1) we are able to restrict the number of possible operators enormously. In this way, we can make predictions about proton decay completely independent of what the short distance physics may be!

Incidentally, this may be construed as an argument for colored quarks. Before the invention of quarks, we were able to write down a dimension four operator of the form $P\bar{e}\pi$, and we must arbitrarily decree that the dimensionless coefficient in front of this operator to be ridiculously small, and since we don't understand why this coefficient is so small, we dignify this ignorance by elevating it to a principle, known as baryon conservation. It is color that forces us to go from a dimension four operator to a dimension six operator. Physics has progressed: a small dimensionless coefficient has been replaced by the ratio (proton mass)^2/M^2_*. We can go on and study any exotic processes involving the known quarks and leptons, by systematically writing down,[25] in order of increasing scaling dimension, all possible operators allowed by SU(3) x SU(2) x U(1).

Given an effective long distance field theory, we can of course try to "induce" what the short distance physics might be. In general, of course, many possible short distance physics may give rise to the same long distance physics: this remark has been elevated to the "principle of universality. Given the long distance physics, we can arrive at the correct short distance physics only be astutely combining experimental observations and inspired guesses guided by general principles and esthetic considerations. Such is the history of physics. Currently, this enterprise is represented by string theory.

An infinitely more modest example involves a possible Majorana mass for the neutrino. The relevant long distance effective Lagrangian, as restricted by SU(3) x SU(2) X U(1), has the form

$$L_{eff} = f\tilde{\psi}_L C \psi_L \tilde{\varphi}\, \varphi$$

where ψ_L and φ_L represent the left-handed lepton doublet and the Higgs doublet respectively. Here $\tilde{\varphi}$ denotes, as usual, $\varepsilon_{ij}\ \varphi_j$. When the neutral componet of φ acquires a vacuum expectation value, the neutrino gets a Majorana mass. Can we "induce" the short distance physics responsible? In other words, can we replace effectively the dimension five operator above by operators with dimension four or less. One attempt is represented by the following[26]

$$\mathcal{L} = g\tilde{\psi}_L C \psi_L h^+ + M\, \tilde{\varphi}\, \varphi\, h^-$$

where h represents a charged SU(2) singlet Higgs. The neutrino Majorana mass is now calculable (in the technical sense!) in terms of the (unknown g and M. Short distance physics of course tells us more (again, the progress of physics!). In this particular case, we can learn something about the flavor of the neutrino Majorana mass.[27]

In some sense, physics is possible only if the effect of a particle of mass M on the low energy or long distance effective theory vanishes as M tends to infinity. Thus, for Schwinger to calculate the g factor of the electron, he didn't have to know the mass of the top quark. There is however a conceptually important exception to this perfectly reasonable expectation. Suppose that after the field for the massive particle is removed, the resulting theory becomes non-renormalizable. Then the low energy theory will remember the massive particle: its absence would be missed even at arbitrarily low energies.

For example, in the standard model the left handed top quark appears in a doublet $(t\ b)_L$. Removing the top field would render the theory non-renormalizable. Now consider[28] the Feynman diagram in which a Z boson couples to a top quark loop which connects to an external up or down quark line via two gluons. This process introduces an isoscalar contribution to the neutral current of the up and down quarks. A simple analysis shows that indeed this radiative correction goes like log m_t for large top quark mass m_t. Note that whether or not the theory left behind after the removal of the heavy field is renormalizable or not depends on the theory being considered. Thus, for Schwinger, the relevant theory was quantum electrodynamics in which the top quark is represented as just another charged field. Removing it leaves the theory perfectly renormalizable.

Another striking example is given by the electric dipole moment of the neutron[29] and electron.[30] Consider the diagram in which a photon couples to a top quark loop which connects to an external electron line via a photon and a Higgs field. Again, we see that this contribution to the electric dipole moment of the electron grows like $\log m_t$. One instructive way of looking at this process is to consider an effective Lagrangian for the coupling of a Higgs field to two photons, which in fact consists the dimension five operator $\varphi F_{\mu\nu}F^{\mu\nu}$. When we insert this effective Lagrangian into a calculation of the electric dipole moment of the electron, we would obtain a (logarithmically) divergent answer since this Lagrangian is not renormalizable. The relevant Feynman integral has to be cut off at the energy scale beyond which the low energy effective Lagrangian ceases to be valid, and this scale is set by precisely the top quark mass.

In condensed matter physics, we are not interested in ultraviolet divergences since these are always cut off by the lattice. As an example, consider the chiral spin state[31,21]. Start with a single particle hopping on a square lattice. We can look at the "Feynman freshman physics" book for example and see that the energy of the particle is related to its momentum by $E = -(\cos k_x + \cos k_y)$; the spectrum, typical of band structure theory, reflects the square symmetry of the lattice. Now suppose every time the electron goes around a square plaquette on the lattice it acquires a factor of (-1). Another way of saying this is that there is a magnetic flux of π threading through each plaquette. The energy spectrum becomes[32,33,34] $E = \pm\sqrt{\cos^2 k_x + \cos^2 k_y}$, still a messy looking expression. But if we want to study a half-filled system, that is, if we fill the band with fermions up to $E = 0$ in accordance with the Pauli exclusion principle, and if we are only interested in physics at long distance and time, then we expand around a point where E vanishes. Writing $k_x = \frac{\pi}{2} + q_x$ and $k_y = \frac{\pi}{2} + q_y$, we find $E = \pm\sqrt{q_x^2 + q_y^2}$. This is a most remarkable result: in a system which is not even rotational invariant we obtain a Lorentz invariant dispersion!

Further, when we allow the particle to hop along the diagonal as well, such that when the particle hops around a triangle its wave function acquires a phase of i, a gap opens up between the upper and lower band. The effective low energy theory describing a half-filled system, that is for energies low compared to the gap, then reads

$$\mathcal{L} = \bar{\psi}\,(i\partial\!\!\!/ - m)\,\psi + \ldots\ldots.$$

We have discovered the Dirac Lagrangian! Finally, introducing the phase degree of freedom contained in the fermion creation and annihilation operator on the lattice, we arrive at the gauge theory

$$\mathcal{L} = \overline{\psi}\,(i\partial\!\!\!/ + a - m)\,\psi$$

For physics below the energy scale m, we can integrate out the fermion field and thus obtain

$$\mathcal{L} = \frac{1}{4\pi}\; 2\varepsilon^{\mu\nu\lambda} a_\mu \partial_\upsilon a_\lambda + \ldots\ldots,$$

the famous Chern-Simon term much talked about by physicists and mathematicians.[35] From here we can proceed to a discussion of fractional statistics and semion superconductivity, but that is another story.[36] One can also go on to discuss extensions to 3 + 1 dimensional spacetime and to non-abelian flux and so on.[37,38]

The remarkable emergence of a relativistic Dirac Lagrangian from a lattice theory without even rotational invariance naturally prompts speculations on whether the observed quarks and leptons can emerge in this way also.[39] In a sense, that is what lattice gauge theory[40] is all about.

We do not want to give the impression that low energy effective theories are easy to write down. The difficulty is that the relevant degrees of freedom may be quite different from those present in the short distance theory. In quantum chromodynamics, for example, we have hadrons in the low energy theory, not quarks. In superconductivity, we have the Ginzburg-Landau field or the Cooperon. In the example just given, the relevant low energy degree of freedom turns out to be a gauge potential governed by Chern-Simons dynamics.

A striking recent example of this phenomenon is the effective field theory of quantum Hall fluid, which after all, simply consists of electrons interacting and moving in a 2 + 1 dimensional spacetime with a perpendicular magnetic field. The microscopic physics is described by a trial wave function of the $N \sim 10^{23}$ electrons proposed by Laughlin. Here I cannot possibly give you anything more than just the flavor of the effective field theory approach.

What I will do here is to argue what the effective field theory must be from general principles.[41] The conservation of the electromagnetic current $\partial^\mu J_\mu = 0$ in (2+1)-dimensional spacetime tells us that the curent can be written as the curl of a vector potential, then reads

$$J^\mu = \frac{1}{2\pi}\,\varepsilon^{\mu\upsilon\lambda}\,\partial_\upsilon\, a_\lambda$$

We now note that when we transform a_μ by $a_\mu \rightarrow a_\mu - \partial_\mu \Lambda$, the current is unchanged. We did not go looking for a gauge potential, the gauge potential came looking for us! There is no place to hide. The existence of a gauge potential follows from completely general considerations.

Let us now try to write down a low energy effective local Lagrangian in terms of operators of lowest possible dimensions, noting that parity and time reversal are broken by the external magnetic field. Since gauge invariance forbids the dimension 2 term $a_\mu a^\mu$ in the Lagrangian, the simplest possible term is in fact the dimension 3 Chern-Simons term $\varepsilon^{\mu\upsilon\lambda} a_\mu \partial^\upsilon a_\lambda$. Thus, the Lagrangian (density) is simply

$$\mathcal{L} = \frac{k}{4\pi} \varepsilon^{\mu\upsilon\lambda} a_\mu \partial^\upsilon a_\lambda + \ldots$$

To determine the dimensionless coefficient k we couple the system to an "external" or "additional" electromagnetic gauge potential A_μ, thus now have

$$\mathcal{L} = \frac{k}{4\pi} \varepsilon^{\mu\upsilon\lambda} a_\mu \partial_\upsilon a_\lambda + \frac{1}{2\pi} \varepsilon^{\mu\upsilon\lambda} A_\mu \partial_\upsilon a_\lambda = \frac{k}{4\pi} \varepsilon^{\mu\upsilon\lambda} a_\mu \partial_\upsilon a_\lambda \frac{1}{2\pi} \varepsilon^{\mu\nu\lambda} a_\mu \partial_\upsilon A_\lambda$$

Integrating out a we obtain

$$\mathcal{L}_{eff} = -\frac{1}{4\pi k} \varepsilon^{\mu\upsilon\lambda} A_\mu \partial_\upsilon A_\lambda.$$

The electromagnetic current that flows in response to A_μ is thus

$$J^\mu \equiv -\frac{\delta \mathcal{L}_{eff}}{\delta A_\mu} = \frac{1}{2\pi k} \varepsilon^{\mu\upsilon\lambda} \partial_\upsilon A_\lambda$$

Looking at the time component of this equation, we recognize 1/k as the ratio of the density of electrons to the magnetic field. To study the elementary excitations in the system, we simply couple the current of these excitations to the gauge potential a. Proceeding in this way, we can easily recover the classic Laughlin results that the elementary excitations carry fractional charge and statistics. For details and references to the original literature, I refer the reader to some pedagogical lectures I gave last winter.[42]

Another topic I like to mention is that of tunnelling in double-layered quantum Hall systems. First, in the absence of tunneling between the two layers, the current associated with each layer is separately conserved $J_I = \frac{1}{2\pi} \varepsilon\, \partial a_I$, I = 1,2. Thus, we can generalize the effective Lagrangian above to

$$\mathcal{L} = \frac{1}{4\pi} \quad (\sum_{IJ} K_{IJ} a\varepsilon\partial a + \sum_{I} 2A\varepsilon\partial a) + a_1 j_1 + a_2 j_2$$

with a 2-by-2 matrix $K = \begin{pmatrix} 1 & n \\ n & m \end{pmatrix}$ We note in passing that this K matrix is in one-to-one correspondences with a class of wave functions proposed long ago by Halperin[43] to describe double-layered Hall systems.

What happens if K has a zero eigenvalue? Then some linear combination of the gauge fields becomes massless, leading to a gapless mode and some interesting physics.[44]

How is tunneling represented in this picture? When an electron tunnels from one layer to another, the currents $J^{\mu}{}_1 = \frac{1}{2\pi}\varepsilon\partial a_1$ and $J^{\mu}{}_2 = \frac{1}{2\pi}\varepsilon\partial a_2$ are no longer separately conserved. Since electrons are represented by flux quanta, tunneling from the first layer to the second converts flux of type $\varepsilon\partial a_1$ to flux of type $\varepsilon\partial a_2$. Thus tunneling corresponds to a kind of magnetic monopole into which flux of type 1 disappears and out of which flux of type 2 appears. Indeed, more formally, we have

$$\Delta(N_1 - N_2) = \pm 2 = \int dt\, d^2x\, \partial_\mu (J^{\mu}{}_1 - J^{\mu}{}_2) = \frac{1}{2\pi} \int dt\, d^2x\, \partial_\mu(\varepsilon^{\mu\upsilon\lambda}\partial_\upsilon a_{-,\lambda})$$

Suppose we continue from Minkowskian (2+1) dimensional spacetime to Enclidean 3 space. We recognize $\partial_\mu(\varepsilon^{\mu\upsilon\lambda}\partial_\upsilon a_{-,\lambda})$ as $\vec{\nabla} \cdot (\vec{\nabla} \times \vec{A}) = \vec{\nabla} \cdot \vec{B}$ if we identify $a_{-,\lambda}$ as a 3-vector gauge potential $\vec{A}$ in Euclidean space and $\vec{B}$ the corresponding magnetic field. This is precisely a Dirac magnetic monopole, with its flux quantized to be $\pm 4\pi$, just as Dirac said it should be.

Thus, in Euclidean 3 space we have a plasma of magnetic monopoles and anti-monopoles. At the location of each monopole and each anti-monopole there occurs a tunneling event in spacetime. Polyakov[45] showed long ago that in the presence of a monopole plasma the photon acquires a mass with an effective sine-Gordon Lagrangian

$$\mathcal{L}_{eff} = g^2(\partial\theta)^2 + \zeta \cos\theta$$

It is immensely pleasing that some of the most profound concepts in theoretical physics are involved here. The discreteness of the electron leads to Dirac quantization of the magnetic monopoles. The quantization of monopoles leads to an angular variable as the order parameter. The appearance of an angular order parameter immediately reminds us of the Josephson effect in superconductors. Wen and I[44] were thus led to predict that there should be "Josephson-like" effects in double-layered quantum Hall systems. We were careful to use the term "Josephson-like" because the double-layered Hall system is to be sure not a superconductor, and thus the usual discussion of the Josephson effect must be taken over here with great care. A detailed discussion is beyond the scope of these lectures. We refer to the original work[44,46,47] and to the recent literature.[48,49,50]

An interesting probe is provided by applying a magnetic field parallel to the plane of the double-layered Hall system. When an electron tunnels from one layer to another, its wave function now acquires a phase factor

$$e^{\pm ie \int dzA_z} \equiv e^{\pm i\xi(x)} \tag{1}$$

(We denote the coordinate perpendicular to the plane by z and the two-dimensional coordinates in the plane by x.) A monopole is now associated with the phase factor $e^{+i\xi(x_a)}$, and an anti-monopole with the phase factor $e^{-i\xi(x_a)}$. It is not difficult to verify that in the effective Lagrangian the cosine term is now modified to $\cos(\theta + \xi)$. Wen and I[46] considered a random magnetic field and showed that its effect was to reduce the tunneling parameter ζ. Yang et al[48] showed that a uniform magnetic field would drive an interesting commensurate-incommensurate transition.

Again, the non-linear sigma model of Gürsey appears naturally. Here the vector field $\vec{n}$ is an order parameter such that when it is pointing up it indicates electrons in the upper layer, and when it is pointing down, electrons in the lower layer. Thus, if we impose a voltage across the double layered system we simply add to the Lagrangian a term Un_z which says that to have an electron in the upper layer costs more energy than to have it in the lower layer. It is also easy to see that tunneling can be represented by a field driving $\vec{n}$ to point in the x-direction (because $S_x = S_+ + S_-$ raises and lowers electrons between the upper and lower layers). In this way, we arrive at a more involved non-linear sigma model[48]

$$\mathcal{L} = -iz^\dagger \dot{z} - (\nabla\vec{n})^2 - \beta n^2{}_z - \eta n_x$$

The sine- Gordon mentioned earlier can be obtained[47] as an effective low energy Lagrangian from this effective Lagrangian by integrating out the fast mode, namely n_z. As is in general the case, there is a hierarchy of low energy effective Lagrangians, presumably with string theory at the top of the hierarchy.

In closing, I would like to mention some recent work[51,52,53] on random matrix theory. First, a faster-than-lightning review of this venerable subject. In the early fifties, Wigner posed the problem of calculating the distribution of the eigenvalues of an N by N (as $N \to \infty$) hermitean matrix randomly taken from a probability distribution, for example,

$$P(\varphi) = \frac{1}{Z} e^{-N \mathrm{tr} V(\varphi)}$$

where V is a polynomial of its argument. The operator measuring the density of eigenvalues is given by $\hat{\rho}(\mu) = \frac{1}{N} \mathrm{tr}\delta(\mu - \varphi)$. The density of eigenvalues $\rho(\mu) = <\hat{\rho}(\mu)>$ has been known for some fifteen years,[54] and as one might expect, depends on V of course. For V an even polynomial, the density is non-vanishing between $-a$ and a where the width of the spectrum a is a complicated functional of V.

What about the density-density correlation function $\rho_c(\mu,\upsilon) = <\hat{\rho}(\mu)\hat{\rho}(\upsilon)> - <\hat{\rho}(\mu)><\hat{\rho}(\upsilon)>$? This correlation function was determined recently for any V and was found to have wild oscillations, as expected since there are N eigenvalues in the spectrum. It is convenient to think of the eigenvalues as the positions of a gas of atoms living in a one-dimensional space of width 2a. The shord distance physics depends on V in detail.

The surprising discovery[51,53] is that when $\rho_c(\mu,\upsilon)$ is smoothed over these short distance details, it becomes universal when expressed in terms of the obvious scaling variables $x = \mu/a$, $y = \upsilon/a$, that is, we found the smoothed correlation to be[51]

$$\rho_c{}^{smooth}(\mu,\upsilon) = \frac{-1}{2N^2\pi^2a^2} f(x,y)$$

where the function

$$f(x,y) = \frac{1}{(x-y)^2} \frac{(1-xy)}{[(1-x^2)(1-y^2)]^{1/2}}.$$

is universal in the sense that it does not depend on V at all. We find this result rather remarkable since the density $\rho(\mu)$ does depend on V. Even after the Gaussian law of large numbers has been proved to us, it seems perhaps somewhat mysterious that random numbers would "know" about the function e^{-x^2}. In the same way, it appears mysterious that somehow random matrices know about the function f(x,y).

Here the long distance effective theory corresponds to hydrodynamics, and not to a renormalization group flow towards low energy. Indeed, the universality can be derived using hydrodynamics.[55] Alternate derivations have also been given.[56,57]

There is however also an analog[58] of the renormalization group. In the renormalization group approach, we thin out the number of degrees of freedom. Here we can take an N by N matrix, integrate over its last row and column, and obtain an

N - 1 by N - 1 matrix. In this way we can obtain a renormalization group flow to determine the density of eigenvalues. The calculation[52] is particularly "neat" because none of the usual complications appears.

ACKNOWLEDGEMENTS

In reading a history of the Turks, I learned that the word "Turk" first appeared in an ancient Chinese chronicle, which describes the Turks as exceptionally hospitable "milk drinking octonion loving barbarians." I would like to thank my Turkish colleagues for their exceptional hospitality. This work is supported in part by the National Science Foundation under Grant No.PHY89-04035.

References

A note to my colleagues who are not referenced here: I feel that this may not be the appropriate place to give a comprehensive set of references to the recent literature. In my more extensive review article[42] a more complete set of references may be found.

1. F.Gürsey, Nuovo Cimento **16**, 230 (1960).
2. F.Gürsey, Annals of Physics **12**, 91 (1961).
3. F.Gürsey, "Effective Lagrangians in Particle Physics," in Particles, Currents, and Symmetries (Acta Phys. Austriaca, Suppl. V, 185, 1968).
4. P.Chang and F.Gürsey, Phys. Rev. **164**, 1752 (1967).
5. K.Nishijima, Nuovo Cimento **11**, 910 (1959).
6. R.J.Glauber, Phys. Rev. **84**, 395 (1951).
7. M.Gell-Mann and M.Lévy, Nuovo Cimento **16**, 705 (1960).
8. S.Adler and R.Dashen, *Current Algebra* (W.A.Benjamin, Inc. New York, 1968).
9. S.Weinberg, Phys. Rev. Lett. **17**, 168 (1966)
10. S.Adler, B.W.Lee, S.B.Treiman, and A.Zee, Phys. Rev. **D4**, 3497 (1971).
11. Results similar to those of Adler, Lee, Treiman, and Zee were given independently by M.V.Terentive, JETP Letters **14**, 140, 1971.
12. The work of Adler, Lee, Treiman, and Zee was inspired by, but did not agree with, the earlier work of R.Aviv, N.D.Hari Dass, and R.F.Sawyer, Phys. Rev. Lett. 26, 591 (1971), and of T.Yao, Phys. Lett, **35B**, 225 (1071).
13. J.Wess and B.Zumino, Phys. Lett. **37B**, 95 (1971).
14. A.Zee, Phys. Rev. **D6**, 885 (1972).
15. J.Schwinger, Phys.Rev. **82**, 664 (1951).

16. R.Aviv and A.Zee, Phys. Rev. **D5**, 2372 (1972)

17. Aviv and I calculated the electromagnetic current $\bar{\psi}\, Q\, \gamma_\mu \psi$ with Q the charge matrix of the quark fields ψ; there was no reason to consider the baryon current at that time. In our final result (equations (4.44) and (4.46) in Ref. 16) the matrix Q can be simply set to unity for comparison with Goldstone-Wilczek.

18. J.Goldstone and F.Wilczek, Phys. Rev. Lett. **47**, 986 (1981).

19. A.Zee, Phys. Lett, **135B**, 307 (1984).

20. E.Witten, Nucl. Phys. **B223**, 422 (1983).

21. E.Fradkin, *Field Theories of Condensed Matter Systems*, (Addison-Wesley Publishing Co. 1991).

22. X.G.Wen and A.Zee, Phys. Rev. Lett. **61**, 1025 (1988).

23. S.Weinberg, Phys. Rev. Lett. **43**, 1566 (1979).

24. F.Wilczek and A.Zee, Phys. Rev. Lett, 43 1571 (1979); Phys. Lett. **88B**, 311 (1979).

25. A.Weldon and A.Zee, Nucl. Phys. **173B**, 269 (1980).

26. A.Zee, Phys. Lett. 93B, 389 (1980).

27. For a review and references to earlier work, see A.Y.Smirnov and Z.Tao, Trieste preprint 1994.

28. J.C.Collins, F.Wilczek, and A.Zee, Phys. Rev. **D18**, 242 (1978).

29. S.Weinberg, Phys. Rev. Lett. **63**, 2333 (1989).

30. S.M.Barr and A.Zee, Phys. Rev. Lett. **65**,21 (1990).

31. X.G. Wen, F.Wilczek, and A.Zee, Phys. Rev. **B39**, 11413 (1990).

32. I.Affleck and J.B.Marston, Phys. Rev. **B37**, 3774 (1988).

33. G.Kotliar, Phys. Rev. **B37**, 3664 (1988).

34. L.B.Ioffe and A.I.Larkin, Phys. Rev. **B39**, 8988 (1989).

35. *Physics and Mathematics of Anyons*, edited by S.S.Chern et al., World Scientific Publishing 1990.

36. A.Zee, *"Semionics", in high Temperature Superconductivity*, edited by K.Bedell, D. Coffey, D.Pines and J.R.Schrieffer, Addison-Wesley 1990.

37. A.Zee, Int.J.Mod. Physic **B5**, 529 (1991), Z.Kunszt and A.Zee, Phys. Rev. B44, 6842 (1991).

38. A.M.Tikofsky, S.B.Libby, and R.B.Laughlin, Nucl. Phys. **B413**, 579 (1994).

39. A.Zee, "Emergence of spinor from flux and lattice hopping," in *M.A.B.Beg Memorial Volume*, edited by A.Ali and P.Hoodbhoy, World Scientific Publishing 1990.

40. J.Kogut and L.Susskind, Phys. Rev. **D11**, 395 (1975).

41. J.Fröhlich and A.Zee, Nucl. Phys. **B364**, 517 (1991).

42. A.Zee, "Field theory of quantum Hall Fluids", to appear in the proceedings of the South African School on Field Theory and Condensed Matter System, Tsitsikamma, South Africa, 1994, to be published by Springer-Verlag.
43. B.I.Halperin, Phys. Hevl. Acta. **56**, 75 (1983).
44. X.G. Wen and A.Zee, Phys. Rev. Lett. **69**, 1811 (1992).
45. A.Polyakov, Nucl. Phys. **B120**, 429 (1977).
46. X.G. Wen and A.Zee, Phys. Rev. **B47**, 2265 (1993).
47. X.G. Wen and A.Zee, "A Phenomenological Study of Interlayer Tunnelling in Double-Layered Quantum Hall Systems", MIT-ITP preprint, 1994.
48. K.Yang, K.Moon, L.Zheng, A.H.MacDonald, L.Zheng, D.Yoshioka, and S.C. Zhang, Phys. Rev. Lett. **72**, 732 (1994).
49. K.Moon, H.Mori, K.Yang, S.M.Girvin, A.H.MacDonald, L.Zheng, D.Yoshioka, and S.C.Zhang, Indiana preprint 1994.
50. I.Ichinose and T.Ohbayashi, U.Tokyo preprint 1994.
51. E.Brézin and A.Zee, Nucl. Phys. **402(FS)**, 613 (1993).
52. E.Brézin and A.Zee, Compt. Rend.Acad. Sci. **317**, 735 (1993).
53. E.Brézin and A.Zee, Phys. Rev. **E49**, 2588 (1994).
54. E.Brézin, C.Itzykson, G.Parisi, and J.B.Zuber, Comm. Math. Phys. **59**, 35 (1978).
55. C.W.J.Beenakker, to appear in Nucl. Phys. **B**.
56. B.Eynard, hep-th/9401165, to appear Nucl. Phys. **B**.
57. P.J.Forrester, to be published.
58. E.Brézin and J.Zinn-Justin, Phys. Lett, **B288**, 54 (1992).

Gürsey's Chiral Model and Its Modifications

Valerii I. Sanyuk

Theoretical Physics Department
Peoples' Friendship University of Russia
Miklukho-Maklay Str. 6, Moskow 117198, Russia
e-mail: sanyuk@udn.msk.su

Up to date available results on the theory of critical points for (3+1)-dimensional G-invariant functionals, occurring in nontrivial topological modifications of Gürsey's σ-model, are listed. In particular we deal with the Skyrme model and the $\vec{n}$-field (or Faddeev) model and their gauge generalizations. It is shown that these functionals are bounded from below and attain their minima in the classes of G-invariant fields, namely on hedgehog or toron configurations.

1. Introduction

In wide range of Gürsey's contributions to Particle Physics, the nonlinear σ-model, invented in paper [1], deserves its special place and still proves to be attractive both from physical and mathematical points of view[1]. From the physical point one might refer σ-models to a universal type due to their widest range of applications: Particle Physics, String Theory, Nematic Liquid Crystals and Ferromagnets, Nonlinear Elasticity and HT Superconductivity, and more. This topic to be discussed at any length requires a special paper, here we would rather turn to those mathematical features of (3+1)-dimensional σ-models and their modifications, which one can observe in the light of the critical points theory.

Universality of Gürsey's σ-model approach was demonstrated one more time during numerous attempts to derive a mesonic Lagrangian as the low-energy limit of QCD, undertaken in 1980s (see e.g. [2], Ch. 9.3 and references there in). There were reproduced different terms of the Lagrangian looked for, but in all derivations Gürsey's σ-model term occurred as the inevitable constituent. Actually it is not a surprise, since this term is just a kinetic energy density in a functional space, when in the general setting field variables of the model (chiral fields, order parameters, or whatever) one regards as mappings

$$\phi(\vec{x}, t) : R^3 \otimes R^1 \to \Phi, \tag{1}$$

from the Minkowskian space-time $R^3 \otimes R^1$ into *a field manifold* Φ.

In this note (without details) there would be listed answers to the following initial questions of *a theory of critical points*:

1. Whether $H[\phi]$ is bounded from below?

2. Does $H[\phi]$ attains its minimum?

[1] This Gürsey's article was published the same year and in the same volume of *Nuovo Cimento* as the known Gell-Mann–Lévy paper on the linear σ-model, but one issue ahead.

3. Does $H[\phi]$ have critical points?

4. What is the structure of critical points?

Here $H[\phi]$ is a descriptive functional of a model, and we will focus our attention on topologically nontrivial (3+1)-dimensional functional of the Skyrme model (when $\Phi = S^3 \simeq SU(2)$), and that of the $\vec{n}$-field (or Faddeev) model ($\Phi = S^2$), and their gauge generalizations, which all together might be considered as topologically nontrivial extensions of earlier Gürsey's ideas.

2. Elements of the G-Invariant Critical Points Theory

We regard as *topologically nontrivial σ-models* those, where fields (1) are endowed with a nontrivial topological index (degree of mapping, Hopf index, Chern-Pontryagin number etc.). For such models the positive answer to the first question is almost straightforward, since descriptive functionals, say $H[\phi]$, are bounded from below through the corresponding topological index (or charge) Q. Actually, this is the case for the class of G –invariant models (where G – is a compact group), which admit existence of topological charges, and, in particular, for chiral models. For the latter models the field manifolds Φ are usually a sphere S^{n-1}, or a compact group G, or a homogeneous space G/H. A vacuum state corresponds to a fixed point $\varphi_\infty \in \Phi$, and therefore field configurations, which might be applied in physics, are to satisfy natural boundary conditions $\varphi \to \varphi_\infty$ as $x \to \infty$. In this way we come to the effective compactification of the domain R^3 into S^3, and this allows one to classify physical states by means of the third homotopy group $\pi_3(M)$.

In particular, in the well-described case of Skyrme model (see e.g. [2], [3]) $\pi_3(SU(2)) = \pi_3(S^3) = Z$, and the topological charge Q is the degree of mapping $Q = \deg(S^3 \to S^3)$. Using the left invariant vector fields (left chiral currents) $l_\mu = U^\dagger \partial_\mu U, \quad \mu = 0,1,2,3$; on $SU(2)$ one can construct the topological charge

$$Q = -\frac{1}{24\pi^2}\int d^3x \epsilon^{ijk} \mathrm{Tr}\,(l_i \cdot l_j,\, l_k)\,, \tag{2}$$

together with the Lagrangian density in the Skyrme model

$$L = -\frac{1}{4\lambda^2}\mathrm{Tr}\, {l_\mu}^2 + \frac{\epsilon^2}{16}\mathrm{Tr}\,[l_\mu, l_\nu]^2, \tag{3}$$

with ϵ, λ being physical constants of the model. The expressions (2) and (3) amount to the lower estimate for the Hamiltonian H through the topological charge Q:

$$H > 6\sqrt{2}\pi^2(\epsilon/\lambda)|Q|, \tag{4}$$

In view of (4) one concludes that the minimum of H can be realized in the given homotopy class i.e. for $Q = N$, and even more. The field configurations, which realize the lower bounds in the right hand side of (4), would be stable in the Lyapunov sense.

To get answers to the second question, as listed in Introduction, and partially to the forth one, we consider special classes of equivariant field configurations $\phi_0(\vec{x})$, defined by the condition

$$\phi_0(\vec{x}) = \hat{T}_g \phi_0(g^{-1}\vec{x}), \quad g \in G, \tag{5}$$

with $\hat{T}_g$ being the representation operator of the group G. For more details see e.g. [2], [3]. In accord with the Coleman-Palais principle, extremals of a G-invariant functional in the class of invariant fields are in fact true extremals.

For (3+1)-dimensional chiral models two groups G are of special interest, namely

$$\begin{aligned} G_1 &= \text{diag}\,[SO(3)_I \otimes SO(3)_S], \\ G_2 &= \text{diag}\,[SO(2)_I \otimes SO(2)_S], \end{aligned}$$

with $SO(3)_I$ being the group of isospin rotations and $SO(3)_S$ denoting the group of spatial rotations. We claim that these are the only maximal compact continuous groups which possess the equivariant fields with nontrivial topological indices in (3+1)-dimensional models. Taking chiral fields in the form

$$U = \exp(i(\vec{n}\,\vec{\tau})\Theta),$$

where $\vec{\tau}-$ are Pauli matrices, and $\vec{n}$ is a unit vector, one finds for G_1-invariant (or spherically-symmetric) fields

$$\vec{n} = \vec{r}/r,\ \Theta = \Theta(r), \tag{6}$$

known in literature as the *hedgehog or skyrmion configuration* [4].

In the case of G_2-invariant (or axially-symmetric) configurations we obtain the following representation for spherical angles β, γ, for the unit vector $\vec{n}$, and for the chiral angle variable Θ:

$$\Theta = \Theta(r,\vartheta);\ \beta = \beta(r,\vartheta);\ \gamma = k\alpha;\ k \in z, \tag{7}$$

where r, ϑ, $\alpha -$ are spherical coordinates. We will refer to this configurations as to *toron* ones.

The special role of G_i-invariant fields in searching for minima of G-invariant functionals comes out from the following

Theorem. *Let a G_i-invariant field ϕ_0, $i = 1,2$, realizes the minimum of the Hamiltonian H, restricted to the class of G_i-invariant fields. Assume the functional $H[\phi]$ to be convex with respect to the derivatives $\partial_i\phi$ at the point $\phi = \phi_0$. Then the field ϕ_0 realizes* the true minimum *of $H[\phi]$.*

The proof one can find in [2] or [3]. It means that in search of minimal (say, energy) field configurations for convex G-invariant functionals it is possible to reduce the variational problem $\delta H[\phi] = 0$ to that for G-invariant fields only: $\delta H[\phi_{\text{inv}}] = 0$. It should be mentioned that abovelisted results might be enforced or checked up by *Direct Minimization in Extended Phase Space Procedure*[2] (when the fields and their derivatives are considered as independent variables), combined with *Spherical Rearrangement Method.* Finally, one needs to check whether the obtained minimal configurations would satisfy the equation of motion. A detailed description of these methods in application to the Skyrme model one can find in Ref. 2, Ch. 5.

These methods were used to prove that in the first homotopy class of the Skyrme model with $Q = 1$ *the absolute minimum of energy functional is attainable.* As a

---

[2] A generalization on functional spaces of the *Gelfand-Zetlin "Valley" Method*[5].

result the variational problem was reduced to the minimization of the simplified Hamiltonian

$$H[\Theta] = \int_0^\infty dr \left[\Theta'^2(r^2/2 + 2\sin^2\Theta) + \sin^2\Theta + \sin^4\Theta/r^2\right]$$

in the class of functions $\Theta(r)$ with : $\Theta(0) = N\pi$, $\Theta(\infty) = 0$, when $Q = N$. The existence of minimal field configurations was proved by direct methods of the calculus of variations [4], [6] and it was shown that only for the skyrmion configuration with $Q = 1$ all the conditions on the minimal energy configurations are fulfilled. It was also proved that the obtained minimizer (the skyrmion) is an infinitely smooth function (from the class $C^\infty(0,\infty)$) satisfies the reduced equation of motion. Thus in the first homotopy class of the Skyrme model we have a sort of complete theory of critical points, in the light of questions listed in Introduction.

3. Critical Points for Modified σ-Models

Using almost the same mashinery it is possible to get all positive answers to the gauge modification of the Skyrme model, which was developed in order to account the contribution of vector ρ-mesons into the structure of baryon [7], [8]. To this end it is introduced a triplet of vector fields A^a_μ, $a = 1,2,3$, with the corresponding matrix $A_\mu = \frac{i}{2}\tau^a A^a_\mu$. Chiral and vector fields are related by means of the covariant derivative

$$D_\mu U = \partial_\mu U - [A_\mu, U],$$

and it leads to the following covariant generalization of the topological charge:

$$Q = -\frac{1}{48\pi^2}\varepsilon^{ijk}\int dx \,\mathrm{Sp}\,\{L_i[L_j, L_k] + 3F_{ij}(L_k + R_k)\}, \tag{8}$$

where the gauge generalizations of the left and right chiral currents, and that of the Yang-Mills field tensor are used:

$$L_i = U^+ D_i U, \quad R_i = D_i U \cdot U^+, \quad F_{ik} = \partial_i A_k - \partial_k A_i - [A_i, A_k].$$

The Lagrangian density of the gauged Skyrme model takes the form

$$\begin{aligned} \mathcal{L} = &-\tfrac{1}{4\lambda^2}\,\mathrm{Sp}\,L_\mu^2 + \tfrac{\varepsilon^2}{16}\,\mathrm{Sp}\,[L_\mu, L_\nu]^2 + \\ &+\tfrac{1}{2e^2}\,\mathrm{Sp}\,F_{\mu\nu}^2 + \tfrac{m_\rho^2}{2\lambda^2}(A^a_\mu)^2 - \tfrac{2m_\pi^2}{\lambda^2}\sin^2\tfrac{\Theta}{2}, \end{aligned} \tag{9}$$

where e stands for the vector field interaction constant, and m_ρ is the ρ-meson mass. The corresponding Hamiltonian is estimated from bellow through the topological charge (8) as follows:

$$H > \frac{12\pi^2\varepsilon\,|\,Q\,|}{\lambda(2 + 9e^2\varepsilon^2)^{1/2}}.$$

To see at what field configurations the energy minimum is attainable we still perform the already mentioned direct minimization in the extended phase space together with the subsequent spherical rearrangement. Already at the next step we find that all minimization conditions holds only for $|Q| = 1$, when $\Theta_{\max} = \pi$. In this way one comes to the conclusion that the minimum (if any) would be attained on G_1 –invariant field configurations. To prove this one needs also to examine the structure of the introduced vector field, which comes in fact from the gauge- and G_1 –invariance of the minimization conditions, and in fact we find that these conditions hold for the G_1 –invariant gauge field $\widetilde{A}_i$,:

$$\widetilde{A}_i^a = f_1\, P_i^a + f_2\, x^a x^i/r^2 + f_3\, \varepsilon^{iak} x^k/r,$$

where $P_i^a = \delta_i^a - x^a x^i/r^2$, and functions f_i are spherically symmetric ones. The attainability of the minimal configuration (or, in other words, the existence of regular solutions) was proved by direct variational methods [8] under the following boundary conditions:

$$\Theta(0) = N\pi, \quad \Theta(\infty) = 0, \quad u(0) = u(\infty) = 1.$$

Thus *the absolute minimum of energy in the gauge Skyrme model* in the first homotopy class (for $|Q| = 1$) is realized in the form of G_1 –invariant critical point.

A similar analysis of critical points in higher homotopy classes ($|Q| = N > 1$) of the Skyrme model and for all Q_H in Faddeev model by direct minimization in the extended phase space of the corresponding functionals leads to axially-symmetric (G_2 –invariant) configurations. Under some conditions these configurations are at least local minimizers of energy functionals. But G_1-invariant configurations, as it is clear from the previous section are also minimizers of energy and one has to compare in this case values of energy, which correspond to these two types of critical points. Actually one has to turn to numerical calculations for two-skyrmion interactions, since analytical results are not available yet. In this way one finds that G_1-invariant critical points would be saddle points of the functional, and that absolute minima will correspond to G_2-invariant critical points (torons) [2], [3], [6] and [9].

In the already mentioned Faddeev σ-model the n-field variable takes its values on the S^2-sphere and are subject to the boundary conditions: $n^a \to \delta_3^a$ as $r \to \infty$. Therewith they are classified with the help of the homotopy group $\pi_3(S^2) = z$, and characterized by the Hopf index Q_H. The latter is defined as the number of connectedness of $\vec{b}$ –lines, related with the n-field through the relation

$$f_{ik} = \varepsilon_{iks} b_s = 2\varepsilon_{abc} \partial_i n^a \partial_k n^b n^c.$$

Then the Hopf index might be expressed as

$$Q_H = -(8\pi)^{-2} \int d^3x (\vec{a} \cdot \vec{b}), \quad \vec{b} = \mathrm{rot}\vec{a}. \tag{10}$$

From (10) and the structure of Faddeev's model Lagrangian

$$\mathcal{L}_F = -\frac{\varepsilon^2}{4} f_{\mu\nu}^2 + \lambda^2 (\partial_\mu n^a)^2 \tag{11}$$

one can derive the following estimate

$$E > \mu |\, Q_H \,|^{3/4}, \quad \mu = \text{const}.$$

It is convenient to use the Hopf map $h: \, S^3 \to S^2$, putting $\vec{n} = \psi^+ \vec{\tau} \psi$, with $\psi^T = (\cos A + i \sin A \cos B, \, \sin A \sin B \, e^{iC})$, in order to obtain the structure of energy minimizers at fixed Q_H. As a result the structure of Faddeev's Hamiltonian reminds that of Skyrme's Hamiltonian and one can use similar tools for its minimization [10]. As a result it comes out that energy minimizers have toroidal structure (torons) as following:

$$\beta = \beta(\rho, \, z); \;\; \gamma = m\alpha + v(\rho, \, z); \;\; m \in z,$$

where ρ, α, z are cylindrical coordinates. The proof of existence of torons by direct variational methods would be given in [11]. Analogous results hold for the gauge Faddeev's model.

Acknowledgements

It is pleasure to express my deep gratitude to Prof. M. Serdaroglu and Dr. G. Aktas for the invitation and warm hospitality provided during Gürsey Memorial Conference

References

1. F. Gürsey, *Nuovo Cim.* **16** (1960) 230-240.
2. V.G. Makhankov, Yu.P. Rybakov, and V.I. Sanyuk , *The Skyrme Model: Fundamentals, Methods, Applications* (Springer-Verlag, Heidelberg-Berlin-New York, 1993) 265 pp.
3. V.G. Makhankov, Yu.P. Rybakov, and V.I. Sanyuk, *Usp.Fiz.Nauk* **162** (1992) 1-62
4. Yu.P. Rybakov and V.I. Sanyuk, Niels Bohr Inst. preprint NBI-HE-81-49 (1981)
5. I.M. Gelfand, and M.L. Zetlin, *Dokl. AN SSSR* (1961) **137** 295-298
6. Yu.P. Rybakov and V.I. Sanyuk, *Int.J.Mod.Phys.* **A7** (1992) 3235-3264
7. L.D. Faddeev, *Lett. Math. Phys.* (1976) **14** 289-293
8. A. Kundu, Yu.P. Rybakov, and V.I. Sanyuk, *Indian J. Pure Appl. Phys.* (1979) **17** 673-677
9. Yu. P. Rybakov, in *Problems of High Energy Physics and Field Theory.* Proc. of X Workshop (Nauka, Moscow, 1988), 349-355
10. V. I. Sanyuk, in *Topological Phases in Quantum Theory.* Proc. of Dubna 1988 Int. Workshop (World Scientific, Singapore, 1988) 316-332
11. V.G. Makhankov, Yu.P. Rybakov, and V.I. Sanyuk, *Usp.Fiz.Nauk* **165** (1995) in preparation.

Instability of a Nielsen-Olesen Vortex Embedded in the Electroweak Theory

Samuel W. MacDowell[1] and Ola Törnkvist[2]

[1] Yale University, Sloane Physics Laboratory, New Haven, CT 06520, USA
[2] NORDITA, Blegdamsvej 17, DK-2100 Copenhagen Ø, Denmark

Abstract. The stability of an abelian (Nielsen-Olesen) vortex embedded in the electroweak theory against W production is investigated in a gauge defined by the condition of a single-component Higgs field. The model is characterized by the parameters $\beta = (\frac{M_H}{M_Z})^2$ and $\gamma = \cos^2\theta_{\rm w}$ where $\theta_{\rm w}$ is the weak mixing angle. It is shown that the equations for W's in the background of the Nielsen-Olesen vortex have no solutions in the linear approximation. A necessary condition for the nonlinear equations to have a solution in the region of parameter space where the abelian vortex is classically unstable is that the W's be produced in a state of angular momentum m such that $0 > m > -2n$. The integer n is defined by the phase of the Higgs field, $\exp(in\varphi)$. It is shown that, in the region of parameter space (β, γ) where the nonlinear equations have a solution with energy lower than that of the abelian vortex, this vortex is a saddle point of the energy in the space of classical field configurations. Solutions for a set of values of the parameters β and γ in this region were obtained numerically for the case $-m = n = 1$. The possibility of existence of a stationary state for $n = 1$ with W's in the state $m = -1$ was investigated. The boundary conditions for the Euler-Lagrange equations required to make the energy finite cannot be satisfied at $r = 0$. For these values of n and m the possibility of a finite-energy stationary state defined in terms of distributions is discussed.

Keywords. electroweak vortex, cosmic string, W condensation

Introduction

It has been shown that the Nielsen-Olesen abelian vortex [1] can be embedded [2, 3] in the electroweak SU(2) × U(1) gauge theory [4, 5] in the form of an azimuthal Z field $Z_\varphi(\rho)$ and a lower component of the Higgs field $\Phi_2 = \Phi(\rho)\exp(in\varphi)$, where (ρ, φ) are polar coordinates of the position vector $\boldsymbol{\rho}$ perpendicular to the vortex. The embedded vortex, hereafter denoted the Z_{NO} vortex, is a tube of confined flux of the Z field strength, which reaches a high value at the center of the vortex.

It is known from previous works that, in a strong uniform magnetic field, the electroweak vacuum develops an instability through the interaction of the magnetic field with the anomalous magnetic moment of the W boson, leading to the formation of a W condensate [6, 7, 8]. The magnetic moment interacts similarly with a Z field; hence an instability with ensuing W production can occur if the Z field strength is sufficiently high within a region large enough compared to the Compton wavelength of the W boson. A measure of these conditions is provided, respectively, by the two parameters $\beta = (M_H/M_Z)^2$ and $\gamma = (M_W/M_Z)^2 \equiv \cos^2\theta$, where M_H, M_Z, M_W are the masses of the Higgs, Z and W bosons and θ is the Weinberg mixing angle. One finds qualitatively that the possibility of instability increases with higher β and higher γ.

A quantitative investigation of the stability of the Z_{NO} vortex for $n = 1$ was performed by James, Perivolaropoulos and Vachaspati [9]. They found numerically that the solution becomes unstable beyond a certain line in the parameter space (β, γ). Their analysis was supplemented with an elegant analytical estimate by Perkins [10], according to which the Z_{NO} solution is unconditionally unstable for $\gamma > .19$. In particular, the points (β, γ) corresponding to the physical value of the Weinberg angle are inside the region of instability.

In this report we have investigated the problem in a gauge which maintains the simple structure of the Z_{NO} vortex, defined by the condition that the upper component of the Higgs field vanishes, rather than the gauge used by James et al. [*op. cit.*] and Achúcarro et al. [11] which allows for a two-component Higgs field in the presence of W bosons. These two gauges are actually inequivalent since (for $n = 1$) the gauge invariant quantity $(\Phi_1^\dagger\Phi_1 + \Phi_2^\dagger\Phi_2)$, at $\rho = 0$, is zero in our gauge but non-zero in theirs.

In the first section an ansatz for the fields is presented that preserves the cylindrical symmetry of the energy density. The Euler-Lagrange equations are obtained and the boundary conditions are established.

In Section 2 we consider the equations for W bosons in the background of the configuration of Higgs and Z fields as given by the Z_{NO} vortex. It is shown that, in this gauge, these equations have no solutions in the linearized form. On the other hand for particular angular momenta of the W's, specified by the condition $-2n < m < 0$ on the phase $\exp(im\varphi)$ of the polar components of the W field, the set of nonlinear equations may admit a solution in a certain domain of the space of parameters (β, γ). It is shown that, in the region of parameter space where the nonlinear equations have a solution and the energy, calculated to lowest order in the W field, is smaller than that for the Z_{NO} vortex, this vortex is not a local minimum but a saddle point in the space of classical field configurations.

In Section 3, these equations were solved numerically for the case $n = 1$, $m = -1$ and a set of values of the parameters $\beta = .5, 1$ and $\gamma = .25, .5, 1$. The energy was computed and found in each case to be lower than that for the Z_{NO} vortex. As remarked already, the field configurations considered here are unrelated by any gauge transformation to those considered in Refs. [9, 11]. Hence the instability regions for the two cases may be different.

The existence of a stationary state with W's, as suggested in Refs. [10, 12, 13], is considered in Section 4. For $m = -1$, one would expect that there exists an analytic solution of the Euler-Lagrange equations for all fields. We found, however, that the boundary conditions cannot be satisfied at $r = 0$. A discussion is given of the possibility of a vortex state with W's defined in terms of distributions.

1 Nonabelian Vortex

We shall investigate the problem of stability of the Z_{NO} vortex in a gauge fixed by the condition that the Higgs field Φ has a zero upper component and a lower component $\Phi(\rho)\exp(in\varphi)$. For the vortex of the two-dimensional abelian theory, n is a topological winding number defined in terms the total flux of the U(1) gauge-field strength [1]. In a non-abelian model n can no longer be defined in a gauge invariant way. In the above chosen gauge it is given by

$$n = \frac{1}{2\pi i\Phi_0^2}\int (d\Phi^\dagger \wedge d\Phi)$$

in the notation of differential forms, where Φ_0 is the magnitude of the Higgs field at infinity. This expression is invariant only under the electromagnetic U(1) gauge group.

The Z_{NO} vortex contains an azimuthal Z field Z_φ. One can easily show that, if a radial component depending on the ρ coordinate alone is added to the Z field, the action increases. Therefore, the vortex solution can only be modified by the inclusion of a W field and an electromagnetic gauge potential. The latter can be chosen purely azimuthal by virtue of the residual electromagnetic gauge invariance.

Let g, g' be the coupling constants for the groups SU(2) and U(1) respectively. They are related to the Weinberg angle θ and the electromagnetic charge e by $g\sin\theta = g'\cos\theta = e$. The physical gauge fields are related to the gauge potentials $\mathbf{V}^a$ and $\mathbf{V}'$ associated with the groups SU(2) and U(1) by

$$\begin{aligned} \mathbf{A} &= \mathbf{V}'\cos\theta + \mathbf{V}^3\sin\theta \ , \\ \mathbf{Z} &= -\mathbf{V}'\sin\theta + \mathbf{V}^3\cos\theta \ , \\ \mathbf{W} &= \tfrac{1}{\sqrt{2}}(\mathbf{V}^1 - i\mathbf{V}^2) \ . \end{aligned}$$

Let us introduce a dimensionless vector

$$\mathbf{r} = \rho\Phi_0 g/(\sqrt{2}\cos\theta) \equiv \rho M_Z \ ,$$

with polar coordinates r, φ, and a set of functions s, X, Y, Z defined by

$$\Phi = \Phi_0 s(r)e^{in\varphi} \ ,$$

$$\tfrac{1}{\sqrt{2}}V^3_\varphi\cos\theta = \Phi_0 Y(r) \ , \qquad \tfrac{1}{\sqrt{2}}V'_\varphi\sin\theta = \Phi_0 X(r) \ ,$$
$$\tfrac{1}{\sqrt{2}}Z_\varphi = \Phi_0(Y - X) = \Phi_0 Z(r) \ , \qquad \tfrac{1}{\sqrt{2}}A_\varphi = \Phi_0(Y\tan\theta + X\cot\theta) \ .$$

In order to preserve the vortex cylindrical symmetry the **W** field must be of the form

$$\mathbf{W}\cos\theta = \Phi_0[u(r)\mathbf{e}_r + iv(r)\mathbf{e}_\varphi]\exp(im\varphi) \ .$$

It can be shown that the functions u and v may be chosen real without loss of generality. It is also convenient to use a set of auxiliary fields

$$\begin{aligned} y &= Y - \tfrac{m}{2r} \\ x &= X - \tfrac{m}{2r} - \tfrac{n}{r} \\ z &= Z + \tfrac{n}{r} = y - x \end{aligned}$$

and the parameters

$$\beta = \left(\frac{M_H}{M_Z}\right)^2, \quad \gamma = \left(\frac{M_W}{M_Z}\right)^2 = \cos^2\theta \ .$$

The energy density in terms of these fields and the new variables $\mathbf{r}$ takes the form

$$\begin{aligned} \mathcal{H} &= \Phi_0^2 \left\{ (s')^2 + ((y-x)s)^2 + \frac{1}{4}\beta(s^2-1)^2 + \frac{1}{\gamma}(v' + \frac{v}{r} + 2yu)^2 \right. \\ &+ \left. \frac{1}{\gamma}\left(y' + \frac{y}{r} - 2uv\right)^2 + \frac{1}{1-\gamma}\left(x' + \frac{x}{r}\right)^2 + (u^2+v^2)s^2 \right\} . \end{aligned} \quad (1.1)$$

The vortex energy per unit length is then given by $\int \mathcal{H}\, d^2\mathbf{r}$. The expression for $\mathcal{H}$ is invariant under the combined substitutions $y \longrightarrow -y$, $x \longrightarrow -x$, $v \longrightarrow -v$ (charge conjugation invariance) so it is sufficient to consider positive values of n.

The Euler-Lagrange equations for the fields are

$$s'' + \tfrac{s'}{r} - [u^2 + v^2 + (y-x)^2 + \tfrac{\beta}{2}(s^2-1)]s = 0 \quad (1.2)$$

$$x'' + \tfrac{x'}{r} - \tfrac{x}{r^2} + (1-\gamma)(y-x)s^2 = 0 \quad (1.3)$$

$$y'' + \tfrac{y'}{r} - \tfrac{y}{r^2} - 2(2v'u + vu' + \tfrac{1}{r}vu) - 4yu^2 - \gamma(y-x)s^2 = 0 \quad (1.4)$$

$$v'' + \tfrac{v'}{r} - \tfrac{v}{r^2} + 2(2y'u + yu' + \tfrac{1}{r}yu) - 4vu^2 - \gamma v s^2 = 0 \quad (1.5)$$

$$v'y - vy' + (2y^2 + \tfrac{\gamma}{2}s^2 + 2v^2)u = 0 \ . \quad (1.6)$$

We shall also study the equations for W's produced in the background of the Z_{NO} vortex. For this purpose we have to consider the Euler-Lagrange equations for the fields u, v, and A_φ, with s, z fixed at their Z_{NO} values. It is convenient to use y, u, v as new independent variables, so that $x = y - z$. The equations for u and v will be the same, (1.5) and (1.6), but the equation for y must be modified:

$$y'' + \frac{y'}{r} - \frac{y}{r^2} - 2(1-\alpha)(2v'u + vu' + \frac{vu}{r} + 2yu^2) - \gamma(y-x)s^2 = 0, \quad (1.7)$$

where $\alpha = \gamma$ if A_φ is allowed to vary or $\alpha = 1$ if A_φ is kept at its Z_{NO} value $A_\varphi = 0$. This equation coincides with the previous equation (1.4) if one sets $\alpha = 0$. Therefore the new equation can be used in the three cases with the appropriate values of α. Note that in all cases α is a non-negative parameter.

From the last three equations one obtains the integrability condition

$$\frac{d}{dr}[ru(\gamma s^2 + 4\alpha v^2)] + 2r(\gamma x s^2 + 4\alpha y u^2)v = 0 \ . \quad (1.8)$$

The full system of equations can be reduced to 4 independent second-order differential equations by solving (1.6) for u and (1.8) for u'.

In order to obtain solutions with a finite energy per unit length of the vortex, the following boundary conditions are imposed.

Boundary conditions near $r = 0$:

$$s = s_0 r^n, \quad x = -\tfrac{2n+m}{2r} + x_0 r, \quad y = -\tfrac{m}{2r} + y_0 r, \quad v = v_0 r^k, \quad u = u_0 r^k \quad (1.9)$$

with $k > -\frac{1}{2}$. Inserting these into the equations, one finds that solutions exist near the origin in the following three cases.

$$\begin{array}{llll} \text{a)} & m = 0, & k = 1, & (1+n)u_0 = nv_0 \ . \\ \text{b)} & m = k+1, & k = 0,1,2\ldots, & u_0 = v_0 \ . \\ \text{c)} & m = -k-2n-1, & k = 0,1,2,\ldots, & mu_0 = -(m+2n)v_0 \ . \end{array} \tag{1.10}$$

These are the boundary conditions valid for the case $\alpha = 0$ corresponding to the variational problem for all the fields or for the same equations linearized with respect to u, v. For the case of the new equations with $\alpha > 0$, the boundary conditions will still be the same for values of n, m such that $k > n$. On the other hand, assuming that $k \leq n$, one finds that the equations near $r = 0$ can only be satisfied if

$$\begin{array}{lll} \text{d)} & m = 0, \quad k = 1, \quad n = 1, & (\gamma s_0^2 + 4\alpha v_0^2)u_0 = \frac{\gamma}{2} s_0^2 v_0 \ . \\ \text{e)} & m \neq 0, \quad k = 0, & v_0 = mu_0 \ . \end{array} \tag{1.11}$$

We remark however that, for $k = 0$, continuity of $\mathbf{W}$ at the origin restricts the value of m to be $m = \pm 1$.

Boundary conditions near $r = \infty$:

Depending on the values of β and γ, different terms in the asymptotic equations are responsible for the leading exponential behavior at large r. Write $s = 1 - f$. Assuming that $0 < \gamma \leq 1$, we consider first the simplest case:

i) $\beta \leq 4\gamma$, $4\gamma > 1$

The asymptotic field equations are, to leading order,

$$\begin{array}{ll} 1.2' : & f'' + \frac{f'}{r} - \beta f = 0 \\ 1.3' : & z'' + \frac{z'}{r} - \frac{z}{r^2} - z = 0 \\ 1.6' : & (v' + \frac{v}{r})\frac{y_1}{r} + (2(\frac{y_1}{r})^2 + \frac{1}{2}\gamma)u = 0 \\ 1.7' : & y'' + \frac{y'}{r} - \frac{y}{r^2} - \gamma z = 0 \\ 1.8' : & u' + \frac{u}{r} + 2y_1\frac{v}{r} = 0 \ , \end{array}$$

where equation (1.3') was obtained by subtracting (1.3) from (1.7). Differentiating (1.8'), and using (1.6') and (1.8') to eliminate v' and v, one obtains an asymptotic equation for u,

$$u'' + 3\frac{u'}{r} + \frac{u}{r^2} - (4(\frac{y_1}{r})^2 + \gamma)u = 0 \ . \tag{1.12}$$

From equations (1.2', 1.3', 1.7', 1.12) and (1.8'), one finds the following approximate solutions for large r:

$$\begin{aligned} f(r) &\sim f_1 K_0(\sqrt{\beta}\, r) \\ z(r) &\sim z_1 K_1(r) \\ y(r) &\sim \tfrac{y_1}{r} + \gamma z \\ u(r) &\sim -v_1 \tfrac{2y_1}{r} K_{|2y_1|}(\sqrt{\gamma}\, r) \\ v(r) &\sim v_1 \tfrac{d}{dr} K_{|2y_1|}(\sqrt{\gamma}\, r)\ . \end{aligned} \tag{1.13}$$

The parameter y_1 is related to the total flux of the electromagnetic field,

$$\oint_\infty \mathbf{A} \cdot d\rho = \frac{2\pi}{e}(2y_1 + 2\gamma n + m)\ .$$

For values of β, γ which do not satisfy condition (i) above, the expressions for z (or y) and f must be modified as follows.

ii) If $4\gamma \leq 1$,

then the asymptotic expression for z in the set of equations (1.2–1.6), or for y in the set of equations (1.5–1.7), has to be modified. Making use of the Green function for the Laplace operator in two dimensions let us define:

$$\begin{aligned} \zeta(r) &= -\frac{1}{2\pi}\int K_0(|\mathbf{r}-\mathbf{r}'|)\cos(\varphi-\varphi')\, j(r')\, d^2\mathbf{r}'\ , \\ j(r) &\equiv 2\left(2v'(r)u(r) + v(r)u'(r) + \tfrac{1}{r}v(r)u(r) + \tfrac{2y_1}{r}[u(r)]^2\right). \end{aligned} \tag{1.14}$$

Then, for the set of equations (1.2–1.6), one would have

$$z = z_1 K_1(r) + \zeta(r)\ , \tag{1.15}$$

while for the set (1.5–1.7) one would have

$$y = \frac{y_1}{r} + \gamma z_1 K_1(r) + (1-\alpha)\zeta(r)\ . \tag{1.16}$$

iii) If $\beta \geq 4\gamma$ or $\beta \geq 4$,

then the expression for f has to be modified. Similarly to case (ii) one finds

$$f(r) = f_1 K_0(\sqrt{\beta}\, r) + \frac{1}{2\pi}\int K_0(\sqrt{\beta}|\mathbf{r}-\mathbf{r}'|)\left(z(r')^2 + u(r')^2 + v(r')^2\right) d^2\mathbf{r}'\ . \tag{1.17}$$

In the integrands of (1.14, 1.17), u and v are given by their asymptotic expressions (1.13), and in equation (1.17), z is given by (1.13) or (1.15) as prescribed by the value of γ.

The result (1.17) means that the leading asymptotic behavior of the Higgs field is determined, in case (iii), not by the Higgs mass but by twice the W boson mass or, in the limit of zero W fields, by twice the Z mass. This finding agrees with the expressions obtained in a recent reanalysis of the Nielsen-Olesen vortex [14].

The asymptotic expressions involve four parameters, f_1, z_1, y_1 and v_1. Together with the four boundary parameters s_0, x_0, y_0 and v_0 at $r = 0$, the number of unknowns equals the rank of the system of differential equations. Therefore, if a solution to the equations exists, then all parameters would be determined by imposing the respective boundary conditions at $r = 0$ and at a large value $r = r_1 \gg 1$.

We shall now investigate the question of existence of a solution with these boundary conditions. Let us introduce the functions $V = v/y$ and $U = ru(\gamma s^2 + 4\alpha v^2)$. The equations for U and V are

$$V' + 2(1 + \tfrac{1}{y^2}(v^2 + \tfrac{\gamma}{4}s^2))u = 0 \ , \tag{1.18}$$

$$U' + 2r(\gamma x s^2 + 4\alpha y u^2)v = 0 \ . \tag{1.19}$$

Multiplying (1.18) by U, (1.19) by V, and adding the equations one obtains

$$(UV)' + 2r\left\{\left([1 + \tfrac{1}{y^2}(v^2 + \tfrac{\gamma}{4}s^2)]\,(\gamma s^2 + 4\alpha v^2) + 4\alpha v^2\right)u^2 + \tfrac{x}{y}\gamma s^2 v^2\right\} = 0 \ . \tag{1.20}$$

The behavior of $UV = ruv(\gamma s^2 + 4\alpha v^2)/y$ as $r \to 0$ and $r \to \infty$ is as follows:

$$r \to 0: \qquad UV \sim \begin{cases} r^2, & m \neq 0, \quad k = 0, \quad \alpha \neq 0 \\ r^{(2n+2)}, & m \neq 0, \quad k = 0, \quad \alpha = 0 \\ r^{(2n+2k+2)}, & m \neq 0, \quad k > 0 \\ r^{(2n+2)}, & m = 0, \quad k = 1 \ . \end{cases}$$

$$r \to \infty: \qquad UV \sim \exp(-2\sqrt{\gamma}\, r) \ . \tag{1.21}$$

If the solution to the equations is such that y does not have a finite zero, then integrating (1.20) from 0 to ∞ one obtains

$$\int 2r\,dr\left([1+\frac{1}{y^2}(v^2+\frac{\gamma}{4}s^2)]\,(\gamma s^2+4\alpha v^2)+4\alpha v^2\right)u^2+\int 2r\,dr\,\frac{x}{y}\gamma s^2 v^2 = 0. \tag{1.22}$$

As $r \to \infty$, since $z \to 0$ exponentially, yx is positive. Therefore, under the assumption that yx does not change sign, the integrands in both integrals in (1.22) are positive definite. We have thus proven the following theorem.

Theorem 1 *Any solution of the field equations* $(1.5 - 1.8)$ *with nonvanishing fields* u, v *must be such that the product* yx *has at least one zero in the open interval* $(0, \infty)$.

We remark here that the theorem is valid also for solutions of the equations linearized with respect to u and v. The derivation in this case parallels that of equations (1.18–1.21) above and leads to an equation like (1.22), except that the v^2 terms are absent from the first integral.

2 Static Solutions for W's in the Background of the Z_{NO} Vortex.

Recall that, for the Z_{NO} vortex, the electromagnetic vector potential is zero. In terms of our auxiliary fields x and y, this translates into the condition $\gamma x + (1-\gamma)y = -\frac{2\gamma n+m}{2r}$. One then obtains

$$x = (\gamma - 1)z - \frac{2\gamma n+m}{2r} \ , \tag{2.1}$$

$$y = \gamma z - \frac{2\gamma n+m}{2r} \ , \tag{2.2}$$

where $z - \frac{n}{r} \equiv Z$ is the vector potential of the Nielsen-Olesen vortex solution.

We shall first investigate the possibility of a perturbative solution about the Z_{NO} vortex. This means that one should look for solutions of the set of equations (1.2–1.8) to lowest order in the fields u, v. Since the source for A_φ is proportional to $j(r)$ given by equation (1.14), perturbations in the electromagnetic field do not contribute to this order. Perturbations in s and z will also be quadratic in u, v. The equations for s, x, y will then be the same as for the Z_{NO} vortex. The equations for u, v are (1.5) and (1.6), linearized with respect to u and v. The boundary conditions for solutions of the linearized equations are the same as the boundary conditions (1.9) and (1.13) of the exact equations.

Let us denote by $\mathcal{Z}_{NO}$ the point in function space corresponding to the Nielsen-Olesen field configuration of the Z_{NO} vortex. We shall establish a condition on the possible solutions in the Z_{NO} background. For this purpose we need the following lemmas.

Lemma 1 *The function $z = y - x$ of $\mathcal{Z}_{NO}$ is positive definite for all values of the parameter β.*

In fact near $r = 0$, $z \approx \frac{n}{r}$ is positive. As $r \to \infty$, z goes exponentially to zero. Then if z were to change sign it would have to go through a negative minimum. The equation for z in the case of the Z_{NO} vortex is

$$z'' + \frac{z'}{r} - \frac{z}{r^2} - zs^2 = 0 \ . \tag{2.3}$$

At a negative minimum of z this gives

$$z'' = (\frac{1}{r^2} + s^2)z < 0 \ . \tag{2.4}$$

But this is the condition for a maximum. Therefore z cannot have a negative minimum, hence it cannot change sign.

Lemma 2 *For $\mathcal{Z}_{NO}$ the value of z_0 in the expansion $z = \frac{n}{r} + z_0 r + \ldots$ near the origin is negative.*

In fact the equation for z can be written

$$\frac{d}{dr}\Big(\frac{1}{r}\frac{d}{dr}(rz)\Big) - zs^2 = 0 \ . \tag{2.5}$$

Integrating from 0 to ∞ one obtains

$$z_0 = -\frac{1}{2}\int_0^\infty zs^2 dr \ . \tag{2.6}$$

Since z is positive definite it follows that $z_0 < 0$. For $\beta = 1$, $z_0 = -1/4$.
We are now ready to prove the following theorem.

Theorem 2 *Any solution of the field equations* $(1.5 - 1.6)$, *with nonvanishing* u, v *and the fields* (s, x, y) *fixed at their* $\mathrm{Z_{NO}}$ *vortex values*, or *these same equations linearized with respect to* u *and* v, *must satisfy the condition* $-2n < m < 0$.

As was already pointed out, Theorem 1 holds true for solutions of the linearized equations as well as of the exact equations. Consider now the asymptotic expressions for y and x,

$$\begin{aligned}
\text{As } r \to 0: \quad & x = -\tfrac{2n+m}{2r} + (\gamma - 1)z_0 r + \mathcal{O}(r^3) \ , \\
& y = -\tfrac{m}{2r} + \gamma z_0 r + \mathcal{O}(r^3) \ , \\
\text{As } r \to \infty: \quad & x = y = -\tfrac{2\gamma n+m}{2r} \ .
\end{aligned}$$

Recall that $z_0 < 0$ for all values of (β, γ).
Assume that the bound on m is not satisfied. Then we have the following two cases:

1) $m \geq 0$. Then x and y are both negative as $r \to 0$ and at large r.
2) $m \leq -2n$. Then x and y are both positive as $r \to 0$ and at large r.

A possible exception occurs for $m = -2n$ and $\gamma = 1$. In this case $x(r) \equiv 0$ and the condition (1.22) for the existence of a solution would require $u(r) \equiv 0$. But then the energy would always increase for any non-vanishing configuration of $v(r)$. For this reason this case is excluded from consideration. In all other cases, if x or y were to change sign, they would have to go through a positive maximum and a negative minimum. But x and y satisfy respectively the equations

$$y'' + \frac{y'}{r} - \frac{y}{r^2} - \gamma zs^2 = 0 \ , \tag{2.7}$$

$$x'' + \frac{x'}{r} - \frac{x}{r^2} - (\gamma - 1)zs^2 = 0 \ , \tag{2.8}$$

where, as already shown, z is positive definite. Then, at a positive maximum of y, one would have

$$y'' = \frac{y}{r^2} + \gamma zs^2 > 0 \ , \tag{2.9}$$

which is the condition for a minimum. At a negative minimum of x

$$x'' = \frac{x}{r^2} + (\gamma - 1)zs^2 < 0 \ , \tag{2.10}$$

which is the condition for a maximum. Therefore neither y nor x changes sign and, by Theorem 1, no solution exists, which proves the theorem.

Let us now apply Theorem 2 to perturbative solutions about the Z_{NO} vortex. The linearized equations lead to the same constraints (1.10) on m, n, and k as the exact equations for all fields. According to these constraints, a solution of the equations near $r = 0$ is possible only for $m \geq 0$ or $m \leq -2n - 1$. But Theorem 2 tells us that no global solution exists for these values of m. We thus arrive at the following result:

Theorem 3 *In the one-component Higgs gauge, a perturbative solution of the Euler-Lagrange equations about the $\mathcal{Z}_{NO}$ configuration does not exist for any values of β and γ, ($\beta > 0$, $0 \leq \gamma \leq 1$).*

Our field ansatz is ill-defined for $\gamma = 0$, but in this case the physical Z field is aligned with the U(1) hypercharge field and does not couple at all to the W bosons.

An analysis of the variation of the energy at $\mathcal{Z}_{NO}$, done in Ref. [9] for $n = 1$, shows that the energy decreases, in a region of the parameter space (β, γ), for a perturbation in the W field with values $m = -1$ and $k = 0$. This value of k implies that the W production is concentrated at the core of the vortex; this is natural since there the Z field strength takes its maximal value, the Higgs field is minimal, and the vacuum instability due to the anomalous magnetic moment of the W boson [6, 7, 8] is most pronounced. As we have seen, for $n = 1$ the values of m for which the boundary conditions at $r = 0$ for the equations (1.5–1.8) linearized with respect to u, v admit a solution, exclude the values $m = -1, -2$. Nevertheless, within the region (β, γ) of instability of the Z_{NO} vortex, one expects the energy to have a minimum for some W configuration with $m = -1$ and the fields s, z fixed at their $\mathcal{Z}_{NO}$ values. This minimum would be a solution of the exact equations (1.5) and (1.6) with boundary conditions (1.11.e) and (1.13).

Before proceeding with the investigation of these equations, we shall establish the following result.

Theorem 4 *If, for some values of β, γ, the equations (1.5–1.7) with s, z given by their $\mathcal{Z}_{NO}$ values, admit a solution such that the energy of the corresponding state, calculated in the quadratic approximation in the W field, is lower than that for the Z_{NO} vortex, then for these values of β and γ the Z_{NO} vortex is a saddle point in the space of field configurations.*

We shall prove the theorem for the case in which A_φ is allowed to vary and y satisfies equation (1.7) with $\alpha = \gamma$. A similar proof can be given for the other case ($y \equiv y_{NO}$ and $\alpha = 1$).

Suppose that for some values of β, γ, we have a solution of the equations (1.5–1.7) corresponding to the production of W's and an electromagnetic potential A_φ in the Z_{NO} background. Let u, v, y be the functions corresponding to this solution and write $y = y_{NO} + \bar{y}$, $x = x_{NO} + \bar{y}$, where x_{NO}, y_{NO} are the values of x, y in the $\mathcal{Z}_{NO}$ configuration, given by (2.1, 2.2). Let $\mathcal{E}_0$ be the energy for the Z_{NO} vortex (with $\Phi_0^2 = 1$ for simplicity) and let us break up the energy difference $\delta\mathcal{E} = \mathcal{E} - \mathcal{E}_0$ into three terms:

$$\mathcal{E}_1 = \frac{2\pi}{\gamma}\int rdr\,[(v' + \frac{v}{r} + 2y_{NO}u)^2 - 4(y'_{NO} + \frac{y_{NO}}{r})uv + \gamma(u^2 + v^2)s^2]\,, \quad (2.11)$$

$$\mathcal{E}_2 = \frac{2\pi}{\gamma}\int rdr\,[4(v'\overline{y} - \overline{y}'v)u + 4(2y_{\rm NO}\overline{y} + v^2)u^2 + \frac{1}{1-\gamma}(\overline{y}' + \frac{\overline{y}}{r})^2]\ , \qquad (2.12)$$

$$\mathcal{E}_3 = \frac{2\pi}{\gamma}\int rdr\,(2\overline{y}u)^2\ . \qquad (2.13)$$

Since $\overline{y}$ is already second order in (u,v), then $\mathcal{E}_1$ is the lowest order energy shift which is assumed to be negative, $\mathcal{E}_1 < 0$.

Consider configurations of the fields $\overline{y}_\lambda = \lambda\overline{y}, u_\lambda = \sqrt{\lambda}u, v_\lambda = \sqrt{\lambda}v$, where λ is a scaling factor. The energy corresponding to these configurations with s, z at their $\mathcal{Z}_{\rm NO}$ values will be given by

$$\mathcal{E}(\lambda) = \mathcal{E}_0 + \lambda\mathcal{E}_1 + \lambda^2\mathcal{E}_2 + \lambda^3\mathcal{E}_3\ . \qquad (2.14)$$

This is a cubic polynomial in λ with extrema at

$$\lambda_\pm = \frac{1}{3\mathcal{E}_3}\Big(- \mathcal{E}_2 \pm \sqrt{\mathcal{E}_2^2 - 3\mathcal{E}_3\mathcal{E}_1}\Big)\ . \qquad (2.15)$$

Since $\mathcal{E}_1 < 0$, and $\mathcal{E}_3 > 0$, only λ_+ is a positive root corresponding to a minimum of $\mathcal{E}(\lambda)$. Thus in the interval $0 < \lambda < \lambda_+$ the energy $\mathcal{E}(\lambda)$ will be smaller than $\mathcal{E}(0) = \mathcal{E}_0$. Therefore there will be configurations of arbitrarily small fields $\overline{y}_\lambda, u_\lambda, v_\lambda$ for which the energy becomes smaller than $\mathcal{E}_0$. This proves the theorem.

Now, by construction, $\mathcal{E}(\lambda)$ has an extremum at $\lambda = 1$. Therefore the value of λ at the minimum of $\mathcal{E}(\lambda)$ is $\lambda_+ = 1$. Inserting this value in the expression (2.15), one obtains

$$\mathcal{E}_1 = -(2\mathcal{E}_2 + 3\mathcal{E}_3)\ . \qquad (2.16)$$

This is a very useful result for numerical computations since it provides a good check on the precision of the calculation of the energy shift $\delta\mathcal{E}$,

$$\delta\mathcal{E} \equiv \mathcal{E}_1 + \mathcal{E}_2 + \mathcal{E}_3 = \frac{1}{2}(\mathcal{E}_1 - \mathcal{E}_3) = -(\mathcal{E}_2 + 2\mathcal{E}_3)\ . \qquad (2.17)$$

3 Numerical Solutions

In this section we present solutions of the set of equations (1.5–1.6) for W's in the presence of the fields s, x, y given by their $\mathcal{Z}_{\rm NO}$ values. The second-order equation (1.5) can be written in the form of two first-order equations for the functions rv and $F \equiv v' + \frac{v}{r} + 2uy$, as

$$(rv)' = r(F - 2uy)\ , \qquad (3.1)$$

$$F' = (\gamma s^2 + 4u^2)v - 2(y' + \frac{y}{r})u\ . \qquad (3.2)$$

Here $y = \gamma Z - \frac{m}{2r}$, where Z and s are given by the Nielsen-Olesen vortex solution and $m = -1$. Equation (1.6) was used to solve algebraically for u in terms of v and F,

$$u = 2\frac{(y' + \frac{y}{r})v - yF}{\gamma s^2 + 4v^2}\ . \qquad (3.3)$$

β	γ	v_0	v_1	$\delta\mathcal{E}$
.5	.25	$-.092750$	.06860281	$-.00028$
.5	.5	$-.211671$	.41657161	$-.0106$
1	.5	$-.237408$	.35632791	$-.0133$
.5	1	$-.377157$	1.04195850	$-.0846$
1	1	$-.418631$	.79309235	$-.1003$

Table 1: Boundary values and change of energy due to W-boson condensation in the fixed background of the Z_{NO} vortex, for a selection of parameters β and γ.

The system of equations was first propagated from $r_1 = 18.5$ down to $r_0 = 0$. The boundary condition $v_0 = -u_0$ imposed at this point allowed for a precise determination of v_1, while v_0 was determined with less precision. Values of these two parameters, which determine the boundary conditions at $r = 0$ and at $r = \infty$, are listed in Table 1 for several different values of β and γ. Next, using these boundary values, the equations were propagated from r_1 and from $r_0 = 0$ to a matching point in the interval $(1.5, 2.5)$. The values of u and v agreed within a precision of 10^{-2} at matching points in this interval. Both u and v vary monotonically without changing sign.

Finally, since this is a non-perturbative solution, one has to check whether the energy of these states with W's is actually lower than that of the Z_{NO} vortex. We computed the energy difference between each of these states and the Z_{NO} vortex and found in each case $\delta\mathcal{E} < 0$. Hence in the sector $(\beta > .5,\ \gamma > .25)$ we find that the Z_{NO} vortex is unstable against W production. The results for $\delta\mathcal{E}$ in units of Φ_0^2 (energy per unit length of vortex) are also given in Table 1. The precision given for the parameters v_0, v_1, is that required to obtain the precision in the energy shift $\delta\mathcal{E}$. However the accuracy of these numbers is sensitive to the accuracy in the determination of the input functions s, z as solutions of the Z_{NO} vortex.

In each of the cases reported here the relations (2.17) (with $\overline{y} = 0$, $\mathcal{E}_2 = 0$) were satisfied within the accuracy obtained for $\delta\mathcal{E}$.

The line in parameter space along which the equations considered in this section cease to have a solution with $\mathcal{E}_0 < 0$, provides an upper bound on the boundary line Γ for the stability region of the Z_{NO} vortex. An upper bound for this curve has also been obtained by Klinkhamer and Olesen [15] using a different approach.

4 Field Configuration of Minimal Energy with W Fields

Since the Z_{NO} vortex with $n = 1$ is unstable in a region of the parameter space (β, γ) with respect to W production in a state of angular momentum $m = -1$, and because the energy is bounded from below, one expects that there exists some configuration of the fields, with W's in such a state, for which the energy is a minimum. The Euler-Lagrange equations (1.2–1.8), however, do not admit a solution for $m = -1$ with the boundary conditions required to make the energy finite. This seems to preclude the existence of a stationary state with W's at least defined in the space of differentiable functions. A minimum energy state may nevertheless exist in the sense of a distribution.

In order to investigate this possibility we propose to study a model obtained from Weinberg-Salam (WS) model with the following modifications:

i) Set $\mathbf{W} = \sqrt{\epsilon}\,\mathbf{W}'$

ii) Add to the energy density a term

$$\frac{1}{\gamma}\epsilon(1-\epsilon)(i\mathbf{W}'^{\dagger} \times \mathbf{W}')^2$$

This term is positive definite for $0 < \epsilon < 1$. In the new model the boundary conditions at $r = 0$ can be satisfied for $m = -1$.

The model coincides with WS for $\epsilon = 1$. In the instability region of parameter space, the Euler-Lagrange equations for the fields have a solution which, as $\epsilon \to 0$, approaches that found in the WS model for W's in the Z_{NO} vortex background. As ϵ increases from 0 to 1, either of the following possibilities may occur:

1) The equations do not have a solution for $\epsilon > \epsilon_{\max} < 1$.
2) As $\epsilon \to 1$ the solutions approach almost everywhere the configuration in the vacuum state and the energy approaches zero.
3) As $\epsilon \to 1$ one obtains a sequence of solutions which has no limit but the energy has a definite limit greater than zero.

In cases 1) and 2) a stable vortex with W's and a finite energy does not exist. In case 3) the existence of a stable vortex depends on establishing the stability of the solutions as $\epsilon \to 1$.

Feza Gürsey in Memoriam

At this conference one of us (S.W.M.) presented a special contribution in memory of Feza Gürsey, which will be published elsewhere in these proceedings. The other (O.T.) wishes to express in this space his feelings of gratitude to Feza Gürsey for his inspired teaching of quantum field theory at Yale University and for the many ways graduate students have benefitted, through the years, from his vibrant personality.

References

[1] H. B. Nielsen and P. Olesen, Nucl. Phys. **B61** (1973), 45.

[2] Y. Nambu, Nucl. Phys. **B130** (1977), 505.

[3] T. Vachaspati, Phys. Rev. Lett. **68** (1992), 1977; *ibid.* **69** (1992), 216.

[4] S. Weinberg, Phys. Rev. Lett. **19** (1967), 1264.

[5] A. Salam in *Elementary Particle Theory*, edited by N. Svartholm (Almqvists Förlag AB, Stockholm, 1968) p. 367.

[6] J. Ambjørn and P. Olesen, Nucl. Phys. **B315** (1989), 606; Nucl. Phys. **B330** (1990), 193; Int. J. Mod. Phys. A **5** (1990), 4525.

[7] S. W. MacDowell and O. Törnkvist, Phys. Rev. D **45** (1992), 3833.

[8] Nils Ola Törnkvist, Yale University Ph. D. Thesis (Nov. 1993) RX-1493, Microfiche UMI-94-15879-mc.

[9] M. James, L. Perivolaropoulos, and T. Vachaspati, Phys. Rev. D **46** (1992), R5232; Nucl. Phys. **B395** (1993), 534.

[10] W. B. Perkins, Phys. Rev. D **47** (1993), R5224.

[11] A. Achúcarro, R. Gregory, J. A. Harvey, and K. Kuijken, Phys. Rev. Lett. **72** (1994), 3646.

[12] O. Törnkvist, Yale Preprint YCTP-P11-92 (Apr. 1992), HEP-PH/9204235.

[13] P. Olesen, Niels Bohr Institute preprint NBI-HE-93-58 (Oct. 1993), HEP-PH/9310275.

[14] L. Perivolaropoulos, Phys. Rev. D **48** (1993), 5961; *ibid.* note to appear.

[15] F. R. Klinkhamer and P. Olesen, Nucl. Phys. **B422** (1994), 227.

Proper-time Methods in the Presence of Non-constant Background Fields

Alan Chodos

Center for Theoretical Physics, Yale University, P.O. Box 208120, New Haven, Connecticut 06520-8120

Abstract: A formalism is developed to enable the construction of the effective action and related quantities in QED for the case of time-varying background electric fields. Some examples are studied and evidence is sought for a possible transition to a phase in which chiral symmetry is spontaneously broken.

The proper-time formalism was used a long time ago by Schwinger [1] to compute the Green function for an electron propagating in the presence of a background electromagnetic field. Although the formalism is general, explicit evaluation of the propagator, and of the associated effective action, was possible only for the case of fields uniform in space and constant in time.

Over the intervening decades, attempts have been made [2] to compute the corrections to Schwinger's results for the case of varying fields. These take the form of a derivative expansion in the fields, but even the first non-trivial corrections turn out to be quite unwieldy, and are, moreover, restricted to fields that do not vary too rapidly (else the higher terms in the expansion must be included).

More recently, there has been cause for a new look at this problem. The motivation is the strange results of the GSI heavy-ion scattering experiments [3], in which mysterious narrow peaks are seen in the energy spectra of emitted e^+e^- pairs. Among the many theoretical ideas that have been advanced, I wish to concentrate on one proposed explanation [4,5]: that the heavy ions create a very strong and rapidly varying electromagnetic field, which then induces a phase transition in QED to a vacuum in which chiral symmetry is spontaneously broken. The observed e^+e^- peaks are due to the decay of positronium - like states in the new phase of QED.

To study this possibility, we employ a proper-time representation for the vaccum expectation value of $\bar{\psi}\psi$, which is an order parameter for this transition. This representation is [5]

$$\langle\bar{\psi}\psi\rangle = m\int_0^\infty d\tau e^{-m^2\tau}\, U(x,\tau) \tag{1}$$

where

$$U(x,\tau) = \mathrm{tr}\langle x|e^{-H\tau}|x\rangle \ . \tag{2}$$

Here τ is the proper-time (continued to imaginary values) and m is the electron mass. U is the trace of a quantum-mechanical matrix element for which the Hamiltonian is $(\gamma\cdot\pi)^2$, with

$$\pi_\mu = p_\mu - eA_\mu(x) \ . \tag{3}$$

The dynamical degrees of freedom are the four coordinates $x_\mu(\tau)$, and p_μ are the associated canonical momenta. $A_\mu(x)$ is the potential that encodes the background field. We are working in Euclidean space, so the γ_μ are Hermitian and H is positive. The trace in eq. (2) is over the indices carried by the γ matrices.

The signal for the spontaneous breakdown of chiral symmetry is that the limit $m\to 0$ of $\langle\bar\psi\psi\rangle$ should not vanish. Because of the explicit factor of m in eq. (1), one requires the integral to diverge as $m\to 0$. In fact, one easily sees [6] that if the large-τ behavior of $U(x,\tau)$ is $\tau^{-\frac{1}{2}}$, $\langle\bar\psi\psi\rangle$ will remain finite and non-zero. If the falloff is more rapid, $\langle\bar\psi\psi\rangle$ will vanish, indicating that there is no spontaneous chiral symmetry breaking.

In a recent paper, Caldi and Vafaeisefat [7] have computed $U(x,\tau)$ numerically using Monte Carlo simulation techniques. For this purpose, it is convenient to recast $U(x,\tau)$ as a path integral:

$$U(x,\tau) = \mathrm{tr}\int \mathcal{D}x_\mu T e^{-S} \ , \tag{4}$$

$$S = \int_0^\tau dt L(x,\dot x) \ , \tag{5}$$

$L(x,\dot x) = \frac{1}{4}\dot x_\mu \dot x_\mu + ie\dot x_\mu A_\mu(x) - e/2\ \sigma_{\mu\nu} F_{\mu\nu}(x)$. The symbol T in eq. (4) denotes τ-ordering.

Caldi and Vafaeisefat look initially at background electric field configurations pointing in one direction only, for which the magnitude varies in time and in the one spatial variable. In particular, they consider

$$\vec{E} = (f(x,t), 0,0) \tag{6}$$

with

$$f(x,t) = eE[\cosh^2(x/W_s)]^{-1} \exp(-t^2/2\ W_t^2). \tag{7}$$

They find, for $W_s^{-1} = W_t = 3$ (in units where eE = 1) that at suitably chosen values for x_μ, $U(x,\tau)$ exhibits the desired $\tau^{-1/2}$ falloff. It is of interest to explore these questions in a more analytic fashion. To this end, we look at a configuration in which the electric field depends only on time (which is, of course, Euclidean time, and which we call x_0): $\vec{E} = (f(x_0), 0,0)$.

Making use of standard manipulations, one obtains

$$U(x,\tau) = \frac{1}{4\pi^2\tau} \sum_{\sigma=\pm 1} \int_{-\infty}^{\infty} d\lambda \int \mathcal{D}x_0 e^{-\int_0^\tau d\tau' L_\sigma} \tag{8}$$

where

$$L_\sigma = \frac{1}{4}\dot{x}_0^2 + (eF - \lambda)^2 + e\sigma f, \qquad \frac{dF}{dx_0} = f . \tag{9}$$

Thus we are summing and integrating over a family of one-dimensional quantum-mechanical models defined by the Hamiltonians

$$H_\sigma = p^2 + V_\sigma(x_0) \ , \tag{10}$$

$$V_\sigma = [eF(x_0) - \lambda]^2 + e\sigma f(x_0) \ . \tag{11}$$

For the purpose of quantitative analysis, it is convenient to re-express $U(x,\tau)$ as

$$U(x,\tau) = \frac{1}{4\pi^2\tau} \sum_{\sigma=\pm 1} \int_{-\infty}^{\infty} d\lambda \langle x|e^{-H_\sigma \tau}|x\rangle \ , \tag{12}$$

and then to insert this in eq. (1), and perform the τ integral after division by m and differentiation with respect to m^2. One obtains

$$I(m) = -\frac{\partial}{\partial m^2}\left[\frac{\langle\bar{\psi}\psi(x)\rangle}{m}\right] = \frac{1}{4\pi^2}\sum_\sigma \int d\lambda \langle x|\frac{1}{H_\sigma + m^2}|x\rangle \ . \tag{13}$$

We then have

$$I(m) = \frac{1}{4\pi^2}\int_{-\infty}^{\infty} d\lambda \sum_{\sigma=\pm 1} \psi_>^{(\sigma)}(x)\, \psi_<^{\sigma}(x). \tag{14}$$

where each ψ obeys the homogeneous equation

$$[-\frac{\partial^2}{\partial x^2} + V_\sigma + m^2]\psi = 0 \quad , \tag{15}$$

subject to the boundary condition that $\psi_>$ ($\psi_<$) is well-behaved as $x \to \infty$ ($x \to -\infty$), and where the Wronskian condition

$$\psi_> \frac{\partial \psi_<}{\partial x} - \psi_< \frac{\partial \psi_>}{\partial x} = 1 \tag{16}$$

is imposed. Our computational strategy is to choose f(x) so that $\psi_>$ and $\psi_<$ can be computed explicitly [8], to insert them in eq. (14), and to determine therefrom the behavior of I(m) as $m \to 0$. If $\langle\bar{\psi}\psi\rangle$ indeed tends to a finite, non-zero value, we expect $I(m) \sim 1/m^3$. Any less singular behavior will be evidence that $\langle\bar{\psi}\psi\rangle$ is tending to zero.

As an illustrative example, we can choose $eF(x) = \gamma \tanh\beta x$. This yields a model that is exactly solvable quantum mechanically [8]. We do not reproduce the formulas here, since they are complicated and not particularly illuminating. When inserted into eq. (14), they yield $1/m^2$ behavior, i.e. no evidence for chiral symmetry breaking.

It is possible to study other exactly solvable quantum mechanical models as well. In all the cases we have examined, the singularity at small m of I(m) is no worse than $1/m^2$.

This result is not in conflict, of course, with the numerical work of Caldi and Vafaeisefat, since their field configurations depend on at least two variables. In deciding how to proceed, one can think of a number of possibilities: (i) extend the search among the one-dimensional models in the hope that an as yet undiscovered class will yield the sought-for $1/m^3$ behavior; (ii) introduce new analytic techniques that will enable one to study the two-variable case. This will permit direct comparison with the Monte Carlo results; (iii) find a way to extract the small m behavior of I(m) (or equivalently the large τ behavior of $U(x,\tau)$) without first having to compute the full functional forms of I(m) or $U(x,\tau)$. This would lead to enormous simplifications not only of the analytic work but also of the Monte Carlo calculations, where the large τ behavior is extracted by computing $U(x,\tau)$ for several values of τ and finding the slope of the best-fitting straight line on a log-log plot.

As new data from GSI and Argonne are reported, one expects the relevance of the ideas upon which the present work is based either to wax or to wane. If the former, it will be interesting to see whether new insight can in fact be gained about the mechanism whereby time- and space-varying background fields induce a chiral phase transition in QED.

A longer version of this paper, with more detail and numerous illuminating comments, is available from hep-ph.

Acknowledgements

I am especially grateful to Daniel Caldi, Andras Kaiser, David Owen and Saeid Vafaeisefat, each of whom contributed substantially to various parts of the work described in this paper. I am also grateful for illuminating conversations with Charles Sommerfield. I am pleased that this work touches on the subject of exactly solvable quantum mechanical models, because this is a subject in which Feza was interested and to which he made significant contributions in collaboration with Franco Iachello and Yoram Alhassid. Finally I wish to express my deepest thanks to Meral Serdaroğlu and the rest of the organizers at Boğaziçi University, to whom the success of the first Gürsey Memorial Conference is in large measure due. The research discussed in this paper was supported in part by the U.S. Department of Energy grant #DEFG0292ER40704.

References

1. J. Schwinger, Phys. Rev. **82**, 664 (1951).
2. L.H. Chan, Phys. Rev. Lett. **54**, 1222 (1985); **57**, 1199 (1986); Phys. Rev. **D38**, 3739 (1988); Charles Sommerfield (private communication); H.W. Lee, P.Y. Pac and H.K. Shin, Phys. Rev. **D40**, 4202 (1989).
3. P. Salabura, et al., Phys. Lett. **B245**, 153 (1990), and references therein; I. Koenig, et al., Z. Phys. **A346**, 153 (1993), and references therein.
4. L.S. Celenza, V.K. Mishra, C.M. Shakin and K.F. Liu, Phys. Rev. Lett. **57**, 55 (1986); L.S. Celenza, C.R. Ji and C.M. Shakin, Phys. Rev. **D36**, 2144 (1987); D.G. Caldi and A. Chodos, Phys. Rev. **D36**, 2876 (1987); Y.J. Ng and Y. Kikuchi, Phys. Rev. **D36**, 2880 (1987).
5. D.G. Caldi, A. Chodos, K. Everding, D.A. Owen and S. Vafaeisefat, Phys. Rev. **D39**, 1432 (1989).
6. T. Banks and A. Casher, Nucl. Phys. **B169**, 103 (1980).
7. D.G. Caldi and S. Vafaeisefat, Phys. Lett. **B287**, 185 (1992).
8. For a discussion of exactly solvable quantum systems, see e.g. Y. Alhassid, F. Gürsey and F. Iachello, Phys. Rev. Lett. **50**, 873 (1983); for a discussion of the connection between solvable quantum systems and supersymmetry, see e.g. F. Cooper, J.N. Ginocchio and A. Khare, Phys. Rev. **D36**, 2458 (1987).

On Abelian Bosonization of Free Fermi Fields in Three Space Dimensions

Charles M. Sommerfield

Center for Theoretical Physics, Yale University, New Haven, Connecticut, USA

Abstract. One of the methods used to extend two-dimensional bosonization to four space-time dimensions involves a transformation to new spatial variables so that only one of them appears kinematically. The problem is then reduced to an Abelian version of two-dimensional bosonization with extra "internal" coordinates. On a formal level, putting these internal coordinates on a finite lattice seems to provide a well-defined prescription for calculating correlation functions. However, in the infinite-lattice or continuum limits, certain difficulties appear that require very delicate specification of all of the many limiting procedures involved in the construction.

Introduction

After the initial success of bosonization in 1+1 dimensions [1,2] Alan Luther presented a heuristic formula [3] for bosonization of free, massless relativistic fermions in 3+1 dimensions. A configuration-space transform [4] was later used by H. Aratyn [5] to put this into a prettier but no less heuristic form. For other approaches to this problem the reader is referred to References [6] and [7].

The basic idea of the method described here is to view field theories in 3+1 dimensions as 1+1 dimensional theories with an internal, non-kinetic degree of freedom. The main subject of this paper is to examine carefully the limiting procedures involved in bosonizing a 1+1 dimensional free fermion with an internal degree of freedom that takes values that are in a continuum.

Tomographic transform

We begin by reviewing the transformation to a 1+1 dimensional theory both for bosons and for fermions. We then present the standard Abelian technique for bosonizing the 1+1 dimensional fermion when the internal degrees of freedom lie on a finite lattice. There follows a careful analysis of the continuum limit of this lattice in the context of a simplified model.

The tomographic transform of a free scalar field $\phi(\mathbf{x}, t)$ with mass m in 3+1 dimensions is defined as

$$\tilde{\phi}(y, \mathbf{n}, t) = \frac{1}{2\pi} \int d^3\mathbf{x}\, \partial_y \delta(y - \mathbf{n}\cdot\mathbf{x})\phi(\mathbf{x}, t).$$

Here $\mathbf{n}$ is a unit vector and $d^2\mathbf{n}$ an element of solid angle in the direction of $\mathbf{n}$.

The field equation satisfied by $\tilde{\phi}(y,\mathbf{n},t)$ and its equal time commutation relations are essentially those of a 1+1 dimensional scalar field, also free and with mass m.

The corresponding transform for a fermion field is

$$\tilde{\psi}^a(y,\mathbf{n},t) = \frac{1}{2\pi}\int d^3\mathbf{x}\,\partial_y\delta(y-\mathbf{n}\cdot\mathbf{x})\sum_\alpha u_\alpha^{\dagger a}(\mathbf{n})\psi_\alpha(\mathbf{x},t).$$

It satisfies the 1+1 dimensional Dirac equation for a right-moving fermion field, (that is, depending only on $y-t$) and has the expected equal-time anticommutation relations. Here $u^a(\mathbf{n})$, $a=1,2$, are orthonormal four-component spinors that satisfy $(\boldsymbol{\alpha}\cdot\mathbf{n})u^a(\mathbf{n}) = u^a(\mathbf{n})$ where $\boldsymbol{\alpha}$ are the Dirac alpha matrices.

The strategy is to use standard Abelian 1+1 dimensional bosonization to construct the right-moving $\tilde{\psi}^a(y,\mathbf{n},t)$, in terms of the right-moving part of the transformed boson $\tilde{\phi}(y,\mathbf{n},t)$, assuming, to begin with, that $\mathbf{n}$ is restricted to a lattice. This is given formally, with the t dependence suppressed, as

$$\tilde{\phi}_r^a(y,\mathbf{n}) = \tfrac{1}{2}[\tilde{\phi}^a(y,\mathbf{n}) - \tfrac{1}{2}\int_{-\infty}^{\infty} dy'\;\epsilon(y-y')\,\partial_t\tilde{\phi}^a(y',\mathbf{n})],$$

where $\epsilon(y-y') = (y-y')/|y-y'|$. It follows that

$$\langle 0|\tilde{\phi}_r^a(y,\mathbf{n})\tilde{\phi}_r^b(y',\mathbf{n}')|0\rangle = -\frac{1}{4\pi}\delta^{ab}\delta(\mathbf{n},\mathbf{n}')\ln\left(\mu[\alpha - i(y-y')]\right)$$

where μ and α are infrared and ultraviolet cutoffs, respectively.

We first take $\mathbf{n}$ to be discretely distributed on some lattice of directions and reinterpret the Dirac delta function as a Kronecker delta function. Replacing the pair a and $\mathbf{n}$ by a single index A, we find the formal bosonization expression

$$\tilde{\psi}^A(y) = \frac{1}{\sqrt{2\pi\alpha}}e^{i\tilde{\phi}_r^A(y)}K^A$$

where

$$K^A = \exp\frac{i}{2}\sqrt{\pi}\sum\nolimits_C \epsilon^{AC}[\phi_r^C(\infty) - \phi_r^C(-\infty)]$$

and where $\epsilon^{AB} = -\epsilon^{BA}$, with $|\epsilon^{AB}|^2 = 1$. The Klein factors K^A are needed to make sure that fermion fields with different indices anticommute rather than commute. In what follows we will ignore these factors. Their only effect is to correct an occasional sign to its proper value.

The correlation functions for the fermion field are determined from the matrix elements of their bosonic representation in the bosonic vacuum. The two-point function is computed as

$$\langle 0|\tilde{\psi}^A(y)\tilde{\psi}^{\dagger B}(y')|0\rangle = \frac{1}{2\pi}\frac{\mu}{\{\mu[\alpha + i(y-y')]\}^{\delta^{AB}}}.$$

If $A = B$ we get the known fermion function. If $A \neq B$ then this vanishes in the limit $\mu \to 0$, so that the result is indeed proportional to δ^{AB}. When reinterpreted in terms of a continuous $\mathbf{n}$ we do get the correct two-point function for the transformed fermion field:

$$\langle 0|\tilde{\psi}^a(y,\mathbf{n})\tilde{\psi}^{\dagger b}(y',\mathbf{n}')|0\rangle = \frac{1}{2\pi}\frac{\delta^{ab}\delta(\mathbf{n},\mathbf{n}')}{\alpha - i(y-y')}.$$

All of the other fermion correlation functions come out properly as well.

To test the consistency of the bosonization formula we must consider the important fermion bilinear operators in the 3+1 dimensional theory such as the Poincaré group generators and the chiral charge operators. If Ω is one of these, does the sequence $\Omega[\psi_{3+1}] \leftrightarrow \Omega[\tilde{\psi}_{1+1}] \leftrightarrow \Omega[\tilde{\phi}_{1+1}] \leftrightarrow \Omega[\phi_{3+1}]$ make sense? Trouble arises when we consider the rotation and boost generators. The proper treatment of the continuum limit of the lattice plays a fundamental role here.

Simplified model and smoothing functions

In order to get at the heart of the problem of the continuum limit we will treat a simple case in which the the internal variable is one-dimensional. We thus consider a single component 1+1 dimensional massless chiral fermion field $\psi(x,u)$ that depends on a continuous internal variable u whose domain is the real line. The bosonization of the fermion field in the Dirac equation $(\partial_t + \partial_x)\psi(x,u) = 0$ will be in terms of a right-moving chiral massless boson field $\phi_r(x,u)$. Assuming for the moment that u is a discrete variable, we write, up to Klein factors, $\psi(x,u) = (2\pi\alpha)^{-1/2}\exp[i\sqrt{4\pi}\phi_r(x,u)]$ and obtain for the fermion two-point function

$$\langle 0|\psi(x,u)\psi^\dagger(x',u')|0\rangle = \frac{1}{2\pi}\frac{\delta_{u,u'}}{\alpha - i(x-x')}$$

For the bosonization procedure to work, it is clear that the exponent that appears in the evaluation of the fermion two-point function must be associated with a Kronecker delta function, rather than a Dirac delta function so that the logarithmic singularity in the bosonic two-point function exponentiates to a simple pole or a simple zero. We need a more analytic method of converting one type delta function to the other. Returning to a continuous u, we introduce a new boson field $\Phi(x,u) = \int du' f(u-u')\phi_r(x,u')$. We take $f(u)$ to be a real, even function. The Dirac delta function is replaced in the commutation relations among the Φ operators, as well as in the bosonic two-point function, by the convolution $g(u-u') = \int du''\, f(u-u'')f(u''-u')$. It is thereby softened. We will choose $f(u)$ so that $g(u) \geq 0$, $g(0) = 1$, and $\epsilon(u)g'(u) < 0$ if $g(u) \neq 0$. To make sure that $g(u)$ goes to 0 very quickly as $|u|$ increases we take it to be a function of u/λ and eventually take the scale parameter λ to zero. This simulates the properties of a Kronecker delta function.

We define the field $\hat{\psi}(x,u) = C(2\pi\alpha)^{-1/2}\exp[i\sqrt{4\pi}\Phi(x,u)]$ which will be the candidate for the canonical fermion field ψ in the appropriate limit. Here C

is some constant, to be specified later, that depends on μ and on the choice of $g(u)$. We then find for the fermionic two-point function

$$\langle 0|\hat{\psi}(x,u)\hat{\psi}^\dagger(x',u')|0\rangle = |C|^2\mu^{[1-g(u,u')]}\frac{1}{2\pi}\frac{1}{[\alpha - i(x-x')]^{g(u-u')}}.$$

As the infra-red cutoff μ goes to zero, this vanishes unless $u = u'$. Choosing C to depend on μ so that C diverges suitably in this limit, we obtain the desired Dirac delta function and the same answer as in the discrete case, with the Kronecker delta replaced by the Dirac delta.

There is a surprise when we consider the four-point function. One must analyze the behavior of the quantity $|C|^4\mu^{(2+g_{12}+g_{34}-g_{13}-g_{23}-g_{14}-g_{24})}$ as $\mu \to 0$. Here $g_{ij} = g(u_i - u_j)$ with u_1 and u_2 assigned to the fermion fields and u_3 and u_4 to the adjoint fields. In order to obtain the correct behavior when all four internal indices are close we find that we cannot choose $f(u)$ such that $g'(0) = 0$. In fact $g(u)$ must have a cusp at $u = 0$. Thus Gaussian functions for $f(u)$ and $g(u)$ are ruled out. A triangular pulse is acceptable for $g(u)$ and this corresponds to a rectangular pulse for $f(u)$.

Fermion bilinears

We go on to discuss the fermion bilinears. This is where we had trouble in the discrete case. A generic form of such a bilinear operator is $[\psi^\dagger(x_2,u_2), \psi(x_1,u_1)]$ where the arguments are allowed to come together, possibly after various derivatives have been taken. This is the nature of the spatially point-split structure of the charge operator, and of the generators of translations in x, t and u. We proceed to evaluate it using operator-product expansion methods (which are exact in this model). We have

$$[\hat{\psi}_2^\dagger, \hat{\psi}_1] = -\frac{|C|^2}{2\pi}\mu^{(1-g_{12})}({R_{12}}^{-g_{12}} - {R_{21}}^{-g_{21}}){:}e^{i\sqrt{4\pi}(\Phi_1-\Phi_2)}{:}.$$

Here $R_{12} = \alpha - i(x_1 - x_2)$. The normal ordering is with respect to the bosonic creation and annihilation operators. We introduce $u = \frac{1}{2}(u_1+u_2)$, $v = u_1 - u_2$, $x = \frac{1}{2}(x_1+x_2)$ and $\xi = x_1 - x_2$, and obtain

$$[\hat{\psi}_2^\dagger, \hat{\psi}_1] = -\frac{|C|^2}{2\pi}\mu^{[1-g_{12}]}[R(\xi)^{-g(v)} - R(-\xi)^{-g(v)}]\Omega(\xi,v)$$

where $\Omega(\xi,v) = {:}\exp i\sqrt{4\pi}[\Phi(x+\frac{1}{2}\xi, u+\frac{1}{2}v) - \Phi(x-\frac{1}{2}\xi, u-\frac{1}{2}v)]{:}$. Taking v and ξ small, (while keeping $\alpha \ll \xi$), and sending μ to 0 we discover that $[\hat{\psi}_2^\dagger, \hat{\psi}_1] \to -i\delta(v)\Omega(\xi,0)/(\pi\xi)$.

The chiral charge operator Q is given in terms of the canonical fermions by

$$\hat{Q} = \frac{1}{2}\int dx\, du_1\, du_2\, \delta(u_1-u_2)[\psi^\dagger(x,u_2), \psi(x,u_1)].$$

We choose to smooth out the u delta function and replace it by $N_\lambda g(u_1 - u_2)$ where N_λ is chosen so that $N_\lambda \int du_1\, g(u_1 - u_2) = 1$. Then $N_\lambda g(u_1 - u_2)$ becomes a Dirac delta function as $\lambda \to 0$. We construct the candidate chiral charge $\hat{Q}$ in terms of $\hat{\psi}$ as $\hat{Q} = \lim_{\xi\to 0} N_\lambda \int dx\, du\, dv\, g(v) \frac{1}{2}[\hat{\psi}_2^\dagger, \hat{\psi}_1]$ so that

$$Q = \lim_{\xi\to 0} \frac{-iN_\lambda g(0)}{4\pi\xi} \int dx\, du\, [\Omega(\xi, 0) - \Omega(-\xi, 0)] = \frac{N_\lambda}{\sqrt{\pi}} \int dx\, du\, \frac{\partial}{\partial x}\Phi(x, u).$$

The charge density in u space is thus $\hat{Q}(u) = N_\lambda \pi^{-1/2}[\Phi(\infty, 0) - \Phi(-\infty, 0)]$. We then have $[\hat{\psi}(x, u), \hat{Q}(u')] = N_\lambda g(u - u')\hat{\psi}(x, u)$ which becomes the correct canonical relation as $\lambda \to 0$.

The canonical operator $J = -i\frac{1}{4} \int dx\, du\, [\psi^\dagger(x, u), \frac{\partial}{\partial u}\psi(x, u)] + (\text{h. c.})$ generates translations in u. It is the analogue of the rotation and boost operators in the tomographic representation of the 3+1 dimensional case. If we were to follow the procedure used for the chiral charge we would find that when expressed in terms of bosonic fields, the candidate generator would vanish. The remedy for this is to back up a step and delay taking μ to zero. Then the spreading function used to define the bilinear, (which need not be the same as what we used for the chiral charge, namely $N_\lambda g(u)$), can be taken to depend on μ in just such a way as to yield the correct answer in terms of canonical boson fields in the limit.

Acknowledgment. Part of this work was done in collaboration with Ramesh Abhiraman. It was partially supported by the United States Department of Energy under grant DE-FG02-92ER 40704.

References

[1] S. Coleman, *Phys. Rev. D* **11** (1975) 2088.

[2] S. Mandelstam, *Phys. Rev. D* **11** (1975) 3026.

[3] A. Luther, *Phys. Rev. B* **19** (1979) 320.

[4] C. M. Sommerfield, in "Symmetries in Particle Physics," A. Chodos, I. Bars and C.-H. Tze, eds., Plenum Press (1984) pp. 127-140.

[5] H. Aratyn, *Nuc. Phys. B* **227** (1983) 172.

[6] E. C. Marino, *Phys. Lett.* **B263** (1991) 63.

[7] A. Kovner and P. S. Kurzepa, *Phys. Lett.* **B328** (1994) 506.

Internal Spin Structure of the Nucleon

Vernon W. Hughes
Yale University, New Haven, Connecticut 06520

Keywords. Deep inelastic scattering, spin dependence, muon nucleon scattering, spin structure functions.

The talk I gave at the Gursey Memorial Symposium in Istanbul, Turkey in June, 1994 was on the topic of the spin-dependent structure functions of the nucleon as determined from the recent measurements at CERN by the Spin Muon Collaboration for which I am the spokesman. I had previously reviewed this topic at the International Workshop on Deep Inelastic Scattering (DIS) and Related Subjects in Eilat, Israel in February, 1994, and then updated that review in July, 1994 for the Workshop Proceedings. That review, which is similar to my talk at the Gursey Memorial Symposium, is being published by World Scientific in the volume entitled **International Workshop on Deep Inelastic Scattering and Related Subjects** with Aharon Levy as editor.[1]

Here I merely summarize briefly the status of our experimental information on the internal spin structure of the nucleon, relate it to some theoretical considerations and remark on some important future experiments.

Feza Gursey did not work actively on the specific topic of the internal spin structure of the nucleon. However, he made important contributions to field theory for strong interactions and to its symmetries, including the development with L. Radicati of the SU(6) symmetry which incorporates spin 1/2 for the quarks. He also represented in his own work, and greatly encouraged at Yale, the spirit of adventure in exploring new fundamental topics.

Since the mid-1970's when experiments on the internal spin structure of the nucleon began at Yale and SLAC, many substantial measurements have been made of spin-dependent asymmetries in high energy electron and muon scattering from protons, deuterons, and ^{3}He. For the proton Fig. 1 shows the spin-dependent structure function $g_1^p(x)$ determined from the inclusive, deep inelastic scattering experiments done with a proton target[2,3,4,5]. These experiments measure the virtual photon-proton asymmetry $A_1(x, Q^2) = (d\sigma_A - d\sigma_P)/(d\sigma_A + d\sigma_P)$, in which $d\sigma_A(d\sigma_P)$ designates the differential scattering cross section when the lepton and proton spins are antiparallel (parallel), and where Q^2 is the four-momentum transfer squared and $x = Q^2/2M\nu$ is the Bjorken scaling variable with $\nu =$

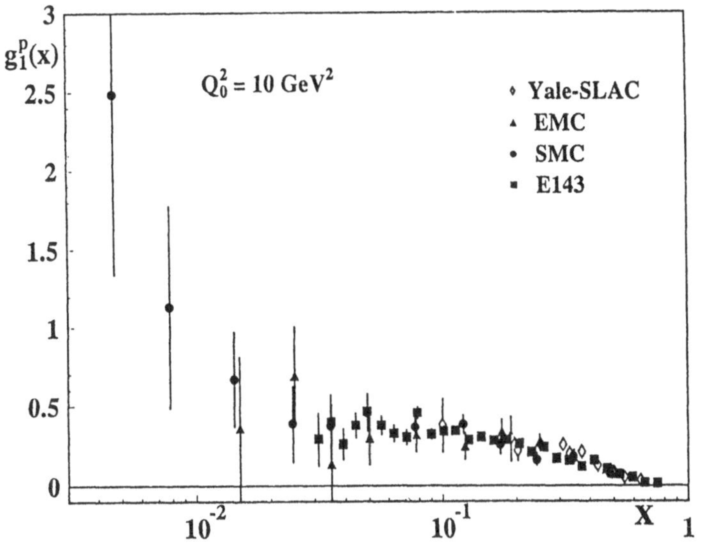

Figure 1: The proton spin-dependent structure function $g_1^p(x)$ as determined from several experiments. Yale-SLAC Ref. 2, EMC Ref. 3, SMC Ref. 4, E143 Ref. 5.

energy loss of scattred lepton and M = nucleon mass. The spin structure function $g_1^p(x)$ is given by:

$$g_1^p(x, Q^2) = A_1^p(x) F_1^p(x, Q^2) \tag{1}$$

in which $F_1^p(x, Q^2)$ is the proton spin-independent structure function.

Within the experimental errors no dependence of $A_1(x, Q^2)$ on Q^2 is observed and A_1 is taken to be independent of Q^2. The QCD evolution of $g_1^p(x, Q^2)$ to a common value Q_0^2 is made with experimental data on $F_1^p(x, Q^2)$. From the figure it is clear that the data from the different experiments are in good agreement.

The only data yet published on the spin-dependent asymmetry for the deuteron A_1^d are shown in Fig. 2.[6] Data are also available for ^3He from which the spin structure function $g_1^n(x, Q^2)$ of the neutron is determined.[7] Within the next 3 to 6 months considerable additional data should be published on p and d asymmetries from SLAC E143 and on d from CERN SMC.

The most fundamental use of this information on proton and neutron spin structure functions is to test the basic Bjorken polarization sum rule[8] which is a model-independent consequence of quantum chromodynamics. This remarkable sum rule relates nucleon quark wavefunctions determined from high energy experiments to low energy neutron beta decay:

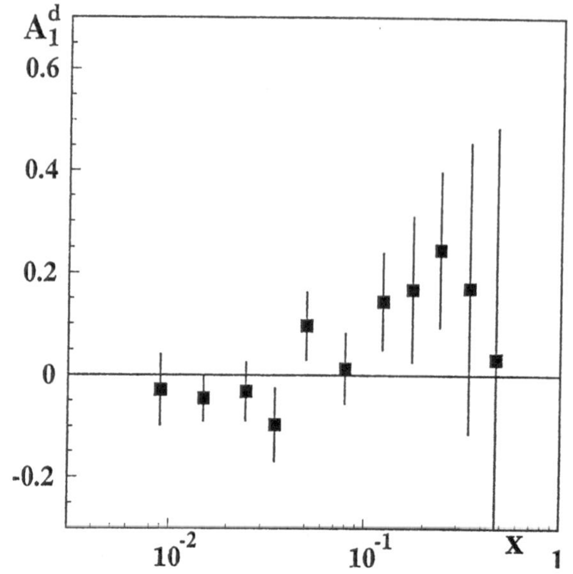

Figure 2: Virtual photon-deuteron asymmetry as measured in Ref. 6.

$$\Gamma_1^p - \Gamma_1^n \equiv \int_0^1 [g_1^p(x) - g_1^n(x)]dx = \frac{1}{6} \mid \frac{g_A}{g_V} \mid = 0.209(1) \qquad (2)$$

in the scaling limit of high $Q^2(Q^2 = \infty)$ in which $g_V(g_A)$ are the vector (axial-vector) coupling constants for neutron beta decay. From the published data, including the extrapolations of $g(x)$ to $x = 0$, a test of the Bjorken sum rule has been made for $Q_0^2 = 5$ GeV^2, where perturbative

QCD contributions change the RHS of Eq. (2) to 0.185(4). The experimental value is in agreement with this theoretical prediction within about 12% at the one standard deviation level.

Additional model-dependent sum rules due to Ellis and Jaffe[9] predict values for Γ_1^p and Γ_1^n individually, which are based on the assumptions that the strange quarks in the nucleon have zero net polarization and that $SU(3)_F$ symmetry applies for the baryon-octet decays. For the proton the data shown in Fig. 1 determine a value for Γ_1^p which is about 2 standard deviations below the Ellis-Jaffe value including the perturbative QCD corrections. For the neutron the situation is less clear[10] because of the small value of Γ_1^n and the relatively large size of the experimental errors and of the perturbative QCD corrections.

Important implications of the violation of the Ellis-Jaffe sum rule for the proton are that only the fraction 0.3 ± 0.1 of the angular momentum of the nucleon is carried by the quarks and that the strange quarks have a negative polarization $\Delta s = -0.10 \pm 0.05$. These conclusions about the internal spin structure of the nucleon are not yet understood but it has been suggested that they may arise from the polarization of the gluons or from orbital angular momentum of the quarks.[11]

Particularly important future experiments include the extension of the measurement of inclusive asymmetries in polarized lepton-nucleon deep inelastic scattering to lower x and high Q^2 which should be possible at HERA with both polarized electron and proton colliding beams. Also the measurement of asymmetry for semi-inclusive polarized deep inelastic scattering in which the J/ψ final state is detected could in principle determine the gluon contribution to the nucleon spin.

Research support in part by the U.S. Department of Energy.

REFERENCES

1. *Internal Spin Structure of the Nucleon from Experiments of the Spin Muon Collaboration* V.W. Hughes (To be published in the **International Workshop on Deep Inelastic Scattering (DIS) and Related Subjects**, ed. by A. Levy.
2. SLAC E-80, M.J. Alguard, et al., *Phys. Rev. Lett.* **37** (1976) 1261; ibid. **41** (1978) 70; M.R. Bergström et al., *Phys. Rev. Lett.* **45** (1980) 2000; SLAC E-130 G. Baum et al., *Phys. Rev. Lett.* **51** (1983) 1135.
3. EMC, J. Ashman et al., *Phys. Lett.* **B206** (1988) 364; *Nucl. Phys.* **B328** (1989) 1.
4. SMC, D. Adams et al., *Phys. Lett.* **B329** (1994) 399.
5. SLAC E-143, K. Abe, et al., *SLAC PUB.* **6508** August, 1994.
6. SMC, B. Adeva et al., *Phys. Lett.* **B302** (1993) 533.
7. SLAC E142, D.L. Anthony et al., *Phys. Rev. Lett.* **71** (1993) 959.
8. J.D. Bjorken, *Phys. Rev.* **148** (1966) 1467; *Phys. Rev.* **D1**(1970) 1376.
9. J. Ellis and R.L. Jaffe, *Phys. Rev.* **D9** (1974) 1444; **D10** (1974) 1669.
10. SMC, B. Adeva et al., *Phys. Lett.* **B320** (1994) 400.
11. G. Altarelli and W.J. Stirling, *Particle World* **1** (1989) No. 2, 40.

Strings and Membranes in Electron Theory

A.O. Barut

Physics Department, University of Colorado, Boulder , CO.80309,
and
ICPAM , International Center for Physics and Applied Mathematics,
P.K. 126 , Edirne , TURKEY

Abstract. We discuss some results in the extended models of the electron to achieve a finite self energy .

1. INTRODUCTION

The point model of the electron has the simplicity that is characterized by only two parameters , mass *m* and charge *e* The anomalous magnetic moment is derived from self energ, although there may be a very small intrinsic magnetic moment. One may think that an extended model of the electron which is probably more realistic would have additional parameters. A model of the electron as an oscillating charged shell, originally due to Lees (1) and Dirac (2), with a surface tension has an equilibrium state which again is characterised by two parameters *e* and *m* only. We discuss some remarkable features of this model which can also be applied to other tructures, and to gravitational systems .We shall emphasizethe problem of self-energy and radiation reaction .

2. THE SPINNING WORDLINE

The realistic spinless particle is described by the well-known action

$$I\left(x^{\mu}, A_{\mu}\right) = \int dx\left[\int d\tau\left(m\sqrt{\dot{x}^2} + eA_{\mu}(x)\dot{x}^{\mu}\right)\delta(x - X(\tau)) - \frac{1}{4}\sqrt{|g|}F_{\mu\nu}F^{\mu\nu} + \sqrt{|g|}R\right] \tag{1}$$

The last term is the gravitational coupling which we shall neglect in the following and work in flat space.

From (1) we obtain: (1°) the Lorentz-force equation, (2°) the Maxwell equation

$$F^{\mu\nu}{}_{,\nu} = -j^{\mu} = -e\int d\tau\, \dot{x}^{\mu}\delta(x - X) \tag{2}$$

(3°) the radiation reaction term in the Lorentz-Dirac equations , namely,

$$\frac{2}{3}e^2(\dddot{X}_{\mu} + (\ddot{X})^2\dot{X}_{\mu}) \tag{3}$$

The latter is obtained by writing $F^{\mu\nu} = F^{\mu\nu}_{ext} + F^{\mu\nu}_{self}$, and by evaluating the value of $F^{\mu\nu}_{self}$ on the particle, by a limiting process and by mass renormalization .By eliminating $A_\mu(x)$ from the action, using the solution of the Maxwell equation (2) we have also an equivalent action between particles ,

$$I\left(x_i^\mu\right)=\sum_i \int d\tau_i\, m_i\left(\dot{x}_i^\mu \dot{x}_i^\nu g_{\mu\nu}\right)^{1/2}+\frac{1}{2}\int \sum_{i,j} e_i e_j \delta\left(\left(x_i-x_j\right)^2\right)\dot{x}_i^\mu \dot{x}_j^\mu g_{\mu\nu} d\tau_i d\tau_j \qquad (4)$$

where we use a symmetric Green's function .

A third form of the action is in phase space

$$I(x\,,\,A)=\int d\tau(p_\mu \dot{x}^\mu - H) \qquad (5)$$

with $H=\Pi_\mu \dot{x}^\mu$, $\quad \Pi_\mu = p_\mu - eA_\mu$

The generalization of this theory to a relativistic spinning electron involves new internal spin co-ordinates. A form of this which gives, when quantized, precisely the Dirac electron, makes use of internal classical c-number spinor variables $z(\tau)$ and itsconjugate $\bar{z}(\tau)=z^+(\tau)\gamma^0$ .In terms of these variables the action (5) is generalized to (3)

$$I(x,\,A;\,z)=\int d\tau\,(p_\mu \dot{x}^\mu - i\lambda\bar{z}\dot{z} - H) \qquad (6)$$

Here $\qquad H=\Pi_\mu \bar{z}\gamma^\mu z$

It has been shown that this classical theory provides a definite model for the Dirac electron which was invented originally without a classical counterpart. It is in fact a classical model of a relativistic moving spin which was thought to have no classical explanation .

3.RELATIVISTIC CHARGED MEMBRANE

Now we work with surfaces of the form $X^\mu(\xi^\alpha)$, where $\xi^\alpha=(\tau\,,\sigma)$ are the points of the membrane (in general, p-brane) and is the time parameter, all Lorentz invariant body-fixed co-ordinate, and X^μ are the space fixed co-ordinates.

The minimally coupled minimal surface action is,

$$I\left(X^\mu,\,A_\mu\right)=\int dx\left[\int d\xi\left\{\kappa\sqrt{|f|}+e^A\partial_A x^\mu A_\mu(x)\right\}\delta(x-X)-\frac{1}{4}F_{\mu\nu}F^{\mu\nu}\right] \qquad (7)$$

Here indices $A=0,\,\ldots\,,\,d-1$, run over the membrane, $|f|=\det f_{AB}$, $f_{AB}=\dfrac{\partial X^\mu}{\partial \xi^A}\dfrac{\partial X_\mu}{\partial \xi^B}$ the induced metric.Again we obtain (1^0) the Lorentz force equation for the membrane

$$\kappa(\sqrt{|f|}x^{\mu}_{,A})_{,A} = -e^{B}\frac{\partial X^{\nu}}{\partial \xi^{B}}F^{\mu}_{\nu} \tag{8}$$

and (2°) the Maxwell's equations

$$F^{\mu\nu}_{,\nu} = -j^{\mu} = \int d\xi\, e^{A}\frac{\partial x^{\mu}}{\partial \xi^{A}}\delta(x-X) \tag{9}$$

which are clearly the generalizations of the particle equations .

Again eliminating the potential $A_{\mu}(x)$ we have the equivalent form of the action

$$I(X^{\mu}(\xi)) =$$

$$\kappa\int d\xi\sqrt{|f|} + \frac{1}{2}\int d\xi d\xi'\delta([x(\xi)-X(\xi')]^{2})e^{A}(\xi)\,e^{B}(\xi')X^{\mu}_{,A}(\xi)\,X_{\mu,B}(\xi') \tag{10}$$

Also this action can be generalized to spinning membrane with the introduction of spin variables on the membrane, $z(\xi)$, which we write now in the phase pace form

$$I(X, A; z) = \int d\xi(p^{A}_{\mu}X^{\mu}_{,A} - i\lambda\bar{z}\gamma^{A}D_{A}z - H) \tag{11}$$

where $D_{A}z = \frac{\partial z}{\partial \xi^{A}} - \Gamma_{A}z$ with Γ_{A} being the spin connection on the membrane and

$$H = \frac{\gamma_{AB}}{\sqrt{|\gamma|}}\Pi^{A}_{\mu}\bar{z}\gamma^{\mu}z \tag{12}$$

$$\Pi^{A}_{\mu} = p^{A}_{\mu} - e^{A}A_{\mu} + i\bar{z}\gamma^{A}\Gamma_{\mu}z \tag{13}$$

4. CHARGED SHELL

Now we consider the specific model of a charged shell, which can be parametrized in spherical co-ordinates as follows

$$X^{\mu}(\xi) = (\tau,\ r(\tau)\sin\theta\cos\varphi,\ r(\tau)\sin\theta\sin\varphi,\ r(\tau)\cos\varphi)$$

This is a co-ordinate condition representing a shell with radial motion only, $r = r(\tau)$, τ is an "observers time". More generally, for a moving shell we may put

$$X^{\mu'}(\xi) = L^{\mu}{}_{\nu}(\tau)X^{\nu}(\xi) \tag{14}$$

which may be represent accelerating and rotating membranes, characterized by the Lorentz parameters.

We can now explicitly evaluate

$$\gamma_{AB} = X^{\mu}_{,A}X_{\mu,B} \tag{15}$$

the induced metric turns out to be

$$ds^2 = (1-\dot{r}^2)dr^2 - r^2 d\theta^2 - r^2 \sin^2\theta \, d\varphi^2 \qquad (16)$$

with
$$\gamma = \det \gamma_{AB} = (1-\dot{r}^2) r^4 \sin^2\theta \qquad (17)$$

The quation of motion for the radial co-ordinate under the action of a minimal urface and minimal interaction is thus

$$\frac{d}{d\tau}\left(\frac{\dot{r}}{\sqrt{1-\dot{r}^2}}\right) + \frac{2}{r\sqrt{1-\dot{r}^2}} = \frac{qE}{kr^2} = \frac{q^2}{kr^4} \qquad (18)$$

where the electric field just on the outside of the surface is a Coulomb field with total charge

$$q = \int d\sigma_A \, e^A = 4\pi r^2 e_0 \qquad (19)$$

The potential and field just outside are

$$A_0 = \frac{q}{r} \quad , \quad \bar{E} = \frac{q^2}{r}\hat{r} \qquad (20)$$

and on the surface itself

$$E(r=R) = \frac{1}{2}\frac{q}{R^2}\hat{r} \qquad (21)$$

just like a conductor .

The action reduces to

$$I(r(\tau)) = \int d\tau \left[4\pi\kappa r^2 \left(1-\dot{r}^2\right)^{1/2} + \frac{q^2}{2r} \right] \qquad (22)$$

and the Hamiltonian of the radial motion is

$$H = \sqrt{(4\pi\kappa r^2)^2 + p^2} + \frac{q^2}{2r} \qquad (23)$$

The equilibrium condition can be obtained from the equation of motion putting $\dot{r}=0$ $(\ddot{r}=0)$, namely

$$r_0^3 = \frac{q^2}{2\kappa} \qquad (24)$$

The total mass of the membrane can be evaluated

(1) either by the Hamiltonian of the membrane

or (2) by the Hamiltonian of the self field of the membrane around it Both give of cource the same result .

REFERENCES

1) G. Lees , Philos. Mag. 28 (1939) 385

2) P.A. Dirac , Proc. Roy. Soc. (London) A 268 , (1962) 57

3) A.O. Barut and N. Zanghi , Phys. Rev. Lett 52 (1984) 2009
A.O. Barut and M. Pavsic , Int. J. Mod. Phys. Letters A 7 1992) 1381

4) A.O. Barut and M. Pavsic , Phys. Lett. B 306 (1993) 49

5) A.O. Barut and M. Pavsic , Phys. Lett. B 331 (1994) 45

Quark and Lepton Masses and the Higgs Field in Fifth Dimension

M.Arik

Department of Physics and Center for Turkish-Balkan Physics Research and Applications, Boğaziçi University, Bebek, Istanbul, Turkey

Keywords. Kaluza-Klein theory, standard model, Higgs field, quark and lepton masses, noncanonical quantization, quantum groups, q-oscillators

In the standard model [1], quarks and leptons are basically massless due to the fact that the parity violating weak interaction is carried by a gauge field and gauge invariance requires chiral fermions. The masses arise through interactions with the Higgs field described by a lagrangian density $\mathcal{L}_i = g_i \Phi \bar{\Psi}_i \Psi_i$. The vacuum expectation value of the Higgs field $\langle \Phi \rangle$ is constant throughout space and time and for each quark or lepton field Ψ_i the coupling g_i introduces a free parameter which gives rise to the mass $m_i = g_i \langle \Phi \rangle$.

The Higgs field is necessary for spontaneous breaking of gauge symmetry, giving rise to the massive weak bosons $W^{\pm}, Z$ and short range weak interactions. Any acceptable theory in which quark and lepton masses are calculable should agree with the standard model in some limit. In this talk I would like to present a model which has two basic ingredients. The first is that space-time is five dimensional [2] and the vacuum expectation value of the Higgs field depends on the macroscopic fifth dimension through a one dimensional solitonic solution of the classical field equations. This produces an effective potential for all particles in the theory which interact with the Higgs field and the four dimensional world sits at the minimum of this potential. The second ingredient involves non-canonical quantization and yields an exponential mass spectrum.

It has often been remarked that the quark and lepton mass spectrum should be described by an exponential mass formula. As an example, the empirical three-parameter mass formula [3]

$$m_{2n} = M(q_1^{2n} + q_2^{2n}) \ , \ m_{2n+1} = M((2q_1)^{2n+1} + (2q_2)^{2n+1}) \qquad (1)$$

when fitted to the quark spectrum for $n = 0, 1, 2, 3, 4, 5$ gives the masses

$$m_n = 13MeV,\ 6MeV,\ 256MeV,\ 1.35GeV,\ 5.0GeV,\ 174GeV \qquad (2)$$

for the quark sequence d, u, s, c, b, t and predicts a b' quark at $98GeV$. For this specific fit m_c, m_b and m_u/m_d have been used as input. Possibly, interpolated up to $\alpha = 2$, (10) can be interpolated up to $\alpha = 4$; with $\alpha = 4$ perhaps corresponding to an exponential mass spectrum. I will now show that such an interpretation indeed follows provided that the canonical quantization procedure is "corrected" by higher order terms in the Planck constant $\hbar$.

The mass squared spectrum provided by (6) is given by the eigenvalues of the hamiltonian

$$H = -\partial_5{}^2 + g^2 F^2 = \{a,\, a^\dagger\} \tag{11}$$

where the annihilation and creation operators

$$a = (\partial_5 + gF)/\sqrt{2}\,,\ a^\dagger = (-\partial_5 + gF)/\sqrt{2} \tag{12}$$

by using (4) satisfy

$$[a,\, a^\dagger]^2 = (gF')^2 = 2g^2 V(F)\,. \tag{13}$$

For a conventional Higgs potential $V(\Phi)$ the right hand side becomes

$$[a,\, a^\dagger]^2 = 2g^2 V_0 + g^2\mu^2 F^2 + g^2\lambda^2 F^4\,. \tag{14}$$

I propose to change the right hand side of this commutator by

$$F^2 \rightarrow -(\hbar^2/g^2)\partial_5{}^2 + F^2\,. \tag{15}$$

Since the classical potential approximation is valid only for low momenta, such a modification does not alter the qualitative behaviour of the mass spectrum for $\lambda = 0$. However for $\lambda \neq 0$ when the classical potential picture gives an instability, this procedure will yield a meaningful mass spectrum. Thus (14) is replaced by

$$\begin{aligned} [a,\, a^\dagger]^2 &= 2g^2 V_0 + \mu^2 H + (\lambda^2/g^2) H^2 \\ H &= \{a,\, a^\dagger\}\,. \end{aligned} \tag{16}$$

The generalized oscillator algebra (16) describes the harmonic oscillator for $\lambda = \mu = 0$ and the q-oscillator [5] for $\mu^4 = 8\lambda^2 V_0$. As far as representations in a Hilbert space are concerned it is identical to the SU(2) Lie algebra for $\lambda = 0$, $\mu^2 < 0$ and to the SU(1,1) Lie algebra for $\lambda = 0$, $\mu^2 > 0$. For $\lambda \neq 0$ it describes [6] the SUq(2) or SUq(1,1) deformed Lie algebra. The compact case of SU(2) or SUq(2) is capable of yielding a finite spectrum and implies that V_0, an irrelevant parameter in the standard interpretation should be quantized.

such a spectrum is related to two parameter deformed Lie algebras and to Fibonacci oscillators [4].

For a simplified model, consider the 5-dimensional lagrangian density

$$\mathcal{L} = \frac{1}{2}\partial_M \Phi \partial^M \Phi - V(\Phi) + \frac{1}{2}\partial_M \Psi \partial^M \Psi - \frac{1}{2} g^2 \Phi^2 \Psi^2 \tag{3}$$

where $M = 0, 1, 2, 3, 5$ with a scalar matter field Ψ. The classical equations of motion admit a solution with $\Psi = 0$ and $\Phi = F(x_5)$ where

$$\partial_5^2 F = V'(F) \ , \ (\partial_5 F)^2 = 2V(F) \ . \tag{4}$$

Expanding the field Ψ in terms of a complete set of states $u_n(x_5)$ gives

$$\begin{aligned} \Psi &= \sum_n \Psi_n(x) u_n(x_5) \ , \quad x = (x_0, x_1, x_2, x_3) \qquad (5) \\ \mathcal{L}_\Psi &= \frac{1}{2} \sum_{n,m} (\partial_\mu \Psi_m \partial^\mu \Psi_n u_m u_n + \Psi_m \Psi_n u_m {\partial_5}^2 u_n - g^2 \Psi_m \Psi_n u_m F^2 u_n) \end{aligned}$$

where $\mathcal{L}_\Psi$ denotes the portion of the lagrangian quadratic in the Ψ_n. The states $u_n(x_5)$ should be chosen to satisfy

$$(-{\partial_5}^2 + g^2 F^2) u_n = {m_n}^2 u_n \tag{6}$$

and if $U_{eff} = (gF(x_5))^2$ is a well behaved confining potential, this procedure will yield a complete set of orthonormal states $u_n(x_5)$. The effective lagrangian density in four dimensions will then be given by

$$\mathcal{L}_4 = \int \mathcal{L}_5 dx_5 = \frac{1}{2} \sum_n (\partial_\mu \Psi_n \partial^\mu \Psi_n - {m_n}^2 {\Psi_n}^2) + \cdots \tag{7}$$

yielding a set of four dimensional matter fields $\Psi_n(x)$ with masses m_n. For a potential $V(F) \sim F^\alpha$, (4) gives

$$F \ \sim \ (x_5)^{2/(2-\alpha)} \tag{8}$$

which is confining for $\alpha < 2$. For $\alpha = 2$ the classical field equation (4) has the solution

$$F \ \sim \ cosh(x_5) \tag{9}$$

which again is a confining potential. The WKB solution to (6) yields

$$ {m_n}^2 \ \sim \ n^{4/(4-\alpha)} \tag{10}$$

For $\alpha > 2$, F is not confining and the conventional interpretation would say that an instability arises. However note that although (8) can be

Denoting the eigenvalues of $a^\dagger a$ by ϵ_n the algebra (16) yields

$$\epsilon_n = Aq^n + Bq^{-n} + C \tag{17}$$

$$\begin{aligned} q + q^{-1} &= 2\,(g^2 + \lambda^2)/(g^2 - \lambda^2) \\ C &= -(g^2\mu^2)/(4\lambda^2) \\ 4AB &= g^2(\mu^4/4\lambda^2 - 2V_0)/(q^{1/2} - q^{-1/2})^2 \,. \end{aligned} \tag{18}$$

Then the eigenvalues of H are given by

$$m_n{}^2 = \epsilon_n + \epsilon_{n+1} \,. \tag{19}$$

To handle fermions one can use the five dimensional lagrangian density

$$\mathcal{L}_\Psi = i\bar{\Psi}\Gamma^M\partial_M\Psi - g\Phi\bar{\Psi}\Gamma^6\Psi \tag{20}$$

with $\Gamma^\mu = \gamma^\mu \otimes \sigma_3$, $\Gamma^5 = 1 \otimes \sigma_2$, $\Gamma^6 = 1 \otimes \sigma_1$, $\Psi = \sum_n \Psi_n(x) \otimes u_n(x_5)$, where Ψ_n is a four component and u_n is a two component spinor. Then

$$H = i\Gamma^5\partial_5 + gF(x_5)\Gamma^6 = \begin{pmatrix} 0 & \sqrt{2}a \\ \sqrt{2}a^\dagger & 0 \end{pmatrix}, H^2 = \begin{pmatrix} 2aa^\dagger & 0 \\ 0 & 2a^\dagger a \end{pmatrix} \tag{21}$$

and the fermion mass spectrum is given by

$$m_n{}^2 = 2\epsilon_n \,. \tag{22}$$

As things stand, (22) is not successful in describing the observed quark and lepton masses. An acceptable mass formula should probably contain two parameters q_1 and q_2 as in (1). Moreover the model predicts a spectrum also for the weak bosons in conflict with current experimental results. Nevertheless, I believe that the basic ideas presented in this talk are relevent for an understanding of quark and lepton masses.

References

[1] S. Weinberg, Rev. Mod. Phys. **52**, 515 (1980); A. Salam, *ibid.* **52**, 525 (1980); S.L. Glashow, *ibid.* **52**, 539 (1980).
[2] T. Kaluza, Sitzungber. Preuss. Akad. Wiss. (Berlin) Math. Phys. **K1**, 966 (1921); O. Klein, Z. Phys. C **37**, 895 (1926).
[3] E. Arik, unpublished.
[4] R. Chakrabarti and R. Jagannathan, J .Phys. A **24** L711 (1991); M. Arik et.al., Z. Phys. C **55**, 89 (1992).
[5] M. Arik and D.D. Coon, J. Math. Phys. **17**, 524 (1975).
[6] M. A. Karaca, Ph.D. thesis, Istanbul Technical University (1994), unpublished.

Regularization and Quantization of Higher Dimensional Current Algebras

Gabriele Ferretti
Institute of Theoretical Physics, Chalmers University of Technology
S-41296 Göteborg, Sweden. ferretti@fy.chalmers.se

Abstract. We present some recently discovered infinite dimensional Lie algebras that can be understood as extensions of the algebra $\mathrm{Map}(M, \mathbf{g})$ of maps from a compact p-dimensional manifold M to some finite dimensional Lie algebra $\mathbf{g}$.

Keywords. Current algebra, pseudo–differential operators, Radul cocycle

1 Higher Dimensional Current Algebras

When considering higher dimensional manifolds we first encounter the fact that, contrary to the one dimensional case, $\mathrm{Map}(M, \mathbf{g})$ has an infinite number of central extensions. Roughly speaking, for any loop one can draw on the manifold (even contractable loops) and for any two elements $\lambda, \xi \in \mathrm{Map}(M, \mathbf{g})$, there is a two cocycle given by evaluating the pull-back of the form $\mathrm{tr}\ (\lambda d\xi)$ on such loop. These are essentially all the possible central extensions one can have. However, none of these extensions are interesting in relation to the study of higher dimensional current algebras precisely because of their "one dimensional" nature.

We must therefore look for more general kinds of extensions (abelian or even non abelian) and this always means adding extra fields to the system. Also, if we are aiming at an algebra that does not depend on the detailed geometry of the manifold M, (e.g. its metric), it is a natural guess to consider adding differential forms.

The algebra we are proposing is made of pairs (X, z), where X represents the free sum $X = X^{(0)} + (X^{(2)} + X^{(4)} + \cdots X^{([p])})$, $X^{(0)} \in \mathrm{Map}(M, \mathbf{g})$ and $X^{(2)}, X^{(4)}, \cdots$ are $\mathbf{g}$ valued differential forms on M of even degree $2, 4, \cdots$. The symbol $[p]$ denotes the largest even integer less or equal than $p = \dim(M)$ and z is a complex number. In the following, $\mathbf{g}$ is taken to be either $\mathbf{gl}(n)$ or $\mathbf{u}(n)$.

The Lie product in such an algebra can be written concisely as

$$[(X;z),(Y;w)] = ([X,Y] + [dX, dY]; \frac{i}{(2\pi)^p} \int_M \mathrm{tr}\ (XdY)), \tag{1}$$

where the following conventions have been made: First, the wedge product is understood in all the commutators of forms on the r.h.s.; second, forms of degree larger that p, arising from the wedge products, all vanish and, finally, the integral on the r.h.s. vanishes on forms of degree not equal to p.

This algebra arises in many different physical situations: In one dimension it reduces to an Affine Kac–Moody algebra; in two dimensions, it is the current algebra of planar QCD in the parity preserving phase [1]. The three dimensional case is the most interesting because, there, (1) is related to the Mickelsson–Faddeev–Shatasvili algebra [2, 3] arising in anomalous chiral gauge theories, the scalar $X^{(0)}$ being the infinitesimal gauge transformation and the two form $X^{(2)}$ being dual to the gauge potential. Finally, let me remark that, in more than three dimensions, (1) becomes a non abelian extension of Map$(M, \mathbf{g})$ and it has been shown in [4] that it is connected to the bosonic sector of the p–brane.

2 Regularization of the Algebra

We will now examine the explicit realization of (1) as algebra of operators in some Fock space. For definitiveness, let us focus on the three dimensional case, where one can write the Lie product of $\widehat{\text{Map}}(M, \mathbf{g})$ as

$$[(X^{(0)} + X^{(2)}; z), (Y^{(0)} + Y^{(2)}; w)] = ([X^{(0)}, Y^{(0)}] + [X^{(0)}, Y^{(2)}] + [X^{(2)}, Y^{(0)}] + [dX^{(0)}, dY^{(0)}]; \frac{i}{(2\pi)^3} \int_M \text{tr}\ (X^{(0)} dY^{(2)} + X^{(2)} dY^{(0)})). \tag{2}$$

The simplest and most obvious choice for the Fock space is a fermionic Fock space $\mathcal{F}$ constructed by considering Weyl spinor fields $\Psi(x)$ at equal time $t = 0$. Using these fields one can easily construct the charge densities (i.e. time components of the currents):

$$J^a(x) = \frac{i}{2} : \Psi^\dagger(x) M^a \Psi(x) : \quad , \tag{3}$$

Where M^a are antihermitian generators of $\mathbf{g}$. If we tried to set $J(X^{(0)}) = \int_M J^a(x) X^{(0)a}(x) d^3x$, we would immediately run into the following problem: contrary to the one dimensional case, the operator $J(X^{(0)})$ creates states of infinite norm out of the ordinary vacuum.

The basic idea , recently proposed by Mickelsson [5, 6] is to regularize the argument of J in a way that, on one hand, preserves the algebra and, on the other hand, makes the norms converge. This can be achieved by replacing the scalar and the two form by appropriate Pseudo–Differential operators (PSDO). Recall that a PSDO is defined through its symbol, ie., a function in phase space. The product of two symbols a and b is the symbol of the composition of the two PSDO and it is denoted by $a * b$.We denote the regularized algebra as $\theta(X^{(0)} + X^{(2)})$:

$$\theta(X^{(0)} + X^{(2)}) = X^{(0)} + S_\mu \frac{\partial X^{(0)}}{\partial x^\mu} + \frac{1}{2} S_{\mu\nu} \frac{\partial}{\partial x^\mu} \frac{\partial X^{(0)}}{\partial x^\nu} + \frac{1}{2} A_{\mu\nu} X^{(2)}_{\mu\nu}, \tag{4}$$

where

$$S_\mu = \frac{\epsilon_{\mu\nu\rho} p_\nu \sigma_\rho}{2p^2} \tag{5}$$

and the other two terms can be expressed as functions of S_μ as

$$\begin{aligned} S_{\mu\nu} &= \frac{1}{2}(S_\mu S_\nu + S_\nu S_\mu) - \frac{i}{2}(\frac{\partial S_\mu}{\partial p^\nu} + \frac{\partial S_\nu}{\partial p^\mu}) \\ A_{\mu\nu} &= \frac{1}{2}(S_\mu S_\nu - S_\nu S_\mu) - \frac{i}{2}(\frac{\partial S_\mu}{\partial p^\nu} - \frac{\partial S_\nu}{\partial p^\mu}). \end{aligned} \tag{6}$$

It can be checked by an explicit calculation that (4) satisfies all the above requirements up to irrelevant PSDO of degree -3.

Also, the central term in (1) can be understood as a (properly normalized) twisted Radul cocycle [6 – 8]

$$c_R(\theta(X^{(0)}+X^{(2)}),\theta(Y^{(0)}+Y^{(2)})) = \frac{i}{(2\pi)^3}\int_M \mathrm{tr}\,(X^{(0)}dY^{(2)}+X^{(2)}dY^{(0)}), \tag{7}$$

which is also independent on PSDO of degree -3.

A more extended version of this paper can be found in [9]. Work is also in progress to gain a better understanding of how the central term can arise from second quantization and on the possible representation theory of (1).

Acknowledgments

This work was done in collaboration with the following people: M. Cederwall, J. Mickelsson, B.E.W. Nilsson, and A. Westerberg. I also would like to thank the organizers of this conference for giving me the opportunity to present this talk and for their kind hospitality during my stay in Istanbul.

References

[1] G.Ferretti and S.G. Rajeev, Phys. Rev. Lett. 69 (1992) 2033. **hep-th/9207039**
[2] J. Mickelsson, Lett. Math. Phys. 7 (1983) 45.
[3] L. Faddeev and S. Shatasvili, Theor. Math. Phys. 60 (1984) 770.
[4] M. Cederwall, G. Ferretti, B.E.W. Nilsson and A. Westerberg, Nucl. Phys. B424 (1994) 97. **hep-th/9401027**
[5] J. Mickelsson, in "Proceedings of the XXII Int. Conf. on Differential Geometric Methods in Theoretical Physics", Ixtapa, Mexico (1993). **hep-th/9311170**
[6] J. Mickelsson, Royal Inst. Tech. Preprint (1994). **hep-th/9404093**
[7] A.O. Radul, Phys. Lett. B265 (1991) 86.
[8] M. Wodzicki, in "Lecture Notes in Mathematics", Berlin, Germany (1985).
[9] G. Ferretti, Chalmers Preprint **hep-th/9406177**

Fermion Masses and Mixing from an Extra Gauge Symmetry

E. Papageorgiu

Laboratoire de Physique Théorique et Hautes Energies Université de Paris XI, Bâtiment 211, 91405 Orsay, France

Looking for simplicity could mean looking for a symmetry. To this most successful undertaking Feza Gürsey has made as we have heard from previous speakers outstanding contributions. Following the same line of thought one may hope to find one day an underlying symmetry principle for the long list of mass parameters which characterize the elementary particle spectrum in our days and enter the Standard Model (SM) as free parameters. As a first step one can look for regularities in terms of some observable. We know for example that the observed mass and mixing hierarchies of the fermion spectrum can be described in terms of the *Wolfenstein* parameter $\lambda \simeq 0.22$ which, to a good approximation, gives the Cabbibo mixing $|V_{us}|$. Taking into account the experimental uncertainties that can be as large as a factor of two one finds the following mass patterns for the up and down quarks:

$$m_u : m_c : m_t \sim (\frac{\lambda^8}{\chi^3} : \frac{\lambda^4}{\chi^3} : 1) \times m_t \tag{1}$$

$$m_d : m_s : m_b \sim (\frac{\lambda^4}{\chi} : \frac{\lambda^2}{\chi} : 1) \times m_b \, . \tag{2}$$

The parameter $\chi \sim 0.7$ is a rough estimate of the radiative corrections (from the heavy top quark) to the mass ratios when they run from the unification scale $M_G = 10^{16}$ GeV down to the electroweak scale according to the renormalisation group equations of the minimal supersymmetric standard model (MSSM). This effect is particularly felt in the up-quark sector. The regularity of the spectra in terms of λ becomes more transparent at the very high scales. Notice also that the mass ratios in the up and down quark sector at M_G are related through a $\lambda^2 \leftrightarrow \lambda$ transformation. Compared to these almost equidistant quark-mass levels the charged lepton masses are perceived rather as an anomaly

$$m_e : m_\mu : m_\tau \sim (\lambda^5 : \lambda : 1) \times m_\tau \tag{3}$$

even at the unification scale where apart from the unification of gauge couplings one also has partial unification of Yukawa couplings and the mass relations:

$$m_\tau \simeq m_b \qquad \text{and} \qquad m_e \cdot m_\mu \simeq m_d \cdot m_s \quad . \tag{4}$$

Given the fact that at low energies one has only 13 observables (six quark and three lepton masses, the three mixing angles and the CP violating phase of the CKM matrix) one cannot fix the entries of the quark and lepton mass matrices M_u, M_d and M_e at M_G, even by assuming that the latter are symmetric. This has led to a further assumption, namely that some of the entries are zero [1,2] while the others are given in powers of λ. In the quark sector the maximum number of zeros that one can have is five (counting together those in M_u and M_d, but without counting symmetric entries twice) and there are five different pairs of $M_u - M_d$ textures at M_G that lead to masses and mixings which are compatible with the present-day experimental values [2]. Some extra consideration is therefore needed to single out a unique solution. In fact, the zeros in the mass matrices can be thought off as relics of a new symmetry that is not "family-blind", in which case the non-zero entries correspond to correction terms related to the symmetry breaking. I will discuss this "old" idea in the light of a new type of unification which consists in extending the gauge group of the standard model by a horizontal $U(1)_X$ factor whose anomalies can be cancelled by a Green-Schwarz mechanism [3,4].

Let us assume the existence of a family-dependent $U(1)_X$ gauge symmetry at M_{Planck}, with respect to which the quarks and leptons carry charges α_i and a_i respectively, where $i = 1, 2, 3$ is the generation index. We first consider the up quark mass matrix. Given the predominant role played by the top quark it is not unnatural to start with a rank-one matrix and make a choice for the charges such that only the (3,3) renormalizable coupling $t^c t h_1$ is allowed. This fixes the charge of the light Higgs h_1 to -2α ($\alpha \equiv \alpha_3$). We expect the other entries to be generated by higher-dimension operators which may occur at the string compactification level and contain combinations of scalar fields some of which acquire vacuum expectation values (vev's) leading to spontaneous symmetry breaking and large masses for the non-observable part of the spectrum. For simplicity let us assume a pair of singlet fields $\sigma_\pm$ developping equal (vev's) along a "D-flat" direction and carrying opposite charges ± 1. They can give rise to higher-order couplings $q_i^c h_1 (\frac{<\sigma>}{M})^{|2\alpha - \alpha_i - \alpha_j|} q_j$. Notice that when the exponent is positive (negative) only the field σ_+ (σ_-) can contribute. The new scale $\mathcal{E} = \frac{<\sigma>}{M}$ which enters in the quark mass matrix is the ratio of the symmetry breaking scale to the scale that governs these higher-dimension operators, and could be the string unification scale $M_S \simeq 10^{18}$ GeV or M_{Planck}. If $\mathcal{E}$ is a small number one finds two universal hierarchy patterns in the generated texture:

$$M_x \sim \begin{pmatrix} \mathcal{E}^{2|x_1|} & \mathcal{E}^{|x_1+x_2|} & \mathcal{E}^{|x_1|} \\ \mathcal{E}^{|x_1+x_2|} & \mathcal{E}^{2|x_2|} & \mathcal{E}^{|x_2|} \\ \mathcal{E}^{|x_1|} & \mathcal{E}^{|x_2|} & 1 \end{pmatrix} \qquad |x_{1,2}| = |\alpha - \alpha_{1,2}|, \qquad (5)$$

namely $m_{11} \sim m_{13}^2$ and $m_{22} \sim m_{23}^2$. The choice $|x_2| = 1$ and $|x_1| = 4$ or $|x_1| = 2$ leads to:

$$M_u^F \sim \begin{pmatrix} \mathcal{E}^8 & \mathcal{E}^3 & \mathcal{E}^4 \\ \mathcal{E}^3 & \mathcal{E}^2 & \mathcal{E} \\ \mathcal{E}^4 & \mathcal{E} & 1 \end{pmatrix} \quad \text{or} \quad M_u^G \sim \begin{pmatrix} \mathcal{E}^4 & \mathcal{E}^3 & \mathcal{E}^2 \\ \mathcal{E}^3 & \mathcal{E}^2 & \mathcal{E} \\ \mathcal{E}^2 & \mathcal{E} & 1 \end{pmatrix}. \qquad (6)$$

If $\mathcal{E}$ is of order λ^2 the two textures above correspond to the phenomenologically acceptable *Ansätze à la Fritzsch* or *à la Giudice* for the up quark mass matrix. The generation of other acceptable textures having a zero in the (2,2) or the (2,3) entry (but not in both entries simultaneously) necessitates a more complicated mechanism involving extra singlets and mixing with heavy Higgses.
The next task is, given the up quark matrices of eq.(6), to construct an acceptable down quark matrix. The assumption of symmetric mass matrices and the $SU(2)_L$ symmetry require the equality between the charges of the up and down quarks. Assuming again that only the (3,3) renormalizable coupling is allowed leads to the other light Higgs h_2 carrying the same charge as h_1. This means that this $U(1)_X$ is anomalous and needs a cancellation mechanism which will be discussed at the end of the talk. Notice that the choice of a particular texture for M_u has already fixed the texture of M_d in terms of some new scale $\mathcal{E}'$ which has to be of order λ to give the correct mass spectrum of eq.(2). The origin of this difference in scale $\mathcal{E}' \sim \mathcal{E}^{1/2}$ is yet unknown.
Thus if one insists in generating the up and down quark mass matrices through the same (simple) mechanism that led to eq.(5) one can only generate textures *à la Fritzsch* containing four zeros in total. In order to generate the five-zero textures of ref.[2] some sort of extension is needed. In many compactification schemes there are among other things additional pairs of Higgs fields which acquire masses at the scale of symmetry breaking. One can thus envisage the possibility of mixing between the light Higgses $h_{1,2}$ with heavy Higgses $H_{1,2}$ whose charge we denote by -2β. Then one can also have couplings $q_i^c H(\frac{<\sigma>}{M})^{|2\beta-\alpha_i-\alpha_j|}q_j$ which give rise to the following texture:

$$M_z \sim \begin{pmatrix} \mathcal{E}^{2|z_1|} & \mathcal{E}^{|z_1+z_2|} & \mathcal{E}^{|z_1+z|} \\ \mathcal{E}^{|z_1+z_2|} & \mathcal{E}^{2|z_2|} & \mathcal{E}^{|z_2+z|} \\ \mathcal{E}^{|z_1+z|} & \mathcal{E}^{|z_2+z|} & 1+\mathcal{E}^{2|z|} \end{pmatrix}, \tag{7}$$

whith $|z_i| = |\beta - \alpha_i|$, and $|z| = |\beta - \alpha|$. When the difference between the light- and heavy-Higgs charges is larger than between the charges of the heavy Higgs and the quarks this automatically gives suppressed (1,3) and (2,3) mass entries. Taken together, the textures of eqs.(5,8) can generate any acceptable up or down quark texture [4]. In particular, the *Fritzsch Ansatz* is obtained also from $|x_2| = 2$ and $|z_1 + z_2| = 6$.
What about the lepton sector? Assuming simply the gauge symmetries of the SM the $U(1)_X$ charges a_i of the leptons are not related to those of the quarks. Allowing however the coupling $\tau^c \tau h_2$ leads to $a_3 = \alpha$ and to $m_\tau = m_b$ unification. Another constraint comes from the second mass relation of eq.(4) which means that we should be looking for textures for which det M_e = det M_d and therefore for combination of charges that satisfy: $a_1 + a_2 = \alpha_1 + \alpha_2$. Since the early days of grand unification it is known that in order to obtain also for the first two generations acceptable mass relations between the charged leptons and down quarks the (2,2) entry of M_d should be multiplied by a factor of minus three, the other entries of M_e and M_d been equal. In this way, though there can be no explanation of the factor minus three in our approach, in terms of the scale $\mathcal{E}'$ and its various powers, M_e becomes identical with M_d.

Again as a consequence of the $SU(2)_L$ symmetry and our symmetric *Ansatz* the lefthanded and righthanded neutrinos, ν_i and N_i, become charged under the $U(1)_X$ with the same charges a_i as the charged leptons. Obviously the presence of the N_i's implies a larger symmetry than what has been assumed so far, but also the assumption of symmetric mass matrices can find its justification only in the context of a left-right symmetric theory. Assuming that the same mechanism which generated masses for the quarks and the leptons generates also Dirac masses for the neutrinos we are automatically led to another well known GUT equality: $M_\nu^D = M_u$. On the other hand, Majorana mass terms M_R^{ij} need not be generated in the same way. In compactified string models, due to the absence of large Higgs representations, righthanded neutrinos donot get tree-level masses, so all entries in M_R are due to nonrenormalizable operators, and nothing is a priori known concerning the existence of a possible hierarchy in this sector. The only constraints on the texture of M_R come from the requirement that the seesaw-suppressed masses of the ordinary neutrinos should be below the experimental limits. For this, M_R has to be a nonsingular matrix and its scale should be well above the electroweak scale. Therefore in addition to the operators that generated the textures of eqs.(5,8) one will need at least an extra piece to set the Majorana mass scale. Common examples are operators containing the heavy Higgses which have been used for generating the texture in eq.(8), namely $N_i^c H H N_j$, whose scale is of $\mathcal{O}(M_G^2/M_S)$ multiplied for some orbifold suppression factor. In this case $\alpha_i + \alpha_j = 4\beta$ leads to $M_R^{ij} \neq 0$. One can always fix the charges of the light and heavy Higgses relative to each other such that there is at least one entry which is different from zero. One can then check whether the charge relations which lead to the *Fritzsch Ansatz* are compatible with the generation of a nonsingular M_R. One finds that starting only from the texture M_x in eq.(5) this is not the case, unless, some of the Majorana entries are generated perturbatively in a similar way as the entries in the other mass matrices. This again implies a hierarchy of righthanded neutrino scales. In contrast, starting from the full scenario and the alternative combination of charges which generates the *Fritzsch Ansatz* one finds in the absence of hierarchy three different M_R textures containing three or four zeros [4] and leading to interesting predictions for neutrino oscillation experiments [5].

Let us finally comment on the cancellation mechanism of the mixed $SU(2)^2U(1)_X$ anomaly, induced by the charge of the light Higgses, via a Green-Schwarz (GS) term in 4D string theories. Notice first that the usual anomaly cancellation condition would require the Higgs fields to be either neutral or to have opposite charges, but these two cases are obviously incompatible with the generation of realistic mass patterns. In contrast, in the GS mechanism one can cancel the mixed anomalies of a $U'(1)$ with $SU(3)$, $SU(2)$ and $U(1)_Y$ by an appropriate shift of the axion in the dilaton multiplet if the anomaly coefficients are in the ratio $A_3 : A_2 : A_1 = 1 : 1 : 5/3$ which symptomatically leads to the successful $sin^2\theta_W = 3/8$ result of the canonical gauge coupling unification [6]. In the case of the horizontal $U(1)_X$ anomaly cancellation can be thus achieved at the expense of adding such a family-independent $U'(1)$ component [3].

References

[1] H. Fritzsch, Phys. Lett. **B** 70(1977)436; Phys. Lett. **B** 73(1978)317; Nucl. Phys. **B** 155(1979)189; J. Harvey, D. Reiss and P. Ramond, Phys. Lett. **B** 92(1980)309; Nucl. Phys. **B** 199(1982)223; S. Dimopoulos, L.J. Hall and S. Raby, Phys. Rev. Lett. 68(1992)1984; Phys. Rev. **D** 45(1992)4192; G. F. Giudice, Mod. Phys. Lett. **A**7(1992)2429. S. Dimopoulos, L.J. Hall and S. Raby, Phys. Rev. **D** 47(1993)3697; L.J. Hall, **UCB**-PTH-92-22.

[2] P. Ramond, R.G. Roberts and G.G. Ross, Nucl. Phys. **B** 406(1993)19.

[3] L.E. Ibáñez and G.G. Ross, Phys. Lett. **B** 332(1994)100.

[4] E. Papageorgiu, Orsay Preprint 94-40, *to appear in Zeit. Phys.* **C** .

[5] E. Papageorgiu, Orsay Preprint 93-45, *to appear in Zeit. Phys.* **C** ; *ibid* 93-51, *to appear in Phys. Lett.* **B** .

[6] L.E. Ibáñez, Phys. Lett. **B** 303(1993)55.

E_6 Model Manifestations in Z and τ Decays

T. M. Aliev[1], S. F. Sultansoy[2] and O. Yılmaz[1]

[1] Physics Department, Middle East Technical University, Ankara, Turkey
[2] Physics Department, Ankara University, Ankara, Turkey

Abstract. Possible manifestations of the E_6 model in $Z \to \ell^+\ell^-$ and in τ-decays are investigated. The differences between the Standard Model with four families and E_6 Model predictions are discussed. It is shown that the experimental study of spin asymmetry in the process $e^+e^- \to Z \to \tau^+\tau^-$ will give direct information on the lepton universality violation due to mixing between ordinary and exotic leptons predicted by E_6 model, and that the investigation of the charged final lepton energy spectrum in τ-decays is an effective tool for establishing new physics predicted by E_6 model.

Keyword. E_6 model

In this report we discuss possible manifestations of the new particles predicted by E_6 model [1] (see also [2]) in τ and Z decays. Here we restrict ourselves considering only leptonic sector.

This investigation is motivated by the fact that at LEP more than 10^7Z boson is produced per year (this number will increase in future ~2.5 times) and proposed standard type c-τ factories will produce 3×10^7 $\tau^+\tau^-$ pairs per year and this number may be increased by a factor more than 10 at linac-type c-τ factories [3].

The classification according to the electroweak SU(2)×U(1) group of a fermion sector of the one E_6 family has the following form

$$\begin{pmatrix} \nu'_\tau \\ \tau'^- \end{pmatrix}_L , \quad \nu'_{\tau R} , \quad \tau'^-_R , \begin{pmatrix} N'^0_\tau \\ T'^-_\tau \end{pmatrix}_L , \quad \begin{pmatrix} N'^0_\tau \\ T'^-_\tau \end{pmatrix}_R , \quad n_\tau \qquad (1)$$

where $N'^{\,o}_\tau$ and T'^- are new heavy leptons, n_τ is a heavy Majorona neutrino. Primes denote possible mixings between leptons with same charge and chirality. As one can see from Eq. (1) low energy phonemenology of E_6 model does not contain "trivial" quark-lepton symmetry in difference from SM. The similar extentions of lepton sector take place also for other lepton generations. As a result, large number of observable mixings and phases appear. Below we will ignore new phases and consider special cases for mixing angels.

The mixing states in the weak multiplets of Eq. (1) are related to their mass eigenstates in the following way,

$$\begin{aligned}
\tau'^{-}_{L,R} &= \tau^{-}_{L,R}\cos\phi_{L,R} + T^{-}_{\tau\ L,R}\sin\phi_{L,R}\ ,\\
T'^{-}_{\tau\ L,R} &= -\tau^{-}_{L,R}\sin\phi_{L,R} + T^{-}_{\tau\ L,R}\cos\phi_{L,R}\ ,\\
\nu'_{\tau L,R} &= \nu_{\tau L,R}\cos\theta_{L,R} + N^{0}_{\tau L,R}\sin\theta_{L,R}\ ,\\
N'^{0}_{\tau\ L,R} &= -\nu_{\tau L,R}\sin\theta_{L,R} + N^{0}_{\tau L,R}\cos\theta_{L,R}\ ,
\end{aligned} \tag{2}$$

where $\phi_L \neq \phi_R$.

Firstly we consider the mixing in the charged lepton sector. Using Eqs. (1) and (2) we get the following Lagrangian containing only usual leptons,

$$\mathcal{L} = \frac{g\ cos\phi_L}{2\sqrt{2}}\bar{\nu}_\tau O_\alpha \tau W_\alpha + h.c. + \frac{\bar{g}}{2}\{\frac{1}{2}\bar{\nu}_\tau O_\alpha \nu_\tau + \bar{\tau}\gamma_\alpha[-\frac{1}{2}(1+x_R) + 2x_w - \frac{1}{2}(1-x_R)\gamma_5]\tau\}Z_\alpha \tag{3}$$

where $x_R = sin^2\phi_R$, $O_\alpha = \gamma_\alpha(1+\gamma_5)$, $x_w = sin^2\theta_w$ $(g = \bar{g}cos\theta_w$, $e = gsin\theta_w)$.

Let us focus on the possible manifestations of $\tau - \mu - e$ universality violation in the neutral current sector due to the new interaction in Eq. (3). Note that a unique opportunity for the investigation of the structure of neutral currents will appear in the Z-resonance region with longitudinally polarized e^+e^- beams, which become accessible at LEP and SLC and therefore a number of polarization effects appears. The most promising one is the spin asymmetry which defined as

$$A^s_\tau = 2\ \frac{a\ (1+cos^2\theta) + b\ cos\theta}{c\ (1+cos^2\theta) + d\ cos\theta} \tag{4}$$

Explicit forms of a, b, c, and d are found in [4].

Calculations show that the Standard Model predicts positive value for A^s at all θ (note that $\sin^2\theta_w(m_Z) = 0.2316 \pm 0.0017$), which becomes negligible at $\theta \geq 140^o$. but E_6 Model lead to essetially different predictions for A^s at $\theta \geq 120^o$. After calculations we get that the number of events which lie between 120° and 170° is about 1/3 of the total number of $\tau^+\tau^-$ pairs. Therefore the experimental study of spin asymmetry will give valuable information on the possible mixing between usual leptons and new charged heavy leptons.

Let consider simultaneously mixing between one generation leptons with the same charge and chiralities for discussing the deviation of experimental value of τ-lepton lifetime from the Standard Model (SM) with three families prediction ($\sim 2.3\ \sigma$). Note that for explanation of the τ lifetime "lengthening" in context of the fourth SM family was discussed in [5]. We reconsider this problem in the framework of E_6 model.

Note that, mixings between different generations can lead to rare lepton number violating decays at tree level. We will come back to discussion this point at the end of this work.

Taking into account the mixings in Eq. (2) it can be easily shown that ($\bar{\nu}_\tau$ τ)-current has the following modification:

$$\bar{\nu}_\tau\gamma_\alpha(1-\gamma_5)\tau \to \cos(\theta_L-\phi_L)\bar{\nu}_\tau\gamma_\alpha(1-\gamma_5)\tau + \sin\theta_R\sin\phi_R\bar{\nu}_\tau\gamma_\alpha(1+\gamma_5)\tau \quad (5)$$

(However, these differences do not appear in physical processes, so we can take $\theta_L = \theta_R = \theta$, $\phi_L = \phi_R = \phi$). In result for τ-lifetime we get

$$\tau = \tau_0[\cos^2(\theta-\phi) + \sin^2\theta sin^2\phi]^{-1} \quad (6)$$

Here τ_0 is the τ-lifetime in SM with three famlies. Note that ϕ=0 corresponds to SM with four families. So, modification of τ lifetime in E_6 and SM with four families is similar, but the consequences in these two models are dramatically different. Indeed, Eq.(4) gives that τ-lifetime in SM with four families has only an increase, however, in E_6 model it increases or decreases, depending on values of mixing angles. To see the difference between SM and E_6 model predictions consider the normalized energy spectrum of the charged final lepton

$$\frac{1}{\Gamma_\ell}\frac{d\Gamma_\ell}{dx} = 4\{3(1-x) + \frac{2\rho}{3}(4x-3)\}x^2 \quad (7)$$

Here $\Gamma_\ell = 1/\tau$, Michel parameter ρ is defined as $\rho = 3(1-\alpha)/4$ with $\alpha = \frac{\sin^2\theta\sin^2\phi}{\cos^2(\theta-\phi)+\sin^2\theta\sin^2\phi}$.

Experimental value $\rho = 0.727 \pm 0.033$ [7] leads that $\alpha \leq 0.075$. Numerical analysis show that difference between SM and E_6 model predictions in the normalized energy spectrum of the charged final lepton for various values of parameter α (α=0 corresponds to the SM case) is very small up to x=0.75. Above x=0.75, SM and E_6 model predictions are noticeable difference [6].

We would like to make some comments on Z-boson decays in framework of the considered E_6 model. Mixing between $\nu_{\tau R}$ and $N_{\tau}{}^0_R$ allows new decay mode of Z boson, namely, $Z \to \nu_{\tau R}\ \bar{\nu}_{\tau\ R}$. It is easy to see that new $Z\nu_{\tau R}\ \bar{\nu}_{\tau\ R}$ vertex is proportional to $\sin^2\theta$. Therefore, if the number of light neutrinos obtained in experiments is larger than three, this fact indicates the existence of the mixing between the ordinary and exotic neutral leptons. From current experimental data $N_\nu = 3.04 \pm 0.04$ [7] one can obtain the restriction on mixing angle as $\sin^4\theta < 0.08$.

Now we would like to briefly discuss the $e-\mu-\tau$ universality and lepton number violations in Z-decays. W consider two possibilities: 1. there are mixings between first generation heavy charged leptons and all ordinary leptons, 2. there are mixings between ordinary and heavy charged leptons in each generations only. It is obvious that the mixings between left components do not lead to FCNC. Therefore we restrict ourselves only right components.

Let us first consider possibility;

$$\begin{pmatrix} N \\ E'^- \end{pmatrix}_R, \quad e'^-_R, \quad \mu'^-_R, \quad \tau'^-_R \quad (8)$$

where

$$E' = c_1E^- + s_1c_2e^- + s_1s_2c_3\mu^- + s_1s_2s_3\tau^- \quad (9)$$

From Eq. (9) one can easily obtain Lagrangian for the flaver changing neutral currents at tree level (note that such currents absent in SM),

$$\mathcal{L} = \frac{e}{2sin2\theta_w}\{s_1^2c_2c_3s_2[\bar{\mu}\gamma_\mu(1+\gamma_5)e + h.c.] + s_1^2c_2s_2s_3[\bar{\tau}\gamma_\mu(1+\gamma_5)e + h.c.]$$
$$+s_1^2s_2^2c_2s_3[\bar{\tau}\gamma_\mu(1+\gamma_5)\mu + h.c.]\}Z_\mu \qquad (10)$$

Suppose that $c_1 \sim c_2 \sim c_3 \sim 1$. (Otherwise we have unwantedly large FCNC contribution to the process $e^+e^- \to \ell^+\ell^-$ ($\ell = \mu, \tau$)).

From Eq. (10), we have

$$Br(Z \to \tau\mu) = s_2^2 Br(Z \to \tau e) = s_2^2 s_3^2 Br(Z \to \mu e) \qquad (11)$$

From Eq. (11) it follows that we can get the best limit only $s_1^2 s_2$ combination. Using the existing experimental limit for Br($Z \to \tau e$) $< 8.7 \times 10^{-6}$ [8], we have $s_1^2 s_2 \leq 3 \times 10^{-3}$

Let us consider second possibility;

$$\begin{pmatrix} N_e^0 \\ E'^- \end{pmatrix}_R, \; e'^-_R, \begin{pmatrix} N_\mu^0 \\ M'^- \end{pmatrix}_R, \; \mu'^-_R, \begin{pmatrix} N_\tau^0 \\ T'^- \end{pmatrix}_R, \; \tau'^-_R,$$

In this case, neutral current of the ordinary leptons looks like

$$\mathcal{L} = \frac{\bar{g}}{2}\bar{\ell}\gamma_\alpha[V^\ell + A^\ell\gamma_5]\ell Z_\alpha \qquad (12)$$

where $V^\ell = -\frac{1}{2}(1 + x_R^\ell) + 2x_w$, $A^\ell = -\frac{1}{2}(1 - x_R^\ell)$, and $x_R^\ell$is the mixing angle between corresponding heavy and ordinary leptons.

Using the SM prediction $\Gamma_0 = \Gamma(Z \to \ell^+\ell^-) \simeq 83.5$ MeV (for $x_w \simeq 0.23$) and experimental results $\Gamma_e = 83.86 \pm 0.30$ MeV, $\Gamma_\mu = 83.78 \pm 0.40$ MeV, $\Gamma_\tau = 83.50 \pm 0.45$ MeV, we obtain $f(x_R^e) = 1.004 \pm 0.004$, $f(x_R^\mu) = 1.003 \pm 0.005$, and $f(x_R^\tau) = 1.000 \pm 0.005$. Here $f(x_R) = 1 + 4x_R + 8x_R^2$. From these results we get the following restrictions on x_R, $x_R^e < 0.01$, $x_R^\mu < 0.01$, and $x_R^\tau < 0.005$.

In conclusion, future precise measurements of above mentioned processes will provide essential information about possible mixings between ordinary and "new heavy leptons" predicted by E_6 model.

References

[1] Gürsey, F., Ramond, P. and Sikivie, P. (1976): A universal gauge theory model based on E_6. Phys. Lett. B60, 177-180;
Gürsey, F., Serdaroğlu, M. (1981): E_6 gauge field theory model revisited. Il Nuovo Cimento 65A, N3, 337-353.

[2] Hewett, J. L., Rizzo, T.G. (1989): Low-energy phenomenology of superstring-inspired E_6 models. Phys. Rep. 183, N5&6 193-381.

[3] Sultansoy, S. F. (1993): Regional project for elementary particle physics: Linac-Ring type c-τ factory. Turkish J. of Phys. Doğa **17**, N8, 591-602.

[4] Aliev, T. M., Sultansoy, S. F., Yılmaz, O. (1992): $\tau - \mu - e$ nonuniversality as indication of ordinary and exotic lepton mixing. Phys. Lett. **B 291**, 106-108.

[5] Ma, E., Pakvasa, S., Tuan, S. F. (1992): τ nonuniversality? Particle World **3**, 27-29.

[6] Aliev, T. M., Sultansoy, S. F., Yılmaz, O. (1993): Looking for E_6 model manifestations in τ decays. Phys. Rev. **D47** N7, 2879-2881.

[7] Particle Data Group (1992): Review of particle properties. Phys. Rev.**D 45**, N11, part II-III.

[8] Shevchenko, L. (1994): Search for lepton flavour violation in Z decays. Talk given in Int. Conf. on High Energy Physics, Glasgow.

Anomaly and Exotic Statistics in One Dimension

Fuad Saradzhev

Department of Physics, Marmara Research Centre,
TUBITAK, 41470 Gebze, TURKEY

Abstract. We show that the physical quantum chiral Schwinger model (CSM) can be formulated in two equivalent ways : either in terms of the fermionic matter degrees of freedom moving in a linearly rising electric field, or in terms of matter fields with exotic statistics.

Keywords. Anomaly,exotic statistics,gauge fields

1 Introduction

In the CSM [1], a 2dim $U(1)$ gauge field is coupled to a chiral massless fermionic field, for example, to the right-handed one. We assume that the $U(1)$ gauge field diminishes rather rapidly when $|x| \to \infty$. Then (i) the dynamics of physical degrees of freedom of the model is governed by the interaction of gauge and right-handed fermionic fields in a large, but finite "volume" $L = \int_{-L/2}^{L/2} dx$, $L \gg 1$, and (ii) the gauge field acquires a global physical degree of freedom represented by

$$\Gamma(A) = \exp\{ieLb\}, \quad b \equiv \frac{1}{L} \int dx A_1.$$

The right-left asymmetric matter content of the CSM shows itself at the quantum level: the local gauge symmetry is realized projectively. The physical states are gauge-invariant only up to a phase and defined as those which are annihilated by the modified Gauss law generator [2]

$$\hat{\tilde{G}}\Phi_{\mathrm{phys}} = (\hat{G} + \frac{e^2}{4\pi} A_1)\Phi_{\mathrm{phys}}, \tag{1}$$

where $\hat{G} \equiv \partial_1 \hat{\pi}_1 + e\hat{j}_R$ is the standard Gauss law generator.

The quantum Hamiltonian compatible with the modified constraint 1, i.e. commuting with $\hat{\tilde{G}}$, in the temporal gauge $A_0 = 0$ has the form:

$$\hat{H} = \int dx \frac{1}{2}(\hat{\pi}_1 + \frac{e^2}{4\pi}\partial_1^{-1} A_1)^2 + \hat{\mathcal{H}}_\psi - e\hat{j}_R A_1,$$

where $\hat{\mathcal{H}}_\psi \equiv i : \hat{\psi}_L^\star \partial_1 \hat{\psi}_L : -i : \hat{\psi}_R^\star \partial_1 \hat{\psi}_R :$, while the operator ∂_1^{-1} is given by

$$(\partial_1^{-1} f)(x) = \frac{1}{2}\int dz \epsilon(x-z) f(z).$$

2 Physical Quantum CSM

We solve the constraint 1 in two steps. In the first step we make a transition to the physical degrees of freedom of the right-handed fermions. Namely, we perform a canonical transformation V_R over the right-handed fermionic field:

$$\hat{\psi}_R \rightarrow \hat{\psi}'_R = V_R^{-1} \hat{\psi}_R V_R \equiv \Omega(x; A)\hat{\psi}_R(x).$$

If $V_R \equiv \exp\{ie \int dx \hat{j}_R(x)\lambda(x; A)\}$, then $\Omega(x; A) = \exp\{ie\lambda(x; A)\}$. We find V_R from the condition

$$V_R^{-1} \hat{\tilde{G}} V_R = \partial_1 \hat{\pi}_1 + C(A), \tag{2}$$

where $C(A)$ is an operator independent of $\hat{\psi}_R$. Eq. 2 specifies that the local gauge symmetry of the transformed Hamiltonian $\hat{H}' = V_R^{-1} H V_R$ must be independent of the fermionic variables, while the transformed physical states $\Phi'_{\text{phys}} = V_R^{-1}\Phi_{\text{phys}}$ must obey the condition

$$(\partial_1 \hat{\pi}_1 + C(A))\Phi'_{\text{phys}} = 0.$$

After certain manipulations, we find from Eq. 2 that $\Omega(y; A)$ satisfies the following equation:

$$\frac{\partial}{\partial x}\frac{\delta \Omega(y; A)}{\delta A_1(x)}\Omega^{-1}(y; A) = -ie\delta(x-y). \tag{3}$$

A general solution to this equation is

$$\Omega(y; A) = \exp\{ie\partial_1^{-1} A_1(y)\} \cdot \Omega_0(y; A),$$

where $\Omega_0(y; A) = \exp\{ie\lambda_0(y; A)\}$ is a solution to the corresponding homogeneous equation. With any selection of Ω that is a solution to Eq. 3 we obtain the Hamiltonian $\hat{H}'$ in which the right-handed fermions are physical. We choose $\Omega_0(y; A) = \exp\{-ieyb\}$, then $C(A) = \frac{e^2}{4\pi}b$.

In the second step, we eliminate the local symmetry of the gauge field, too. We use for the transformed physical states the following ansatz:

$$\Phi'_{\text{phys}} = \exp\{-\frac{ie^2}{4\pi}b \int dx x A_1\} \cdot \Phi(b, \psi_R).$$

On the states $\Phi(b,\psi_R)$,

$$\hat{\pi}_1(x)\Phi(b,\psi_R) = \frac{1}{L}\hat{\pi}_b\Phi(b,\psi_R), \quad \hat{\pi}_b \equiv -i\frac{d}{db},$$

and the "physical" Hamiltonian with the local gauge symmetry completely removed is

$$\begin{aligned} \hat{H}_{\text{phys}} &= \int dx\hat{\mathcal{H}}_\psi + \hat{\mathcal{H}}_b - \frac{e^2}{2}\int dxdy\hat{\jmath}_{R,N}(x)D(x,y)\hat{\jmath}_{R,N}(y), \\ \hat{\mathcal{H}}_b &\equiv \frac{1}{2L^2}\hat{\mathrm{p}}_b^2 + \frac{e^2}{4\pi}b^2 + e\cdot\frac{e^2}{4\pi}bx^2\cdot\hat{\jmath}_{R,N}(x), \end{aligned} \tag{4}$$

where

$$D(x,y) \equiv \frac{1}{2}|x-y| + \frac{xy}{L}$$

and

$$\hat{\mathrm{p}}_b(x) \equiv \hat{\pi}_b - \frac{L}{2}\mathcal{E}(x).$$

We see that b not only represents the physical degrees of freedom for the gauge field, but also creates in "volume" L a background electric field $\mathcal{E}(x) \equiv -\frac{e^2}{\pi}xb$ in which the physical degrees of freedom of the model are moving [3].

3 Exotization

We can formally decouple the matter and gauge-field degrees of freedom by introducing the exotic statistics matter field [4]. Indeed, let us define the composite field

$$\hat{\tilde{\psi}}_R(x) = \exp\{i\frac{\pi}{L}x - i\frac{2\pi}{eL}\hat{\mathrm{p}}_b(x)\}\cdot\hat{\psi}_R(x). \tag{5}$$

The field $\hat{\tilde{\psi}}_R(x)$ has the commutation relations:

$$\begin{aligned} \hat{\tilde{\psi}}^\star_R(x)\hat{\tilde{\psi}}_R(y) &+ e^{iF(x,y)}\hat{\tilde{\psi}}_R(y)\hat{\tilde{\psi}}^\star_R(x) = \delta(x-y), \\ \hat{\tilde{\psi}}_R(x)\hat{\tilde{\psi}}_R(y) &+ e^{-iF(x,y)}\hat{\tilde{\psi}}_R(y)\hat{\tilde{\psi}}_R(x) = 0, \end{aligned} \tag{6}$$

where $F(x,y) \equiv \Delta\cdot(x-y)$, $\Delta \equiv \frac{2\pi}{L}$. The commutation relations 6 are indicative of an exotic statistics of $\hat{\tilde{\psi}}_R(x)$.

Let us construct the Fock representation of the commutation relations 6. We expand $\hat{\psi}_R(x)$ in the Fourier integral

$$\hat{\psi}_R(x) = \frac{1}{\sqrt{2\pi}}\int_{-\infty}^{\infty} dqe^{iqx}\hat{a}_q, \tag{7}$$

where $\hat{a}^\star_q$, $\hat{a}_q$ are fermionic creation and annihilation operators for a right-handed particle of momentum q. The corresponding Fock vacuum is defined as

$$\hat{a}_q|0\rangle = 0 \quad \text{for} \quad q \geq 0, \quad \hat{a}^\star_q|0\rangle = 0 \quad \text{for} \quad q < 0.$$

Using 5 and 7, we get the Fourier expansion for the exotic field:

$$\hat{\tilde{\psi}}_R = \frac{1}{\sqrt{2\pi}} \int_{-\infty}^{\infty} dq e^{iqx} \hat{\tilde{a}}_q,$$

where

$$\hat{\tilde{a}}_q \equiv \exp\{-i\frac{2\pi}{eL}\hat{\pi}_b\} \cdot \hat{a}_{q+eb}.$$

It can be verified that $\hat{\tilde{a}}_q$, $\hat{\tilde{a}}_q^{\star}$ fulfil the following commutation relation algebra:

$$\begin{aligned} \hat{\tilde{a}}_q \hat{\tilde{a}}_p^{\star} &+ \hat{\tilde{a}}_{p-\Delta}^{\star} \hat{\tilde{a}}_{q-\Delta} = \delta(q-p), \\ \hat{\tilde{a}}_q \hat{\tilde{a}}_p &+ \hat{\tilde{a}}_{p-\Delta} \hat{\tilde{a}}_{q+\Delta} = 0. \end{aligned} \tag{8}$$

We introduce next the new Fock vacuum $|\tilde{0}\rangle$ defined as:

$$\hat{\tilde{a}}_q|\tilde{0}\rangle = 0 \quad \text{for} \quad q \geq 0, \quad \hat{\tilde{a}}_q^{\star}|\tilde{0}\rangle = 0 \quad \text{for} \quad q < \Delta,$$

and denote the normal ordering with respect to $|\tilde{0}\rangle$ by $\vdots\vdots$. The old creation and annihilation operators act on the new Fock vacuum by the rule:

$$\hat{a}_q|\tilde{0}\rangle = 0 \quad \text{for} \quad q \geq eb, \quad \hat{a}_q^{\star}|\tilde{0}\rangle = 0 \quad \text{for} \quad q < eb,$$

i.e. the background charge is incorporated in the new Fock vacuum.

Using the exotic composite field introduced, we rewrite $\hat{H}_{\text{phys}}$ in the compact form with matter and gauge-field degrees of freedom decoupled:

$$\hat{H}_{\text{phys}} = \frac{1}{2L}\hat{\pi}_b^2 + \int dx \hat{\tilde{\mathcal{H}}}_{\psi} - \frac{e^2}{2} \int dx dy \hat{\tilde{j}}_{R,N}(x) D(x,y) \hat{\tilde{j}}_{R,N}(y),$$

$$\hat{\tilde{\mathcal{H}}}_{\psi} \equiv i : \hat{\psi}_L^{\star} \partial_1 \hat{\psi}_L : -i \vdots \hat{\tilde{\psi}}_R^{\star} \partial_1 \hat{\tilde{\psi}}_R \vdots .$$

References

[1] R. Jackiw and R. Rajaraman, Phys.Rev.Lett. **54**, 1219 (1985).

[2] F.M. Saradzhev, Int.J.Mod.Phys. **A6**, 3823 (1991).

[3] F.M. Saradzhev, Phys.Lett. **B278**, 449 (1992).

[4] F.M. Saradzhev. Phys.Lett. **B324**, 192 (1994).

Samuel MacDowell
Closing Remarks
Gürsey Memorial Conference I
June 6-10, 1994

As this conference draws to a close, I would like to add a few words to express my personal and the gratitude of all of us who came here, for the privilege and the pleasure of participating in this tribute to our dear and beloved friend and colleague Feza Gürsey.

We are grateful to the Turkish Scientific and Technological Research Council and its director Professor Namık Kemal Pak for sponsoring this conference, the first of a series of biannual conferences in honor of this great man.

More important is the fact that this memorial was the initiative of Feza's own students for whom, as has been said here, he was a role model, an inspiring teacher and guide, and a caring friend. This is the best evidence that the ideals and convictions of Feza about the importance of science and the role it plays in the cultural and technological development of a society will live on.

It is very fitting that this conference is being held in this beautiful city of Istanbul, the city that at heart Feza loved and considered his city. Many times have I listened to him talk with longing and tenderness about Istanbul, its treasures, its museums, its historical past and its beauty. I would like to thank Professor John Freely for giving us a delightful and expert guided tour of the old city and the Bosphorus.

It is also very fitting that this conference is taking place in this great University. As was said in the opening address by your Rector Professor Üstün Ergüder, Boğaziçi University was the place Feza had planned to spend most of his time after retiring from Yale. I know that he was looking forward with great expectations to the work that he would have liked to carry on here. In spite of the fact that his life can be said to have been very rich and productive, and that he left a tremendous legacy of accomplishments, he had still many projects for the years ahead that unfortunately could not be fulfilled. Some of this work was to be done alone and some with Meral, the dynamic secretary of this Conference, who was carefully preparing herself for this forthcoming collaboration with Feza.

I am glad to say, and I think my collegues would agree, that this conference has achieved well its objectives.

The main topic on Strings was covered by a large number of high level papers. In his talk, Gregg Zuckerman made a rather thorough survey of many of the topics, showing how they are linked together, from a mathematical perspective. It is worth noticing that springing from the basic Virasoro algebra of conformal Field Theories a great variety of algebraic structures have been uncovered or constructed.

Discussed here were algebras of Vertex operators, W-algebras which are peculiar extensions of the Virasoro algebra, closing on non-linear terms, Kac Moody algebras super symmetries and the algebra of infinitesimal invariant transformations of off-shell supergravity.

There were contributions describing recent developments in Superconformal Field Theories and Superstrings, Topological Field Theories, Quantum groups and 2-D Gravity and an interesting application of Hopf algebra leading to yet another perhaps more general framework of Quantum Theory incorporating a new duality principle. I emphasize this side of the conference because Feza himself was a pioneer in introducing and applying to Physics new algebraic structures, in particular, the algebra of octonions.

On the other hand, however, many other subjects related to Feza's work on symmetries, have been also submitted to this conference. I shall mention just a couple that are typical examples of the far reaching influence of his work.

We had a very nice review by Tony Zee on Symmetry and Effective Field Theories. Here the seminal work was Feza's paper on the chiral non-linear sigma model which has become a paradigm for the description of effective interactions especially at low energies. We had also a report by Franco Iachello on the solutions of problems in systems with dynamical symmetry.

It is impossible to do justice to all the authors of quality papers presented here. I will just say that this conference provided a forum for the discussion of their work in a very friendly and family-like atmosphere, much like one would have expected had it been organized by Feza himself.

Although for myself some of the talks were at a level beyond my comprehension, I sensed, especially from the discussions and the questions raised, that there were some very fruitful exchanges here.

On a more personal note, I would like to say that I see this conference also as a tribute to our dear lady Suha Gürsey. I am sure that Feza would have felt very pleased in sharing with her this kind of tribute.

Finally I would like to express my own gratitude and the gratitude of all of us participants in this conference for the warm hospitality extended to us by the staff, who so tirelessly prepared this conference.

Gülen Aktaş, Ömür Akyüz, Cihan Saçlıoğlu and Meral Serdaroğlu - I can tell them that their untiring efforts were rewarded by a very successful and well-organized conference. We immensely enjoyed this meeting, this opportunity to hear so many interesting talks and this reunion with so many friends and colleagues.

I would like to ask the audience to give to the organizers a warm round of applause.

LIST OF PARTICIPANTS

ACAR(FIRAT), F.Gülay	İTÜ
AÇIKTEPE, Tevfik	İTÜ
ADAGİDELİ, İnanç	METU
AKKAYA, Aydın	BÜ
AKTAŞ, Gülen	BÜ
AKYÜZ, Ömür	BÜ
ALIEV, Alikram	TÜBİTAK
ARGYRES, Philip	IAS, Princeton
ARIK, Engin	BÜ
ARIK, Metin	BÜ
AYDIROĞLU, Eser	EMU
BARS, İthzak	USC, USA
BARUT, Asım	University of Colorado,USA
BOUWKNEGT, Peter	USC, USA
BOZ, Müge	METU
CACERES Elena	University of Texas, USA
CEDERWALL, Martin	Chalmers University Sweden
CHAICHIAN, Mesud	University of Helsinki, Finland
CHODOS, Alan	Yale University, USA
DAYI, Ö. Faruk	TÜBİTAK
DEMİR, Durmuş Ali	METU
DICK, Rainer	University of Munich, Germany
DOMOKOS, Gabor	John Hopkins Univ., USA
DOMOKOS, S. Kovesi	John Hopkins University, USA
DURU, İsmail Hakkı	TÜBİTAK
EVANS, Mark	Rockefeller University, USA
EVANS, Jonathan	CERN, Switzerland
FAİNBERG, Viladmir	METU
FERAPONTOV, Eugene	TÜBİTAK
FERRETTI, Gabriele	Chalmers University, Sweden
FRE, Pietro	SISSA, Trieste, Italy
FREELY, John	BÜ
FREUND, Peter	University of Chicago, USA
GABAY, Vili	Balkan Physics Center
GERVAIS, Jean-Loup	ENS, France
GÜLMEZ, Erhan	BÜ
GÜNAYDIN, Murat	Penn State University, USA
GÜNGÖRMEZ Meltem	İTÜ
GÜRSEY, Suha	
GÜRSEY, Yusuf	METU
GÜVEN, Rahmi	BÜ
HACINLIYAN, Avadis	BÜ
HANSOY, Cenger	İTÜ

HIGUCHI, Atsushi	University of Berne, Switzerland
HORTAÇSU, Mahmut	İTÜ
HUGHES, Vernon	Yale University, USA
IACHELLO, Francesco	Yale University, USA
IKEDA, İlhan	TÜBİTAK
ISBERG, Jan	Kings College, London. England
KALAYCI, Jan	İTÜ
KAPLAN, Zuhal	BÜ
KARADAYI, Hasan	İTÜ
KARASU, Atalay	METU
KAYA, Reyhan	TÜBİTAK
KHURİ, Ramzi	CERN, SWITZERLAND
KOCA, Mehmet	Çukurova University
MACDOWELL, Samuel	Yale University, USA
MAJID, Shahn	Cambridge University, UK
MARCHILDON, Louis	University of Quebec, Canada
MC CLOUD Paul	Kings College, England
MELNIKOV, Vitaly	SVRC, Moskow, Russia
NERGİS, Serdar	BÜ
NEYZİ, Fahrunissa	BÜ
NICOLAI, Hermann	University of Hamburg, Germany
NIEMİ, Antti	Uppsala University, Sweden
NUTKU, Yavuz	TÜBİTAK
OGIEVETSKY, Victor	Bonn University, Germany
OĞUZ, Ömer	BÜ
OOGURI, Hirosi	Kyoto University, Japan
ÖNENGÜT, Gülsen	Çukurova University
ÖZBEK, Haluk	İTÜ
ÖZDEMİR, Neşe	İTÜ
ÖZER, Hakkı Tuncay	İTÜ
PAK, Namik Kemal	TUBITAK
PAPAGEORGİU, Elena	LPTHE, Paris, France
PELTS, Gregory	Rockefeller University, USA
PILCH, Krzysztof	USC, USA
RAMOND, Pierre	University of Florida, USA
SAÇLIOĞLU, Cihan	BÜ
SANALAN, Yalçın	AEK
SANYUK, Valerii	PFU, Moskow
SARADZHEV, Fuad	TÜBİTAK
SAYGILI, Kamuran	BÜ
SCHWANDER, Thorsten	Hannover University, Germany
SEMİZ, İbrahim	University of North Carolina, USA
SERDAROĞLU, Meral	BÜ

SEZGIN, Ergin	Texas A &M USA
SOMMERFIELD, Charles	Yale University,. USA
TAORMINA, Anne	University of Durham, UK
TURAN, Gürsevil	METU
VISWANATHAN, K.S.	Simon Fraser University, Canada
VON PROEYEN, Antoine	K.U., Leuven, Belgium
WALİ Kameshwar	Syracuse University, USA
WESTERBERG, Anders	ITP, Sweden
YILDIZ, Ali	BÜ
YILMAZ, İhsan	Ege University, Turkey
YILMAZ, Osman	METU
YÖRÜK, Engin	BÜ
ZEE, Antony	ITP, U.C. Santa Barbara, USA
ZUCKERMAN, Gregg	Yale University, USA

Lecture Notes in Physics

For information about Vols. 1–409
please contact your bookseller or Springer-Verlag

Vol. 408: V. J. Martínez, M. Portilla, D. Sàez (Eds.), New Insights into the Universe. Proceedings, 1991. XI, 298 pages. 1992.

Vol. 409: H. Gausterer, C. B. Lang (Eds.), Computational Methods in Field Theory. Proceedings, 1992. XII, 274 pages. 1992.

Vol. 410: J. Ehlers, G. Schäfer (Eds.), Relativistic Gravity Research. Proceedings, VIII, 409 pages. 1992.

Vol. 411: W. Dieter Heiss (Ed.), Chaos and Quantum Chaos. Proceedings, XIV, 330 pages. 1992.

Vol. 412: A. W. Clegg, G. E. Nedoluha (Eds.), Astrophysical Masers. Proceedings, 1992. XX, 480 pages. 1993.

Vol. 413: Aa. Sandqvist, T. P. Ray (Eds.); Central Activity in Galaxies. From Observational Data to Astrophysical Diagnostics. XIII, 235 pages. 1993.

Vol. 414: M. Napolitano, F. Sabetta (Eds.), Thirteenth International Conference on Numerical Methods in Fluid Dynamics. Proceedings, 1992. XIV, 541 pages. 1993.

Vol. 415: L. Garrido (Ed.), Complex Fluids. Proceedings, 1992. XIII, 413 pages. 1993.

Vol. 416: B. Baschek, G. Klare, J. Lequeux (Eds.), New Aspects of Magellanic Cloud Research. Proceedings, 1992. XIII, 494 pages. 1993.

Vol. 417: K. Goeke P. Kroll, H.-R. Petry (Eds.), Quark Cluster Dynamics. Proceedings, 1992. XI, 297 pages. 1993.

Vol. 418: J. van Paradijs, H. M. Maitzen (Eds.), Galactic High-Energy Astrophysics. XIII, 293 pages. 1993.

Vol. 419: K. H. Ploog, L. Tapfer (Eds.), Physics and Technology of Semiconductor Quantum Devices. Proceedings, 1992. VIII, 212 pages. 1993.

Vol. 420: F. Ehlotzky (Ed.), Fundamentals of Quantum Optics III. Proceedings, 1993. XII, 346 pages. 1993.

Vol. 421: H.-J. Röser, K. Meisenheimer (Eds.), Jets in Extragalactic Radio Sources. XX, 301 pages. 1993.

Vol. 422: L. Päivärinta, E. Somersalo (Eds.), Inverse Problems in Mathematical Physics. Proceedings, 1992. XVIII, 256 pages. 1993.

Vol. 423: F. J. Chinea, L. M. González-Romero (Eds.), Rotating Objects and Relativistic Physics. Proceedings, 1992. XII, 304 pages. 1993.

Vol. 424: G. F. Helminck (Ed.), Geometric and Quantum Aspects of Integrable Systems. Proceedings, 1992. IX, 224 pages. 1993.

Vol. 425: M. Dienes, M. Month, B. Strasser, S. Turner (Eds.), Frontiers of Particle Beams: Factories with $e^+ e^-$ Rings. Proceedings, 1992. IX, 414 pages. 1994.

Vol. 426: L. Mathelitsch, W. Plessas (Eds.), Substructures of Matter as Revealed with Electroweak Probes. Proceedings, 1993. XIV, 441 pages. 1994

Vol. 427: H. V. von Geramb (Ed.), Quantum Inversion Theory and Applications. Proceedings, 1993. VIII, 481 pages. 1994.

Vol. 428: U. G. Jørgensen (Ed.), Molecules in the Stellar Environment. Proceedings, 1993. VIII, 440 pages. 1994.

Vol. 429: J. L. Sanz, E. Martínez-González, L. Cayón (Eds.), Present and Future of the Cosmic Microwave Background. Proceedings, 1993. VIII, 233 pages. 1994.

Vol. 430: V. G. Gurzadyan, D. Pfenniger (Eds.), Ergodic Concepts in Stellar Dynamics. Proceedings, 1993. XVI, 302 pages. 1994.

Vol. 431: T. P. Ray, S. Beckwith (Eds.), Star Formation and Techniques in Infrared and mm-Wave Astronomy. Proceedings, 1992. XIV, 314 pages. 1994.

Vol. 432: G. Belvedere, M. Rodonò, G. M. Simnett (Eds.), Advances in Solar Physics. Proceedings, 1993. XVII, 335 pages. 1994.

Vol. 433: G. Contopoulos, N. Spyrou, L. Vlahos (Eds.), Galactic Dynamics and N-Body Simulations. Proceedings, 1993. XIV, 417 pages. 1994.

Vol. 434: J. Ehlers, H. Friedrich (Eds.), Canonical Gravity: From Classical to Quantum. Proceedings, 1993. X, 267 pages. 1994.

Vol. 435: E. Maruyama, H. Watanabe (Eds.), Physics and Industry. Proceedings, 1993. VII, 108 pages. 1994.

Vol. 436: A. Alekseev, A. Hietamäki, K. Huitu, A. Morozov, A. Niemi (Eds.), Integrable Models and Strings. Proceedings, 1993. VII, 280 pages. 1994.

Vol. 437: K. K. Bardhan, B. K. Chakrabarti, A. Hansen (Eds.), Non-Linearity and Breakdown in Soft Condensed Matter. Proceedings, 1993. XI, 340 pages. 1994.

Vol. 438: A. Pękalski (Ed.), Diffusion Processes: Experiment, Theory, Simulations. Proceedings, 1994. VIII, 312 pages. 1994.

Vol. 439: T. L. Wilson, K. J. Johnston (Eds.), The Structure and Content of Molecular Clouds. 25 Years of Molecular Radioastronomy. Proceedings, 1993. XIII, 308 pages. 1994.

Vol. 440: H. Latal, W. Schweiger (Eds.), Matter Under Extreme Conditions. Proceedings, 1994. IX, 243 pages. 1994.

Vol. 441: J. M. Arias, M. I. Gallardo, M. Lozano (Eds.), Response of the Nuclear System to External Forces. Proceedings, 1994, VIII, 293 pages. 1995.

Vol. 443: D. Koester, K. Werner (Eds.), White Dwarfs. Proceedings, Kiel, Germany 1994. XII, 348.

Vol. 444: A. O. Benz, A. Krüger (Eds.), Coronal Magnetic Energy Releases. Proceedings, 1994. X, 293 pages. 1995.

Vol. 446: V. Rivasseau (Ed.), Constructive Physics. Results in Field Theory, Statistical Mechanics and Condensed Matter Physics. Proceedings, 1994. X, 338 pages. 1995.

Vol. 447: G. Aktas, C. Saçlıoğlu, M. Serdaroğlu (Eds.), Strings and Symmetries. Proceedings, Istanbul, Turkey 1994. XIV, 389 pages. 1995.

Vol. 448: P. L. Garrido, J. Marro (Eds.), 3rd Granada Lectures in Computational Physics. Proceedings, Granada, Spain 1994. XIV, 346 pages. 1995.

New Series m: Monographs

Vol. m 1: H. Hora, Plasmas at High Temperature and Density. VIII, 442 pages. 1991.

Vol. m 2: P. Busch, P. J. Lahti, P. Mittelstaedt, The Quantum Theory of Measurement. XIII, 165 pages. 1991.

Vol. m 3: A. Heck, J. M. Perdang (Eds.), Applying Fractals in Astronomy. IX, 210 pages. 1991.

Vol. m 4: R. K. Zeytounian, Mécanique des fluides fondamentale. XV, 615 pages, 1991.

Vol. m 5: R. K. Zeytounian, Meteorological Fluid Dynamics. XI, 346 pages. 1991.

Vol. m 6: N. M. J. Woodhouse, Special Relativity. VIII, 86 pages. 1992.

Vol. m 7: G. Morandi, The Role of Topology in Classical and Quantum Physics. XIII, 239 pages. 1992.

Vol. m 8: D. Funaro, Polynomial Approximation of Differential Equations. X, 305 pages. 1992.

Vol. m 9: M. Namiki, Stochastic Quantization. X, 217 pages. 1992.

Vol. m 10: J. Hoppe, Lectures on Integrable Systems. VII, 111 pages. 1992.

Vol. m 11: A. D. Yaghjian, Relativistic Dynamics of a Charged Sphere. XII, 115 pages. 1992.

Vol. m 12: G. Esposito, Quantum Gravity, Quantum Cosmology and Lorentzian Geometries. Second Corrected and Enlarged Edition. XVIII, 349 pages. 1994.

Vol. m 13: M. Klein, A. Knauf, Classical Planar Scattering by Coulombic Potentials. V, 142 pages. 1992.

Vol. m 14: A. Lerda, Anyons. XI, 138 pages. 1992.

Vol. m 15: N. Peters, B. Rogg (Eds.), Reduced Kinetic Mechanisms for Applications in Combustion Systems. X, 360 pages. 1993.

Vol. m 16: P. Christe, M. Henkel, Introduction to Conformal Invariance and Its Applications to Critical Phenomena. XV, 260 pages. 1993.

Vol. m 17: M. Schoen, Computer Simulation of Condensed Phases in Complex Geometries. X, 136 pages. 1993.

Vol. m 18: H. Carmichael, An Open Systems Approach to Quantum Optics. X, 179 pages. 1993.

Vol. m 19: S. D. Bogan, M. K. Hinders, Interface Effects in Elastic Wave Scattering. XII, 182 pages. 1994.

Vol. m 20: E. Abdalla, M. C. B. Abdalla, D. Dalmazi, A. Zadra, 2D-Gravity in Non-Critical Strings. IX, 319 pages. 1994.

Vol. m 21: G. P. Berman, E. N. Bulgakov, D. D. Holm, Crossover-Time in Quantum Boson and Spin Systems. XI, 268 pages. 1994.

Vol. m 22: M.-O. Hongler, Chaotic and Stochastic Behaviour in Automatic Production Lines. V, 85 pages. 1994.

Vol. m 23: V. S. Viswanath, G. Müller, The Recursion Method. X, 259 pages. 1994.

Vol. m 24: A. Ern, V. Giovangigli, Multicomponent Transport Algorithms. XIV, 427 pages. 1994.

Vol. m 25: A. V. Bogdanov, G. V. Dubrovskiy, M. P. Krutikov, D. V. Kulginov, V. M. Strelchenya, Interaction of Gases with Surfaces. XIV, 132 pages. 1995.

Vol. m 26: M. Dineykhan, G. V. Efimov, G. Ganbold, S. N. Nedelko, Oscillator Representation in Quantum Physics. IX, 279 pages. 1995.

Vol. m 27: J. T. Ottesen, Infinite Dimensional Groups and Algebras in Quantum Physics. IX, 218 pages. 1995.

Vol. m 28: O. Piguet, S. P. Sorella, Algebraic Renormalization. IX, 134 pages. 1995.

Vol: m 30: A. J. Greer, W. J. Kossler, Low Magnetic Fields in Anisotropic Superconductors. VII, 161 pages. 1995.

Springer-Verlag and the Environment

We at Springer-Verlag firmly believe that an international science publisher has a special obligation to the environment, and our corporate policies consistently reflect this conviction.

We also expect our business partners – paper mills, printers, packaging manufacturers, etc. – to commit themselves to using environmentally friendly materials and production processes.

The paper in this book is made from low- or no-chlorine pulp and is acid free, in conformance with international standards for paper permanency.